T0206327

Biodiesel Fuels Based on Edible and Nonedible Feedstocks, Wastes, and Algae

Biodiesel Fuels Based on Edible and Nonedible Feedstocks, Wastes, and Algae

Science, Technology, Health, and Environment

Edited by
Ozcan Konur

CRC Press
Taylor & Francis Group
Boca Raton London New York

CRC Press is an imprint of the
Taylor & Francis Group, an **informa** business

First edition published 2021
by CRC Press
6000 Broken Sound Parkway NW, Suite 300, Boca Raton, FL 33487-2742

and by CRC Press
2 Park Square, Milton Park, Abingdon, Oxon, OX14 4RN

ISBN: 978-0-367-45615-3 (hbk)
ISBN: 978-0-367-70503-9 (pbk)
ISBN: 978-0-367-45620-7 (ebk)

Typeset in Times
by SPi Global, India

Contents

PART V   Edible Oil-based Biodiesel Fuels

Part VI   Nonedible Oil-based Biodiesel Fuels

Part VII   Waste Oil-based Biodiesel Fuels

Part VIII   Algal Biodiesel Fuels

Preface

Crude oils have been primary sources of energy and fuels, such as petrodiesel. However, significant public concerns about the sustainability, price fluctuations, and adverse environmental impact of crude oils have emerged since the 1970s. Thus, biooils and biooil-based biodiesel fuels have emerged as alternatives to crude oils and crude oil-based petrodiesel fuels, respectively, in recent decades. Nowadays, although petrodiesel fuels are still used extensively, biodiesel fuels are being used increasingly as petrodiesel–biodiesel blends in the transportation and power sectors. Therefore, there has been great public interest in the development of environment and human-friendly and sustainable petrodiesel and biodiesel fuels. However, it is necessary to reduce the total cost of biodiesel production by reducing the feedstock cost through the improvement of biomass and lipid productivity. It is also necessary to mitigate the adverse impact of petrodiesel fuels on the environment and human health.

Although there have been over 1,500 reviews and book chapters in this field, there has been no review of the research as a representative sample of all the population studies done in the field of both petrodiesel and biodiesel fuels. Thus, this second volume on feedstock-specific biodiesel fuels provides a representative sample of all the population studies in this field. The major research fronts are determined from the sample and population paper-based scientometric studies. Table 1.1 in volume 1 presents information on the major and secondary research fronts arising from these studies.

There are four secondary research fronts regarding the second volume on feedstock-based biodiesel fuels: edible oil-based biodiesel fuels, nonedible oil-based biodiesel fuels, waste oil-based biodiesel fuels, and algal oil-based biodiesel fuels.

As seen from Table 1.1 in volume 1 there is a substantial correlation between the distribution of research fronts in this handbook and the 100-most-cited sample papers. Papers on feedstock-based biodiesel fuels are over-represented compared to the population papers: Handbook chapters (33%), sample papers (30%) v. population papers (18.3%).

Table 1.2 in volume 1 extends Table 1.1 in volume 1 and provides more information on the content of this handbook, including chapter numbers, paper references, primary and secondary research fronts, and finally the titles of the papers presented.

The data presented in the tables and figures show that a small number of authors, institutions, funding bodies, journals, keywords, research fronts, subject categories, and countries have shaped the research in this field.

The findings show the importance of the progression of efficient incentive structures for the development of research in this field, as in other fields. It further seems that, although the research funding is a significant element of these incentive structures, it might not be a sole solution for increasing the incentives for research. On the other hand, it seems there is more to do to reduce the significant gender deficit in this field as in other fields of science and technology.

The information provided on nanotechnology applications suggests that there is ample scope for the expansion of advanced applications in this research field.

The research on feedstocks for biodiesel fuels first focused on edible oils as a first generation. However, public concerns about competition with foods based on these feedstocks and the adverse impact on ecological diversity and deforestation have resulted in the exploration of nonedible oil-based biodiesel fuels as a second generation. Due to the ecological and cost benefits of treating wastes, waste oil-based biodiesel fuels as a third generation have emerged. Furthermore, following a seminal paper by Chisti in 2007 and other influential review papers, the research has focused recently on algal oil-based biodiesel fuels. Since the cost of feedstocks in general constitutes 85% of total biodiesel production costs, research has focused more on improving biomass and lipid productivity. Furthermore, since water, CO_2, and nutrients (primarily N and P) have been major ingredients for algal biomass and lipid production, research has also intensified in the use of wastewaters and flue gasses for algal biomass production to reduce the ecological burdens and production costs.

Thus, this handbook is a valuable source for stakeholders primarily in the research fields of Energy Fuels, Chemical Engineering, Environmental Sciences, Biotechnology and Applied Microbiology, Physical Chemistry, Petroleum Engineering, Environmental Engineering, Multidisciplinary Chemistry, Thermodynamics, Analytical Chemistry, Mechanical Engineering, Agricultural Engineering, Marine Freshwater Biology, Green Sustainable Science Technology, Applied Chemistry, Multidisciplinary Geosciences, Microbiology, Multidisciplinary Materials Science, Mechanics, Toxicology, Multidisciplinary Sciences, Biochemistry and Molecular Biology, Water Resources, Plant Sciences, Multidisciplinary Engineering, Transportation Science Technology, Geochemistry and Geophysics, Food Science Technology, Ecology, Public Environmental Occupational Health, Meteorology and Atmospheric Sciences, Electrochemistry, and Biochemical Research Methods.

This handbook is also particularly relevant in the context of biomedical sciences for Public and Environmental Occupational Health, Pharmacology, Immunology, Respiratory System, Allergy, Genetics Heredity, Oncology, Experimental Medical Research, Critical Care Medicine, General Internal Medicine, Cardiovascular Systems, Physiology, Medicinal Chemistry, and Endocrinology and Metabolism.

Ozcan Konur

Acknowledgements

This handbook was a multi-stakeholder project from its inception to its publication. CRC Press and Taylor & Francis Group were the major stakeholders in financing and executing it. Marc Gutierrez was the executive editor. Eighty-three authors have kindly contributed chapters despite the relatively low level of incentives, compared to journals. As stated in many chapters, a small number of highly cited scholars have shaped the research on both biodiesel and petrodiesel fuels. The contribution of all these and other stakeholders is greatly acknowledged.

Editor's Biography

Ozcan Konur, as both a materials scientist and social scientist by training, has focused on the bibliometric evaluation of research in the innovative high-priority research areas of algal materials and nanomaterials for energy and fuels as well as for biomedicine at the level of researchers, journals, institutions, countries, and research areas, including the social implications of the research conducted in these areas.

He has also researched extensively in the development of social policies for disadvantaged people on the basis of disability, age, religious beliefs, race, gender, and sexuality at the interface of science and policy.

He has edited a book titled *Bioenergy and Biofuels* (CRC Press, January 2018) and a handbook titled *Handbook of Algal Science, Technology, and Medicine* (Elsevier, April 2020).

Contributors

Caleb Acquah
University of Ottawa, Canada

B. Ashok
Vellore Institute of Technology, India

Nagaraj R. Banapurmath
KLE Technological University,
India

J. E. Castanheiro
University of Evora, Portugal

Madhu Sudan Reddy Dandu
Department of Mechanical Engineering
Sree Vidyanikethan Engineering
College, Tirupati
India

Michael K. Danquah
University of Tennessee, USA

Sharanabasava V. Ganachari
KLE Technological University, India

Aran Incharoensakdi
Chulalongkorn University; Royal
Society of Thailand, Thailand

G. Janani
PSG College of Technology,
India

Jaison Jeevanandam
Curtin University, Malaysia

Shankha Koley
Indian Institute of Technology
Kharagpur, India

Ozcan Konur
Formerly, Ankara Yildirim Beyazit
University, Turkey

Saravanan Krishnan
Indian Institute of Technology Madras,
India

Pratap S. Kulkarni
Proudhadevaraya Institute of
Technology, India

N. Keerthi Kumar
S.D.M. College of Engineering and
Technology, Dharwad
India

R. Vinoth Kumar
National Institute of Technology
Andhra Pradesh, India

Nirupama Mallick
Indian Institute of Technology
Kharagpur, India

I. Ganesh Moorthy
Kamaraj College of Engineering and
Technology, India

K. Nanthagopal
Vellore Institute of Technology, India

S. Niju
PSG College of Technology, India

Sohrab Rohani
Western University, Canada

K.V.V. Satyannarayana
National Institute of Technology
Andhra Pradesh, India

Muhammad Nurunnabi Siddiquee
University of Alberta, Canada

Randeep Singh
Indian Institute of Technology
Guwahati, India

Ramachandran Sivaramakrishnan
Chulalongkorn University, Thailand

Sashi Sonkar
Indian Institute of Technology
Kharagpur, India

Manzoore Elahi Soudagar
University of Malaya, Malaysia

Charles K. Westbrook
Lawrence Livermore National
Laboratory, USA

V.S. Yaliwal
S.D.M. College of Engineering and
Technology, India

Part V

Edible Oil-based Biodiesel Fuels

21 Edible Oil-based Biodiesel Fuels

A Scientometric Review of the Research

Ozcan Konur

CONTENTS

21.1 INTRODUCTION

Crude oils have been primary sources of energy and fuels, such as petrodiesel (Busca et al., 1998; Khalili et al., 1995; Rogge et al., 1993; Schauer et al., 1999). However, significant public concerns about the sustainability, price fluctuations, and adverse environmental impact of crude oils have emerged since the 1970s (Ahmadun et al.,

2009; Atlas, 1981; Babich and Moulijn, 2003; Kilian, 2009; Perron, 1989). Thus, biooils have emerged as an alternative to crude oils in recent decades (Bridgwater and Peacocke, 2000; Czernik and Bridgwater, 2004; Gallezot, 2012; Mohan et al., 2006). In this context, edible oil-based biodiesel fuels (Darnoko and Cheryan, 2000; Freedman et al., 1986; Kusdiana and Saka, 2001; Noureddini and Zhu, 1997; Saka and Kusdiana, 2001) have also emerged as a viable alternative to crude oil-based petrodiesel fuels (Busca et al., 1998; Khalili et al., 1995; Rogge et al., 1993; Schauer et al., 1999). Nowadays, both petrodiesel fuels and edible oil-based biodiesel fuels are being used extensively at the global scale (Konur, 2021a–ag).

However, for the efficient development of the research in this field, it is necessary to develop efficient incentive structures for the primary stakeholders and to inform these stakeholders about the research (Konur, 2000, 2002a–c, 2006a–b, 2007a–b; North, 1991a–b).

Scientometric analysis offers ways to evaluate the research in a respective field (Garfield, 1955, 1972). This method has been used to evaluate research in a number of fields (Konur, 2011, 2012a–n, 2015, 2016a–f, 2017a–f, 2018a–b, 2019a–b). However, there has been no current scientometric study of this field.

This chapter presents a study of the scientometric evaluation of the research in this field using two datasets. The first dataset includes the 100-most-cited papers ($n = 100$ sample papers) whilst the second set includes population papers ($n =$ over 2,650 papers) published between 1980 and 2019. This complements the chapters on crude oil-based petrodiesel fuels and other biooil-based biodiesel fuels.

The data on the indices, document types, authors, institutions, funding bodies, source titles, 'Web of Science' subject categories, keywords, research fronts, and citation impact are presented and discussed.

21.2   MATERIALS AND METHODOLOGY

The search for the literature was carried out in the 'Web of Science' (WOS) database in January 2020. It contains the 'Science Citation Index-Expanded' (SCI-E), the Social Sciences Citation Index' (SSCI), the 'Book Citation Index-Science' (BCI-S), the 'Conference Proceedings Citation Index-Science' (CPCI-S), the 'Emerging Sources Citation Index' (ESCI), the 'Book Citation Index-Social Sciences and Humanities' (BCI-SSH), the 'Conference Proceedings Citation Index-Social Sciences and Humanities' (CPCI-SSH), and the 'Arts and Humanities Citation Index' (A&HCI).

The keywords for the search of the literature are collated from the screening of abstract pages for the first 500 highly cited papers. This keyword set is provided in the Appendix.

Two datasets are used for this study. The highly cited 100 papers comprise the first dataset (sample data set, $n = 100$ papers) whilst all the papers form the second dataset (population data set, $n =$ over 2,650 papers).

The data on the indices, document types, publication years, institutions, funding bodies, source titles, countries, 'Web of Science' subject categories, citation impact, keywords, and research fronts are collated from these datasets. The key findings are provided in the relevant tables and figure, supplemented with explanatory notes in

the text. The findings are discussed and a number of conclusions are drawn and a number of recommendations for further study are made.

21.3 RESULTS

21.3.1 INDICES AND DOCUMENTS

There are over 3,200 papers in this field in the 'Web of Science' as of January 2020. This original population dataset is refined for the document type (article, review, book chapter, book, editorial material, note, and letter) and language (English), resulting in over 2,650 papers comprising over 83.5% of the original population dataset.

The primary index is the SCI-E for both the sample and population papers. Of the population papers, 95.1% are indexed by the SCI-E database. Additionally 3.5, 4.3, and 0.3% of these papers are indexed by the CPCI-S, ESCI, and BCI-S databases, respectively. The papers on the social and humanitarian aspects of this field are relatively negligible with 1.1 and 0.0.% of the population papers indexed by the SSCI and A&HCI, respectively.

Brief information on the document types for both datasets is provided in Table 21.1. The key finding is that articles are the primary documents for both the sample and population papers, whilst reviews form only 2% of the sample papers.

21.3.2 AUTHORS

Brief information about the nine-most-prolific authors with at least three sample papers each is provided in Table 21.2. Around 330 and 6,700 authors contribute to the sample and population papers, respectively.

The most-prolific authors are 'Farooq Anwar', 'Bryan R. Moser', 'Umer Rashid', and 'Wenlei Xie' with four sample papers each, working primarily on 'biodiesel fuel production' and 'biodiesel fuel properties'. These top four authors have the most impact with a 9.9% publication surplus altogether.

On the other hand, a number of authors have a significant presence in the population papers: 'Hassan H. Masjuki', 'M. Abul Kalam', 'Vlada B. Veljkovic', 'J. Vladimir

TABLE 21.1
Document Types

	Document Type	Sample Dataset (%)	Population Dataset (%)	Difference (%)
1	Article	97	98.4	−1.4
2	Review	2	1.2	0.8
3	Book chapter	0	0.3	−0.3
4	Proceeding paper	4	3.9	0.1
5	Editorial material	0	0.0	0.0
6	Letter	0	0.3	−0.3
7	Book	0	0.0	0.0
8	Note	1	0.2	0.8

TABLE 21.2
Authors

	Author	Sample Papers (%)	Population Papers (%)	Surplus (%)	Institution	Country	Edible Oils	Research Front
1	Anwar, Farooq	4	0.3	3.7	Univ. Agr. Faisalabad	Pakistan	Rape, moringa, cotton, sunfloweer	Biodiesel production, biodiesel properties
2	Moser, Bryan R.	4	0.7	3.3	Dept. Agr.	USA	Moringa, camelina, sunflower, soy	Biodiesel production, biodiesel properties
3	Rashid, Umer	4	0.6	3.4	Univ. Agr. Faisalabad	Pakistan	Rape, moringa, cotton, sunflower	Biodiesel production, biodiesel properties
4	Xie, Wenlei	4	0.5	3.5	Henan Univ. Technol.	China	Soy	Biodiesel production
5	Bunyakiat, Kunchana	3	0.2	2.8	Chulalongkorn Univ.	Thailand	Palm, coconut	Biodiesel production
6	Du, Wei	3	0.4	2.6	Tsinghua Univ.	China	Soy, rape	Biodiesel production
7	Kusdiana, Dadan	3	0.1	2.9	Kyoto Univ.	Japan	Rape	Biodiesel production
8	Liu, Dehua	3	0.4	2.6	Tsinghua Univ.	China	Soy, rape	Biodiesel production
9	Saka, Shiro	3	0.2	2.8	Kyoto Univ.	Japan	Rape	Biodiesel production

Oliveira', 'Olivera S. Stamenkovic', 'M. A. Fazal', 'Bryan R. Moser', 'A. S. Md. Abdul Haseeb', 'G. Nagarajan', 'Antonio G. Souza', 'Ieda dos Santos', 'Frantisel Skopal', 'Ajay K. Dalai', 'Hasanah M. Ghazali', and 'Martin Hajek' with at least 0.5% of the population papers each.

The most-prolific institution for these top authors is 'Kyoto University', 'Tsinghua University', and the 'University of Agriculture, Faisalabad' with two authors each. In total, five institutions house these top authors.

It is notable that none of these top researchers are listed in the 'Highly Cited Researchers' (HCR) in 2019 (Clarivate Analytics, 2019; Docampo and Cram, 2019).

The most-prolific country for these top authors is China with three. The other prolific countries are Japan and Pakistan with two authors each. In total, five countries contribute to these top papers.

There are two key research fronts for these top researchers: 'biodiesel fuel production' and 'biodiesel fuel properties' with nine and three authors, respectively. On the other hand, these top authors focus on eight edible oils. The most-prolific oil is 'rapeseed oil' with six authors. The other prolific oils are 'soybean oil', 'moringa oil', 'sunflower oil', and 'cotton seed oil' with at least two authors each.

It is further notable that there is a significant gender deficit among these top authors as only one of them is female (Lariviere et al., 2013; Xie and Shauman, 1998).

21.3.3 PUBLICATION YEARS

Information about the publication years for both datasets is provided in Figure 21.1. This figure shows that 5, 8, 71, and 16% of the sample papers and 2.0, 4.9, 20.0, and 72.8% of the population papers were published in the 1980s, 1990s, 2000s, and 2010s, respectively.

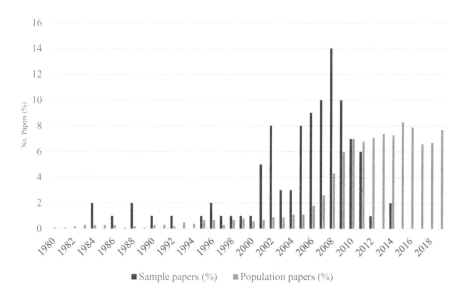

■ Sample papers (%)   ■ Population papers (%)

FIGURE 21.1   Research output between 1980 and 2019.

Similarly, the most-prolific publication years for the sample dataset are 2008, 2007, 2009, and 2006 with 14, 10, 10, and 9 papers, respectively. On the other hand, the most-prolific publication years for the population dataset are 2015, 2016, 2019, and 2013 with 8.3, 7.9, 7.7, and 7.4% of the population papers each. It is notable that there is a sharply rising trend for the population papers, particularly in the late 2000s and a significant consolidation in the 2010s.

21.3.4 INSTITUTIONS

Brief information on the top 10 institutions with at least 3% of the sample papers each is provided in Table 21.3. In total, around 130 and 1,900 institutions contribute to the sample and population papers, respectively.

These top institutions publish 45.0 and 12.9% of the sample and population papers, respectively. The top institution is the US 'Department of Agriculture' with eight sample papers and a 6% publication surplus. This top institution is closely followed by 'Tsinghua University' of China with seven sample papers and a 6.9% publication surplus. The other top institutions are the 'Scientific Research National Center' of France, the 'Chulalongkorn University' of Thailand, 'Henan University of Technology' of China, the 'University of Malaya' of Malaysia, and the 'University of Agriculture, Faisalabad' of Pakistan with four sample papers each.

The most-prolific countries for these top institutions are China, Malaysia, and the USA with two institutions each. In total, seven countries cover these institutions.

The institutions with the most impact are 'Tsinghua University' and the US 'Department of Energy' with 6.9 and 6.0% publication surpluses, respectively. On the other hand, the institutions with the least impact are the 'University of Sains Malaysia' and the 'University of Malaya' with at least a –3.3% publication surplus each.

It is notable that some institutions have a heavy presence in the population papers: the 'Malaysian Palm Oil Board', the 'University of Putra Malaysia', 'Anna

TABLE 21.3
Institutions

	Institution	Country	No. of Sample Papers (%)	No. of Population Papers (%)	Difference (%)
1	Depart. Agric.	USA	8	2.0	6.0
2	Tsinghua Univ.	China	7	1.1	6.9
3	Sci. Res. Natl. Ctr.	France	4	1.2	2.8
4	Chulalongkorn Univ.	Thailand	4	1.5	2.5
5	Henan Univ. Technol.	China	4	0.7	3.3
6	Univ. Malaya	Malaysia	4	3.7	–3.3
7	Univ. Agric. Faisalabad	Pakistan	4	0.3	3.7
8	Iowa State Univ.	USA	3	0.4	2.6
9	Kyoto Univ.	Japan	3	0.2	2.8
10	Univ. Sains Malaysia	Malaysia	3	1.9	1.1

University', the 'Indian Institute of Technology' of India, the 'Superior Council of Scientific Research' of Spain, the 'University of Sao Paulo', the 'State University of Campinas', and the 'Federal University of Ceara' of Brazil, with at least a 1% presence in the population papers each.

21.3.5 FUNDING BODIES

Brief information about the top four funding bodies with at least 2% of the sample papers each is provided in Table 21.4. It is significant that only 19.0 and 54.2% of the sample and population papers declare any funding, respectively.

The top funding body is the 'Ministry of Science' of Serbia, funding 3% of the sample papers and a 3.4% publication surplus. This top funding body is closely followed by the 'Higher Education Commission' of Pakistan, the 'National Natural Science Foundation' of China, and the 'University of Malaya' of Malaysia with two sample papers each. It is notable that all these funding bodies are from developing countries.

It is notable that some top funding agencies have a heavy presence in the population studies. Some of them are the 'National Council for Scientific and Technological Development', 'CAPES', the 'Sao Paulo State Research Foundation', and 'Science, Technology, and Innovation' of Brazil, the 'Thailand Research Fund', the 'European Union', the 'Ministry of Energy Science Technology Environment and Climate Change' of Malaysia, the US 'Department of Energy', and the 'University of Sains Malaysia' with at least 1% of the population papers each. Similarly, most of these prolific funding bodies are from developing countries.

21.3.6 SOURCE TITLES

Brief information about the top 16 source titles with at least two sample papers each is provided in Table 21.5. In total, 32 and over 530 source titles publish the sample and population papers, respectively. On the other hand, these top 16 journals publish 84.0 and 38.9% of the sample and population papers, respectively.

The top journal is 'Fuel', publishing 18 sample papers with a 9.1% publication surplus. This top journal is followed by the 'Journal of the American Oil Chemists Society' and 'Bioresource Technology' with ten sample papers each. The other top journals are 'Energy Fuels', 'Renewable Energy', 'Energy Conversion and

TABLE 21.4
Funding Bodies

	Institution	Country	No. of Sample Papers (%)	No. of Population Papers (%)	Difference (%)
1	Ministry. Science	Serbia	3	0.7	2.3
2	Higher Ed. Comm.	Pakistan	2	0.4	1.6
3	Natl. Natr. Sci. Found.	China	2	4.0	1.6
4	Univ. Malaya	Malaysia	2	1.6	0.4

TABLE 21.5
Source Titles

	Source Title	Wos Subject Category	No. of Sample Papers (%)	No. of Population Papers (%)	Difference (%)
1	Fuel	Ener. Fuels, Eng. Chem.	18	8.9	9.1
2	Journal of the American Oil Chemists Society	Chem. Appl., Food Sci. Technol.	10	5.2	4.8
3	Bioresource Technology	Agr. Eng., Biot. Appl. Microb., Ener. Fuels	10	3.3	6.7
4	Energy Fuels	Ener. Fuels, Eng. Chem.	5	3.5	1.5
5	Renewable Energy	Green Sust. Sci. Technol., Ener. Fuels	5	3.5	1.5
6	Energy Conversion and Management	Therm., Ener. Fuels, Mechs.	5	2.9	2.1
7	Fuel Processing Technology	Che. Appl., Ener. Fuels, Eng. Chem.	5	2.7	2.3
8	Industrial Engineering Chemistry Research	Eng. Chem.	4	1.6	2.4
9	Journal of Molecular Catalysis B Enzymatic	Bioch. Mol. Biol., Chem. Phys.	4	0.6	3.4
10	Enzyme and Microbial Technology	Biot. Appl. Microb.	4	0.3	3.7
11	Biomass Bioenergy	Agr. Eng., Biot. Appl. Microb., Ener. Fuels	3	2.4	0.6
12	Applied Catalysis A General	Chem. Phys., Env. Sci.	3	1.0	2.0
13	Applied Energy	Ener. Fuels, Eng. Chem	2	1.3	0.7
14	Renewable Sustainable Energy Reviews	Green Sust. Sci. Technol., Ener. Fuels	2	0.7	1.3
15	Chemical Engineering Journal	Eng., Env., Eng. Chem.	2	0.7	1.3
16	Journal of Molecular Catalysis A Chemical	Chem. Phys.	2	0.3	1.7

Management', and 'Fuel Processing Technology' with at least five sample papers each.

Although these journals are indexed by 14 subject categories, the top categories are 'Energy Fuels' and 'Engineering Chemical' with nine and six journals, respectively. The other prolific subject categories are 'Biotechnology and Applied Microbiology' and 'Chemistry Physics' with three journals each and 'Agricultural Engineering', 'Chemistry Applied', and 'Green and Sustainable Science and Technology' with two journals each.

The journals with the most impact are 'Fuel', 'Bioresource Technology', and the 'Journal of the American Oil Chemists Society' with at least a 4.8% publication surplus each. On the other hand, the journals with the least impact are 'Biomass Bioenergy', 'Applied Energy', 'Renewable Sustainable Energy Reviews', and the 'Chemical Engineering Journal' with at least a 0.6% publication surplus each.

It is notable that some journals have a heavy presence in the population papers. Some of them are 'Energy Sources Part A Recovery Utilization and Environmental Effects', 'Energy', 'Industrial Crops and Products', the 'Journal of Cleaner Production', the 'Journal of Thermal Analysis and Calorimetry', the 'European Journal of Lipid Science and Technology', 'Applied Biochemistry and Biotechnology', and the 'Journal of Chemical and Engineering Data' with at least a 0.8% presence in the population papers each.

21.3.7   COUNTRIES

Brief information about the top 12 countries with at least three sample papers each is provided in Table 21.6. In total, 30 and over 100 countries contribute to the sample and population papers, respectively.

The top country is the USA, publishing 22.0 and 11.8% of the sample and population papers, respectively. China follows the USA with 15.0 and 9.7% of the sample

TABLE 21.6
Countries

	Country	No. of Sample Papers (%)	No. of Population Papers (%)	Difference (%)
1	USA	22	11.8	10.2
2	China	15	9.7	5.3
3	Japan	7	2.6	4.4
4	Malaysia	7	10.7	−3.7
5	Turkey	7	5.9	1.1
6	Thailand	6	3.8	2.2
7	France	5	2.1	2.9
8	Pakistan	5	1.6	3.4
9	Brazil	4	12.0	−8.0
10	India	4	10.1	−6.1
11	Spain	4	4.0	0.0
12	Serbia	3	1.4	1.6
	Europe-3	12	7.5	4.5
	Asia-7	48	50.5	−2.5

and population papers, respectively. The other prolific countries are Japan, Malaysia, and Turkey, publishing seven sample papers each.

On the other hand, the European and Asian countries represented in Table 21.6 publish altogether 12 and 48% of the sample papers, whilst they publish 7.5 and 50.5% of the population papers, respectively.

It is notable that the publication surplus for the USA and these European and Asian countries is 10.2, 4.5, and −2.5%, respectively. On the other hand, the countries with the most impact are the USA, China, and Japan with 10.2, 5.3, and 4.4% publication surpluses, respectively. Furthermore, the countries with the least impact are Brazil, India, and Malaysia with −8.0, −6.1, and −3.7% publication deficits, respectively.

It is also notable that some countries have a heavy presence in the population papers. The major producers of these papers are Canada, Iran, the UK, South Korea, Taiwan, Italy, Indonesia, the Czech Republic, Germany, Colombia, Greece, Poland, Saudi Arabia, Australia, and Romania with at least 1.0% of the population papers each.

21.3.8 'WEB OF SCIENCE' SUBJECT CATEGORIES

Brief information about the top 12 'Web of Science' subject categories with at least four sample papers each is provided in Table 21.7. The sample and population papers are indexed by 20 and over 90 subject categories, respectively.

For the sample papers, the top subject is 'Energy Fuels' with 58.0 and 43.6% of the sample and population papers, respectively. This top subject category is followed by 'Engineering Chemical' with 40.0 and 33.8% of the sample and population papers, respectively. The other prolific subjects are 'Biotechnology Applied

TABLE 21.7
Web of Science Subject Categories

	Subject	No. of Sample Papers (%)	No. of Population Papers (%)	Difference (%)
1	Energy Fuels	58	43.6	14.4
2	Engineering Chemical	40	33.8	6.2
3	Biotechnology Applied Microbiology	18	10.5	7.5
4	Chemistry Applied	16	11.9	4.1
5	Agricultural Engineering	15	9.0	6.0
6	Food Science Technology	12	9.9	2.1
7	Chemistry Physical	11	8.2	2.8
8	Green Sustainable Science Technology	9	8.0	1.0
9	Thermodynamics	8	9.9	−1.9
10	Biochemistry Molecular Biology	6	3.1	2.9
11	Environmental Sciences	6	9.1	−3.1
12	Mechanics	6	3.6	2.4
13	Engineering Environmental	4	5.0	−1.0

Microbiology', 'Chemistry Applied', 'Agricultural Engineering', 'Food Science Technology', and 'Chemistry Physical' with at least 11 sample papers each.

It is notable that the publication surplus is most significant for 'Energy Fuels', 'Biotechnology Applied Microbiology', 'Engineering Chemical', and 'Agricultural Engineering' with 14.4, 7.5, 6.2, and 6.0% publication surpluses, respectively. On the other hand, the subjects with least impact are 'Environmental Sciences', 'Thermodynamics', and 'Engineering Environmental' with at least a –3.1% publication deficit each. This latter group of subject categories are under-represented in the sample papers.

Additionally, some subject categories have a heavy presence in the population papers: 'Chemistry Multidisciplinary', 'Engineering Mechanical', 'Agronomy', 'Chemistry Analytical', 'Nutrition Dietetics', 'Engineering Multidisciplinary', 'Materials Science Multidisciplinary', 'Multidisciplinary Sciences', 'Transportation Science Technology', and 'Agriculture Multidisciplinary' have at least a 1% presence in the population papers each.

21.3.9   CITATION IMPACT

These sample papers received about 22,600 citations as of January 2020. Thus, the average number of citations per paper is about 225.

21.3.10   KEYWORDS

Although a number of keywords are listed in the Appendix for the datasets related to this field, some of them are more significant for the sample papers.

The most-prolific keyword for the set related to edible oils is 'soy*' with 32 papers. This top keyword is followed by 'palm*', 'sunflower', and 'rape*' with 19, 16, and 14 sample papers, respectively. There are also two 'canola' papers related to rapeseed. The other prolific keywords are 'cotton*', 'brassica', 'coconut*', 'fish oil*', 'rice bran', and 'camelina' with at least two sample papers each.

On the other hand, the most-prolific keyword related to diesel fuels is '*diesel' with 78 sample papers. This top keyword is followed by 'transester*' and 'methyl ester*' with 32 and 10 sample papers, respectively. The other prolific keywords are 'compression ignition engine', 'methanolysis', 'hydrotreat*', 'hydroprocess*', 'alcoholysis', and 'ethyl-ester' with at least two sample papers each.

21.3.11   RESEARCH FRONTS

Brief information about the key research fronts is provided in Table 21.8. There are two major topical research fronts for these sample papers: 'edible oil-based biodiesel production' (Antolin et al., 2002; Darnoko and Cheryan, 2000; Kusdiana and Saka, 2001; Noureddini and Zhu, 1997; Saka and Kusdiana, 2001) and 'properties and characterization of edible oil-based biodiesel fuels' (Benjumea et al., 2008; Clark et al., 1984; Dunn, 2005; Gelbard et al., 1995; Labeckas and Slavinskas, 2006) with 84 and 35 sample papers, respectively.

TABLE 21.8
Research Fronts

Research front	No. of Sample Papers (%)
Topical research fronts	
Biodiesel production	84
Biodiesel characterization and properties	35
Biomass-based research fronts	
Soybean oil-based biodiesel fuels	32
Palm oil-based biodiesel fuels	19
Rapeseed oil-based biodiesel fuels	16
Sunflower oil-based biodiesel fuels	16
Brassica-Camelina oil-based biodiesel fuels	6
Cotton seed oil-based biodiesel fuels	5
Fish oil-based biodiesel fuels	4
Rice bran oil-based biodiesel fuels	3
Other edible oil-based biodiesel fuels	9

On the other hand, there are nine research fronts for the edible oils used for biodiesel production. The most-prolific research front is 'soybean oil-based biodiesel fuels' with 32 sample papers (Freedman et al., 1986; Noureddini and Zhu, 1997). This top research front is followed by 'palm oil-based biodiesel fuels' (Darnoko and Cheryan, 2000; Jitputti et al., 2006), 'rapeseed oil-based biodiesel fuels' (Kusdiana and Saka, 2001; Saka and Kusdiana, 2001), and 'sunflower oil-based biodiesel fuels' (Antolin et al., 2002; Soumanou and Bornscheuer, 2003) with 19, 16, and 16 sample papers, respectively. These top four research fronts form 83% of the sample papers in total.

The other prolific research fronts are 'Brassica-Camelina oil-based biodiesel fuels' (Dorado et al., 2004), 'cotton seed oil-based biodiesel fuels' (Nabi et al., 2009), 'fish oil-based biodiesel fuels' (Indarti et al., 2005), 'rice bran oil-based biodiesel fuels' (Zullaikah et al., 2005), and 'other edible oil-based biodiesel fuels' (Banapurmath et al., 2008) with six, five, four, three, and nine sample papers, respectively.

21.4 DISCUSSION

The size of the research in this field has increased to over 2,650 papers as of January 2020. It is expected that the number of the population papers in this field will exceed 7,500 papers by the end of the 2020s.

The research has developed more in the technological aspects of this field, rather than the social and humanitarian pathways, as evidenced by the negligible number of population papers in the indices of the 'Web of Science', SSCI and A&HCI.

The article types of documents are the primary documents for both datasets and reviews are over-represented by 30.4% in the sample papers (Table 21.1). Thus, the contribution of reviews by only 2% of the sample papers in this field is highly exceptional (cf. Konur, 2011, 2012a–n, 2015, 2016a–f, 2017a–f, 2018a–b, 2019a–b).

Nine authors from five institutions have at least three sample papers each (Table 21.2). Three of these authors are from China. The other prolific countries are Japan and Pakistan with two authors.

These authors focus on 'biodiesel fuel production' and 'biodiesel fuel properties'. There is significant 'gender deficit' among these top authors as only one of them is female (Lariviere et al., 2013; Xie and Shauman, 1998).

The population papers have built on the sample papers, primarily published in the 2000s and in the 2010s (Figure 21.1). Following a rising trend, particularly in the late 2000s, it is expected that the number of papers will reach 7,500 by the end of the 2020s, nearly tripling the current size.

The engagement of the institutions in this field at the global scale is significant as around 130 and 1,900 institutions contribute to the sample and population papers, respectively.

Ten top institutions publish 45.0 and 12.9% of the sample and population papers, respectively (Table 21.3). The top institution is the US 'Department of Agriculture' and 'Tsinghua University' of China with eight and seven sample papers, respectively. The other top institutions are the 'Scientific Research National Center' of France, the 'Chulalongkorn University' of Thailand, 'Henan University of Technology' of China, the 'University of Malaya' of Malaysia, and the 'University of Agriculture, Faisalabad' of Pakistan with four sample papers each.

The most-prolific countries for these top institutions are China, Malaysia, and the USA. It is notable that some institutions with a heavy presence in the population papers are under-represented in the sample papers.

It is significant that only 19.0 and about 54.2% of the sample and population papers declare any funding, respectively. It is notable that all the top funding bodies are from developing countries (Table 21.4). It is further notable that some top funding agencies for the population studies do not enter this top funding body list.

However, the lack of Chinese funding bodies in this table is notable. This finding is in contrast with the studies showing heavy research funding in China, where the NSFC is the primary funding agency (Wang et al., 2012).

The sample and population papers are published by 32 and over 500 journals, respectively. It is significant that the top 16 journals publish 84.0 and 38.9% of the sample and population papers, respectively (Table 21.5).

The top journal is 'Fuel', publishing 18 sample papers with a 9.1% publication surplus. The other top journals are the 'Journal of the American Oil Chemists Society', 'Bioresource Technology', 'Energy Fuels', 'Renewable Energy', 'Energy Conversion and Management', and 'Fuel Processing Technology' with at least five sample papers each.

The top categories for these journals are 'Energy Fuels', 'Engineering Chemical', 'Biotechnology and Applied Microbiology', and 'Chemistry Physics'.

It is notable that some journals with a heavy presence in the population papers are relatively under-represented in the sample papers.

In total, 30 and over 100 countries contribute to the sample and population papers, respectively. The top country is the USA publishing 22.0 and 11.8% of the sample and population papers, respectively (Table 21.6). This finding is in line with the studies arguing that the USA is not losing ground in science and technology (Leydesdorff and Wagner, 2009).

China follows the USA with 15.0 and 9.7% of the sample and population papers, respectively. The other prolific countries are Japan, Malaysia, and Turkey, publishing seven sample papers each. These findings are in line with the studies showing huge research incentives in these countries (Kumar and Jan, 2014; Urata, 1990).

On the other hand, the European and Asian countries represented in this table publish altogether 12 and 48% of the sample papers and 7.5 and 50.5% of the population papers, respectively. These findings are in line with the studies showing that European countries have superior publication performance in science and technology (Bordons et al., 2015; Youtie et al., 2008).

On the other hand, the European and Asian countries represented in this table publish altogether 12 and 48% of the sample papers and 7.5 and 50.5% of the population papers, respectively.

It is notable that the publication surplus for the USA and these European and Asian countries is 10.2, 4.5, and –2.5%, respectively.

It is further notable that China has a significant publication surplus (5.3%). This finding is in line with China's efforts to be a leading nation in science and technology (Guan and Ma, 2007; Youtie et al., 2008; Zhou and Leydesdorff, 2006).

On the other hand, the countries with the most impact are the USA, China, and Japan with 10.2, 5.3, and 4.4% publication surpluses, respectively. Furthermore, the countries with the least impact are Brazil, India, and Malaysia with –8.0, –6.1, and –3.7% publication deficits, respectively (Bhattacharya et al., 2012; Glanzel et al., 2006)

It is also notable that some countries have a heavy presence in the population papers. The major producers of these papers are Canada, Iran, the UK, South Korea, Taiwan, Italy, Indonesia, the Czech Republic, Germany, Colombia, Greece, Poland, Saudi Arabia, Australia, and Romania with at least 1.0% of the population papers each (Braun et al., 1991; Fu and Ho, 2015; Huang et al. 2006; Leydesdorff and Zhou, 2005; Moin et al., 2005).

The sample and population papers are indexed by 20 and over 90 subject categories, respectively.

For the sample papers, the top subject is 'Energy Fuels' with 58.0 and 43.6% of the sample and population papers, respectively (Table 21.7). This top subject category is followed by 'Engineering Chemical' with 40.0 and 33.8% of the sample and population papers, respectively. The other prolific subjects are 'Biotechnology Applied Microbiology', 'Chemistry Applied', 'Agricultural Engineering', 'Food Science Technology', and 'Chemistry Physical' with at least 11 sample papers each.

It is notable that the publication surplus is most significant for 'Energy Fuels', 'Biotechnology Applied Microbiology', 'Engineering Chemical', and 'Agricultural Engineering' with 14.4, 7.5, 6.2, and 6.0% publication surpluses, respectively. On the other hand, the subjects with least impact are 'Environmental Sciences', 'Thermodynamics', and 'Engineering Environmental' with at least a –3.1% publication deficit each. This latter group of subject categories are under-represented in the sample papers.

These sample papers received about 22,600 citations as of January 2020. Thus, the average number of citations per paper is about 226. Hence, the citation impact of these top 100 papers in this field is significant.

Although a number of keywords are listed in the Appendix for the datasets related to this field, some of them are more significant for the sample papers.

The most-prolific keyword for the set related to edible oils is 'soy*' with 32 papers. This top keyword is followed by 'palm*', 'sunflower', and 'rape*' with 19, 16, and 14 sample papers, respectively. There are also two 'canola' papers related to rapeseed. The other prolific keywords are 'cotton*', 'brassica', 'coconut*', 'fish oil*', 'rice bran', and 'camelina' with at least two sample papers each.

On the other hand, the most-prolific keyword related to diesel fuels is '*diesel' with 78 sample papers. This top keyword is followed by 'transester*' and 'methyl ester*' with 32 and 10 sample papers, respectively. The other prolific keywords are 'compression ignition engine', 'methanolysis', 'hydrotreat*', 'hydroprocess*', 'alcoholysis', and 'ethyl-ester' with at least two sample papers each. As expected, these keywords provide valuable information about the pathways of research in this field.

There are two major topical research fronts for these sample papers: 'edible oil-based biodiesel production' and 'edible oil-based biodiesel properties' with 84 and 35 sample papers, respectively (Table 21.8).

On the other hand, there are nine research fronts for the edible oils used for biodiesel production. The most-prolific research front is 'soybean oil-based biodiesel fuels' with 32 sample papers. This top research front is followed by 'palm oil-based biodiesel fuels', 'rapeseed oil-based biodiesel fuels', and 'sunflower oil-based biodiesel fuels' with 19, 16, and 16 sample papers, respectively. These top four research fronts form 83% of the sample papers in total.

The other prolific research fronts are 'brassica-camelina oil-based biodiesel fuels', 'cotton seed oil-based biodiesel fuels', 'fish oil-based biodiesel fuels', 'rice bran oil-based biodiesel fuels', and 'other edible oil-based biodiesel fuels' with six, five, four, three, and nine sample papers, respectively.

The key emphasis in these research fronts is the exploration of the structure–processing–property relationships of edible oil biomass, edible biooils, and edible oil-based biodiesel fuels (Cheng and Ma, 2011; Konur and Matthews, 1989; Rogers and Hopfinger, 1994; Scherf and List, 2002).

21.5 CONCLUSION

This chapter has mapped the research on edible oil-based biodiesel fuels using a scientometric method.

The size of over 2,650 population papers shows the public importance of this interdisciplinary research field. However, it is significant that the research has developed more in the technological aspects than in social and humanitarian pathways.

Articles dominate both the sample and population papers. The population papers, primarily published in the 2010s, build on these sample papers, primarily published in the 2000s.

The data presented in the tables and figure show that a small number of authors, institutions, funding bodies, journals, keywords, research fronts, types of edible oils, subject categories, and countries have shaped the research in this field.

It is notable that the authors, institutions, and funding bodies from the USA, China, Japan, Malaysia, and Turkey dominate the research in this field. Furthermore, it is also notable that some countries have a heavy presence in the population papers. The major producers of the population papers are Canada, Iran, the UK, South Korea, Taiwan, Italy, Indonesia, the Czech Republic, Germany, Colombia, Greece, Poland, Saudi Arabia, Australia, and Romania with at least 1.0% of the population papers each. Brazil, India, and Malaysia are under-represented significantly in the sample papers.

These findings show the importance of the progression of efficient incentive structures for the development of the research in this field as in other fields. It seems that the USA and European countries (such as France and Serbia) have efficient incentive structures for the development of the research in this field, contrary to Brazil, India, Malaysia, Canada, Iran, the UK, South Korea, Taiwan, Italy, Indonesia, the Czech Republic, Germany, Colombia, Greece, Poland, Saudi Arabia, Australia, and Romania.

It further seems that, although research funding is a significant element of these incentive structures, it might not be a sole solution for increasing the incentives for research as is the case in Brazil, India, Malaysia, Canada, Iran, the UK, South Korea, Taiwan, Italy, Indonesia, the Czech Republic, Germany, Colombia, Greece, Poland, Saudi Arabia, Australia, and Romania.

On the other hand, it seems there is more to do to reduce the significant gender deficit in this field, as in other fields of science and technology (Lariviere et al., 2013; Xie and Shauman, 1998).

The data on the research fronts, keywords, source titles, and subject categories provide valuable evidence for the interdisciplinary (Lariviere and Gingras, 2010; Morillo et al. 2001) nature of the research in this field.

There is ample justification for the broad search strategy employed in this study due to the interdisciplinary nature of this research field, as evidenced by the top subject categories. The search strategy employed in this study is in line with those employed for related and other research fields (Konur, 2011, 2012a–n, 2015, 2016a–f, 2017a–f, 2018a–b, 2019a–b). It is particularly noted that only 58.0 and 43.6% of the sample and population papers are indexed by the 'Energy Fuels' subject category, respectively.

There are two major topical research fronts for these sample papers: 'edible oil-based biodiesel fuel production' and 'properties and characterization of edible oil-based biodiesel fuels' (Table 21.8). On the other hand, there are nine research fronts for the edible oils used for biodiesel production. The most-prolific research front is 'soybean oil-based biodiesel fuels' with 32 sample papers. This top research front is followed by 'palm oil-based biodiesel fuels', 'rapeseed oil-based biodiesel fuels', and 'sunflower oil-based biodiesel fuels' with 19, 16, and 16 sample papers, respectively. These top four research fronts form 83% of the sample papers in total.

It is recommended that further scientometric studies are carried out for each of these research fronts, building on the pioneering studies in these fields.

ACKNOWLEDGMENTS

The contribution of the highly cited researchers in the fields of edible oil-based biodiesel fuels is greatly acknowledged.

21.A APPENDIX

The keyword set for the edible oil-based biodiesel fuels
Syntax: (1 AND 2) OR 3

21.A.1. EDIBLE OIL-RELATED KEYWORDS

TI=("cooking oil" or edible or cotton* or soy* or sunflower* or rape* or palm or corn or olive* or opium or canola or coffee or peanut* or coconut* or cynara or artichoke or brassica or hazelnut* or helianthus or "fish oil*" or groundnut* or elaeis or cocos or babassu or pequi or caryocar or dende or linseed or flax* or camelina or candlenut* or aleurites or moringa or mustard or sesam* or safflower* or apricot or "rocket seed*" or "yellow horn" or paradise or soapnut or eruca or walnut or andiroba or cumaru or tall or cerbera or mango or pumpkin or cucurbita or okra or pistacia or almond or date or "rice bran*" or arachis or hibiscus or terebinth or colza) NOT TI=(waste* or "non-edible*" or soapstock* or frying or used or stover or shell* or *cob* or cake* or spent or meal).

21.A.2 DIESEL FUELS-RELATED KEYWORDS

TI=(*diesel or transester* or "trans-ester*" or "methyl-ester*" or "compression ignition engine*" or methanolysis or "ci engine*" or hydrotreat* or hydroprocessing or alcoholysis or "ethyl-ester*" or ethanolysis or fame or hydrodeoxygenation or hydrocracking or "thermal cracking").

21.A.3 CROSS-SUBJECT KEYWORDS

TI=("methyl-soyate").

REFERENCES

Ahmadun, F. R., A. Pendashteh, and L. C. Abdullah, et al. 2009. Review of technologies for oil and gas produced water treatment. *Journal of Hazardous Materials* 170:530–551.
Antolin, G., F. V. Tinaut, and Y. Briceno. 2002. Optimisation of biodiesel production by sunflower oil transesterification. *Bioresource Technology* 83:111–114.
Atlas, R. M. 1981. Microbial degradation of petroleum hydrocarbons: An environmental perspective. *Microbiological Reviews* 45:180–209.
Babich, I. V. and J. A. Moulijn. 2003. Science and technology of novel processes for deep desulfurization of oil refinery streams: A review. *Fuel* 82:607–631.
Banapurmath, N. R., P. G. Tewari, and R. S. Hosmath. 2008. Performance and emission characteristics of a DI compression ignition engine operated on Honge, Jatropha and sesame oil methyl esters. *Renewable Energy* 33:1982–1988.

Benjumea, P., J. Agudelo, and A. Agudejo. 2008. Basic properties of palm oil biodiesel-diesel blends. *Fuel* 87:2069–2075.

Bhattacharya, S., Shilpa, and M. Bhati. 2012. China and India: The two new players in the nanotechnology race. *Scientometrics* 93:59–87.

Bordons, M., B. Gonzalez-Albo, J. Aparicio, and L. Moreno. 2015. The influence of R&D intensity of countries on the impact of international collaborative research: Evidence from Spain. *Scientometrics* 102:1385–1400.

Braun, T., W. Glanzel, and A. Schubert. 1991. The bibliometric assessment of UK scientific performance—Some comments on Martin's "reply". *Scientometrics* 20:359–362.

Bridgwater, A. V. and G. V. C. Peacocke. 2000. Fast pyrolysis processes for biomass. *Renewable & Sustainable Energy Reviews* 4:1–73.

Busca, G., L. Lietti, G. Ramis, and F. Berti. 1998. Chemical and mechanistic aspects of the selective catalytic reduction of NO_x by ammonia over oxide catalysts: A review. *Applied Catalysis B-Environmental* 18:1–36.

Cheng, Y. Q. and E. Ma. 2011. Atomic-level structure and structure–property relationship in metallic glasses. *Progress in Materials Science* 56:379–473.

Clarivate Analytics. 2019. *Highly cited researchers: 2019 Recipients.* Philadelphia, PA: Clarivate Analytics. https://recognition.webofsciencegroup.com/awards/highly-cited/2019/ (accessed January, 3, 2020).

Clark, S. J., L. Wagner, M. D. Schrock, and P. G. Piennaar. 1984. Methyl and ethyl soybean esters as renewable fuels for diesel engines. *Journal of the American Oil Chemists Society* 61:1632–1638.

Czernik, S. and A. V. Bridgwater. 2004. Overview of applications of biomass fast pyrolysis oil. *Energy & Fuels* 18:590–598.

Darnoko, D. and M. Cheryan. 2000. Kinetics of palm oil transesterification in a batch reactor. *Journal of the American Oil Chemists Society* 77:1263–1267.

Docampo, D. and L. Cram. 2019. Highly cited researchers: A moving target. *Scientometrics* 118:1011–1025.

Dorado, M. P., E. Ballesteros, F. J. Lopez, and M. Mittelbach. 2004. Optimization of alkali-catalyzed transesterification of *Brassica carinata* oil for biodiesel production. *Energy & Fuels* 18:77–83.

Dunn, R. O. 2005. Effect of antioxidants on the oxidative stability of methyl soyate (biodiesel). *Fuel Processing Technology* 86:1071–1085.

Freedman, B., R. O. Butterfield, and E. H. Pryde. 1986. Transesterification kinetics of soybean oil. *Journal of the American Oil Chemists Society* 63:1375–1380.

Fu, H. Z. and Y. S. Ho. 2015. Highly cited Canada articles in Science Citation Index Expanded: A bibliometric analysis. *Canadian Social Science* 11:50.

Gallezot, P. 2012. Conversion of biomass to selected chemical products. *Chemical Society Reviews* 41:1538–1558.

Garfield, E. 1955. Citation indexes for science. *Science* 122:108–111.

Garfield, E. 1972. Citation analysis as a tool in journal evaluation. *Science* 178:471–479.

Gelbard, G., O. Bres, R. M. Vargas, F. Vielfaure, and U. F. Schuchardt. 1995. [1]H nuclear-magnetic resonance determination of the yield of the transesterification of rapeseed oil with methanol. *Journal of the American Oil Chemists Society* 72:1239–1241.

Glanzel, W., J. Leta, and B. Thijs. 2006. Science in Brazil. Part 1: A macro-level comparative study. *Scientometrics* 67: 67–86.

Guan, J. C. and N. Ma. 2007. China's emerging presence in nanoscience and nanotechnology: A comparative bibliometric study of several nanoscience 'giants'. *Research Policy* 36: 880–886.

Huang, M. H., H. W. Chang, D. Z. Chen. 2006. Research evaluation of research-oriented universities in Taiwan from 1993 to 2003. *Scientometrics* 67: 419–435.

Indarti, E., M. I. A. Majid, R. Hashim, and A. Chong. 2005. Direct FAME synthesis for rapid total lipid analysis from fish oil and cod liver oil. *Journal of Food Composition and Analysis* 18:161–170.

Jitputti, J., B. Kitiyanan, and P. Rangsunvigit. 2006. Transesterification of crude palm kernel oil and crude coconut oil by different solid catalysts. *Chemical Engineering Journal* 116: 61–66.

Khalili, N. R., P. A. Scheff, and T. M. Holsen. 1995. PAH source fingerprints for coke ovens, diesel and gasoline-engines, highway tunnels, and wood combustion emissions. *Atmospheric Environment* 29: 533–542.

Kilian, L. 2009. Not all oil price shocks are alike: Disentangling demand and supply shocks in the crude oil market. *American Economic Review* 99: 1053–1069.

Konur, O. 2000. Creating enforceable civil rights for disabled students in higher education: An institutional theory perspective. *Disability & Society* 15:1041–1063.

Konur, O. 2002a. Access to Nursing Education by disabled students: Rights and duties of nursing programs. *Nurse Education Today* 22: 364–374.

Konur, O. 2002b. Assessment of disabled students in higher education: Current public policy issues. *Assessment and Evaluation in Higher Education* 27:131–152.

Konur, O. 2002c. Access to employment by disabled people in the UK: Is the Disability Discrimination Act working? *International Journal of Discrimination and the Law* 5: 247–279.

Konur, O. 2006a. Participation of children with dyslexia in compulsory education: Current public policy issues. *Dyslexia* 12: 51–67.

Konur, O. 2006b. Teaching disabled students in Higher Education. *Teaching in Higher Education* 11: 351–363.

Konur, O. 2007a. A judicial outcome analysis of the Disability Discrimination Act: A windfall for the employers? *Disability & Society* 22:187–204.

Konur, O. 2007b. Computer-assisted teaching and assessment of disabled students in higher education: The interface between academic standards and disability rights. *Journal of Computer Assisted Learning* 23: 207–219.

Konur, O. 2011. The scientometric evaluation of the research on the algae and bio-energy. *Applied Energy* 88: 3532–3540.

Konur, O. 2012a. Evaluation of the research on the social sciences in Turkey: A scientometric approach. *Energy Education Science and Technology Part B: Social and Educational Studies* 4:1893–1908.

Konur, O. 2012b. Prof. Dr. Ayhan Demirbas' scientometric biography. *Energy Education Science and Technology Part A: Energy Science and Research* 28:727–738.

Konur, O. 2012c. The evaluation of the biogas research: A scientometric approach. *Energy Education Science and Technology Part A: Energy Science and Research* 29: 1277–1292.

Konur, O. 2012d. The evaluation of the educational research: A scientometric approach. *Energy Education Science and Technology Part B: Social and Educational Studies* 4:1935–1948.

Konur, O. 2012e. The evaluation of the global energy and fuels research: A scientometric approach. *Energy Education Science and Technology Part A: Energy Science and Research* 30: 613–628.

Konur, O. 2012f. The evaluation of the research on the Arts and Humanities in Turkey: A scientometric approach. *Energy Education Science and Technology Part B: Social and Educational Studies* 4:1603–1618.

Konur, O. 2012g. The evaluation of the research on the biodiesel: A scientometric approach. *Energy Education Science and Technology Part A: Energy Science and Research* 28:1003–1014.

Konur, O. 2012h. The evaluation of the research on the bioethanol: A scientometric approach. *Energy Education Science and Technology Part A: Energy Science and Research* 28:1051–1064.

Konur, O. 2012i. The evaluation of the research on the biofuels: A scientometric approach. *Energy Education Science and Technology Part A: Energy Science and Research* 28: 903–916.

Konur, O. 2012j. The evaluation of the research on the biohydrogen: A scientometric approach. *Energy Education Science and Technology Part A: Energy Science and Research* 29: 323–338.

Konur, O. 2012k. The evaluation of the research on the microbial fuel cells: A scientometric approach. *Energy Education Science and Technology Part A: Energy Science and Research* 29: 309–322.

Konur, O. 2012l. The scientometric evaluation of the research on the production of bioenergy from biomass. *Biomass and Bioenergy* 47: 504–515.

Konur, O. 2012m. The scientometric evaluation of the research on the deaf students in higher education. *Energy Education Science and Technology Part B: Social and Educational Studies* 4:1573–1588.

Konur, O. 2012n. The scientometric evaluation of the research on the students with ADHD in higher education. *Energy Education Science and Technology Part B: Social and Educational Studies* 4:1547–1562.

Konur, O. 2015. Current state of research on algal biodiesel. In *Marine Bioenergy: Trends and Developments*, S. K. Kim, and C. G. Lee, ed., 487–512. Boca Raton, FL: CRC Press.

Konur, O. 2016a. Scientometric overview in nanobiodrugs. In *Nanoarchitectonics for Smart Delivery and Drug Targeting*, A. M. Holban and A.M. Grumezescu, ed., 405–428. Amsterdam: Elsevier.

Konur, O. 2016b. Scientometric overview regarding nanoemulsions used in the food industry. In *Emulsions: Nanotechnology in the Agri-Food Industry*, A. M. Grumezescu, ed., 689–711. Amsterdam: Elsevier.

Konur, O. 2016c. Scientometric overview regarding the nanobiomaterials in antimicrobial therapy. In *Nanobiomaterials in Antimicrobial Therapy,* A. M. Grumezescu, ed., 511–535. Amsterdam: Elsevier.

Konur, O. 2016d. Scientometric overview regarding the nanobiomaterials in dentistry. In *Nanobiomaterials in Dentistry,* A. M. Grumezescu, ed., 425–453. Amsterdam: Elsevier.

Konur, O. 2016e. Scientometric overview regarding the surface chemistry of nanobiomaterials. In *Surface Chemistry of Nanobiomaterials,* A. M. Grumezescu, ed., 463–486. Amsterdam: Elsevier.

Konur, O. 2016f. The scientometric overview in cancer targeting. In *Nanoarchitectonics for Smart Delivery and Drug Targeting*, A. M. Holban and A. Grumezescu, ed., 871–895. Amsterdam; Elsevier.

Konur, O. 2017a. Recent citation classics in antimicrobial nanobiomaterials. In *Nanostructures for Antimicrobial Therapy*, A. Ficai and A. M. Grumezescu, ed., 669–685. Amsterdam: Elsevier.

Konur, O. 2017b. Scientometric overview in nanopesticides. In *New Pesticides and Soil Sensors,* A. M. Grumezescu, ed. 719–744. Amsterdam: Elsevier.

Konur, O. 2017c. Scientometric overview regarding oral cancer nanomedicine. In *Nanostructures for Oral Medicine*, E. Andronescu, A. M. Grumezescu, ed., 939–962. Amsterdam: Elsevier.

Konur, O. 2017d. Scientometric overview regarding water nanopurification. In *Water Purification,* A. M. Grumezescu, ed., 693–716. Amsterdam: Elsevier.

Konur, O. 2017e. Scientometric overview in food nanopreservation. In *Food Preservation,* A. M. Grumezescu, ed., 703–729. Amsterdam: Elsevier.

Konur, O. 2017f. The top citation classics in alginates for biomedicine. In *Seaweed Polysaccharides: Isolation, Biological and Biomedical Applications*, J. Venkatesan, S. Anil, S. K. Kim, ed., 223–249. Amsterdam: Elsevier.

Konur, O. 2018a. Scientometric evaluation of the global research in spine: An update on the pioneering study by Wei et al. *European Spine Journal* 27: 525–529.

Konur, O. 2018b. Bioenergy and biofuels science and technology: Scientometric overview and citation classics. In *Bioenergy and Biofuels*, O. Konur, ed., 3–63. Boca Raton: CRC Press.

Konur, O. 2019a. Cyanobacterial bioenergy and biofuels science and technology: A scientometric overview. In *Cyanobacteria: From Basic Science to Applications*, ed. A. K. Mishra, D. N. Tiwari and A. N. Rai, 419–442. Amsterdam: Elsevier.

Konur, O. 2019b. Nanotechnology applications in food: A scientometric overview. In *Nanoscience for Sustainable Agriculture*, ed. R. N. Pudake, N. Chauhan, and C. Kole, 683–711. Cham: Springer.

Konur, O., ed. 2021a. *Handbook of Biodiesel and Petrodiesel Fuels: Science, Technology, Health, and Environment*. Boca Raton, FL: CRC Press.

Konur, O., ed. 2021b. *Handbook of Biodiesel and Petrodiesel Fuels: Science, Technology, Health, and Environment. Volume 1. Biodiesel Fuels: Science, Technology, Health, and Environment*. Boca Raton, FL: CRC Press.

Konur, O., ed. 2021c. *Handbook of Biodiesel and Petrodiesel Fuels: Science, Technology, Health, and Environment. Volume 2. Biodiesel Fuels based on the Edible and Nonedible Feedstocks, Wastes, and Algae: Science, Technology, Health, and Environment*. Boca Raton, FL: CRC Press.

Konur, O., ed. 2021d. *Handbook of Biodiesel and Petrodiesel Fuels: Science, Technology, Health, and Environment. Volume 3. Petrodiesel Fuels: Science, Technology, Health, and Environment*. Boca Raton, FL: CRC Press.

Konur, O. 2021e. Biodiesel and petrodiesel fuels: Science, technology, health, and environment. In *Handbook of Biodiesel and Petrodiesel Fuels: Science, Technology, Health, and Environment. Volume 1. Biodiesel Fuels: Science, Technology, Health, and Environment*, ed. O. Konur. Boca Raton, FL: CRC Press.

Konur, O. 2021f. Biodiesel and petrodiesel fuels: A scientometric review of the research. In *Handbook of Biodiesel and Petrodiesel Fuels: Science, Technology, Health, and Environment. Volume 1. Biodiesel Fuels: Science, Technology, Health, and Environment*, ed. O. Konur. Boca Raton, FL: CRC Press.

Konur, O. 2021g. Biodiesel and petrodiesel fuels: A review of the research. In *Handbook of Biodiesel and Petrodiesel Fuels: Science, Technology, Health, and Environment. Volume 1. Biodiesel Fuels: Science, Technology, Health, and Environment*, ed. O. Konur. Boca Raton, FL: CRC Press.

Konur, O. 2021h Nanotechnology applications in the diesel fuels and the related research fields: A review of the research. In *Handbook of Biodiesel and Petrodiesel Fuels: Science, Technology, Health, and Environment. Volume 1. Biodiesel Fuels: Science, Technology, Health, and Environment*, ed. O. Konur. Boca Raton, FL: CRC Press.

Konur, O. 2021i. Biooils: A scientometric review of the research. In *Handbook of Biodiesel and Petrodiesel Fuels: Science, Technology, Health, and Environment. Volume 1. Biodiesel Fuels: Science, Technology, Health, and Environment*, ed. O. Konur. Boca Raton, FL: CRC Press.

Konur, O. 2021j. Characterization and properties of biooils: A review of the research. In *Handbook of Biodiesel and Petrodiesel Fuels: Science, Technology, Health, and Environment. Volume 1. Biodiesel Fuels: Science, Technology, Health, and Environment*, ed. O. Konur. Boca Raton, FL: CRC Press.

Konur, O. 2021k. Biomass pyrolysis and pyrolysis oils: A review of the research. In *Handbook of Biodiesel and Petrodiesel Fuels: Science, Technology, Health, and Environment.*

Volume 1. Biodiesel Fuels: Science, Technology, Health, and Environment, ed. O. Konur. Boca Raton, FL: CRC Press.

Konur, O. 2021l. Biodiesel fuels: A scientometric review of the research. In *Handbook of Biodiesel and Petrodiesel Fuels: Science, Technology, Health, and Environment. Volume 1. Biodiesel Fuels: Science, Technology, Health, and Environment,* ed. O. Konur. Boca Raton, FL: CRC Press.

Konur, O. 2021m. Glycerol: A scientometric review of the research. In *Handbook of Biodiesel and Petrodiesel Fuels: Science, Technology, Health, and Environment. Volume 1. Biodiesel Fuels: Science, Technology, Health, and Environment,* ed. O. Konur. Boca Raton, FL: CRC Press.

Konur, O. 2021n. Propanediol production from glycerol: A review of the research. In *Handbook of Biodiesel and Petrodiesel Fuels: Science, Technology, Health, and Environment. Volume 1. Biodiesel Fuels: Science, Technology, Health, and Environment,* ed. O. Konur Boca Raton, FL: CRC Press.

Konur, O. 2021o. Edible oil-based biodiesel fuels: A scientometric review of the research. *In Handbook of Biodiesel and Petrodiesel Fuels: Science, Technology, Health, and Environment. Volume 2. Biodiesel Fuels based on the Edible and Nonedible Feedstocks, Wastes, and Algae: Science, Technology, Health, and Environment,* ed. O. Konur. Boca Raton, FL: CRC Press.

Konur, O. 2021p. Palm oil-based biodiesel fuels: A review of the research. In *Handbook of Biodiesel and Petrodiesel Fuels: Science, Technology, Health, and Environment. Volume 2. Biodiesel Fuels based on the Edible and Nonedible Feedstocks, Wastes, and Algae,* ed. O. Konur. Boca Raton, FL: CRC Press.

Konur, O. 2021q. Rapeseed oil-based biodiesel fuels: A review of the research. In *Handbook of Biodiesel and Petrodiesel Fuels: Science, Technology, Health, and Environment. Volume 2. Biodiesel Fuels based on the Edible and Nonedible Feedstocks, Wastes, and Algae,* ed. O. Konur. Boca Raton, FL: CRC Press.

Konur, O. 2021r. Nonedible oil-based biodiesel fuels: A scientometric review of the research. In *Handbook of Biodiesel and Petrodiesel Fuels: Science, Technology, Health, and Environment. Volume 2. Biodiesel Fuels based on the Edible and Nonedible Feedstocks, Wastes, and Algae: Science, Technology, Health, and Environment,* ed. O. Konur. Boca Raton, FL: CRC Press.

Konur, O. 2021s. Waste oil-based biodiesel fuels: A scientometric review of the research. In *Handbook of Biodiesel and Petrodiesel Fuels: Science, Technology, Health, and Environment. Volume 2. Biodiesel Fuels based on the Edible and Nonedible Feedstocks, Wastes, and Algae: Science, Technology, Health, and Environment,* ed. O. Konur. Boca Raton, FL: CRC Press.

Konur, O. 2021t. Algal biodiesel fuels: A scientometric review of the research. In *Handbook of Biodiesel and Petrodiesel Fuels: Science, Technology, Health, and Environment. Volume 2. Biodiesel Fuels based on the Edible and Nonedible Feedstocks, Wastes, and Algae: Science, Technology, Health, and Environment,* ed. O. Konur. Boca Raton, FL: CRC Press.

Konur, O. 2021u. Algal biomass production for biodiesel production: A review of the research. In *Handbook of Biodiesel and Petrodiesel Fuels: Science, Technology, Health, and Environment. Volume 2. Biodiesel Fuels based on the Edible and Nonedible Feedstocks, Wastes, and Algae,* Ed. O. Konur Boca Raton, FL: CRC Press.

Konur, O. 2021v. Algal biomass production in wastewaters for biodiesel production: A review of the research. In *Handbook of Biodiesel and Petrodiesel Fuels: Science, Technology, Health, and Environment. Volume 2. Biodiesel Fuels based on the Edible and Nonedible Feedstocks, Wastes, and Algae,* ed. O. Konur. Boca Raton, FL: CRC Press.

Konur, O. 2021x. Algal lipid production for biodiesel production: A review of the research. In *Handbook of Biodiesel and Petrodiesel Fuels: Science, Technology, Health, and*

Environment. Volume 2. Biodiesel Fuels based on the Edible and Nonedible Feedstocks, Wastes, and Algae, Ed. O. Konur Boca Raton, FL: CRC Press.

Konur, O. 2021y. Crude oils: A scientometric review of the research. In *Handbook of Biodiesel and Petrodiesel Fuels: Science, Technology, Health, and Environment. Volume 3. Petrodiesel Fuels: Science, Technology, Health, and Environment*, ed. O. Konur. Boca Raton, FL: CRC Press.

Konur, O. 2021z. Petrodiesel fuels: A scientometric review of the research. In *Handbook of Biodiesel and Petrodiesel Fuels: Science, Technology, Health, and Environment. Volume 3. Petrodiesel Fuels: Science, Technology, Health, and Environment*, ed. O. Konur. Boca Raton, FL: CRC Press.

Konur, O. 2021aa. Bioremediation of petroleum hydrocarbons in the contaminated soils: A review of the research. In *Handbook of Biodiesel and Petrodiesel Fuels: Science, Technology, Health, and Environment. Volume 3. Petrodiesel Fuels: Science, Technology, Health, and Environment*, ed. O. Konur. Boca Raton, FL: CRC Press.

Konur, O. 2021ab. Desulfurization of diesel fuels: A review of the research. In *Handbook of Biodiesel and Petrodiesel Fuels: Science, Technology, Health, and Environment. Volume 3. Petrodiesel Fuels: Science, Technology, Health, and Environment*, ed. O. Konur. Boca Raton, FL: CRC Press.

Konur, O. 2021ac. Diesel fuel exhaust emissions: A scientometric review of the research. In *Handbook of Biodiesel and Petrodiesel Fuels: Science, Technology, Health, and Environment. Volume 3. Petrodiesel Fuels: Science, Technology, Health, and Environment*, ed. O. Konur. Boca Raton, FL: CRC Press.

Konur, O. 2021ad. The adverse health and safety impact of diesel fuels: A scientometric review of the research. In *Handbook of Biodiesel and Petrodiesel Fuels: Science, Technology, Health, and Environment. Volume 3. Petrodiesel Fuels: Science, Technology, Health, and Environment*, ed. O. Konur. Boca Raton, FL: CRC Press.

Konur, O. 2021ae. Respiratory illnesses caused by the diesel fuel exhaust emissions: A review of the research. In *Handbook of Biodiesel and Petrodiesel Fuels: Science, Technology, Health, and Environment. Volume 3. Petrodiesel Fuels: Science, Technology, Health, and Environment*, ed. O. Konur. Boca Raton, FL: CRC Press.

Konur, O. 2021af. Cancer caused by the diesel fuel exhaust emissions: A review of the research. In *Handbook of Biodiesel and Petrodiesel Fuels: Science, Technology, Health, and Environment. Volume 3. Petrodiesel Fuels: Science, Technology, Health, and Environment*, ed. O. Konur. Boca Raton, FL: CRC Press.

Konur, O. 2021ag. Cardiovascular and other illnesses caused by the diesel fuel exhaust emissions: A review of the research. In *Handbook of Biodiesel and Petrodiesel Fuels: Science, Technology, Health, and Environment. Volume 3. Petrodiesel Fuels: Science, Technology, Health, and Environment*, ed. O. Konur. Boca Raton, FL: CRC Press.

Konur, O. and F. L. Matthews. 1989. Effect of the properties of the constituents on the fatigue performance of composites: A review. *Composites* 20:317–328.

Kumar, S. and J. M. Jan. 2014. Research collaboration networks of two OIC nations: Comparative study between Turkey and Malaysia in the field of 'Energy Fuels', 2009–2011. *Scientometrics* 98: 387–414.

Kusdiana, D. and S. Saka. 2001. Kinetics of transesterification in rapeseed oil to biodiesel fuel as treated in supercritical methanol. *Fuel* 80: 693–698.

Labeckas, G. and S. Slavinskas. 2006. The effect of rapeseed oil methyl ester on direct injection Diesel engine performance and exhaust emissions. *Energy Conversion and Management* 47:1954–1967.

Lariviere, V. and Y. Gingras. 2010. On the relationship between interdisciplinarity and scientific impact. *Journal of the American Society for Information Science and Technology* 61:126–131.

Lariviere, V., C. Ni, Y. Gingras, B. Cronin, and C.R. Sugimoto. 2013. Bibliometrics: Global gender disparities in science. *Nature News* 504: 211–213.

Leydesdorff, L. and P. Zhou. 2005. Are the contributions of China and Korea upsetting the world system of science? *Scientometrics* 63: 617–630.

Leydesdorff, L. and C. Wagner. 2009. Is the United States losing ground in science? A global perspective on the world science system. *Scientometrics* 78: 23–36.

Mohan, D., C. U. Pittman, and P. H. Steele. 2006. Pyrolysis of wood/biomass for bio-oil: A critical review. *Energy & Fuels* 20: 848–889.

Moin, M., M. Mahmoudi, M., and N. Rezaei, N. 2005. Scientific output of Iran at the threshold of the 21st century. *Scientometrics* 62: 239–248.

Morillo, F., M. Bordons, and I. Gomez. 2001. An approach to interdisciplinarity through bibliometric indicators. *Scientometrics* 51: 203–222.

Nabi, M. N., M. M. Rahman, and M. S. Akhter. 2009. Biodiesel from cotton seed oil and its effect on engine performance and exhaust emissions. *Applied Thermal Engineering* 29: 2265–2270.

North, D. C. 1991a. *Institutions, Institutional Change and Economic Performance*. Cambridge, Mass.: Cambridge University Press.

North, D.C. 1991b. Institutions. *Journal of Economic Perspectives* 5: 97–112.

Noureddini, H. and D. Zhu. 1997. Kinetics of transesterification of soybean oil. *Journal of the American Oil Chemists Society* 74:1457–1463.

Perron, P. 1989. The great crash, the oil price shock, and the unit root hypothesis. *Econometrica: Journal of the Econometric Society* 57:1361–1401.

Rogers, D. and A. J. Hopfinger. 1994. Application of genetic function approximation to quantitative structure-activity relationships and quantitative structure-property relationships. *Journal of Chemical Information and Computer Sciences* 34: 854–866.

Rogge, W. F., L. M. Hildemann, M. A. Mazurek, G. R. Cass, and B. R. T. Simoneit. 1993. Sources of fine organic aerosol. 2. Noncatalyst and catalyst-equipped automobiles and heavy-duty diesel trucks. *Environmental Science & Technology* 27: 636–651.

Saka, S. and D. Kusdiana. 2001. Biodiesel fuel from rapeseed oil as prepared in supercritical methanol. *Fuel* 80: 225–231.

Schauer, J. J., M. J. Kleeman, G. R. Cass, and B. R. T. Simoneit. 1999. Measurement of emissions from air pollution sources. 2. C_1 through C_{30} organic compounds from medium duty diesel trucks. *Environmental Science & Technology* 33:1578–1587.

Scherf, U. and E. J. List. 2002. Semiconducting polyfluorenes-towards reliable structure–property relationships. *Advanced Materials* 14: 477–487.

Soumanou, M. M. and U. T. Bornscheuer. 2003. Improvement in lipase-catalyzed synthesis of fatty acid methyl esters from sunflower oil. *Enzyme and Microbial Technology* 33: 97–103.

Urata, H. 1990. Information flows among academic disciplines in Japan. *Scientometrics* 18: 309–319.

Wang, X., D. Liu, K. Ding, K., and X. Wang. 2012. Science funding and research output: A study on 10 countries. *Scientometrics* 91:591–599.

Xie, Y. and K. A. Shauman. 1998. Sex differences in research productivity: New evidence about an old puzzle. *American Sociological Review* 63: 847–870.

Youtie, J, P. Shapira, and A. L. Porter. 2008. Nanotechnology publications and citations by leading countries and blocs. *Journal of Nanoparticle Research* 10: 981–986.

Zhou, P. and L. Leydesdorff. 2006. The emergence of China as a leading nation in science. *Research Policy* 35: 83–104.

Zullaikah, S., C. C. Lai, S. R. Vali, and Y. H. Ju. 2005. A two-step acid-catalyzed process for the production of biodiesel from rice bran oil. *Bioresource Technology* 96:1889–1896.

22 Chemistry of Biodiesel Fuels based on Soybean Oil

Charles K. Westbrook

CONTENTS

22.1   INTRODUCTION

This chapter will review current the understanding of the combustion chemistry of 'soybean diesel' (SBD) fuel, from the large, engine-size scale down to the atomic and molecular scale, and how the two scales interact to provide a tool for understanding at both scales. I reviewed this subject previously (Westbrook, 2013), focusing on the development of a detailed chemical kinetic reaction mechanism that was applicable to biodiesel fuel, consisting of 'methyl esters' (MEs) of vegetable oils. In the years since, steady progress in chemical reaction types and development of experimental studies of biodiesel combustion have enabled a better understanding. The emphasis on soy is appropriate because soybeans are an important agricultural crop, primarily as a feedstock for livestock, but also because SBD fuel has acquired a growing position as a transportation fuel in the United States.

Biodiesel fuels have important advantages over conventional petroleum-based diesel fuel. The two O atoms in each fuel molecule mean that the fuel is approximately 10% oxygenated, and biodiesel fuels contain little or no aromatic compounds, both of which lead to lower CO production and lower sooting than petroleum-based diesel fuel (Westbrook et al., 2006, 2011a–b). Biodiesel fuel contains little or no sulfur and has a higher flash point, faster biodegradation, and greater lubricity than

459

conventional diesel fuel. Perhaps most important, biodiesel qualifies as a sustainable transportation fuel and reduces net emissions of greenhouse gases. Society has experimented with biodiesel fuels enough to merit further attention to the advantages and disadvantages of their use at greater scales.

22.2 HOW BIODIESEL FUELS ARE DIFFERENT FROM CONVENTIONAL DIESEL FUELS

Conventional fuels that drive automobiles, trucks, jet aircraft, and most other practical engines usually begin with petroleum, pumped from the ground and then refined to accommodate the needs of the particular type of engine as required. Many types of hydrocarbon molecules are found in petroleum, and the refining process roughly separates out those chemical species which are best matched to the special needs of each type of engine (Westbrook et al., 2017). Although there are many fuel molecule sizes and compositions in these fuels, they can be grouped into a few general classes as shown in Figure 22.1.

At the molecular scale, nearly all of the performance properties of a transportation fuel are governed by the chemical bonds in the fuel molecules themselves. This includes heat release rates, power production, fuel/oxidizer ignition, chemical pollutant production and emissions, overall combustion efficiency, and eventually economies of energy production. The present chapter is intended to explain how the chemical structure of SBD molecules influences its properties as a diesel fuel and how they differ from diesel fuel produced from petroleum.

Typical petroleum-based fuels used in jet aircraft, diesel, and gasoline engines can be divided into some of these component classes as shown in Figure 22.2 (Westbrook et al., 2011a–b). Similar sorting of diesel fuels (Farrell et al., 2007; Sato et al., 2004) divided them into *n*-paraffins, *iso*-paraffins, 1-ring and 2-ring naphthenes, 1-ring and 2-ring aromatics, and naphtho-aromatics. A feature of transportation fuels from petroleum, including gasoline, diesel fuel, and jet fuel, is that they contain hundreds or thousands of structurally distinct chemical species from each of the classes shown

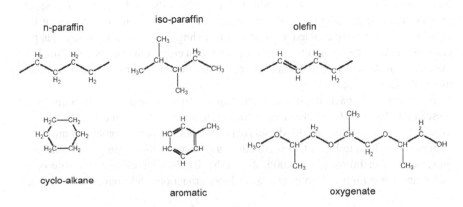

FIGURE 22.1   Some classes of molecular structures found in petroleum-based fuels.

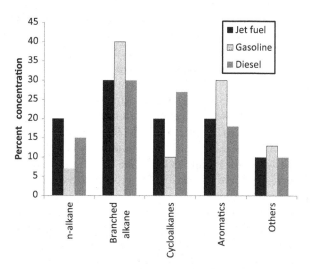

FIGURE 22.2   Relative amounts of fuel classes in petroleum-based transportation fuels.

in Figure 22.1 and many more fuel classes as well. None of the lists of conventional diesel fuel classes shows any significant fraction of oxygenated species.

In contrast, biodiesel fuels, including SBD fuel, are distinctly different from petroleum-based diesel fuels because they consist almost exclusively of mixtures of a small number of unique 'fatty acid methyl ester' (FAME) chemical species, with two O atoms contained in each ME group. These dominant FAME species include methyl palmitate, methyl stearate, methyl oleate, methyl linoleate, and methyl linolenate, shown schematically in Figure 22.3. All of the components have a linear

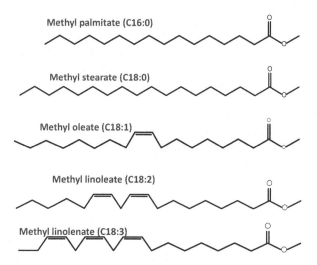

FIGURE 22.3   Structural diagrams of major components of soy and other biodiesel fuels.

('straight-chain') backbone of carbon atoms, with a ME-(C=O)-O-CH3 group at one end of the chain. Four of them have a chain of 17 and one has a chain of 15 C atoms in that chain. Two of these species are saturated, having only C-C single bonds throughout, and the other three components have one, two, and three C=C double bonds, respectively.

A common naming scheme for them is shown in Figure 22.3; for example, methyl oleate is named C18:1 to show it has a chain of 18 C atoms (counting the first C atom in the ME group) and one C=C double bond. The locations of the C=C double bonds in these molecules are at the 9, 12, and 15 positions in the carbon atom chain, where the counting starts at the ME group. As I will show below, the number and location of these C=C double bonds are responsible for a considerable variety of the practical performance properties of these fuels.

The reactivity of these fuels is similar to that of long-chain purely hydrocarbon fuels with a long straight chain of C atoms. There is a great deal known about the kinetics of straight chain saturated fuels, particularly n-alkane and 2-methyl alkane fuels with straight chains as long as 20 C atoms (Sarathy et al., 2011; Westbrook et al., 2009). The long chain of methylene -(CH_2)- groups produces fuels with very small or even negative octane numbers and large 'cetane numbers' (CNs), both indicating rapid reaction at low temperatures (Curran et al., 1998; Lovell, 1934; Murphy et al., 2004; Pollard, 1977). To understand this key property of SBD fuel molecules, it is necessary to highlight the low temperature kinetics of hydrocarbon fuels. Further details can be found in other sources (Simmie, 2003; Westbrook and Dryer, 1981, 1984; Westbrook et al., 2018a–b).

22.3 LOW TEMPERATURE COMBUSTION OF BIODIESEL FUELS

Diesel engines are best understood as 'compression ignition' engines. The combustion cycle begins with nearly room-temperature air confined in a combustion cylinder that is then compressed and heated by piston motion, followed at a time near maximum compression by injection of usually liquid diesel fuel, which vaporizes as it mixes with and is heated by the compressed air until rapid oxidation begins (i.e. ignition). When a biodiesel fuel is being used, the fuel consists largely of the long chain molecules shown in Figure 22.3. The chemical reactions and ignition delay time of this heated, mixed gaseous mixture depend on its temperature, density, and composition.

The main reaction pathways change as the mixture temperature increases as the gases are being mixed and heated. There are several general stages of temperature increase encountered by the fuel/air reactants in diesel fuels. The first range is from end-of-compression temperature to about 600 K, during which a slow reaction occurs that supports only minor changes in the fuel/air mixture composition.

From 600 K to about 850 K, fuel and fuel fragments begin to react, converting much of the fuel to partially processed hydrocarbon molecules along reaction pathways that depend strongly on the molecular structure of the initial fuel molecules. It is primarily in this range of temperatures that important fuel parameters such as the CN influence the rates of key reactions. In this range, considerable conversion of the fuel to intermediate molecules takes place, producing species such as large peroxides, olefins, and other complex structures that are 'waiting' for an initial ignition step. Relatively slow heating of the mixture due to continued piston motion takes place throughout this first stage. Fuels with high cetane ratings experience a preliminary ignition period, with some chemical reaction heat release and temperature growth, but fuels with low cetane and high octane numbers simply continue their relatively slow temperature growth.

In the intermediate stage, details of the molecular structures of the fuels make a difference in the rate of total oxidation and heat release, meaning that some fuels experience an intermediate temperature 'first stage ignition' while other fuels do not, and understanding these processes is the key to prediction of many combustion results. The first of these key reaction stages is the low temperature regime that begins at about 600 K and takes the fuel/air mixtures into their first, incomplete ignition. This is commonly named the 'cool flame' regime.

22.4   COOL FLAMES

In a long chain molecule like those in SBD and in conventional diesel fuel, most of the C-H bonds are at secondary sites in the fuel, with many methylene -(CH_2)- groups. Secondary C-H bonds are relatively weak (i.e. 98 kcal/mol, compared to 101 kcal/mol for primary C-H bonds), making them relatively easy to abstract. Using n-heptane (C_7H_{16}) for illustration, the result of abstraction of a secondary H atom results in large quantities of C_7H_{15} heptyl radical species

$$HH \bullet HHHHH - C - C - C - C - C - C - C - HHHHHHHH \qquad (22.1)$$

which then adds to molecular oxygen O_2 to produce a heptylperoxy radical $C_7H_{15}O_2$

$$OHH \bullet HHHHHHOHHHHH - C - C - C - C - C - C - C - H + O_2$$
$$\rightarrow H - C - C - C - C - C - C - C - HHHHHHHHHHHHHHHH \qquad (22.2)$$

where the • indicates a radical site. The O• then easily abstracts another H atom from a different location within the $C_7H_{15}O_2$ radical to produce a hydroperoxyalkyl radical QOOH:

$$\bullet HOOHHOHHHHHHOHH \bullet HH - C - C - C - C - C - C - C - H$$
$$\rightarrow H - C - C - C - C - C - C - C - HHHHHHHHHHHHHHHH \qquad (22.3)$$

In more general terms, the above reaction sequence is summarized as being originated by H atom abstraction from the fuel (RH) to produce a radical 'R', which then adds to molecular oxygen to produce an alkylperoxy radical 'RO$_2$', followed by an internal isomerization (or H atom transfer) reaction to produce a hydroperoxyalkyl radical 'QOOH'. Since there is a free radical site in the QOOH above, it can add another O_2 to produce a hydroperoxyalkylperoxy

$$HH \bullet OOOHHOHH \bullet HHHOHHOHH - C - C - C - C - C - C - C - H$$
$$+ O_2 \rightarrow H - C - C - C - C - C - C - C - HHHHHHHHHHHHHHHH \qquad (22.4)$$

radical O_2QOOH which can abstract another H atom from within the species via other complex H atom transfer reactions. At every stage in these reaction sequences, the interim species can decompose to smaller fragments, each of which then begins its own family of reactions.

The final step of a lot of these reactions is a relatively complex reaction

$$H \cdot HHOOOOHHOHHOHHHOHHOHH - C - C - C - C - C - C$$
$$-C - H \rightarrow H - C - C - C - C - C - C - C - HHHHHHHHHH \cdot HHHH \quad (22.5)$$

which then decomposes into a metastable intermediate ketohydroperoxide species $C_7H_{14}O_3$ and an OH radical:

$$HHHOOOOHHOHHOHHH\|HHOHH - C - C - C - C - C - C$$
$$-C - H \rightarrow H - C - C - C - C - C - C - C - HHH \cdot HHHHHHHHHH \quad (22.6)$$

This species is temporarily stable enough to remain until the reacting mixture reaches the higher temperatures required to break the O-O double bond in the keto-hydroperoxide species, effectively producing a brief pause or delay in the overall fuel oxidation process.

Overall low temperature oxidation of conventional hydrocarbon fuels begins with the addition of O_2 to the fuel radicals at temperatures of approximately 550–600 K; the subsequent series of reactions is sometimes called a 'cool flame' or, more formally, an 'alkyl peroxy radical isomerization cycle of reactions', or in short, 'low temperature combustion' (LTC). Although not shown in this brief discussion, the cool flame produces many reactive hydroxyl OH radicals, which react rapidly by abstracting an H atom from one of the fuel molecules to produce water (H_2O), releasing substantial amounts of heat and steadily increasing the temperature of the reacting gas mixture.

As the temperature of the gas increases, RO_2 adduct stability decreases and the rate of dissociation of RO_2 back to its $R + O_2$ reactants accelerates. The temperature at which the $R + O_2 = RO_2$ reaction becomes dynamically balanced is called the 'ceiling temperature' (Benson, 1976). Heat release from water production in the cool flame leads to a modest temperature increase of 100 to 200 K, not large when compared with the overall heat release of the full oxidation process, which is close to 2,000 K from reactants to final products. However, even a small temperature increase due to low temperature oxidation reaction pathways leads to an earlier time of ignition of the gas mixture than would occur without that low temperature reactivity. In a chemical sense, low temperature reactivity 'sensitizes' the fuel/air mixture and accelerates its ignition.

A slower reaction phase usually follows the end of the cool flame, but some OH reactivity continues. Soon the reactive mixture reaches a higher temperature of about 900–1,000 K where H_2O_2 and considerable amounts of hydrocarbon hydroperoxides that have been accumulating during the cool flame decompose. The resulting flood of OH ignites a rapid second stage that leads almost instantaneously to the final fuel mixture ignition, an intense stage driven by radical species chain branching from the reaction:

$$H + O_2 = O + OH \tag{22.7}$$

This reaction consumes one radical species (i.e. H atoms) and produces two new radical species (O and OH) which consume all of the remaining fuel and all of the remaining intermediate hydrocarbon species. This particular chain branching reaction has been described as the most important single reaction in all of hydrocarbon combustion (Warnatz, 1983; Westbrook, 2000; Westbrook and Dryer, 1984) and is responsible for many critical combustion phenomena, with biodiesel ignition as a good example of its central role in combustion.

We can illustrate the results of the ignition of a sample of methyl stearate, the C18:0 component of SBD fuel, mixed with a stoichiometric amount of air at 12 atm pressure and initial temperatures from 650 to 1,200 K. The computed ignition delay time results are plotted in Figure 22.4 as a function of inverse temperature.

The shortest ignition delay time is 0.1 ms at about 1,200 K ($1000/T = 0.8$); as the temperature is decreased (i.e. as $1000/T$ increases), the mixture ignites more slowly and the ignition delay time increases in an intuitive way. Surprisingly, however, when the temperature approaches 900 K (i.e. $1000/T = 1.11$), the computed ignition delay time begins to decrease as the temperature is further decreased. That is, the ignition is faster at 800 K than at 900 K, and this counter-intuitive behavior continues until the ignition delay time reaches a minimum of about 3.5 ms at 750 K. As the temperature decreases further, the ignition delay time then increases again as the rate of overall reaction slows, returning to a more intuitive behavior. But over the temperature range from 750 to 850 K, the reaction rate decreases with increasing temperature, a phenomenon that is called 'negative temperature coefficient' (NTC) behavior. NTC behavior is active over only a limited range of temperatures.

All of these conditions exist for methyl stearate and methyl palmitate, the saturated components of SBD fuel, and both of these individual components exhibit significant NTC behavior. The dashed line in Figure 22.4 shows that the same cool flame phenomenon is seen in the ignition of methyl linoleate (C18:2), although the amount of reduction of the ignition delay time by the cool flame is somewhat less than its effect on methyl stearate.

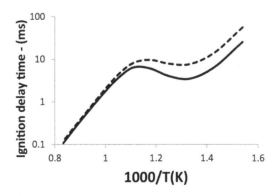

FIGURE 22.4 Computed ignition delay times for methyl stearate in air at 12 atm pressure (solid line) and methyl linoleate (dashed line).

The ability of a fuel to provide a cool flame increases its overall reactivity in engine combustion applications. In such conditions, a fuel/air mix is steadily heated until the cool flame ignites, and most typical fuels will see approximately the same pressure/temperature/time history. After the cool flame period ends, fuels will experience another partial ignition when the temperature reaches about 900–1,000 K when O-O bonds in H_2O_2 and other species, including ketohydroperoxides, are broken. These degenerate chain branching steps provide a pulse of heat due to the oxidation of the large amounts of OH that are produced, so the mixture will then proceed rapidly to the final high temperature step in the ignition process.

22.5   HOW DO THE C=C DOUBLE BONDS MAKE A DIFFERENCE?

These same five MEs shown in Figure 22.3 are the main components of biodiesel fuels produced from soy, rapeseed, sunflower, safflower, linseed, jatropha, cottonseed, corn, palm, and peanut oils, each of which has a unique mixture of the same five components. For example, SBD fuel contains much different fractions of methyl oleate and methyl linoleate than rapeseed oil biodiesel fuel, yet both are acceptable fuels in diesel engines, although the two fuels have other different properties as results of their different compositions.

Measured CNs for each of these biodiesel fuels are shown in Table 22.1, and the last two columns show the average contents of the most widely used biodiesel fuels: SBD fuel commonly used in the United States and rapeseed (i.e. canola) oil commonly used in Europe. The unsaturated SBD components methyl oleate (C18:1), methyl linoleate (C18:2), and methyl linolenate (C18:3) are similar in structure to the saturated component methyl stearate (C18:0), but with one, two, or three C=C double bonds within the long C atom chain. I showed in Figure 22.4 that inclusion of two C=C double bonds slowed the ignition (i.e. longer ignition delay time) of linoleate relative to stearate, so I will examine the effect of the C=C double bond in the oleate and then the multiple C=C bonds in the linoleate and linolenate.

I Start with the structure of the fully saturated methyl stearate:

$$HHHHHHHHHHHHHHHHHHHHOHH-C-C-C-C$$
$$-C-C-C-C-C-C-C-C-C-C-C-C-C-C-O$$
$$-C-HHHHHHHHHHHHHHHHHHHHH \qquad (22.8)$$

Abstraction of an H atom from a secondary site is possible almost anywhere within the chain to produce, for example:

$$HHHHHH\cdot HHHHHHHHHHHOHH-C-C-C$$
$$-C-C-C-C-C-C-C-C-C-C-C-C-C-C$$
$$-C-O-C-HHHHHHHHHHHHHHHHHHHHH \qquad (22.9)$$

which then can react at high temperatures by decomposing to smaller fragments, a familiar reaction sequence in biodiesel and conventional diesel fuels. At low temperatures, however, the stearate radical can add an O_2 molecule and initiate a cool

TABLE 22.1

Composition of Biodiesel and CN values of Fuels made from Transesterification of Selected Oils

	Sunflower	Safflower	Linseed	Jatropha	Cottonseed	Corn	Olive	Beef Tallow	Palm	Peanut	Soy	Rapeseed
Palmitate	7	7	7	4	23	10	13	28	46	11	8	4
Stearate	5	1	2	8	3	4	4	21	4	8	4	1
Oleate	19	13	19	49	20	38	72	47	40	49	25	60
Linoleate	68	78	19	38	53	48	10	3	10	32	55	21
Linolenate	1	0	54	1	1	0	1	1	0	0	8	14
CN	49	50	39	58	51	49	55	58	62	54	47	54

flame series of reactions as discussed above. However, if the stearate species is replaced by methyl oleate,

$$
\begin{aligned}
&\mathrm{H\,H\,H\,H\,H\,H\,H\,H\,H\,H\,H\,H\,H\,H\,H\,O\,H\,H-C-C-C} \\
&\mathrm{-C-C-C-C-C-C=C-C-C-C-C-C-C-C} \\
&\mathrm{-C-O-C-H\,H\,H\,H\,H\,H\,H\,H\,H\,H\,H\,H\,H\,H\,H} \qquad (22.10)
\end{aligned}
$$

with the C=C double bond at the '9' location from the ME group, a large section of the molecule is changed. Focusing on the neighborhood of the C=C double bond, I find

$$
\mathrm{H\,H\,H\,H\,H\,H\,H\,H...-C-C-C-C=C-C-C-C-...H\,H\,H\,H\,H\,H} \qquad (22.11)
$$

I can rewrite this segment and rename some of the H atoms to indicate the nature and strength of some C-H bonds:

$$
\mathrm{s\,s\,a\,v\,v\,a\,s\,s...-C-C-C-C=C-C-C-C-...s\,s\,a\,a\,s\,s} \qquad (22.12)
$$

where 's' represents a conventional secondary C-H bond, which is the main feature of the saturated carbon chains in methyl stearate and methyl palmitate and in most of the remainder of the methyl oleate molecule. When the C=C double bond is inserted into the chain, three major changes occur. First, two H atoms are no longer needed, having been replaced by the double bond. Second, the two H atoms that remain bonded to the two C atoms in the double bond become much 'more strongly bound' in a 'vinylic' C-H bond (indicated above by the 'v'), and third, the four H atoms bound to C atoms adjacent to the double bond become 'more weakly bound' to their C atoms than in conventional secondary bonds, and these new, much weaker C-H bonds are 'allylic' secondary C-H bonds (indicated by 'a' in the above formula). Chemical theory tells us that the C=C double bond attracts (i.e. 'steals') some electron density that would normally be employed to bond the 'a' H atoms to their local C atoms, making those allylic secondary C-H bonds weaker than usual secondary bonds.

As a result, these 'a' H atoms at allylic sites are easier to abstract than those at conventional secondary sites, and much easier to abstract than the much more strongly bonded H atoms at primary or vinylic sites, so radical attack and H atom abstraction takes place preferentially at these weakened secondary allylic sites, producing an allylic radical

$$
\mathrm{s\,s\,a\,v\,v\,\bullet\,s\,s...-C-C-C-C=C-C-C-C-...s\,s\,a\,a\,s\,s} \qquad (22.13)
$$

Intuition might then say that O_2 addition and initiation of cool flames should occur easily at these sites, leading to enhanced RO_2 cool flame reaction and even faster ignition. However, this does not occur. The allylic radical site, which was created because the H atom was so weakly bound there to the fuel molecule, remains a weak location for any atom to stay for very long, and although O_2 is indeed added quickly at these sites, the allylic RO_2 does not 'live' long enough to sustain subsequent H atom transfer reactions. While the 'bond dissociation energy' (BDE) for O_2 at a conventional secondary site is 37.6 kcal/mol, at an allylic site the BDE is only 26.9 kcal/mol, a substantial difference that makes the adduct weak.

Due to the poor bonding capabilities at that allylic site, the O_2 dissociates rapidly back to the allylic radical and the free O_2, and only low temperature kinetics are possible from this beginning site. In this perhaps non-intuitive way, easy abstraction of allylic H atoms actually slows the overall rate of fuel oxidation in the low temperature regime, and the net result is a retarded rather than advanced ignition. The retarding effect of one C=C double bond is not so extreme as to completely quench low temperature cool flame reactivity in the oleate, since the differences in the allylic R-H and R-O_2 BDE values are not prohibitive, so some cool flame behavior originating at these and other sites in the oleate chain can still initiate cool flame behavior.

However, an additional feature occurs for the methyl linoleate and methyl linolenate due to the addition of a second, and then a third, C=C double bond. The location of these additional double bonds is the key to their effects (Westbrook et al., 2013; Yang et al., 2011). In the methyl linoleate, the chain with its two C=C double bonds is

$$ssavvbvvas...-C-C-C-C=C-C-C=C-C-C...ssabas \quad (22.14)$$

and for methyl linolenate, another C=C link in the C atom chain is included with the same connections as in the linoleate. For both the linoleate and linolenate, the key is the introduction of a 'doubly allylic' or 'bis-allylic' pair of C-H bonds represented in these diagrams by the 'b' label for those two H atoms. In the same theoretical terms as above, a pair of C=C double bonds draws so much electron density from the C-b bonds that those H atoms are 'extremely' weakly bound. This can be quantified by remembering that the conventional secondary C-H BDE is about 37.6 kcal/mol. The first C=C double bond reduces the C-a allylic BDE to 26.9 kcal/mol, and addition of another C=C makes the bis-allylic C-b BDE equal to 16.4 kcal/mol.

These energetics create an almost impossible set of conditions for low temperature kinetics due to these interconnected effects. First, the extremely weak C-b bond gives that H atom an enormous advantage for radical abstraction reactions from the fuel. It is literally 'hanging by a thread'. Then, because the huge majority of the fuel radicals will have lost that 'b' H atom, that 'b' site will be the greatly favored site for O_2 addition, denying those O_2 species the potential to bind elsewhere to the fuel molecule and initiate low temperature reaction sequences. And third, even in those rare cases where an unstable RO_2 adduct is formed at the 'b' site, the favored reaction of that adduct will be dissociation back to the R and O_2 species from which it was produced.

Thus, even though the methyl linoleate and methyl linolenate react rapidly via abstraction of their bis-allylic H atoms, very little stable RO_2 can be formed that persists long enough to initiate the alkylperoxy radical isomerization reaction paths that lead to cool flames and significant low temperature chemistry. The net result is that those components of SBD, specifically methyl linoleate and methyl linolenate, which according to Table 22.1 comprise more than 60% of the fuel molecules in the

SBD mixture, do not initiate any low temperature reactions that could produce cool flames and their accompanying low temperature heat release, thereby providing no support for advanced timing of their ignition under diesel engine conditions.

22.6 CETANE NUMBERS FOR BIODIESEL FUELS

The CN is a convenient measure of diesel fuel ignition quality, comparing the time of autoignition of a test fuel/air mixture to the time of autoignition of a standardized mixture of reference fuels. The reference fuels for determining the CN of any type of diesel fuel are the easily ignited *n*-hexadecane (CN=100) and the difficult-to-ignite fuel *iso*-cetane (2,2,4,4,6,8,8-heptamethyl nonane) (CN=15), and the CN or ON is determined by finding a mixture of the reference fuels that ignites in the test process at the same time as the fuel being tested.

Since most biodiesel fuel consists of mixtures of the five MEs in Figure 22.3, we can use a well-tested detailed chemical kinetic reaction mechanism to make computational predictions of realistic ignition simulations. In a computational study (Westbrook et al., 2011a–b, 2013) I used kinetic models for each SBD component to examine the details of its ignition chemistry and see how each component influences the chemistry of the combination into SBD. Our focus was on SBD, but I will also discuss rapeseed ME, also known as canola oil ME. Due to oddities of history and geography, soybeans have been used in the United States as the primary source of biodiesel fuel while rapeseed oil has had the largest impact as diesel fuel in Europe. Historically, peanut oil was the first diesel fuel used by Rudolf Diesel, who invented the first diesel engine in the 1890s, because his first customers at the time had large supplies of peanuts from French farms in North Africa.

Experimentally, methyl stearate has CN = 101, methyl palmitate has CN = 86, methyl oleate has CN = 59, methyl linoleate has CN = 38, and methyl linolenate has CN = 23, so the biodiesel components cover a very wide range of CNs. As shown in Table 22.1, SBD has a CN of 47 and rapeseed oil biodiesel has a CN of 54, indicating that the rapeseed biodiesel ignites more rapidly than the SBD fuel. This difference in ignition delay time reflects the smaller fraction (0.25) of methyl oleate and the higher fraction (0.55) of methyl linoleate in SBD, compared to the higher fraction (0.60) of methyl oleate and the lower fraction (0.21) of methyl linoleate in rapeseed biodiesel.

The kinetic model shows that methyl oleate, with only one C=C double bond, ignites more rapidly than the methyl linoleate, with two C=C double bonds and two resulting bis-allylic C-H bonds. Those bis-allylic C-H bonds produced by the methyl linoleate effectively inhibit chain reaction initiation, and the greater rate of chain initiation in methyl oleate leads to more cool flame activity than in methyl linoleate and therefore to the observed difference in CN between soy and rapeseed biodiesel fuels. The lower rate of low temperature reactivity, as I explained above, is due to the inhibiting effects of the elevated level of bis-allylic radicals in methyl linoleate in SBD. This difference has economic and political impacts. The EU requires a minimum CN of 51 for diesel fuel while the United States recommends a minimum CN of 40–45; these requirements reflect the fact that biodiesel fuel in Europe is produced from rapeseed oil while that produced in the USA is produced primarily from soybean oil, so biodiesel fuel produced in the USA cannot be sold in Europe.

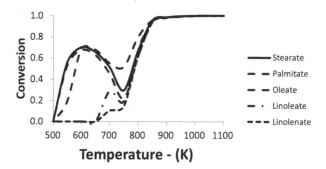

FIGURE 22.5 Reactivity of each biodiesel fuel component at atmospheric pressure in a simulated stirred reactor.

I computed the autoignition of each of the separate ME components of biodiesel fuels, using the detailed chemical kinetic model for the biodiesel fuels shown in Figure 22.5. The saturated biodiesel fuel components, methyl stearate and methyl palmitate, as well as the mono-unsaturated component methyl oleate, show considerable reactivity at reaction temperatures from 500 to 750 K that results from vigorous cool flame LTC. Figure 22.5 shows that the LTC process is essentially complete at a temperature of about 750 K at a pressure of 1 atm, which is used for these model computations; at higher pressures the LTC would continue to about 800 K, due to the effect of pressure on the equilibrium constant of the $R + O_2 = RO_2$ reaction.

The combustion rates shown in Figure 22.5 of the two poly-unsaturated SBD components, methyl linoleate, and methyl linolenate, are effectively zero at temperatures from 500–650 K, with the first small amounts of heat release between 650 and 700 K. These two components then proceed to react vigorously beginning at about 750 K and following a path much like that of the more saturated components. Closer examination of the specific reaction pathways show that at 750 K and higher temperatures, all five components react along paths that are effectively the same, with chain branching from H_2O_2 decomposition

$$H-O-O-H \ \rightarrow \ OH + OH \tag{22.15}$$

at temperatures from 800–900 K, and then from

$$H + O_2 \rightarrow O + OH \tag{22.16}$$

at temperatures above about 1,050 K. These intermediate and high temperature chain branching reaction pathways have been well understood for many years, and the importance of the low temperature reactivity and heat release contributes to this picture by determining when the fuel/oxidizer mixture reaches the temperatures at which the intermediate and high temperature chain branching processes take control of the reactant mixtures. That is, a fuel mixture which exhibits a great deal of cool flame reactivity that significantly increases the reactant mixture results in an earlier onset of the later, full ignition, and such fuels are those with higher CNs, while a mixture with little or no low temperature heat release reaches the conditions for full ignition at a later time and has a lower CN rating.

I next compare how these five MEs, combined in the proportions from Table 22.1 to make simulated SBD and rapeseed biodiesel fuels, react at the same temperatures. Again using a computational stirred reactor simulation of each biodiesel fuel, the plots of reactivity at temperatures from 500 to 1,100 K are shown in Figure 22.6. The SBD (Figure 22.6) exhibits a smaller region of NTC than the rapeseed biodiesel fuel (Figure 22.6b). A real engine test using either biodiesel fuel would traverse either of these two reactivity plots at very nearly the same rate, so the test using 'soy ME' (SME) would experience less early time heat release than the test using 'rapeseed ME' (RME) and would ignite later than the RME fueled test. This is reflected by the higher CN rating of the RME fuel, specifically CN = 54 for the RME fuel and CN = 47 for the SBD fuel, and this difference in CN can be seen to be caused by the reduced residence time of reactive fuel/air in the smaller NTC region in Figure 22.6 than in Figure 22.6b, and this difference is caused in turn by the different component fractions of methyl oleate and methyl linoleate.

As the fuel/air mixture is heated from room temperature, the RME biodiesel fuel begins to react at about 550 K, while combustion of the SBD fuel is delayed until the

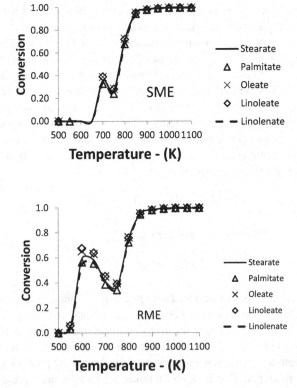

FIGURES 22.6   (a) and (b) Stirred reactor simulations of SME and RME, atmospheric pressure, stoichiometric fuel/air, 0.2% fuel.

temperature reaches about 650 K. Both fuels display a distinct NTC region, but this region is more pronounced in the rapeseed biodiesel fuel, and the reactivity increases rapidly in both fuels from the NTC minimum to complete conversion over a temperature range of about 750 to 850 K.

22.7 HOW BIODIESEL FUEL COMPOSITION AFFECTS ITS PROPERTIES

Biodiesel fuels in general, including SBD, have properties that can make them troublesome for use in practical engines. These characteristics include poor cold-temperature performance, poor oxidative stability, and increased emissions of 'nitrogen oxides' (NO_x), all of which can be addressed by modifying their relative fractions of the basic MEs in the fuel. For example, the fully saturated components such as methyl stearate and methyl palmitate can condense at cold temperatures; the stearate is a solid at normal room temperature but becomes soluble in mixtures with the unsaturated MEs. However, the presence of only one C=C double bond improves the cold-temperature flow properties of methyl oleate relative to methyl stearate (Durrett et al., 2008). In general, biodiesel fuel performance at low temperatures suggests minimizing the fractions of saturated MEs in the fuel mixture.

The presence of bis-allyl C-H bonds in the poly-unsaturated MEs, methyl linoleate, and linolenate, with their low bond dissociation energies discussed above, make fuels with those species very reactive at quite low temperatures, leading to significant oxidation and polymerization in those fuels when stored for extended periods of time. It has been noted (Frankel, 1998) that linoleate esters are 40 times more reactive than oleate esters because of the reactivity of the bis-allyl group in the linoleate, and the number of bis-allylic bonds is greater when the linolenates are present. Even low concentrations of polyunsaturated fatty esters have a disproportionately large effect on the oxidative stability of biodiesel (Knothe, 2009; Knothe and Dunn, 2003). In addition, I have seen earlier that the poly-unsaturated components adversely affect the CN of biodiesel fuels, while Durrett et al. (2008) concluded that biodiesel fuels with high fractions of methyl oleate have excellent characteristics for ignition quality, NO_x emissions, and fuel stability.

Emissions studies of the constituent FAMEs typically present in biodiesel revealed that the saturated MEs methyl palmitate, methyl laurate, and methyl stearate all produce less NO_x than standard diesel, as do the methyl and ethyl esters of hydrogenated soybean oil. On the other hand, progressively increasing the number of double bonds in FAMEs resulted in a corresponding increase in NO_x production. Further, MEs with longer fatty acid chains had lower NO_x emissions (McCormick et al., 2001). Therefore, reducing the unsaturated (particularly the polyunsaturated) fatty acid content of the input oil should result in lower NO_x emissions from the resulting biodiesel. All of these trends suggest that the mono-unsaturated oleate fraction in biodiesel fuels is the only component that has no inherent drawbacks and optimizes the overall performance of the biodiesel fuel.

As a result, an important topic of biodiesel fuel research involves optimization of the composition of a 'native' biodiesel fuel by adding and removing specific ME species. Due to the prominence of SBD, much of this research has occurred with soybeans and the composition of its ME mix. This optimization is accomplished by incrementing the fraction of one or more of the five major components in Figure 22.3, or by adding other MEs not usually found in that particular fuel. This is analogous to the common practice of fuel blending at the refinery level for optimizing the performance characteristics of conventional gasoline or diesel fuels. Usually this involves additions of other desirable MEs such as methyl palmitoleate (C16:1), which has the same good features of its C18:1 analog, methyl oleate. In search of the ideal biodiesel composition, high fractions of monounsaturated MEs (as oleic and palmitoleic MEs), reduced presence of polyunsaturated acids, and limited saturated ME content are recommended.

In this sense, C18:1 and C16:1 are the best-fitting components in terms of oxidative stability and cold weather behavior, among many other properties. Furthermore, genetic engineering may be a valuable tool to design oils presenting the most suitable fatty acid profile to provide high quality biodiesel (Pinzi et al., 2009). Genetic engineering (Buhr et al., 2002) has decreased consumption of the oleate to poly-unsaturated products and increased production of the oleate from the saturated products, resulting in dramatic yields of the oleate fraction to 85%, from its unmodified level of 17.9%, and other positive results.

Recent research has taken place on the composition of vegetable oils, primarily directed towards oils for use in food services. For such markets, social objections to genetic engineering are very important, so the most intriguing advances are being made on selective propagation and other plant modifications that might be called selective evolution. Their impact on fuels for biodiesel applications may follow, if such advances lead to oils better suited for combustion applications, although the amounts being produced today are not yet at the appropriate scale for such applications. The appeal of these oil research approaches is the development of plants that 'naturally' produce an altered distribution of MEs that are more dominated by C16:1 and C18:1 directly (Durrett et al., 2008).

Very recent studies have reported developments of vegetable oils with very high fractions of oleic content. These varieties are developments of systematic selection and do not involve artificial genetic engineering, which avoids political issues. The most recent progress is in soybeans with oleic fractions of 75% and higher, as a replacement of commodity oils used in food applications (Napolitano et al., 2018), but previous similar levels for high-oleic oils have been made for sunflower, palm, safflower, canola, and others. The quantities being produced appear still to be modest, but there is a great deal of enthusiasm in this rapidly growing application field. Much of the focus has been on extending the shelf life of the oils for domestic consumption, but as the sizes of the crops steadily increase, applications for use as biodiesel fuels will develop eventually.

Clearly, the benefits of vegetable oils that have superior nutritional properties and oils that lead to excellent combustion properties as a biodiesel fuel are both valuable, and the present kinetic analysis applies equally to both purposes. The point is that the molecular structures of the oils are moving towards the mono-unsaturated varieties, which have been emphasized above as being superior in both types of applications, food and fuel. The important role of soybeans as a special case, due to the enormous production of that crop, is enabling for both types of applications and that appears likely to continue indefinitely. The analysis above is entirely consistent with the current trends in vegetable oil fractions, and the same analysis predicts that the same trend would benefit the fuels as well.

Much of the groundwork to provide a useful kinetic modeling tool for such fuels is available to respond to any changes that can be made. In addition, that modeling capability could be employed to direct the genetic engineering to prioritize the research directions in fuel sizes and structures that could be most productive in real-world applications.

In a sense, the chemistry of all biodiesel fuels is the same, since it is the chemistry of the same five MEs in Figure 22.3 and related components. It is the relative fractions of each of these components that distinguish these fuels from different plant oils and could predict their optimal properties.

ACKNOWLEDGMENTS

This work was supported by the U.S. Department of Energy, Office of Basic Energy Sciences and the Vehicle Technologies Office, program managers Gurpreet Singh and Kevin Stork, and was performed under the auspices of the U.S. Department of Energy by the Lawrence Livermore National Laboratory under contract DE-AC52-07NA27344.

REFERENCES

Benson, S. W. 1976. *Thermochemical Kinetics: Methods for the Estimation of Thermochemical Data and Rate Parameters*. New York, NY: John Wiley & Sons.

Buhr, T., S. Sato, and F. Ebrahim, et al. 2002. Ribozyme termination of RNA transcripts down-regulate seed fatty acid genes in transgenic soybean. *Plant Journal* 30: 155–163.

Curran, H. J., P. Gaffuri, W. J. Pitz and C. K. Westbrook. 1998. A comprehensive modeling study of *n*-heptane oxidation. *Combustion and Flame* 114: 149–177.

Durrett, T.P., C. Benning, and J. Ohlrogge. 2008. Plant triacylglycerols as feedstocks for the production of biofuels. *Plant Journal* 54: 593–607.

Farrell, J. T., N. P. Cernansky, and F. L. Dryer, et al. 2007. Development of an experimental database and kinetic models for surrogate diesel fuels. *SAE Technical Paper* 2007: SAE-2007-01-0201. Society of Automotive Engineers.

Frankel, E. N. 1998. *Lipid Oxidation*. Dundee, UK: Oily Press.

Knothe G. 2009. Improving biodiesel fuel properties by modifying fatty ester composition. *Energy & Environmental Science* 2: 759-766.

Knothe, G. and R. O. Dunn. 2003. Dependence of oil stability index of fatty compounds on their structure and concentration and presence of metals. *Journal of the American Chemical Society* 80: 1021–1026.

Lovell, W. G., J. M. Campbell, and T. A. Boyd. 1934. Knocking characteristics of hydrocarbons determined from compression ratios at which individual compounds begin to knock under specified conditions. *Industrial & Engineering Chemistry* 26:1105–1108.

McCormick, R.L., M. S. Graboski, T. L. Alleman, and A. M. Herring. 2001. Impact of biodiesel source material and chemical structure on emissions of criteria pollutants from a heavy-duty engine. *Environmental Science & Technology* 35: 1742–1747.

Murphy, M.J., J. D. Taylor and R. L. McCormick. 2004. *Compendium of Experimental Cetane Number Data*. NREL/SR-540-36805. Golden, CO: National Renewable Energy Laboratory.

Napolitano, G. E., Y. Ye, and C. Cruz-Hernandez. 2018. Chemical characterization of a high-oleic soybean oil. *Journal of the American Oil Chemists' Society* 95: 583–589.

Pinzi S., I. L. Garcia, and F. J. Lopez-Gimenez, et al. 2009. The ideal vegetable oil-based biodiesel composition: A review of social, economical and technical implications. *Energy & Fuels* 23: 2325–2341.

Pollard, R. T. 1977. Hydrocarbons. In *Comprehensive Chemical Kinetics*, ed. C. H. Bamford and C. F. H. Tipper, 17: 249–367. New York: Elsevier Scientific Publishing Company.

Sarathy, S.M., C. K. Westbrook, and M. Mehl, et al. 2011. Comprehensive chemical kinetic modeling of the oxidation of 2-methyl alkanes from C_7 to C_{20}. *Combustion and Flame* 158:2338–2357.

Sato, S., Y. Sugimoto, K. Sakanishi, I. Saito and S. Yui. 2004. Diesel quality and molecular structure of bitumen-derived middle distillates. *Fuel* 83: 1915–1927.

Simmie, J. M. 2003. Detailed chemical kinetic models for the combustion of hydrocarbon fuels. *Progress in Energy and Combustion Science* 29: 599–634.

Warnatz, J. 1983. Hydrocarbon oxidation at high temperatures. *Berichte der Bunsengesellschaft fur Physikalische Chemie* 87: 1008–1022.

Westbrook, C. K. and F. L. Dryer. 1981. Chemical kinetics and modeling of combustion processes. *Proceedings of the Combustion Institute* 18: 749–767.

Westbrook, C.K. and F.L. Dryer. 1984. Chemical kinetics modeling of hydrocarbon combustion. *Progress in Energy and Combustion Science* 10: 1–57.

Westbrook, C. K. 2000. Chemical kinetics of hydrocarbon ignition in practical combustion systems. *Proceedings of the Combustion Institute* 28:1563–1577.

Westbrook, C. K., W. J. Pitz and H. J. Curran. 2006. Chemical kinetic modeling study of the effects of oxygenated hydrocarbons on soot emissions from diesel engines. *Journal of Physical Chemistry A* 110, 6912–6922.

Westbrook, C. K., W. J. Pitz, O. Herbinet, H. J. Curran and E. J. Silke. 2009. A comprehensive detailed chemical kinetic reaction mechanism for combustion of *n*-alkane hydrocarbons from *n*-octane to *n*-hexadecane. *Combustion and Flame* 156:181–199.

Westbrook, C. K., C. V. Naik, and O. Herbinet, et al. 2011a. Detailed chemical kinetic reaction mechanisms for soy and rapeseed biodiesel fuels. *Combustion and Flame* 158: 742–755.

Westbrook, C. K., W. J. Pitz, M. Mehl and H. J. Curran. 2011b. Detailed chemical kinetic reaction mechanisms for primary reference fuels for diesel cetane number and spark-ignition octane number. *Proceedings of the Combustion Institute* 33: 185–192.

Westbrook, C. K. 2013. Biofuels Combustion. *Annual Review of Physical Chemistry* 64: 201–219.

Westbrook, C. K., W. J. Pitz, S. M. Sarathy and M. Mehl. 2013. Detailed chemical kinetic modeling of the effects of C=C double bonds on the ignition of biodiesel fuels. *Proceedings of the Combustion Institute* 34: 3049–3056.

Westbrook, C. K., M. Mehl, W. J. Pitz, and M. Sjoberg. 2017. Chemical kinetics of octane sensitivity in a spark-ignition engine. *Combustion and Flame* 175: 2–15.

Westbrook, C. K., M. Sjoberg and N. P. Cernansky. 2018a. A new chemical kinetic method of determining RON and MON values for single component and multicomponent mixtures of engine fuels. *Combustion and Flame* 195: 50–62.

Westbrook, C. K., M. Mehl, W. J. Pitz, S. Wagnon, and K. Zhang. 2018b. Multi-fuel surrogate chemical kinetic mechanisms for real world applications. *Physical Chemistry Chemical Physics* 20: 10588–10606.

Yang, B., C. K. Westbrook, T. A. Cool, N. Hansen and K. Kohse-Hoinghaus. 2011. The effect of carbon-carbon double bonds on the combustion chemistry of small fatty acid esters. *Zeitschrift fur Physikalische Chemie* 225: 1293–1314.

23 Palm Oil-based Biodiesel Fuels

A Review of the Research

Ozcan Konur

CONTENTS

23.1 INTRODUCTION

Crude oils have been primary sources of energy and fuels, such as petrodiesel. However, significant public concerns about the sustainability, price fluctuations, and adverse environmental impact of crude oils have emerged since the 1970s (Ahmadun et al., 2009; Atlas, 1981; Babich and Moulijn, 2003; Kilian, 2009; Perron, 1989). Thus, biooils (Bridgwater et al., 1999; Bridgwater and Peacocke, 2000; Czernik and Bridgwater, 2004; Mohan et al., 2006; Zhang et al., 2007) and biooil-based biodiesel fuels (Chisti, 2007; Hill et al., 2006) have emerged as alternatives to crude oils and crude oil-based petrodiesel fuels, respectively, in recent decades.

Nowadays, although petrodiesel fuels are still used extensively, biodiesel fuels are being used increasingly in the transportation and power sectors (Konur, 2021a–ag). Therefore, there has been great public interest in the development of edible oil-based biodiesel fuels as the first generation of such fuels (Antolin et al., 2002; Freedman et al., 1986; Kusdiana and Saka, 2001; Noureddini and Zhu, 1997; Saka and Kusdiana, 2001). The POBD fuels have been among the most-produced and marketed edible oil-based biodiesel fuels (Al-Widyan and Al-Shyoukh, 2002; Benjumea et al., 2008; Darnoko and Cheryan, 2000; Jitputti et al., 2006; Ozsezen et al., 2009; Sarin et al., 2007).

However, for the efficient progression of the research in this field, it is necessary to develop efficient incentive structures for the primary stakeholders and to inform these stakeholders about the research (Konur, 2000, 2002a–c, 2006a–b, 2007a–b; North, 1991a–b).

Although there have been a number of reviews and book chapters in this field (Chew and Bhatia, 2008; Lam et al., 2009; Mekhilef et al., 2011; Ong et al., 2011; Sumathi et al., 2008; Shuit et al., 2009; Yusoff, 2006), there has been no review of the 25-most-cited articles in this field. Thus, this chapter reviews these most-cited articles by highlighting the key findings of these studies on the production and properties of POBD fuels. Then, it discusses these key findings.

23.2 MATERIALS AND METHODOLOGY

The search for the literature was carried out in the 'Web of Science' (WOS) database in February 2020. It contains the 'Science Citation Index-Expanded' (SCI-E), the Social Sciences Citation Index' (SSCI), the 'Book Citation Index-Science' (BCI-S), the 'Conference Proceedings Citation Index-Science' (CPCI-S), the 'Emerging Sources Citation Index' (ESCI), the 'Book Citation Index-Social Sciences and Humanities' (BCI-SSH), the 'Conference Proceedings Citation Index-Social Sciences and Humanities' (CPCI-SSH), and the 'Arts and Humanities Citation Index' (A&HCI).

The keywords for the search of the literature are collated from the screening of abstract pages for the first 1,000 highly cited papers on edible oil-based biodiesel fuels. These keywords sets are provided in the Appendix of the related chapter (Konur, 2021o).

The 25-most-cited articles are selected for this chapter and the key findings of the review are presented and discussed briefly.

23.3 RESULTS

23.3.1 Palm Oil-based Biodiesel Production

23.3.1.1 Crude Palm Oil-based Biodiesel Production

Darnoko and Cheryan (2000) study the kinetics of palm oil transesterification in a batch reactor in a paper with 436 citations. They produced 'methyl esters' (MEs) by the transesterification of palm oil with methanol in the presence of a catalyst (KOH). They observed that the rate of transesterification increased with temperature, up to 60°C. Higher temperatures did not reduce the time to reach maximal conversion. The conversion of 'triglycerides' (TGs), diglycerides (DGs), and monoglycerides (MGs) was second order, up to 30 min of reaction time. Reaction rate constants for TG, DG, and MG hydrolysis reactions were 0.018–0.191 (wt% min^{-1}), and were higher at higher temperatures and higher for the MC reaction than for TG hydrolysis. Activation energies were 14.7, 14.2, and 6.4 kcal/mol for the TG, DG, and MG hydrolysis reactions, respectively. The optimal catalyst concentration was 1% KOH.

Crabbe et al. (2001) study biodiesel production from crude palm oil and evaluate the butanol extraction and fuel properties in a paper with 243 citations. They focus on the molar ratio of methanol to oil, the amount of catalyst, and the reaction temperature, which affect the yield of the acid-catalyzed production of ME from crude palm oil. They then use the biodiesel as an extractant in batch and continuous acetone-butanol-ethanol fermentation, and analyze its fuel properties and that of the biodiesel–'acetone-butanol-ethanol' (ABE) product mix extracted from the batch culture. The optimized variables – 40:1 methanol/oil (mol/mol) with 5% H_2SO_4 (vol/wt) reacted at 95°C for 9 h – provided a maximum ester yield of 97%. Biodiesel preferentially extracted butanol and enhanced its production in the batch culture from 10 to 12 gl^{-1}. The fuel properties of biodiesel and the biodiesel–ABE mix were comparable to that of No. 2 diesel, but their cetane numbers and the boiling points of the 90% fractions were higher. Therefore, they could serve as efficient No. 2 diesel substitutes. The biodiesel–ABE mixture had the highest cetane number.

Noiroj et al. (2009) study the transesterification of palm oil to biodiesel using KOH loaded on Al_2O_3 and NaY zeolite supports as heterogeneous catalysts in a paper with 241 citations. They optimized the reaction parameters, such as the reaction time, wt% KOH loading, molar ratio of oil to methanol, and amount of catalyst for the production of biodiesel. They observed that the 25 wt% KOH/Al_2O_3 and the 10 wt% KOH/NaY catalysts were the best formula due to their biodiesel yield of 91.07% at temperatures below 70°C within 2–3 h at a 1:15 molar ratio of palm oil to methanol and a catalyst amount of 3–6 wt%. They also observed the leaching of potassium species in both spent catalysts. The amount of leached potassium species of the KOH/Al_2O_3 was somewhat higher compared to that of the KOH/NaY catalyst.

Mootabadi et al. (2010) study the ultrasonic-assisted transesterification of palm oil in the presence of alkaline earth metal oxide catalysts (CaO, SrO, and BaO) in a paper with 165 citations. They performed a batch process assisted by a 20 kHz ultrasonic cavitation to study the effect of the reaction time (10–60 min), alcohol to palm oil molar ratio (3:1–15:1), catalyst loading (0.5–3.0%), and the varying of ultrasonic amplitudes (25–100%). They observed that the activities of the catalysts were mainly related to their basic strength. The catalytic activity was in the sequence of CaO<SrO<BaO. At optimum conditions, 60 min was required to achieve a 95% yield compared to 2–4 h with conventional stirring. Further, the yields achieved in 60 min increased from 5.5 to 77.3% (CaO), 48.2 to 95.2% (SrO), and 67.3 to 95.2% (BaO). A 50% amplitude of ultrasonic irradiation was the most suitable value, and physical changes on the catalysts after the ultrasonic-assisted reaction were successfully elucidated. The BaO catalyst underwent a relatively more severe activity drop in the

catalyst reusability test. Catalyst dissolution was mainly responsible for the activity drop of the reused catalysts, especially with a BaO catalyst.

Hayyan et al. (2010) develop a low cost quaternary ammonium salt-glycerol-based ionic liquid as a solvent for extracting glycerol from biodiesel in a paper with 164 citations. They tested this separation technique on POBD with KOH as a reaction catalyst. They focused on the effect of the 'deep eutectic solvent' (DES):biodiesel ratio and the DES composition on the efficiency of the extraction process. The lab scale purification experiments proved the viability of the separation technique with a best DES:biodiesel molar ratio of 1:1 and a DES molar composition of 1:1 (salt:glycerol). The purified biodiesel fulfilled the EN 14214 and ASTMD 6751 standard specifications for biodiesel fuel in terms of glycerol content. The authors propose a continuous separation process for industrial scale application.

Yee et al. (2009) perform a 'life cycle assessment' (LCA) of palm biodiesel (PBD) to study and validate the popular belief that it is a green and sustainable fuel in a paper with 164 citations. They divided the LCA study into three main stages: agricultural activities, oil milling, and the transesterification process for the production of biodiesel. For each stage, they presented the energy balance and greenhouse gas assessments as these are important data for a technoeconomic and environmental feasibility evaluation of palm biodiesel. They then compared the results obtained for palm biodiesel with rapeseed biodiesel. They found that the utilization of PBD would generate an energy yield ratio of 3.53 (output energy:input energy), indicating a net positive energy generated and ensuring its sustainability. The energy ratio for PBD was more than double that of rapeseed biodiesel, which was estimated as 1.44, thereby indicating that palm oil would be a more sustainable feedstock for biodiesel production as compared to rapeseed oil. Moreover, combustion of PBD was more environment friendly than petrodiesel, as a significant 38% reduction of CO_2 emission could be achieved per liter combusted.

Aranda et al. (2008) study the acid-catalyzed homogeneous esterification reaction for biodiesel production from palm fatty acids in a batch reactor using homogeneous acid catalysts in a paper with 162 citations. They evaluated the effect of the alcohol used, the presence of water, and the type and concentration of the catalysts. They found that methanesulfonic and sulfuric acid were the best catalysts. Reaction with methanol showed greater yields. The presence of water in the reaction medium showed a negative effect in the reaction velocity. They estimate kinetic parameters and perform molecular modeling. They locate protonation of the carboxylic moiety of the fatty acid as a rate determinant step for the reaction.

Chongkhong et al. (2007) study the production of 'fatty acid methyl esters' (FAMEs) from a 'palm fatty acid distillate' (PFAD) having high 'free fatty acids' (FFAs) in a paper with 154 citations. They performed batch esterifications of PFAD. The study parameters were reaction temperatures of 70°C–100°C, molar ratios of methanol:PFAD of 0.4:1–12:1, a quantity of catalysts of 0–5.502% (wt of sulfuric acid/wt of PFAD), and reaction times of 15–240 min. They find that the optimum condition for the 'continuous esterification process' (CSTR) was a molar ratio of methanol:PFAD at 8:1 with a 1.834 wt% of H_2SO_4 at 70°C under its own pressure with a retention time of 60 min. The amount of FFA was reduced from 93 wt% to less than 2 wt% at the end of the esterification process. The FAME was purified by neutralization with 3 M sodium hydroxide in a water solution at a reaction temperature of 80°C for 15 min, followed by a transesterification process with 0.396 M sodium hydroxide in a methanol solution at a reaction temperature of 65°C for 15 min. The final FAME product met with the Thai biodiesel quality standard and ASTM D6751-02.

Boey et al. (2009) use waste mud crab (*Scylla serrata*) shells as a source of calcium oxide to transesterify palm olein into biodiesel in a paper with 153 citations. The main component of the shell was calcium carbonate which transformed into calcium oxide when activated above 700°C for 2 h. The optimal conditions were a methanol:oil mass ratio of 0.5:1; a catalyst amount of 5 wt%; a reaction temperature of 65°C, and a stirring rate of 500 rpm. The waste catalyst performed equally well as laboratory CaO, thus creating another low-cost catalyst source for producing biodiesel. The prepared catalyst was able to be reemployed up to 11 times. They performed statistical analysis using a 'central composite design' to evaluate the contribution and performance of the parameters on biodiesel purity.

Al-Zuhair et al. (2007) study the kinetic mechanism of the production of biodiesel from palm oil using lipase in a paper with 146 citations. The reaction took place in an *n*-hexane organic medium and the lipase used was from *Mucor miehei*. At a constant methanol concentration of 300 molm^{-3}, they observed that initially, as the palm oil concentration increased, the reaction rate increased. However, the initial rate dropped sharply at substrate concentrations larger than 1,250 molm^{-3}. They observed similar behavior for the methanol concentration effect, where, at a constant substrate concentration of 1,000 molm^{-3}, the initial rate of reaction dropped at methanol concentrations larger than 3,000 molm^{-3}. They adopted a 'Ping Pong Bi mechanism' with inhibition by both reactants and developed a mathematical model from a proposed kinetic mechanism and used it to identify the regions where the effect of inhibition by both substrates had arisen. The proposed model equation was essential for predicting the rate of the methanolysis of palm oil in a batch or a continuous reactor and for determining the optimal conditions for the biodiesel product.

Melero et al. (2010) study biodiesel production from crude palm oil containing a high percentage of FFAs over 'sulfonic acid'-functionalized SBA-15 materials ('propyl-SO$_3$H', 'arene-SO$_3$H', 'perfluoro-SO$_3$H') in a paper with 136 citations. They observed that 'sulfonic acid'-modified mesostructured materials were more active than conventional 'ion-exchange sulfonic resins' (Amberlyst-36 and SAC-13) in the simultaneous esterification of FFAs and transesterification of TGs with methanol. They also study the reusability of the catalysts showing high stability for 'propyl-SO$_3$H' and 'arene-SO$_3$H'-modified mesostructured materials.

In contrast, 'ionic-exchange sulfonic acid resins' displayed low-conversion rates, with stronger decay of activity in the second consecutive catalytic run. Interestingly, a 'perfluorosulfonic acid'-functionalized SBA-15 sample yielded a dramatic loss of activity, indicating that Si-O-C bonding is not stable under the reaction conditions as compared with the Si-C bond present in 'propyl-SO$_3$H' and 'arene-SO$_3$H' catalysts. Further functionalization of an 'arene-SO$_3$H' SBA-15 catalyst with hydrophobic trimethylsilyl groups enhanced its catalytic performance. This material was able to produce a yield to FAME of ca.95% in 4 h of reaction with a moderate methanol to oil molar ratio (20:1), 140°C, and a catalyst concentration of 6 wt% referred to the starting oil.

23.3.1.2 Palm Kernel Oil-based Biodiesel Production

Jitputti et al. (2006) study the transesterification of 'crude palm kernel oil' (PKO) and 'crude coconut oil' (CCO) by different solid catalysts in a paper with 356 citations. They used several acidic and basic solids, such as ZrO_2, ZnO, SO_4^{2-}/SnO_2, SO_4^{2-}/ZrO_2, KNO_3/KL zeolite, and KNO_3/ZrO, as heterogeneous catalysts for PKO and CCO transesterification with methanol. They observed that ZnO and SO_4^{2-}/ZrO_2 exhibited the highest activity for both PKO and CCO transesterification. In the case of SO_4^{2-}/ZrO_2, only 1 wt% of this acidic solid was needed to catalyze the reaction, which resulted in FAME content higher than 90%. The spent SO_4^{2-}/ZrO_2 cannot be directly reused for the transesterification. However, this spent catalyst can be easily regenerated and the same activity can be obtained.

Ngamcharussrivichai et al. (2008) study transesterification of PKO with methanol over mixed oxides of Ca and Zn, batchwise at 60°C and 1 atm in a paper with 186 citations. They prepared CaO ZnO catalysts via a conventional coprecipitation of the corresponding mixed metal nitrate solution in the presence of a soluble carbonate salt at near neutral conditions. They find that the mixed oxides possessed relatively small particle sizes and high surface areas, compared to pure CaO and ZnO. Moreover, the combination of Ca and Zn reduced the calcination temperature required for the decomposition of metal carbonate precipitates to active oxides.

They also studied the influences of the Ca:Zn atomic ratio in the mixed oxide catalyst, the catalyst amount, the methanol:oil molar ratio, the reaction time, and the water amount on the ME content. Under suitable transesterification conditions at 60°C (catalyst amount = 10 wt%, methanol:oil molar ratio = 30, reaction time = 1 h), the ME content of > 94% can be achieved over a CaO·ZnO catalyst with a Ca:Zn ratio of 0.25. The mixed oxide can also be applied to the transesterification of palm olein, soybean, and sunflower oils. They further studied the effects of different regeneration methods on the reusability of the CaO·ZnO catalyst.

Benjapornkulaphong et al. (2009) study the transesterification of PKO and CCO with methanol under a heterogeneous catalysis system in a paper with 146 citations. They prepared various Al_2O_3-supported alkali and alkali earth metal oxides via an impregnation method and applied them as solid catalysts. They observed that the supported alkali metal catalysts, $LiNO_3/Al_2O_3$, $NaNO_3/Al_2O_3$, and KNO_3/Al_2O_3, with active metal oxides formed at calcination temperatures of 450–550°C, showed a very high ME (ME) content (> 93%). This was due to a homogeneous catalysis of dissoluted alkali oxides.

On the other hand, $Ca(NO_3)_2/Al_2O_3$ calcined at 450°C yielded an ME content as high as 94% with only a small loss of active oxides from the catalyst, whereas a calcined $Mg(NO_3)_2/Al_2O_3$ catalyst possessed an inactive magnesium-aluminate phase, resulting in very low ME formation. At calcination temperatures of > 650°C, alkali metal- and alkali earth metal-aluminate compounds were formed. Whilst water-soluble alkali metal aluminates that formed over $NaNO_3/Al_2O_3$ and KNO_3/Al_2O_3 were catalytically active, the aluminate compounds on $LiNO_3/Al_2O_3$ and $Ca(NO_3)_2/Al_2O_3$ were less soluble, giving a very low ME content. The suitable conditions for the heterogeneously catalyzed transesterification of PKO and CCO over $Ca(NO_3)_2/Al_2O_3$ were the methanol:oil molar ratio of 65, a temperature of 60°C, and a reaction time of 3 h, with a 10 and 15–20% (w/w) catalyst:oil ratio for PKO and CCO, respectively.

Some important physical and fuel properties of the resultant biodiesel products met the standards of petrodiesel fuel and biodiesel issued by the Department of Energy Business, Ministry of Energy, Thailand.

23.3.1.3 Waste Palm Oil-based Biodiesel Production

Al-Widyan and Al-Shyoukh (2002) performed a transesterification of waste palm oil into biodiesel in a paper with 251 citations. They used H_2SO_4 and different concentrations of HCl and ethanol at different excess levels. They observed that higher catalyst concentrations (1.5–2.25 M) produced biodiesel with a lower specific gravity, γ, in a much shorter reaction time than lower concentrations. H_2SO_4 performed better than HCl at 2.25 M, as it resulted in a lower γ. Moreover, a 100% excess of alcohol effected significant reductions in reaction time and a lower gamma relative to lower excess levels. The best process combination was 2.25 M H_2SO_4 with 100% excess ethanol, which reduced γ from an initial value of 0.916 to a final value of 0.8737 in about 3 h of reaction time. Biodiesel had the behavior of a Newtonian fluid.

Halim et al. (2009) developed an optimal continuous procedure of lipase-catalyzed transesterification of 'waste cooking palm oil' (WPO) in a 'continuous packed bed reactor' to investigate the possibility of large scale production in a paper with 178 citations. They used a 'response surface methodology' (RSM) based on a 'central composite rotatable design' (CCRD) to optimize the two important reaction variables of packed bed height (cm) and substrate flow rate (ml/min). The optimum condition for the transesterification of WPO was a 10.53 cm packed bed height and a 0.57 ml/min substrate flow rate. The optimum predicted FAME yield was 80.3% and the actual value was 79%. The RSM study based on a CCRD was adaptable to the FAME yield studied for the current transesterification system. They also studied the effect of mass transfer in the packed bed reactor and developed models for FAME yield for cases of reaction control and mass transfer control. There was very good compatibility between the mass transfer model and the experimental results obtained from an immobilized lipase packed bed reactor operation, showing that in this case the FAME yield was mass transfer controlled.

23.3.2 Properties of Palm Oil-based Biodiesel Fuels

Sarin et al. (2007) study the properties of Jatropha-palm biodiesel blends in a paper with 349 citations. They note that due to the substantial amount of saturated fats in palm, palm biodiesel has poor low temperature properties. They examined blends of Jatropha and palm biodiesel to study their physicochemical properties and to obtain an optimum mix of them to achieve better low temperature properties with improved oxidation stability.

Benjumea et al. (2008) measured the properties of PBD–diesel blends according to the corresponding ASTM standards in a paper with 279 citations. They evaluate mixing rules as a function of the volume fraction of biodiesel in the blend in order to predict these properties. They use Kay's mixing rule for predicting density, heating value, three different points of the distillation curve (T10, T50, and T90), the 'cloud point', and calculated a 'cetane index', while an Arrhenius mixing rule is used for viscosity. They observed that the absolute average deviations obtained were low, demonstrating the suitability of the used mixing rules. They found that the calculated cetane index of PBD obtained using ASTM D4737 was in better agreement with the reported cetane number than the one corresponding to ASTM D976. This was most likely due to the fact that the former standard takes into account the particular characteristics of the distillation curve.

Ozsezen et al. (2009) study the performance and combustion characteristics of a 'direct injection' (DI) diesel engine fueled with 'waste palm oil methyl ester' (WPBD) and 'canola oil methyl ester' (CBD) in a paper with 246 citations. They performed the experiments at a constant engine speed mode (1,500 rpm) under the full load condition of the engine. They found that when the test engine was fueled with WPBD or CBD, the engine performance slightly weakened; the combustion characteristics slightly changed when compared to petrodiesel fuel. The biodiesels caused reductions in CO, 'unburned hydrocarbon' (UHC) emissions, and 'smoke opacity', but they caused increases in nitrogen oxide (NO_x) emissions.

Crabbe et al. (2001) study biodiesel production from crude palm oil and evaluate butanol extraction and fuel properties in a paper with 243 citations. They focused on the molar ratio of methanol:oil, the amount of catalyst, and the reaction temperature, which affect the yield of acid-catalyzed production of ME (biodiesel) from crude palm oil. They then use the PBD as an extractant in a batch and continuous acetone-butanol-ethanol fermentation, and analyze its fuel properties and that of the biodiesel–ABE product mix extracted from the batch culture. The optimized variables were 40:1 methanol:oil (mol/mol) with 5% H_2SO_4 (vol/wt) reacted at 95°C for 9 h, which gave a maximum ester yield of 97%. Biodiesel preferentially extracted butanol and enhanced its production in the batch culture from 10 to 12 gL^{-1}. The fuel properties of biodiesel and the biodiesel–ABE mix were comparable to that of No. 2 diesel, but their cetane numbers and the boiling points of the 90% fractions were higher. Therefore, they could serve as efficient No. 2 diesel substitutes. The biodiesel–ABE mixture had the highest cetane number.

Liang et al. (2006) study the effect of natural and synthetic antioxidants on the oxidative stability of PBD in a paper with 165 citations. They produced 'crude palm oil methyl esters' (CPBDs) from the transesterification of crude palm oil with minor components such as carotenes and vitamin E. They obtained the 'distilled palm oil methyl esters' (DPBDs) after the recovery of these minor components from the CPBDs.

Although both possessed fuel characteristics which were comparable to those of petrodiesel, CPBD exhibited better oxidative stability (rancimat induction period > 25 h) than DPBD (about 3.5 h). They attribute this to the presence of vitamin E (about 600 ppm), a natural antioxidant in the former. While the DPBDs contain practically no vitamin E (< 50 ppm) and, as a result, they exhibited poor oxidative stability. Thus, the CPBDs met the European standard for biodiesel (EN 14214) which sets a minimum rancimat induction period of 6 h. They further performed a study to enhance the oxidative stability of DPBD in order to meet this standard. They used natural and synthetic antioxidants to investigate their effect on the oxidative stability of DPBD. They observe that both types of antioxidants showed beneficial effects in inhibiting the oxidation of DPBD. Comparatively, they found that the synthetic antioxidants were more effective than the natural antioxidants, as a lower dosage (17 times less) was needed to achieve the minimum rancimat induction period of 6 h.

Yee et al. (2009) performed a 'life cycle assessment' (LCA) of PBD to study and validate the popular belief that PBD is a green and sustainable fuel in a paper with 164 citations. They divided the LCA study into three main stages: agricultural activities, oil milling, and the transesterification process for the production of PBD. For each stage, they present the 'energy balance' and 'greenhouse gas' assessments as these are important data for the technoeconomicl and environmental feasibility evaluation of palm biodiesel. They then compared the results obtained for PBD with rapeseed biodiesel. They found that the utilization of PBD generated an energy yield ratio of 3.53 (output energy/input energy), indicating a net positive energy generated and ensuring its sustainability. The energy ratio for PBD was more than double that of rapeseed biodiesel which was estimated to be only 1.44, thereby indicating that palm oil would be a more sustainable feedstock for biodiesel production as compared to rapeseed oil. Moreover, the combustion of PBD was more environment friendly than petrodiesel as a significant 38% reduction of CO_2 emissionsns could be achieved per liter combusted.

Moser (2008) studied the impact of blending canola, palm, soybean, and sunflower oil biodiesel on the fuel properties of biodiesel in a paper with 153 citations. They prepared single, binary, ternary, and quaternary mixtures of canola (low erucic acid rapeseed), palm, soybean, and sunflower (high oleic acid) oil biodiesel (CME, PME, SME, and SFME, respectively) and measure important fuel properties such as the 'oil stability index' (OSI), the 'cold filter plugging point' (CFPP), the 'cloud point' (CP), the 'pour point' (PP), 'kinematic viscosity' (40°C), 'lubricity', 'acid value' (AV), and 'iodine value' (IV). They observed that the fuel properties of SME were improved through blending with CME, PME, and SFME to satisfy the IV (<120) and OSI (>6 h) specifications contained within EN 14214. SME was satisfactory according to ASTM D6751 with regard to OSI (> 3 h). The CFPP of PME was improved by up to 15°C through blending with CME. They elucidated statistically significant relationships between OSI and IV, OSI and the saturated FAME (SFAME) content, OSI and CFPP, CFPP and IV, and CFPP and SFAME content. However, the only relationship of practical significance was that of CFPP versus SFAME content when the SFAME content was greater than 12 wt%.

Ozsezen and Canakci (2011) studied the performance, combustion, and injection characteristics of a DI diesel engine experimentally when it was fueled with CBD and 'waste (frying) palm oil methyl ester' (WPBD) in a paper with 150 citations. They performed the experiments at constant engine speeds under the full load condition of the engine. They observed that when the test engine was fueled with WPBD or CBD instead of petrodiesel fuel (PD), the 'brake power' reduced by 4–5%, while the 'brake specific fuel consumption' (BSFC) increased by 9–10%. On the other hand, MEs caused reductions in CO by 59–67%, in UHC emissions by 17–26%, in CO_2 by 5–8%, and smoke opacity by 56–63%. However, both MEs produced more NO_x emissions by 11–22% compared with those of the PD over the speed range.

Canakci et al. (2009) predicted the performance and exhaust emissions of a diesel engine fueled with biodiesel produced from WPBD in a paper with 142 citations. They performed the prediction of engine performance and exhaust emissions for five different neural networks to define how the inputs affect the outputs using the bio-diesel blends produced from WPBD. They used PD, B100, and biodiesel blends with PD, which are 50% (B50), 20% (B20), and 5% (B5), to measure the engine perfor-mance and exhaust emissions for different engine speeds at full load conditions. Using the 'artificial neural network' (ANN) model, they predicted the performance and exhaust emissions of a diesel engine for WPBD blends.

They found that the fifth network was sufficient for all the outputs. In this net-work, they took fuel properties, engine speed, and environmental conditions as the input parameters, while they used the values of flow rates, maximum injection pres-sure, emissions engine load, maximum cylinder gas pressure, and thermal efficiency as the output parameters. For all the networks, they applied the learning algorithm called back-propagation for a single hidden layer. They used a 'scaled conjugate gradient' (SCG) and 'Levenberg–Marquardt' (LM) for the variants of the algorithm, and provided the formulations for outputs obtained from the weights. The fifth net-work produced R^2 values of 0.99, and the mean percentage errors were smaller than 5 except for some emissions. They obtained higher mean errors for emissions such as CO, NO_x, and UHC. The complexity of the burning process and the measurement errors in the experimental study caused higher mean errors.

Fattah et al. (2014) study the effect of antioxidant addition to PBD on engine per-formance and emission characteristics in a paper with 138 citations. They produced PBD by transesterification using potassium hydroxide (KOH) as a catalyst. They added two monophenolic antioxidants, '2, 6-di-tert-butyl-4-methylphenol' (BHA) and '2(3)-tert-butyl-4-methoxy phenol' (BHT), at 1,000 ppm concentration to 20% PBD (B20) to study their effect. They observed that the addition of antioxidants increased 'oxidation stability' without causing any significant negative effect on physicochemical properties. BHA showed greater capability to increase the stability of B20. They used a 42 kW, 1.8 L, four-cylinder diesel engine to carry out tests under conditions of constant load and varying speed. They show that B20 and antioxidant-treated B20 produced 0.68–1.02% lower brake power (BP) and 4.03–4.71% higher BSFC compared to petrodiesel. Both of the antioxidants reduced NO_x by a mean of 9.8–12.6% compared to B20. However, they observed, compared to B20, mean increases in CO and hydrocarbon (HC) emissions of 8.6–12.3% and 9.1–12.0%, respectively. The emission levels of the three pollutants were lower than those of petrodiesel. Thus, they conclude that B20 blends with added antioxidant can be used in diesel engines without any modifications.

Lin et al. (2006) study the emissions of 'polycyclic aromatic hydrocarbons' (PAHs), 'carcinogenic potencies' (BaPeq), 'particulate matter' (PM), fuel consump-tion, and energy efficiency of PBD blends from the diesel generator in a steady state in a paper with 135 citations. They tested fuels of B0 (premium diesel fuel), B10 (10% palm biodiesel+90% B0), B20, B30, B50, B75, and B100. They observed that PAH emission decreased with increasing PBD blends due to the small PAH content in biodiesel. The 'mean reduction fraction' of total 'PAH emission factor' (B0 = 1110 μgL^{-1}) from the exhaust of the diesel generator was 13.2, 28.0, 40.6, 54.4, 61.89, and 98.8% for B10, B20, B30, B50, B75, and B100, respectively, compared with B0. The mean reduction fraction of the total BaPeq (B0 = 1.65 μgL^{-1}) from the exhaust of the diesel generator were 15.2, 29.1, 43.3, 56.4, 58.2, and 97.6% for B10, B20, B30, B50, B75, and B100, respectively, compared with B0.

PM emission decreased as the PBD blends increased from 0 to 10%, and increased as the PBD blends increased from 10 to 100% because the soluble organic fraction of the PM emission was high in blends with high PBD content. The BSFC rose with rising PBD blends due to the low gross heat value of PBD. The increasing fraction of BSFC of PBD was lower than those of soy, soapstock, *Brassica carinate*, and rapeseed biodiesel. PBD was the most feasible biodiesel. The best energy efficiency occurred between B10 and B20, close to B15. The curve dropped as the PBD content rose above B20. They conclude that PBD was an oxygenated fuel appropriate for use in diesel engines to promote combustion efficiency and decreased PAH emission. However, adding an excess of PBD to B0 led to incomplete combustion in the diesel-engine generator and inhibited the release of energy in the fuel.

23.4   DISCUSSION

Table 23.1 provides information on the research fronts in this field. As this table shows, the primary research fronts of 'production of POBD fuels' and 'properties of POBD fuels' comprise 64 and 44% of these papers, respectively. In the first group of papers, 'production of crude POBD fuels', 'production of PKO-based biodiesel fuels', and 'production of waste POBD fuels' form 44, 12, and 8% of these papers, respectively.

23.4.1   PALM OIL-BASED BIODIESEL PRODUCTION

23.4.1.1   Crude Palm Oil-based Biodiesel Production

Darnoko and Cheryan (2000) study the kinetics of palm oil transesterification in a batch reactor in a paper with 436 citations. Crabbe et al. (2001) study biodiesel production from crude palm oil and evaluate butanol extraction and fuel properties in a paper with 243 citations. They focus on the molar ratio of methanol:oil, the amount of catalyst, and the reaction temperature, which affect the yield of acid-catalyzed production of ME from crude palm oil. Noiroj et al. (2009) study the transesterification of palm oil to biodiesel using KOH loaded on Al_2O_3 and NaY zeolite supports as heterogeneous catalysts in a paper with 241 citations.

TABLE 23.1
Research Fronts

	Research Front	Papers (%)
1	Production of palm oil-based biodiesel (POBD) fuels	64
1.1	Production of crude POBD fuels	44
1.2	Production of palm kernel oil (PKO)-based biodiesel fuels	12
1.3	Production of waste POBD fuels	8
2	Properties of POBD fuels	44

Mootabadi et al. (2010) study the ultrasonic-assisted transesterification of palm oil in the presence of alkaline earth metal oxide catalysts (CaO, SrO, and BaO) in a paper with 165 citations. Hayyan et al. (2010) develop a low cost quaternary ammonium salt-glycerol-based ionic liquid as a solvent for extracting glycerol from biodiesel in a paper with 164 citations. Yee et al. (2009) perform an LCA of PBD to study and validate the popular belief that palm biodiesel is a green and sustainable fuel in a paper with 164 citations.

Aranda et al. (2008) study the acid-catalyzed homogeneous esterification reaction for biodiesel production from palm fatty acids in a batch reactor using homogeneous acid catalysts in a paper with 162 citations. Chongkhong et al. (2007) study the production of FAMEs from PFAD having high FFAs in a paper with 154 citations. They perform batch esterifications of PFAD. Boey et al. (2009) use waste mud crab (*Scylla serrata*) shells as a source of calcium oxide to transesterify palm olein into biodiesel in a paper with 153 citations.

Al-Zuhair et al. (2007) study the kinetic mechanism of the production of biodiesel from palm oil using lipase in a paper with 146 citations. Melero et al. (2010) study the biodiesel production from crude palm oil containing a high percentage of FFAs over 'sulfonic acid'-functionalized SBA-15 materials ('propyl-SO$_3$H', 'arene-SO$_3$H', 'perfluoro-SO$_3$H') in a paper with 136 citations.

These prolific studies highlight the production of crude POBD fuels.

23.4.1.2   Palm Kernel Oil-based Biodiesel Production

Jitputti et al. (2006) study the transesterification of crude PKO and CCO by different solid catalysts in a paper with 356 citations. Ngamcharussrivichai et al. (2008) study transesterification of PKO with methanol over mixed oxides of Ca and Zn batchwise at 60°C and 1 atm in a paper with 186 citations. Benjapornkulaphong et al. (2009) study transesterification of PKO and coconut oil with methanol under a heterogeneous catalysis system in a paper with 146 citations.

These prolific studies highlight the production of PKO-based biodiesel fuels which have more economic and environmental advantages compared to POBD fuels.

23.4.1.3   Waste Cooking Palm Oil-based Biodiesel Production

Al-Widyan and Al-Shyoukh (2002) perform the transesterification of WPO into biodiesel in a paper with 251 citations. Halim et al. (2009) develop an optimal continuous procedure of lipase-catalyzed transesterification of WPO in a continuous packed bed reactor to investigate the possibility of large scale production in a paper with 178 citations.

These prolific studies highlight the properties of WPO biodiesel fuels, which have more economic and environmental advantages compared to POBD fuels.

23.4.2   Properties of Palm Oil-based Biodiesel Fuels

Sarin et al. (2007) study the properties of Jatropha-palm biodiesel blends in a paper with 349 citations. Benjumea et al. (2008) measure the properties of PBD–diesel blends according to the corresponding ASTM standards in a paper with 279 citations. Ozsezen et al. (2009) study the performance and combustion characteristics of a DI diesel engine fueled with WPBD and CBD in a paper with 246 citations.

Crabbe et al. (2001) study biodiesel production from crude palm oil and evaluate butanol extraction and fuel properties in a paper with 243 citations. They focus on the molar ratio of methanol:oil, the amount of catalyst, and the reaction temperature, which affect the yield of acid-catalyzed production of ME from crude palm oil. Liang et al. (2006) study the effect of natural and synthetic antioxidants on the oxidative stability of PBD in a paper with 165 citations. Yee et al. (2009) perform an LCA of PBD to study and validate the popular belief that PBD is a green and sustainable fuel in a paper with 164 citations.

Moser (2008) study the impact of blending canola, palm, soybean, and sunflower oil biodiesel on the fuel properties of biodiesel in a paper with 153 citations. Ozsezen and Canakci (2011) study the performance, combustion, and injection characteristics of a DI diesel engine experimentally when it was fueled with CBD and 'WPO methyl ester' in a paper with 150 citations. Canakci et al. (2009) predict the performance and exhaust emissions of a diesel engine fueled with biodiesel produced from WPO in a paper with 142 citations.

Fattah et al. (2014) study the effect of antioxidant addition to PBD on engine performance and emission characteristics in a paper with 138 citations. Lin et al. (2006) study the emissions of PAHs, BaPeq, , PM, fuel consumption, and energy efficiency of PBD blends from the diesel generator in a steady state in a paper with 135 citations.

These prolific studies highlight the properties of POBD fuels in the context of the diesel engines.

23.4.3 THE PUBLIC CONCERNS REGARDING THE PRODUCTION OF PALM OIL-BASED BIODIESEL FUELS

There has been some public concerns about the production and utilization of POBD fuels.

One of these concerns emanates from the fact that both palm and palm oils are edible (Al-Shahib and Marshall, 2003; Basiron, 2007; Edem, 2002; Sundram et al., 2003, Tan et al., 2009). Thus, there has been significant competition between biodiesel production from palm oils and the household consumption of palm oils, resulting in significant public concern about food security (Godfray et al., 2010; Lobell et al., 2008; Rosegrant and Cline, 2003; Schmidhuber and Tubiello, 2007).

However, there has been substantial exploration of POBD fuels produced from PKOs (Jitputti et al., 2006; Ngamcharussrivichai et al., 2008; Benjapornkulaphong et al., 2009) and WPOs (Al-Widyan and Al-Shyoukh, 2002; Halim et al., 2009). These versions of biodiesel fuels help also in reducing the cost of these fuels as around 85% of the total cost emanates from the feedstocks (Haas, 2005).

Another of these concerns relates to the fact that oil palm plantations have replaced large areas of forest in Southeast Asia and other places (Carlson et al., 2012, 2013; Fitzherbert et al., 2008; Koh and Wilcove, 2008; Koh et al., 2011). This has resulted in public concern about the destruction of forests (deforestation) (Fitzherbert et al., 2008; Koh and Wilcove, 2008; Koh et al., 2011), the destruction of ecological biodiversity (Fitzherbert et al., 2008; Koh and Wilcove, 2008; Koh et al., 2011), a significant increase in CO_2 emissions (Carlson et al., 2012, 2013), and finally, exploitation of local communities (McCarthy, 2010; Obidzinski et al., 2012, Rist et al., 2010).

All these concerns about the expansion of oil palm plantations has resulted in the exploration of biodiesel production from nonedible oils (Konur, 2021r), waste oils (Konur, 2021s), and algae (Konur, 2021t).

23.5 CONCLUSION

This chapter has presented the key findings of the 25-most-cited article papers in this field. Table 23.1 provides information on the research fronts in this field. As this table shows the primary research fronts of 'production of POBD fuels' and 'properties of POBD fuels' comprise 64 and 44% of these papers, respectively. In the first group of papers, 'production of crude POBD fuels', 'production of PKO-based biodiesel fuels', and 'production of waste POBD fuels' form 44, 12, and 8% of these papers, respectively.

These prolific studies on two complementary research fronts provide valuable evidence on the production and properties of POBD fuels. The significant public concerns about the production and utilization of POBD fuels as a first generation edible oil-based biodiesel fuel has also been noted.

It is recommended that similar studies are carried out for each research front as well.

ACKNOWLEDGMENTS

The contribution of the highly cited researchers in this field is greatly acknowledged.

REFERENCES

Ahmadun, F. R., A. Pendashteh, and L. C. Abdullah, et al. 2009. Review of technologies for oil and gas produced water treatment. *Journal of Hazardous Materials* 170: 530–551.

Al-Shahib, W. and R. J. Marshall. 2003. The fruit of the date palm: Its possible use as the best food for the future? *International Journal of Food Sciences and Nutrition* 54: 247–259.

Al-Widyan, M. I. and A. O. Al-Shyoukh. 2002. Experimental evaluation of the transesterification of waste palm oil into biodiesel. *Bioresource Technology* 85: 253–256.

Al-Zuhair, S., F. W. Ling, and L. S. Jun. 2007. Proposed kinetic mechanism of the production of biodiesel from palm oil using lipase. *Process Biochemistry* 42: 951–960.

Antolin, G., F. V. Tinaut, and Y. Briceno, et al. 2002. Optimisation of biodiesel production by sunflower oil transesterification. *Bioresource Technology* 83:111–114.

Aranda, D. A. G., R. T. P. Santos, N. C. O. Tapanes, A. L. D. Ramos, and O. A. C. Antunes. 2008. Acid-catalyzed homogeneous esterification reaction for biodiesel production from palm fatty acids. *Catalysis Letters* 122: 20–25.

Atlas, R. M. 1981. Microbial degradation of petroleum hydrocarbons: An environmental perspective. *Microbiological Reviews* 45:180–209.

Babich, I. V. and J. A. Moulijn. 2003. Science and technology of novel processes for deep desulfurization of oil refinery streams: A review. *Fuel* 82: 607–631.

Basiron, Y. 2007. Palm oil production through sustainable plantations. *European Journal of Lipid Science and Technology* 109: 289–295.

Benjapornkulaphong, S., C. Ngamcharussrivichai, and K. Bunyakiat. 2009. Al$_2$O$_3$-supported alkali and alkali earth metal oxides for transesterification of palm kernel oil and coconut oil. *Chemical Engineering Journal* 145: 468–474.

Benjumea, P., J. Agudelo, and A. Agudejo. 2008. Basic properties of palm oil biodiesel-diesel blends. *Fuel* 87: 2069–2075.

Boey, P. L., G. P. Maniam, and S. A. Hamid. 2009. Biodiesel production via transesterification of palm olein using waste mud crab (*Scylla serrata*) shell as a heterogeneous catalyst. *Bioresource Technology* 100: 6362–6368.

Bridgwater, A. V. and G. V. C. Peacocke. 2000. Fast pyrolysis processes for biomass. *Renewable & Sustainable Energy Reviews* 4:1–73.

Bridgwater, A. V., D. Meier, and D. Radlein. 1999. An overview of fast pyrolysis of biomass. *Organic Geochemistry* 30:1479–1493.

Canakci, M., A. N. Ozsezen, E. Arcaklioglu, and A. Erdil. 2009. Prediction of performance and exhaust emissions of a diesel engine fueled with biodiesel produced from waste frying palm oil. *Expert Systems with Applications* 36: 9268–9280.

Carlson, K. M., L. M. Curran, and D. Ratnasari, et al. 2012. Committed carbon emissions, deforestation, and community land conversion from oil palm plantation expansion in West Kalimantan, Indonesia. *Proceedings of the National Academy of Sciences of the United States of America* 109: 7559–7564.

Carlson, K. M., L. M. Curran, and G. P. Asner, et al. 2013. Carbon emissions from forest conversion by Kalimantan oil palm plantations. *Nature Climate Change* 3: 283–287.

Chew, T. L. and S. Bhatia. 2008. Catalytic processes towards the production of biofuels in a palm oil and oil palm biomass-based biorefinery. *Bioresource Technology* 99: 7911–7922.

Chisti, Y. 2007. Biodiesel from microalgae. *Biotechnology Advances* 25: 294–306.

Chongkhong, S., C. Tongurai, P. Chetpattananondh, and C. Bunyakan. 2007. Biodiesel production by esterification of palm fatty acid distillate. *Biomass & Bioenergy* 31: 563–568.

Crabbe, E., C. Nolasco-Hipolito, G. Kobayashi, K. Sonomoto, and A. Ishizaki. 2001. Biodiesel production from crude palm oil and evaluation of butanol extraction and fuel properties. *Process Biochemistry* 37: 65–71.

Czernik, S. and A. V. Bridgwater. 2004. Overview of applications of biomass fast pyrolysis oil. *Energy & Fuels* 18: 590–598.

Darnoko, D. and M. Cheryan. 2000. Kinetics of palm oil transesterification in a batch reactor. *Journal of the American Oil Chemists Society* 77: 1263–1267.

Edem, D. O. 2002. Palm oil: Biochemical, physiological, nutritional, hematological, and toxicological aspects: A review. *Plant Foods for Human Nutrition* 57: 319–341.

Fattah, I. M. R., H. H. Masjuki, M. A. Kalam, M. Mofijur, and M. J. Abedin. 2014. Effect of antioxidant on the performance and emission characteristics of a diesel engine fueled with palm biodiesel blends. *Energy Conversion and Management* 79: 265–272.

Fitzherbert, E. B., M. J. Struebig, and A. Morel, et al. 2008. How will oil palm expansion affect biodiversity? *Trends n Ecology & Evolution* 23: 538–545.

Freedman, B., R. O. Butterfield, and E. H. Pryde. 1986. Transesterification kinetics of soybean oil. *Journal of the American Oil Chemists Society* 63:1375–1380.

Godfray, H. C. J., J. R. Beddington, and I. R. Crute, et al., 2010. Food security: The challenge of feeding 9 billion people. *Science* 327: 812–818.

Haas, M. J. 2005. Improving the economics of biodiesel production through the use of low value lipids as feedstocks: Vegetable oil soapstock. *Fuel Processing Technology* 86: 1087–1096.

Halim, S. F. A., A. H. Kamaruddin, and W. J. N. Fernando. 2009. Continuous biosynthesis of biodiesel from waste cooking palm oil in a packed bed reactor: Optimization using response surface methodology (RSM) and mass transfer studies. *Bioresource Technology* 100: 710–716.

Hayyan, M, F. S. Mjalli, M. A. Hashim, and I. M. AlNashef. 2010. A novel technique for separating glycerine from palm oil-based biodiesel using ionic liquids. *Fuel Processing Technology* 91:116–120.

Hill, J., E. Nelson, D. Tilman, S. Polasky, and D. Tiffany. 2006. Environmental, economic, and energetic costs and benefits of biodiesel and ethanol biofuels. *Proceedings of the National Academy of Sciences of the United States of America* 103:11206–11210.

Jitputti, J., B. Kitiyanan, and P. Rangsunvigit, et al. 2006. Transesterification of crude palm kernel oil and crude coconut oil by different solid catalysts. *Chemical Engineering Journal* 116: 61–66.

Kilian, L. 2009. Not all oil price shocks are alike: Disentangling demand and supply shocks in the crude oil market. *American Economic Review* 99: 1053–1069.

Koh, L. P. and D. S. Wilcove. 2008. Is oil palm agriculture really destroying tropical biodiversity? *Conservation Letters* 1: 60–64.

Koh, L. P., J. Miettinen, S. C. Liew, and J. Ghazoul. 2011. Remotely sensed evidence of tropical peatland conversion to oil palm. *Proceedings of the National Academy of Sciences of the United States of America* 108: 5127–5132.

Konur, O. 2000. Creating enforceable civil rights for disabled students in higher education: An institutional theory perspective. *Disability & Society* 15:1041–1063.

Konur, O. 2002a. Access to Nursing Education by disabled students: Rights and duties of nursing programs. *Nurse Education Today* 22: 364–374.

Konur, O. 2002b. Assessment of disabled students in higher education: Current public policy issues. *Assessment and Evaluation in Higher Education* 27: 131–152.

Konur, O. 2002c. Access to employment by disabled people in the UK: Is the Disability Discrimination Act working? *International Journal of Discrimination and the Law* 5: 247–279.

Konur, O. 2006a. Participation of children with dyslexia in compulsory education: Current public policy issues. *Dyslexia* 12: 51–67.

Konur, O. 2006b. Teaching disabled students in Higher Education. *Teaching in Higher Education* 11: 351–363.

Konur, O. 2007a. A judicial outcome analysis of the Disability Discrimination Act: A windfall for the employers? *Disability & Society* 22:187–204.

Konur, O. 2007b. Computer-assisted teaching and assessment of disabled students in higher education: The interface between academic standards and disability rights. *Journal of Computer Assisted Learning* 23: 207–219.

Konur, O., ed. 2021a. *Handbook of Biodiesel and Petrodiesel Fuels: Science, Technology, Health, and Environment.* Boca Raton, FL: CRC Press.

Konur, O., ed. 2021b. *Handbook of Biodiesel and Petrodiesel Fuels: Science, Technology, Health, and Environment. Volume 1. Biodiesel Fuels: Science, Technology, Health, and Environment.* Boca Raton, FL: CRC Press.

Konur, O., ed. 2021c. *Handbook of Biodiesel and Petrodiesel Fuels: Science, Technology, Health, and Environment. Volume 2. Biodiesel Fuels based on the Edible and Nonedible Feedstocks, Wastes, and Algae: Science, Technology, Health, and Environment.* Boca Raton, FL: CRC Press.

Konur, O., ed. 2021d. *Handbook of Biodiesel and Petrodiesel Fuels: Science, Technology, Health, and Environment. Volume 3. Petrodiesel Fuels: Science, Technology, Health, and Environment.* Boca Raton, FL: CRC Press.

Konur, O. 2021e. Biodiesel and petrodiesel fuels: Science, technology, health, and environment. In *Handbook of Biodiesel and Petrodiesel Fuels: Science, Technology, Health, and Environment. Volume 1. Biodiesel Fuels: Science, Technology, Health, and Environment*, ed. O. Konur. Boca Raton, FL: CRC Press.

Konur, O. 2021f. Biodiesel and petrodiesel fuels: A scientometric review of the research. In *Handbook of Biodiesel and Petrodiesel Fuels: Science, Technology, Health, and Environment. Volume 1. Biodiesel Fuels: Science, Technology, Health, and Environment*, ed. O. Konur. Boca Raton, FL: CRC Press.

Konur, O. 2021g. Biodiesel and petrodiesel fuels: A review of the research. In *Handbook of Biodiesel and Petrodiesel Fuels: Science, Technology, Health, and Environment. Volume*

1. Biodiesel Fuels: Science, Technology, Health, and Environment, ed. O. Konur. Boca Raton, FL: CRC Press.

Konur, O. 2021h Nanotechnology applications in the diesel fuels and the related research fields: A review of the research. In *Handbook of Biodiesel and Petrodiesel Fuels: Science, Technology, Health, and Environment. Volume 1. Biodiesel Fuels: Science, Technology, Health, and Environment*, ed. O. Konur. Boca Raton, FL: CRC Press.

Konur, O. 2021i. Biooils: A scientometric review of the research. In *Handbook of Biodiesel and Petrodiesel Fuels: Science, Technology, Health, and Environment. Volume 1. Biodiesel Fuels: Science, Technology, Health, and Environment*, ed. O. Konur. Boca Raton, FL: CRC Press.

Konur, O. 2021j. Characterization and properties of biooils: A review of the research. In *Handbook of Biodiesel and Petrodiesel Fuels: Science, Technology, Health, and Environment. Volume 1. Biodiesel Fuels: Science, Technology, Health, and Environment*, ed. O. Konur. Boca Raton, FL: CRC Press.

Konur, O. 2021k. Biomass pyrolysis and pyrolysis oils: A review of the research. In *Handbook of Biodiesel and Petrodiesel Fuels: Science, Technology, Health, and Environment. Volume 1. Biodiesel Fuels: Science, Technology, Health, and Environment*, ed. O. Konur. Boca Raton, FL: CRC Press.

Konur, O. 2021l. Biodiesel fuels: A scientometric review of the research. In *Handbook of Biodiesel and Petrodiesel Fuels: Science, Technology, Health, and Environment. Volume 1. Biodiesel Fuels: Science, Technology, Health, and Environment*, ed. O. Konur. Boca Raton, FL: CRC Press.

Konur, O. 2021m. Glycerol: A scientometric review of the research. In *Handbook of Biodiesel and Petrodiesel Fuels: Science, Technology, Health, and Environment. Volume 1. Biodiesel Fuels: Science, Technology, Health, and Environment*, ed. O. Konur. Boca Raton, FL: CRC Press.

Konur, O. 2021n. Propanediol production from glycerol: A review of the research. In *Handbook of Biodiesel and Petrodiesel Fuels: Science, Technology, Health, and Environment. Volume 1. Biodiesel Fuels: Science, Technology, Health, and Environment*, ed. O. Konur. Boca Raton, FL: CRC Press.

Konur, O. 2021o. Edible oil-based biodiesel fuels: A scientometric review of the research. *In Handbook of Biodiesel and Petrodiesel Fuels: Science, Technology, Health, and Environment. Volume 2. Biodiesel Fuels based on the Edible and Nonedible Feedstocks, Wastes, and Algae: Science, Technology, Health, and Environment*, ed. O. Konur. Boca Raton, FL: CRC Press.

Konur, O. 2021p. Palm oil-based biodiesel fuels: A review of the research. In *Handbook of Biodiesel and Petrodiesel Fuels: Science, Technology, Health, and Environment. Volume 2. Biodiesel Fuels based on the Edible and Nonedible Feedstocks, Wastes, and Algae*, ed. O. Konur. Boca Raton, FL: CRC Press.

Konur, O. 2021q. Rapeseed oil-based biodiesel fuels: A review of the research. In *Handbook of Biodiesel and Petrodiesel Fuels: Science, Technology, Health, and Environment. Volume 2. Biodiesel Fuels based on the Edible and Nonedible Feedstocks, Wastes, and Algae*, ed. O. Konur. Boca Raton, FL: CRC Press.

Konur, O. 2021r. Nonedible oil-based biodiesel fuels: A scientometric review of the research. In *Handbook of Biodiesel and Petrodiesel Fuels: Science, Technology, Health, and Environment. Volume 2. Biodiesel Fuels based on the Edible and Nonedible Feedstocks, Wastes, and Algae: Science, Technology, Health, and Environment*, ed. O. Konur. Boca Raton, FL: CRC Press.

Konur, O. 2021s. Waste oil-based biodiesel fuels: A scientometric review of the research. In *Handbook of Biodiesel and Petrodiesel Fuels: Science, Technology, Health, and Environment. Volume 2. Biodiesel Fuels based on the Edible and Nonedible Feedstocks, Wastes, and Algae: Science, Technology, Health, and Environment*, ed. O. Konur. Boca Raton, FL: CRC Press.

Konur, O. 2021t. Algal biodiesel fuels: A scientometric review of the research. In *Handbook of Biodiesel and Petrodiesel Fuels: Science, Technology, Health, and Environment. Volume 2. Biodiesel Fuels based on the Edible and Nonedible Feedstocks, Wastes, and Algae: Science, Technology, Health, and Environment*, ed. O. Konur. Boca Raton, FL: CRC Press.

Konur, O. 2021u. Algal biomass production for biodiesel production: A review of the research. In *Handbook of Biodiesel and Petrodiesel Fuels: Science, Technology, Health, and Environment. Volume 2. Biodiesel Fuels based on the Edible and Nonedible Feedstocks, Wastes, and Algae*, Ed. O. Konur. Boca Raton, FL: CRC Press.

Konur, O. 2021v. Algal biomass production in wastewaters for biodiesel production: A review of the research. In *Handbook of Biodiesel and Petrodiesel Fuels: Science, Technology, Health, and Environment. Volume 2. Biodiesel Fuels based on the Edible and Nonedible Feedstocks, Wastes, and Algae*, ed. O. Konur. Boca Raton, FL: CRC Press.

Konur, O. 2021x. Algal lipid production for biodiesel production: A review of the research. In *Handbook of Biodiesel and Petrodiesel Fuels: Science, Technology, Health, and Environment. Volume 2. Biodiesel Fuels based on the Edible and Nonedible Feedstocks, Wastes, and Algae*, ed. O. Konur. Boca Raton, FL: CRC Press.

Konur, O. 2021y. Crude oils: A scientometric review of the research. In *Handbook of Biodiesel and Petrodiesel Fuels: Science, Technology, Health, and Environment. Volume 3. Petrodiesel Fuels: Science, Technology, Health, and Environment*, ed. O. Konur. Boca Raton, FL: CRC Press.

Konur, O. 2021z. Petrodiesel fuels: A scientometric review of the research. In *Handbook of Biodiesel and Petrodiesel Fuels: Science, Technology, Health, and Environment. Volume 3. Petrodiesel Fuels: Science, Technology, Health, and Environment*, ed. O. Konur. Boca Raton, FL: CRC Press.

Konur, O. 2021aa. Bioremediation of petroleum hydrocarbons in the contaminated soils: A review of the research. In *Handbook of Biodiesel and Petrodiesel Fuels: Science, Technology, Health, and Environment. Volume 3. Petrodiesel Fuels: Science, Technology, Health, and Environment*, ed. O. Konur. Boca Raton, FL: CRC Press.

Konur, O. 2021ab. Desulfurization of diesel fuels: A review of the research. In *Handbook of Biodiesel and Petrodiesel Fuels: Science, Technology, Health, and Environment. Volume 3. Petrodiesel Fuels: Science, Technology, Health, and Environment*, ed. O. Konur. Boca Raton, FL: CRC Press.

Konur, O. 2021ac. Diesel fuel exhaust emissions: A scientometric review of the research. In *Handbook of Biodiesel and Petrodiesel Fuels: Science, Technology, Health, and Environment. Volume 3. Petrodiesel Fuels: Science, Technology, Health, and Environment*, ed. O. Konur. Boca Raton, FL: CRC Press.

Konur, O. 2021ad. The adverse health and safety impact of diesel fuels: A scientometric review of the research. In *Handbook of Biodiesel and Petrodiesel Fuels: Science, Technology, Health, and Environment. Volume 3. Petrodiesel Fuels: Science, Technology, Health, and Environment*, ed. O. Konur. Boca Raton, FL: CRC Press.

Konur, O. 2021ae. Respiratory illnesses caused by the diesel fuel exhaust emissions: A review of the research. In *Handbook of Biodiesel and Petrodiesel Fuels: Science, Technology, Health, and Environment. Volume 3. Petrodiesel Fuels: Science, Technology, Health, and Environment*, ed. O. Konur. Boca Raton, FL: CRC Press.

Konur, O. 2021af. Cancer caused by the diesel fuel exhaust emissions: A review of the research. In *Handbook of Biodiesel and Petrodiesel Fuels: Science, Technology, Health, and Environment. Volume 3. Petrodiesel Fuels: Science, Technology, Health, and Environment*, ed. O. Konur. Boca Raton, FL: CRC Press.

Konur, O. 2021ag. Cardiovascular and other illnesses caused by the diesel fuel exhaust emissions: A review of the research. In *Handbook of Biodiesel and Petrodiesel Fuels: Science, Technology, Health, and Environment. Volume 3. Petrodiesel Fuels: Science, Technology, Health, and Environment*, ed. O. Konur. Boca Raton, FL: CRC Press.

Kusdiana, D. and S. Saka. 2001. Kinetics of transesterification in rapeseed oil to biodiesel fuel as treated in supercritical methanol. *Fuel* 80: 693–698.

Lam, M. K., K. T. Tan, K. T. Lee, and A. R. Mohamed. 2009. Malaysian palm oil: Surviving the food versus fuel dispute for a sustainable future. *Renewable & Sustainable Energy Reviews* 13:1456–1464.

Liang, Y. C., C. Y. May, and C. S. Foon, et al. 2006. The effect of natural and synthetic antioxidants on the oxidative stability of palm diesel. *Fuel* 85: 867–870.

Lin, Y. C., W. J. Lee, and H. C. Hou. 2006. PAH emissions and energy efficiency of palm-biodiesel blends fueled on diesel generator. *Atmospheric Environment* 40: 3930–3940.

Lobell, D. B., M. B. Burke, and C. Tebaldi, et al. 2008. Prioritizing climate change adaptation needs for food security in 2030. *Science* 319: 607–610.

McCarthy, J. F. 2010. Processes of inclusion and adverse incorporation: oil palm and agrarian change in Sumatra, Indonesia. *Journal of Peasant Studies* 37: 821–850.

Mekhilef, S., S. Siga, and R. Saidur. 2011. A review on palm oil biodiesel as a source of renewable fuel. *Renewable & Sustainable Energy Reviews* 15:1937–1949.

Melero, J. A., L. F. Bautista, G. Morales, J. Iglesias, and R. Sanchez-Vazquez. 2010. Biodiesel production from crude palm oil using sulfonic acid-modified mesostructured catalysts. *Chemical Engineering Journal* 161: 323–331.

Mohan, D., C. U. Pittman, and P. H. Steele. 2006. Pyrolysis of wood/biomass for bio-oil: A critical review. *Energy & Fuels* 20: 848–889.

Mootabadi, H., B. Salamatinia, S. Bhatia, and A. Z. Abdullah. 2010. Ultrasonic-assisted biodiesel production process from palm oil using alkaline earth metal oxides as the heterogeneous catalysts. *Fuel* 89: 1818–1825.

Moser, B. R. 2008. Influence of blending canola, palm, soybean, and sunflower oil methyl esters on fuel properties of biodiesel. *Energy & Fuels* 22: 4301–4306.

Ngamcharussrivichai, C., P. Totarat and K. Bunyakiat. 2008. Ca and Zn mixed oxide as a heterogeneous base catalyst for transesterification of palm kernel oil. *Applied Catalysis A-General* 341: 77–85.

Noiroj, K., P. Intarapong, A. Luengnaruemitchai, and S. Jai-In. 2009. A comparative study of KOH/Al$_2$O$_3$ and KOH/NaY catalysts for biodiesel production via transesterification from palm oil. *Renewable Energy* 34:1145–1150.

North, D. C. 1991a. *Institutions, Institutional Change and Economic Performance*. Cambridge, Mass.: Cambridge University Press.

North, D.C. 1991b. Institutions. *Journal of Economic Perspectives* 5: 97–112.

Noureddini, H. and D. Zhu. 1997. Kinetics of transesterification of soybean oil. *Journal of the American Oil Chemists Society* 74.1457–1463.

Obidzinski, K., R. Andriani, H. Komarudin, and A. Andrianto. 2012. Environmental and social impacts of oil palm plantations and their implications for biofuel production in Indonesia. *Ecology and Society* 17: 25.

Ong, H. C., T. M. I. Mahlia, H. H. Masjuki, and R. S. Norhasyima. 2011. Comparison of palm oil, *Jatropha curcas* and *Calophyllum inophyllum* for biodiesel: A review. *Renewable & Sustainable Energy Reviews* 15: 3501–3515.

Ozsezen, A. N. and M. Canakci. 2011. Determination of performance and combustion characteristics of a diesel engine fueled with canola and waste palm oil methyl esters. *Energy Conversion and Management* 52:108–116.

Ozsezen, A. N., M. Canakci, A. Turkcan, and C. Sayin. 2009. Performance and combustion characteristics of a DI diesel engine fueled with waste palm oil and canola oil methyl esters. *Fuel* 88: 629–636.

Perron, P. 1989. The great crash, the oil price shock, and the unit root hypothesis. *Econometrica: Journal of the Econometric Society* 57: 1361–1401.

Rist, L., L. Feintrenie, and P. Levang. 2010. The livelihood impacts of oil palm: Smallholders in Indonesia. *Biodiversity and Conservation* 19: 1009–1024.

Rosegrant, M. W. and S. A. Cline. 2003. Global food security: Challenges and policies. *Science* 302:1917–1919.

Saka, S. and D. Kusdiana. 2001. Biodiesel fuel from rapeseed oil as prepared in supercritical methanol. *Fuel* 80: 225–231.

Sarin, R., M. Sharma, S. Sinharay, and R. K. Malhotra. 2007. Jatropha-palm biodiesel blends: An optimum mix for Asia. *Fuel* 86: 1365–1371.

Schmidhuber, J. and F. N. Tubiello. 2007. Global food security under climate change. *Proceedings of the National Academy of Sciences* 104:19703–19708.

Shuit, S. H., K. T. Tan, K. T. Lee, and A. H. Kamaruddin. 2009. Oil palm biomass as a sustainable energy source: A Malaysian case study. *Energy* 34:1225–1235.

Sumathi, S., S. P. Chai, and A. R. Mohamed. 2008. Utilization of oil palm as a source of renewable energy in Malaysia. *Renewable & Sustainable Energy Reviews* 12: 2404–2421.

Sundram, K., R. Sambanthamurthi, and Y. A. Tan. 2003. Palm fruit chemistry and nutrition. *Asia Pacific Journal of Clinical Nutrition* 12: 355–362.

Tan, K. T., K. T. Lee, A. R. Mohamed, and S. Bhatia. 2009. Palm oil: Addressing issues and towards sustainable development. *Renewable & Sustainable Energy Reviews* 13: 420–427.

Yee, K. F., K. T. Tan, A. Z. Abdullah, and K. T. Lee. 2009. Life cycle assessment of palm biodiesel: Revealing facts and benefits for sustainability. *Applied Energy* 86: S189–SS96.

Yusoff, S. 2006. Renewable energy from palm oil: Innovation on effective utilization of waste. *Journal of Cleaner Production* 14: 87–93.

Zhang, Q., J. Chang, T. J. Wang, and Y. Xu. 2007. Review of biomass pyrolysis oil properties and upgrading research. *Energy Conversion and Management* 48: 87–92.

24 Rapeseed Oil-based Biodiesel Fuels

A Review of the Research

Ozcan Konur

CONTENTS

24.1 INTRODUCTION

Crude oils have been primary sources of energy and fuels, such as petrodiesel. However, significant public concern about the sustainability, price fluctuations, and adverse environmental impact of crude oils have emerged since the 1970s (Ahmadun et al., 2009; Atlas, 1981; Babich and Moulijn, 2003; Kilian, 2009; Perron, 1989). Thus, biooils (Bridgwater et al., 1999; Bridgwater and Peacocke, 2000; Czernik and Bridgwater, 2004; Mohan et al., 2006; Zhang et al., 2007) and biooil-based biodiesel fuels (Chisti, 2007; Hill et al., 2006) have emerged as alternatives to crude oils and crude oil-based petrodiesel fuels, respectively, in recent decades. Nowadays, although petrodiesel fuels are still used extensively, biodiesel fuels are being used increasingly in the transportation and power sectors (Konur, 2021a–ag).

Therefore, there has been great public interest in the development of edible oil-based biodiesel fuels as the first generation of biodiesel fuels (Antolin et al., 2002;

Freedman et al., 1986; Kusdiana and Saka, 2001a–b; Noureddini and Zhu, 1997; Saka and Kusdiana, 2001). Rapeseed oil-based biodiesel fuels (ROBD) fuels have been among the most-produced and marketed edible oil-based biodiesel fuels (Gelbard et al., 1995; Gryglewicz, 1999; Kusdiana and Saka, 2001a–b; Labeckas and Slavinskas, 2006; Li et al., 2006; Rashid and Anwar, 2008; Saka and Kusdiana, 2001; Warabi et al., 2004).

However, for the efficient progression of the research in this field, it is necessary to develop efficient incentive structures for the primary stakeholders and to inform these stakeholders about the research (Konur, 2000, 2002a–c, 2006a–b, 2007a–b; North, 1991a–b).

Although there have been a number of reviews and book chapters in this field (Aldhaidhawi et al., 2017; El-Enin et al., 2013; Ge et al., 2017; Milazzo et al., 2013), there has been no review of the 25-most-cited articles. Thus, this chapter reviews these articles by highlighting the key findings of these most-prolific studies on the production and properties of ROBD fuels. Then, it discusses these key findings.

24.2 MATERIALS AND METHODOLOGY

The search for the literature was carried out in the 'Web of Science' (WOS) database in February 2020. It contains the 'Science Citation Index-Expanded' (SCI-E), the Social Sciences Citation Index' (SSCI), the 'Book Citation Index-Science' (BCI-S), the 'Conference Proceedings Citation Index-Science' (CPCI-S), the 'Emerging Sources Citation Index' (ESCI), the 'Book Citation Index-Social Sciences and Humanities' (BCI-SSH), the 'Conference Proceedings Citation Index-Social Sciences and Humanities' (CPCI-SSH), and the 'Arts and Humanities Citation Index' (A&HCI).

The keywords for the search of the literature were collated from the screening of abstract pages for the first 1,000 highly cited papers on edible oil-based biodiesel fuels. These keyword sets are provided in the Appendix of the related chapter (Konur, 2021o).

The 25-most-cited articles are selected for this review and the key findings are presented and discussed briefly.

24.3 RESULTS

24.3.1 PRODUCTION OF RAPESEED OIL-BASED BIODIESEL FUELS

Saka and Kusdiana (2001) study the transesterification reaction of rapeseed oil in supercritical methanol without using any catalyst in a paper with 687 citations. They performed an experiment in the batch-type reaction vessel, preheated to between 350 and 400°C and at a pressure of 35–65 MPa, and with a molar ratio of 1:42 of rapeseed oil to methanol. They observed that in a preheated temperature of 350°C, 240 s of supercritical treatment of methanol was sufficient to convert the rapeseed oil to 'methyl esters' (MEs). Although the prepared MEs were basically the same as those of the common method with a basic catalyst, their yield by the former was higher than that by the latter. In addition, this new supercritical methanol process required a shorter reaction time and a simpler purification procedure because of the unused catalyst.

Kusdiana and Saka (2001a) performed a kinetic study on the free catalyst trans-esterification of rapeseed oil in subcritical and supercritical methanol under different reaction conditions of temperature and reaction time in a paper with 540 citations. They made runs in a bath-type reaction vessel, ranging from 200°C in a subcritical temperature to 500°C in a supercritical state, with different molar ratios of methanol to rapeseed oil to determine rate constants by employing a simple method. They observed that the conversion rate of rapeseed oil to its MEs increased dramatically in the supercritical state, and that a reaction temperature of 350°C was the best condition, with the molar ratio of methanol in rapeseed oil being 42.

Gryglewicz (1999) study a heterogeneous catalyst, in particular calcium compounds, to produce MEs of rapeseed oil in a paper with 379 citations. They observed that the transesterification of rapeseed oil by methyl alcohol could be catalyzed effectively by basic alkaline-earth metal compounds, namely: calcium oxide, calcium methoxide, and barium hydroxide. Calcium catalysts, due to their weak solubility in the reaction medium, were less active than sodium hydroxide. However, calcium catalysts were cheaper and led to decreases in the number of technological stages and the amount of unwanted waste products. They further observed that the transesterification reaction rate could be enhanced by ultrasound as well as by introducing an appropriate reagent into a reactor to promote methanol solubility in the rapeseed oil. Hence, they used tetrahydrofuran as an additive to accelerate the transesterification process.

Gelbard et al. (1995) monitored the formation of 'fatty acid methyl esters' (FAMEs) by the transesterification of rapeseed oil with methanol by ^{1}H 'nuclear magnetic resonance spectroscopy' (NMR) in a paper with 345 citations. They observed that this accurate determination was simpler than chromatographic methods.

Rashid and Anwar (2008) developed an optimized protocol for the production of biodiesel through the alkaline-catalyzed transesterification of rapeseed oil in a paper with 301 citations. They used the reaction variables of the methanol:oil molar ratio (3:1–21:1), a catalyst concentration (0.25–1.50%), temperature (35–65°C), mixing intensity (180–600 rpm), and catalyst type. They followed the evaluation of the trans-esterification process by gas chromatographic analysis of the rapeseed oil FAMEs at different reaction times. They observed that the biodiesel with the best yield and quality was produced at a methanol:oil molar ratio of 6:1, a potassium hydroxide catalyst concentration of 1.0%, a mixing intensity of 600 rpm, and a reaction temperature of 65°C. The yield of the biodiesel produced under optimal condition was 95–96%. They further note that with a greater or lower concentration of KOH or methanol in relation to the optimal values, the reaction either did not fully occur or led to soap formation. They evaluated the quality of the biodiesel produced by the determinations of important properties, such as density, specific gravity, kinematic viscosity, higher heating value, acid value, flash point, pour point, cloud point, combustion point, 'cold filter plugging point' (CFPP), cetane index, ash content, sulfur content, water content, copper strip corrosion value, distillation temperature, and fatty acid composition. The produced biodiesel exhibited fuel properties within the limits prescribed by the latest 'American Standards for Testing Materials' (ASTM) and European EN standards.

Warabi et al. (2004) developed a catalyst-free biodiesel production method with supercritical methanol that allows a simple process and high yield because of the simultaneous transesterification of triglycerides and the methyl esterification of the fatty acids of rapeseed oil in a paper with 300 citations. They set the reaction temperature at 300°C, and used methanol, ethanol, 1-propanol, 1-butanol, or 1-octanol as the reactant. They observed that the transesterification of rapeseed oil was slower in reaction rates than the alkyl esterification of fatty acids for any of the alcohols employed. Furthermore, saturated fatty acids such as palmitic and stearic acids had slightly lower reactivity than that of the unsaturated fatty acids, such as oleic, linoleic, and linolenic.

Li et al. (2006) study the lipase-catalyzed transesterification of rapeseed oils for biodiesel production with a novel organic solvent, 'tert-Butanol', as the reaction medium in a paper with 257 citations. They propose the combined use of 'Lipozyme TL IM' and 'Novozym 435' to catalyze the methanolysis. They observed that the highest biodiesel yield of 95% could be achieved under optimum conditions (tert-butanol:oil volume ratio 1:1; methanol:oil molar ratio 4:1; 3% Lipozyme TL IM and 1% Novozym 435, based on the oil weight; temperature 35°C; 130 rpm, 12 h). There was no obvious loss in lipase activity even after being repeatedly used for 200 cycles with tert-butanol as the reaction medium. They also explored waste rapeseed oil for biodiesel production and found that lipase also showed good stability in this novel system.

Yuan et al. (2008) used waste rapeseed oil with high 'free fatty acids' (FFAs) as feedstock for producing biodiesel in a paper with 192 citations. In the pretreatment step, they reduced the FFAs by a distillation refining method. Then, they produced biodiesel by an alkaline-catalyzed transesterification process, which was designed according to the 2^4 'full-factorial central composite design'. They used the 'response surface methodology' (RSM) to optimize the conditions for the maximum conversion to biodiesel and determined the significance and interaction of the factors affecting biodiesel production. They observed that the catalyst concentration and reaction time were the limiting conditions and that little variation in their values would alter the conversion. At the same time, there was a significant mutual interaction between the catalyst concentration and reaction time. They observed that biodiesel mainly contained six FAMEs. In addition, most of the fuel properties were in reasonable agreement with GB252-2000 and ASTM D6751.

Leclercq et al. (2001) study the transesterification of rapeseed oil in the presence of basic zeolites and related solid catalysts in a paper with 171 citations. They performed transesterification of rapeseed oil with methanol by reflux of methanol over cesium-exchanged NaX faujasites, mixed magnesium-aluminum oxides, magnesium oxide, and barium hydroxide for different methanol-to-oil ratios. Over cesium-exchanged NaX faujasites and mixed magnesium-aluminum oxides, they observed that a long reaction time and a high methanol:oil ratio were required to achieve both high oil conversion and high yields in MEs. However, over a 300 m²/g magnesium oxide, methanol:oil ratios and reaction times were significantly reduced to obtain both high oil conversion and high yield in the MEs, particularly when the hydroxide precursor was calcined at 823 K. Finally, preliminary results with other basic solids such as barium hydroxide showed very high activity and a very high yield in esters.

Verziu et al. (2008) study the catalytic activity for the production of biodiesel from sunflower and rapeseed oils with three morphologically different nanocrystalline MgO materials prepared using simple, green, and reproducible methods in a paper with 161 citations. The nanocrystalline samples studied were MgO (111) nanosheets (MgOI), conventionally prepared MgO (MgOII), and aerogel prepared MgO (MgOIII). They tested them in the transesterification of sunflower and rapeseed vegetable oils at low temperatures, and under the different experimental conditions of autoclave, microwave, and ultrasound. They observed that working with these materials under microwave conditions provided higher conversions and selectivities to methylesters compared to autoclave or ultrasound conditions. Under ultrasound, a leaching of the magnesium was evidenced as a direct consequence of a saponification reaction. These systems also allowed working with much lower ratios of methanol to vegetable oil than reported in the literature for other heterogeneous systems. The activation temperature providing the most active catalysts varied depending on the exposed facet: for MgOI this was 773 K, while for MgOII and MgOIII it was 583 K.

Westbrook et al. (2011) developed a detailed chemical kinetic reaction mechanism for the five major components of soybean biodiesel and rapeseed biodiesel fuels in a paper with 154 citations. These components, methyl stearate, methyl oleate, methyl linoleate, methyl linolenate, and methyl palmitate, were large ME molecules, some with carbon-carbon double bonds; kinetic mechanisms for them as a family of fuels have not previously been available. Of particular importance in these mechanisms were models for 'alkylperoxy radical isomerization' reactions in which a C=C double bond was embedded in the transition state ring. They validated the resulting kinetic model through comparisons made between the predicted results and a relatively small experimental literature. They also used this model in simulations of biodiesel oxidation in a jet-stirred reactor and with intermediate shock tube ignition and oxidation conditions to demonstrate the capabilities and limitations of these mechanisms. They traced the differences in the combustion properties between the two biodiesel fuels to the differences in the relative amounts of the same five ME components.

Peterson and Scarrah (1984) identified a heterogeneous catalyst to selectively produce methyl fatty esters from low erucic rapeseed oil in a paper with 146 citations. They performed most experiments at atmospheric pressure and at approximately the corresponding boiling point temperature of the mixture, 60–63°C. However, they tested the catalytic activity of an anion exchange resin at 200°C and 68 atm (1,000 psig) and at 91°C and 9.2 atm (135 psig). They observed that the most promising catalyst was CaO·MgO. The activities of the catalysts CaO and ZnO were enhanced with the addition of MgO, therefore the transesterification reaction mechanism might be bifunctional. The anion exchange resin catalyst at 200°C and 68 atm generated substantial amounts of both methyl fatty esters and straight-chain hydrocarbons, even though these reactions did not go to completion. At 91°C and 9.2 atm, cracking also occurred but at a substantially reduced rate, and no transesterification was noted.

Sotelo-Boyas et al. (2011) study the hydrocracking of rapeseed oil on three different types of bifunctional catalysts – Pt/H-Y, Pt/H-ZSM-5, and sulfided NiMo/γ-Al_2O_3 – as an alternative way to produce diesel hydrocarbons in a paper with 136 citations. They performed these experiments in a batch reactor over a temperature range of 300–400°C and initial hydrogen pressures from 5 to 11 MPa. The reaction time was limited to 3 h to prevent a high degree of cracking. They observed that the Pt-zeolite catalysts had strong catalytic activity for both cracking and hydrogenation reactions, and therefore a higher severity was required to reach a relatively high oil conversion into liquid hydrocarbons. With dependence on the activity of the acid sites of the catalysts, there was a trade-off between the yield of biodiesel and the degree of isomerization, which had a direct effect on the cold properties of the diesel. Among the three catalysts, hydrocracking on Ni-Mo/γ-Al$_2$O3 gave the highest yield of liquid hydrocarbons in the boiling range of the diesel fraction, i.e. biodiesel, which contained mainly n-paraffins from C$_{15}$ to C$_{18}$, and therefore with poor cold flow properties. For both zeolitic catalysts, the hydrotreating of rapeseed oil produced more iso- than n-paraffins in the boiling range C$_5$ to C$_{22}$, which included significant amounts of both biodiesel and biogasoline. They observed that both pressure and temperature played an important role in the transformation of triglycerides and fatty acids into hydrocarbons.

Kusdiana and Saka (2001b) study the methyl esterification of the FFAs of rapeseed oil as treated in supercritical methanol in a paper with 129 citations. They evaluated the supercritical methanol method without using any catalyst for the reaction of FFAs. As a result, they observed a complete conversion for saturated fatty acids to MES at temperatures above 400°C, whereas for unsaturated fatty acids, a lower temperature of 350°C was appropriate, and a higher temperature resulted in a degradation of the products. Consequently, a conversion of FFAs to MEs was highest, over 95%, when treated at 350°C. Fortunately, this temperature treatment was also most appropriate for the transesterification of triglycerides. Thus, the overall conversion process of rapeseed oil to MEs was adequate at 350°C. They conclude that the supercritical methanol method on biodiesel fuel production, in which the production process became much simpler and increased the total yield due to the MEs produced from FFAs, was superior to the conventional method.

Zeng et al. (2008) study the activation of 'Mg-Al hydrotalcite' catalysts for the transesterification of rapeseed oil in a paper with 127 citations. They prepared these catalysts with different Mg:Al molar ratios. They confirm that the materials had a hydrotalcite structure. They observe that the hydrotalcite catalyst calcined at 773 K with an Mg:Al molar ratio of 3.0 exhibited the highest catalytic activity in the transesterification. In addition, they performed a study for optimizing the transesterification reaction conditions, such as the molar ratio of the methanol to oil, the reaction temperature, the reaction time, the stirring speed, and the amount of catalyst. The optimized parameters – 6:1 methanol:oil molar ratio with 1.5% catalyst (w/w of oil) reacted under a stirring speed of 300 rpm at 65°C for a 4 h reaction – gave a maximum ester conversion of 90.5%.

Yan et al. (2008) use supported CaO/MgO catalysts in the transesterification of rapeseed oil with methanol in a paper with 127 citations. They observed that this supported catalyst showed a higher activity than pure CaO and was easily separated from the product mixture. They then observed that the activity of CaO catalysts was associated with their alkalinity. They optimized the preparation method of this catalyst and studied the reaction parameters. With the CaO/MgO catalyst so obtained, the conversion of rapeseed oil reached 92% at 64.5°C. This supported basic catalyst was easily contaminated by the gaseous poisons in the air, such as O_2, CO_2, and H_2O, and as a result, a thermal treatment was required before the reaction to activate the catalyst.

Dagaut et al. (2007) study the kinetics of the oxidation of 'rapeseed oil methyl ester' (RME) in a jet-stirred reactor in a paper with 124 citations. The RME was a complex mixture of C_{14}, C_{16}, C_{18}, C_{20}, and C_{22} esters. It was preferable to use a surrogate model fuel of simple and well characterized composition for the modeling. Based on the present experimental results, they propose n-hexadecane to represent RME in the computations. The chemical kinetic reaction mechanism consisted of 225 species and 1,841 reversible reactions. The kinetic modeling gave a good description of the experimental results. A very good modeling of the relative importance of the olefins (C_2-C_6) was obtained and the experimental and simulated reactivity of RME were in good agreement. They confirm the effectiveness of using surrogate model fuels for modeling the combustion of complex commercial fuels.

Kulkarni et al. (2007) performed the transesterification of canola oil with methanol, ethanol, and various mixtures of methanol/ethanol, keeping the molar ratio of oil to alcohol at 1:6 and using KOH as a catalyst in a paper with 119 citations. They observed that mixtures of alcohol increased the rate of the transesterification reaction and produced methyl as well as ethyl esters. The increased rate was the result of better solubility of oil in the reaction mixture, due to the better solvent properties of ethanol than methanol and the equilibrium due to methanol. With a 3:3 molar ratio of methanol to ethanol (ME) (3:3)) the amount of ethyl ester formed was 50% that of ME. The properties (acid value, viscosity, density) of all esters including mixed esters were within the limits of ASTM standards. The lubricities of these esters were in the order: ethyl ester > methyl ethyl ester > ME.

24.3.2 Properties of Rapeseed Oil-based Biodiesel Fuels

Labeckas and Slavinskas (2006) performed a comparative bench testing of a four stroke, four cylinder, 'direct injection' (DI), unmodified, naturally aspirated diesel engine when operating on neat RME and its 5%, 10%, 20%, and 35% blends with diesel fuel in a paper with 258 citations. They examined the effects of RME inclusion in petrodiesel fuel on the 'brake specific fuel consumption' (BSFC) of a high speed diesel engine, its 'brake thermal efficiency' (BTE), emission composition changes, and smoke opacity of the exhausts. They observed that the BSFC at maximum torque (273.5 g/kW h) and rated power (281 g/kW h) for RME was higher by 18.7 and 23.2% relative to petrodiesel fuel. It was difficult to determine the RME concentration in the petrodiesel fuel that could be recognized as equally good for all loads and speeds. The maximum BTE varied from 0.356 to 0.398 for RME and from 0.373 to 0.383 for petrodiesel fuel. They obtained the highest fuel energy content(9.36–9.61 MJ/kW h) during operation with oil blend B10, whereas the lowest ones belonged to B35 and neat RME. The maximum NO_x emissions increased proportionally with the mass percentage of oxygen in the biofuel and engine speed, reaching the highest values at the speed of 2,000 min^{-1}, the highest being a 2,132 ppm value for the B35 blend and 2,107 ppm for RME. The CO emissions and visible smoke emerging from the biodiesel over all load and speed ranges were lower by up to 51.6% and 13.5% to

60.3%, respectively. The CO_2 emissions along with the fuel consumption and gas temperature were slightly higher for the B20 and B35 blends and neat RME. The emissions of 'unburned hydrocarbons' (UHC) for all biofuels were low, ranging over 5–21 ppm levels.

Ozsezen et al. (2009) study the performance and combustion characteristics of a DI diesel engine fueled with biodiesels such as 'waste (frying) palm oil methyl ester' (WPME) and 'canola oil methyl ester' (CME) in a paper with 246 citations. In order to determine the performance and combustion characteristics, they performed the experiments at a constant engine speed mode (1,500 rpm) under the full load condition of the engine. They observed that when the test engine was fueled with WPME or CME, the engine performance slightly weakened and the combustion characteristics slightly changed when compared to petrodiesel fuel. The biodiesels caused reductions in CO, UHC emissions, and smoke opacity, but they caused increases in nitrogen oxide (NO_x) emissions.

Tsolakis et al. (2007) study the effects of RME and different petrodiesel/RME blends on diesel engine NO_x emissions, smoke, fuel consumption, engine efficiency, cylinder pressure, and the net heat release rate in a paper with 229 citations. They observed that the combustion of RME as a pure fuel or blended with petrodiesel in an unmodified engine resulted in advanced combustion, reduced ignition delay, and an increased heat release rate in the initial uncontrolled premixed combustion phase. The increased in-cylinder pressure and temperature led to increased NO_x emissions while the more advanced combustion assisted in the reduction of smoke compared to pure petrodiesel combustion. The lower calorific value of RME resulted in increased fuel consumption though the engine's thermal efficiency was not affected significantly. When similar percentages (by volume) of 'exhaust gas recirculation' (EGR) were used in the cases of petrodiesel and RME, NO_x emissions were reduced to similar values, but the smoke emissions were significantly lower in the case of RME. The retardation of the injection timing in the case of pure RME and 50/50 (by volume) blend with petrodiesel resulted in further reduction of NO_x, at a cost of small increases of smoke and fuel consumption.

Mittelbach and Gangl (2001) study the degree of physical and chemical deterioration of biodiesel produced from rapeseed oil and used frying oil under different storage conditions in a paper with 161 citations. They observed that these produced drastic effects when the fuel was exposed to daylight and air. However, there were no significant differences between undistilled biodiesel made from fresh rapeseed oil and used frying oil. The viscosity and neutralization numbers rose during storage owing to the formation of dimers and polymers and to hydrolytic cleavage of MEs into fatty acids. However, even for samples studied under different storage conditions for over 150 d, the specified limits for viscosity and neutralization numbers were not reached. In European biodiesel specifications there is a mandatory limit for oxidative stability, because it may be a crucial parameter for injection pump performance. The value for the induction period of the distilled product was very low. The induction period values for the undistilled samples decreased very rapidly during storage, especially with exposure to light and air.

Ozsezen and Canakci (2011) study the performance, combustion, and injection characteristics of a DI diesel engine experimentally when it was fueled with CME and WPME in a paper with 150 citations. In order to determine the performance and combustion characteristics, they performed the experiments at constant engine speeds under the full load condition of the engine. They observed that when the test engine was fueled with WPME or CME instead of petrodiesel fuel, the brake power reduced by 4–5%, while the BSFC increased by 9–10%. On the other hand, MEs caused reductions in CO by 59–67%, in UHC by 17–26%, in CO_2 by 5–8%, and smoke opacity by 56–63%. However, both MEs produced more nitrogen oxide (NO_x) emissions by 11–22% compared with those of the PD over the speed range.

Moser (2008) prepared single, binary, ternary, and quaternary mixtures of canola (low erucic acid rapeseed), palm, soybean, and sunflower (high oleic acid) oil MEs (CME, PME, SME, and SFME, respectively) and measured important fuel properties in a paper with 153 citations. These properties include an 'oil stability index' (OSI), a , a 'cloud point' (CP), a 'pour point' (PP), 'kinematic viscosity' (40°C), lubricity, 'acid value' (AV), and 'iodine value' (IV). He observed that the fuel properties of SME were improved through blending with CME, PME, and SFME to satisfy the IV (< 120) and OSI (> 6 h) specifications contained within EN 14214. SME was satisfactory according to ASTM D6751, in relation to OSI (> 3 h). The CFPP of PME was improved by up to 15°C through blending with CME. He then elucidates statistically significant relationships between OSI and IV, OSI and saturated FAME (SFAME) content, OSI and CFPP, CFPP and IV, and CFPP and SFAME content. However, the only relationship of practical significance was that of CFPP versus SFAME content when the SFAME content was greater than 12 wt%.

Ozsezen and Canakci (2011) studied the performance, combustion, and injection characteristics of a DI diesel engine experimentally when it was fueled with CME and WPME in a paper with 150 citations. In order to determine the performance and combustion characteristics, they performed the experiments at constant engine speeds under the full load condition of the engine. They observed that when the test engine was fueled with WPME or CME instead of petrodiesel fuel, the brake power reduced by 4–5%, while the BSFC increased by 9–10%. On the other hand, MEs caused reductions in CO by 59–67%, in UHC by 17–26%, in CO_2 by 5–8%, and smoke opacity by 56–63%. However, both MEs produced more nitrogen oxide (NO_x) emissions by 11–22% compared with those of the PD over the speed range.

Tsolakis (2006) studied the effects of the RME fueled diesel engine with the use of EGR on the particle size distribution in a paper with 135 citations. He observes that the combustion of RME significantly improved the engine smoke and total particle mass but increased both NO_x and particle concentration with low aerodynamic diameters (<0.091 μm) when compared to the petrodiesel fueled engine. Although the particle size and mass distribution were not affected significantly by the different EGR additions, the particle total number and mass were increased considerably for both fuels. For the RME fueled engine, the EGR addition reduced the particles in the lowest aerodynamic diameter measured (0.046 μm). The use of EGR better suited the RME combustion, as apart from resulting in the higher NO_x reduction, it maintained the smoke (soot, particulate matter) at relatively low levels. It was challenging to reduce simultaneously the total particle mass without increasing the number of particles at low aerodynamic diameters. Furthermore, the lower RME calorific value compared to PD resulted in increased fuel consumption, although the engine efficiency was not noticeably affected.

24.4 DISCUSSION

Table 24.1 provides information on the research fronts in this field. As this table shows the primary research fronts of 'production of ROBD fuels' and 'properties of ROBD fuels' comprise 72 and 28% of these papers, respectively. Furthermore, some of the papers in the first group have sections on the 'properties and characterization of ROBD fuels'.

24.4.1 RAPESEED OIL-BASED BIODIESEL PRODUCTION

Saka and Kusdiana (2001) study the transesterification reaction of rapeseed oil in supercritical methanol without using any catalyst in a paper with 687 citations. Kusdiana and Saka (2001a) perform a kinetic study in free catalyst transesterification of rapeseed oil in subcritical and supercritical methanol under different reaction conditions of temperatures and reaction times in a paper with 540 citations. Gryglewicz (1999) study a heterogeneous catalyst, in particular calcium compounds, to produce MEs of rapeseed oil in a paper with 379 citations.

Gelbard et al. (1995) monitor the formation of FAMEs by transesterification of rapeseed oil with methanol by ^{1}H nuclear magnetic resonance spectroscopy in a paper with 345 citations. Rashid and Anwar (2008) develop an optimized protocol for the production of biodiesel through alkaline-catalyzed transesterification of rapeseed oil in a paper with 301 citations. Warabi et al. (2004) develop a catalyst-free biodiesel production method with supercritical methanol that allows a simple process and high yield because of simultaneous transesterification of triglycerides and methyl esterification of fatty acids of rapeseed oil in a paper with 300 citations.

Li et al. (2006) study the lipase-catalyzed transesterification of rapeseed oils for biodiesel production with a novel organic solvent, 'tert-Butanol', as the reaction medium in a paper with 257 citations. Yuan et al. (2008) use waste rapeseed oil with high FFAs as feedstock for producing biodiesel in a paper with 192 citations. Leclercq et al. (2001) study the transesterification of rapeseed oil in the presence of basic zeolites and related solid catalysts in a paper with 171 citations.

Verziu et al. (2008) study the catalytic activity for the production of biodiesel from sunflower and rapeseed oils with three morphologically different

TABLE 24.1

Research Fronts

	Research Front	Papers (%)
1	Production of rapeseed oil-based biodiesel (ROBD) fuels	72
2	Properties of ROBD fuels	28

nanocrystalline MgO materials prepared using simple, green, and reproducible methods in a paper with 161 citations. Westbrook et al. (2011) develop a detailed chemical kinetic reaction mechanism for the five major components of soybean biodiesel and rapeseed biodiesel fuels in a paper with 154 citations. Peterson and Scarrah (1984) identify a heterogeneous catalyst to selectively produce methyl fatty esters from low erucic rapeseed oil in a paper with 146 citations.

Sotelo-Boyas et al. (2011) study the hydrocracking of rapeseed oil on three different types of bifunctional catalysts: Pt/H-Y, Pt/H-ZSM-5, and sulfided NiMo/γ-Al$_2$O$_3$ as an alternative way to produce diesel hydrocarbons in a paper with 136 citations. Kusdiana and Saka (2001b) study the methyl esterification of FFAs of rapeseed oil as treated in supercritical methanol in a paper with 129 citations. Zeng et al. (2008) study the activation of 'Mg-Al hydrotalcite' catalysts for the transesterification of rapeseed oil in a paper with 127 citations.

Yan et al. (2008) use supported CaO/MgO catalysts for the transesterification of rapeseed oils with methanol in a paper with 127 citations. Dagaut et al. (2007) study the kinetic of oxidation of RME in a jet-stirred reactor in a paper with 124 citations. Kulkarni et al. (2007) perform the transesterification of canola oil with methanol, ethanol, and various mixtures of methanol/ethanol, keeping the molar ratio of oil to alcohol at 1:6 and using KOH as a catalyst in a paper with 119 citations.

These prolific studies highlight the properties of ROBD fuels.

24.4.2 PROPERTIES OF RAPESEED OIL-BASED BIODIESEL FUELS

Labeckas and Slavinskas (2006) perform a comparative bench testing of a four stroke, four cylinder, DI, unmodified, naturally aspirated diesel engine when operating on neat RME and its 5%, 10%, 20%, and 35% blends with diesel fuel in a paper with 258 citations. Ozsezen et al. (2009) study the performance and combustion characteristics of a DI diesel engine fueled with biodiesels such as WPME and CME in a paper with 246 citations. Tsolakis et al. (2007) study the effects of RME and different petrodiesel/RME blends on the diesel engine's NO$_x$ emissions, smoke, fuel consumption, engine efficiency, cylinder pressure, and net heat release rate in a paper with 229 citations.

Mittelbach and Gangl (2001) study the degree of physical and chemical deterioration of biodiesel produced from rapeseed and used frying oil under different storage conditions in a paper with 161 citations. Moser (2008) prepares single, binary, ternary, and quaternary mixtures of canola (low erucic acid rapeseed), palm, soybean, and sunflower (high oleic acid) oil MEs (CME, PME, SME, and SFME, respectively) in a paper with 153 citations.

Tsolakis (2006) study the effects of the RME fueled diesel engine with the use of EGR on the particle size distribution in a paper with 135 citations. Ozsezen and Canakci (2011) study the performance, combustion, and injection characteristics of a DI diesel engine experimentally when it was fueled with CME and WPME in a paper with 150 citations.

These prolific studies highlight the properties of ROBD fuels.

24.4.3 THE PUBLIC CONCERN REGARDING THE PRODUCTION OF RAPESEED OIL-BASED BIODIESEL FUELS

There has been some public concern about the production and utilization of ROBD fuels.

One of these concerns emanates from the fact that both rapeseed and rapeseed oils are edible (Cumby et al., 2008; He et al., 2013; Koski et al., 2003; Naczk and Shahidi, 1989; Naczk et al., 1998). Thus, there has been significant competition between biodiesel production from rapeseed oils and the household consumption of rapeseed oils, resulting in significant public concern about food security' (Ajanovic, 2011; Godfray et al., 2010; Lam et al., 2009; Lobell et al., 2008; Naylor et al., 2007; Rosegrant and Cline, 2003; Schmidhuber and Tubiello, 2007; Tenenbaum, 2008).

However, there has been substantial exploration of ROBD fuels from waste cooking rapeseed oils (Ozsezen et al., 2009; Yuan et al., 2008). This version of biodiesel fuels helps also in reducing the cost of these fuels as around 85% of their total cost in general emanates from the feedstocks (Haas, 2005; Haas et al., 2006).

Public concern about food security has resulted in the exploration of biodiesel production from nonedible oils (Konur, 2021r), waste oils (Konur, 2021s), and algal oils (Konur, 2021t).

24.5   CONCLUSION

This chapter has presented the key findings of the 25-most-cited article papers in this field. Table 24.1 provides information on the research fronts. As this table shows the primary research fronts of 'production of ROBD fuels' and 'properties of ROBD fuels' comprise 72 and 28% of these papers, respectively.

These prolific studies on two complementary research fronts provide valuable evidence on the production and properties of ROBD fuels. The significant public concern about the production and utilization of ROBD fuels as a first generation fuel has also been noted.

It is recommended that similar studies are carried out for each research front as well.

ACKNOWLEDGMENTS

The contribution of the highly cited researchers in this field is greatly acknowledged.

REFERENCES

Ahmadun, F. R., A. Pendashteh, and L. C. Abdullah, et al. 2009. Review of technologies for oil and gas produced water treatment. *Journal of Hazardous Materials* 170:530–551.
Ajanovic, A. 2011. Biofuels versus food production: Does biofuels production increase food prices? *Energy* 36:2070–2076.
Aldhaidhawi, M., R. Chiriac, and V. Badescu. 2017. Ignition delay, combustion and emission characteristics of diesel engine fueled with rapeseed biodiesel: A literature review. *Renewable & Sustainable Energy Reviews* 73:178–186.

Antolin, G., F. V. Tinaut, and Y. Briceno, et al. 2002. Optimisation of biodiesel production by sunflower oil transesterification. *Bioresource Technology* 83:111–114.

Atlas, R. M. 1981. Microbial degradation of petroleum hydrocarbons: An environmental perspective. *Microbiological Reviews* 45: 180–209.

Babich, I. V. and J. A. Moulijn. 2003. Science and technology of novel processes for deep desulfurization of oil refinery streams: A review. *Fuel* 82: 607–631.

Bridgwater, A. V. and G. V. C. Peacocke. 2000. Fast pyrolysis processes for biomass. *Renewable & Sustainable Energy Reviews* 4:1–73.

Bridgwater, A. V., D. Meier, and D. Radlein. 1999. An overview of fast pyrolysis of biomass. *Organic Geochemistry* 30: 1479–1493.

Chisti, Y. 2007. Biodiesel from microalgae. *Biotechnology Advances* 25: 294–306.

Cumby, N., Y. Zhong, M. Naczk, and F. Shahidi. 2008. Antioxidant activity and water-holding capacity of canola protein hydrolysates. *Food Chemistry* 109: 144–148.

Czernik, S. and A. V. Bridgwater. 2004. Overview of applications of biomass fast pyrolysis oil. *Energy & Fuels* 18: 590–598.

Dagaut, P., S. Gail, and M. Sahasrabudhe. 2007. Rapeseed oil methyl ester oxidation over extended ranges of pressure, temperature, and equivalence ratio: Experimental and modeling kinetic study. *Proceedings of the Combustion Institute* 31: 2955–2961.

El-Enin, S. A., N. K. Attia, N. N. El-Ibiari, G. I. El-Diwani, and K. M. El-Khatib. 2013. In-situ transesterification of rapeseed and cost indicators for biodiesel production. *Renewable & Sustainable Energy Reviews* 18: 471–477.

Freedman, B., R. O. Butterfield, and E. H. Pryde. 1986. Transesterification kinetics of soybean oil. *Journal of the American Oil Chemists Society* 63:1375–1380.

Ge, J. C., S. K. Yoon, and N. J. Choi. 2017. Using canola oil biodiesel as an alternative fuel in diesel engines: A review. *Applied Sciences-Basel* 7: 881.

Gelbard, G., O. Bres, R. M. Vargas, F. Vielfaure, and U. F. Schuchardt. 1995. [1]H nuclear-magnetic-resonance determination of the yield of the transesterification of rapeseed oil with methanol. *Journal of the American Oil Chemists Society* 72: 1239–1241.

Godfray, H. C. J., J. R. Beddington, and I. R. Crute, et al., 2010. Food security: The challenge of feeding 9 billion people. *Science* 327: 812–818.

Gryglewicz, S. 1999. Rapeseed oil methyl esters preparation using heterogeneous catalysts. *Bioresource Technology* 70: 249–253.

Haas, M. J. 2005. Improving the economics of biodiesel production through the use of low value lipids as feedstocks: vegetable oil soapstock. *Fuel Processing Technology* 86:1087–1096.

Haas, M. J., A. J. McAloon, W. C. Yee, and T. A. Foglia. 2006. A process model to estimate biodiesel production costs. *Bioresource Technology* 97: 671–678.

He, R., A. T. Girgih, S. A. Malomo, J. R. Ju, and R. E. Aluko. 2013. Antioxidant activities of enzymatic rapeseed protein hydrolysates and the membrane ultrafiltration fractions. *Journal of Functional Foods* 5: 219–227.

Hill, J., E. Nelson, D. Tilman, S. Polasky, and D. Tiffany. 2006. Environmental, economic, and energetic costs and benefits of biodiesel and ethanol biofuels. *Proceedings of the National Academy of Sciences of the United States of America* 103:11206–11210.

Kilian, L. 2009. Not all oil price shocks are alike: Disentangling demand and supply shocks in the crude oil market. *American Economic Review* 99: 1053–1069.

Konur, O. 2000. Creating enforceable civil rights for disabled students in higher education: An institutional theory perspective. *Disability & Society* 15: 1041–1063.

Konur, O. 2002a. Access to Nursing Education by disabled students: Rights and duties of nursing programs. *Nurse Education Today* 22: 364–374.

Konur, O. 2002b. Assessment of disabled students in higher education: Current public policy issues. *Assessment and Evaluation in Higher Education* 27: 131–152.

Konur, O. 2002c. Access to employment by disabled people in the UK: Is the Disability Discrimination Act working? *International Journal of Discrimination and the Law* 5: 247–279.

Konur, O. 2006a. Participation of children with dyslexia in compulsory education: Current public policy issues. *Dyslexia* 12: 51–67.

Konur, O. 2006b. Teaching disabled students in Higher Education. *Teaching in Higher Education* 11: 351–363.

Konur, O. 2007a. A judicial outcome analysis of the Disability Discrimination Act: A windfall for the employers? *Disability & Society* 22: 187–204.

Konur, O. 2007b. Computer-assisted teaching and assessment of disabled students in higher education: The interface between academic standards and disability rights. *Journal of Computer Assisted Learning* 23: 207–219.

Konur, O., ed. 2021a. *Handbook of Biodiesel and Petrodiesel Fuels: Science, Technology, Health, and Environment.* Boca Raton, FL: CRC Press.

Konur, O., ed. 2021b. *Handbook of Biodiesel and Petrodiesel Fuels: Science, Technology, Health, and Environment. Volume 1. Biodiesel Fuels: Science, Technology, Health, and Environment.* Boca Raton, FL: CRC Press.

Konur, O., ed. 2021c. *Handbook of Biodiesel and Petrodiesel Fuels: Science, Technology, Health, and Environment. Volume 2. Biodiesel Fuels based on the Edible and Nonedible Feedstocks, Wastes, and Algae: Science, Technology, Health, and Environment.* Boca Raton, FL: CRC Press.

Konur, O., ed. 2021d. *Handbook of Biodiesel and Petrodiesel Fuels: Science, Technology, Health, and Environment. Volume 3. Petrodiesel Fuels: Science, Technology, Health, and Environment.* Boca Raton, FL: CRC Press.

Konur, O. 2021e. Biodiesel and petrodiesel fuels: Science, technology, health, and environment. In *Handbook of Biodiesel and Petrodiesel Fuels: Science, Technology, Health, and Environment. Volume 1. Biodiesel Fuels: Science, Technology, Health, and Environment*, ed. O. Konur. Boca Raton, FL: CRC Press.

Konur, O. 2021f. Biodiesel and petrodiesel fuels: A scientometric review of the research. In *Handbook of Biodiesel and Petrodiesel Fuels: Science, Technology, Health, and Environment. Volume 1. Biodiesel Fuels: Science, Technology, Health, and Environment*, ed. O. Konur. Boca Raton, FL: CRC Press.

Konur, O. 2021g. Biodiesel and petrodiesel fuels: A review of the research. In *Handbook of Biodiesel and Petrodiesel Fuels: Science, Technology, Health, and Environment. Volume 1. Biodiesel Fuels: Science, Technology, Health, and Environment*, ed. O. Konur. Boca Raton, FL: CRC Press.

Konur, O. 2021h Nanotechnology applications in the diesel fuels and the related research fields: A review of the research. In *Handbook of Biodiesel and Petrodiesel Fuels: Science, Technology, Health, and Environment. Volume 1. Biodiesel Fuels: Science, Technology, Health, and Environment*, ed. O. Konur. Boca Raton, FL: CRC Press.

Konur, O. 2021i. Biooils: A scientometric review of the research. In *Handbook of Biodiesel and Petrodiesel Fuels: Science, Technology, Health, and Environment. Volume 1. Biodiesel Fuels: Science, Technology, Health, and Environment*, ed. O. Konur. Boca Raton, FL: CRC Press.

Konur, O. 2021j. Characterization and properties of biooils: A review of the research. In *Handbook of Biodiesel and Petrodiesel Fuels: Science, Technology, Health, and Environment. Volume 1. Biodiesel Fuels: Science, Technology, Health, and Environment*, ed. O. Konur. Boca Raton, FL: CRC Press.

Konur, O. 2021k. Biomass pyrolysis and pyrolysis oils: A review of the research. In *Handbook of Biodiesel and Petrodiesel Fuels: Science, Technology, Health, and Environment. Volume 1. Biodiesel Fuels: Science, Technology, Health, and Environment*, ed. O. Konur. Boca Raton, FL: CRC Press.

Konur, O. 2021l. Biodiesel fuels: A scientometric review of the research. In *Handbook of Biodiesel and Petrodiesel Fuels: Science, Technology, Health, and Environment. Volume 1. Biodiesel Fuels: Science, Technology, Health, and Environment*, ed. O. Konur. Boca Raton, FL: CRC Press.

Konur, O. 2021m. Glycerol: A scientometric review of the research. In *Handbook of Biodiesel and Petrodiesel Fuels: Science, Technology, Health, and Environment. Volume 1. Biodiesel Fuels: Science, Technology, Health, and Environment*, ed. O. Konur. Boca Raton, FL: CRC Press.

Konur, O. 2021n. Propanediol production from glycerol: A review of the research. In *Handbook of Biodiesel and Petrodiesel Fuels: Science, Technology, Health, and Environment. Volume 1. Biodiesel Fuels: Science, Technology, Health, and Environment*, ed. O. Konur. Boca Raton, FL: CRC Press.

Konur, O. 2021o. Edible oil-based biodiesel fuels: A scientometric review of the research. In *Handbook of Biodiesel and Petrodiesel Fuels: Science, Technology, Health, and Environment. Volume 2. Biodiesel Fuels based on the Edible and Nonedible Feedstocks, Wastes, and Algae: Science, Technology, Health, and Environment*, ed. O. Konur. Boca Raton, FL: CRC Press.

Konur, O. 2021p. Palm oil-based biodiesel fuels: A review of the research. In *Handbook of Biodiesel and Petrodiesel Fuels: Science, Technology, Health, and Environment. Volume 2. Biodiesel Fuels based on the Edible and Nonedible Feedstocks, Wastes, and Algae*, ed. O. Konur. Boca Raton, FL: CRC Press.

Konur, O. 2021q. Rapeseed oil-based biodiesel fuels: A review of the research. In *Handbook of Biodiesel and Petrodiesel Fuels: Science, Technology, Health, and Environment. Volume 2. Biodiesel Fuels based on the Edible and Nonedible Feedstocks, Wastes, and Algae*, ed. O. Konur. Boca Raton, FL: CRC Press.

Konur, O. 2021r. Nonedible oil-based biodiesel fuels: A scientometric review of the research. In *Handbook of Biodiesel and Petrodiesel Fuels: Science, Technology, Health, and Environment. Volume 2. Biodiesel Fuels based on the Edible and Nonedible Feedstocks, Wastes, and Algae: Science, Technology, Health, and Environment*, ed. O. Konur. Boca Raton, FL: CRC Press.

Konur, O. 2021s. Waste oil-based biodiesel fuels: A scientometric review of the research. In *Handbook of Biodiesel and Petrodiesel Fuels: Science, Technology, Health, and Environment. Volume 2. Biodiesel Fuels based on the Edible and Nonedible Feedstocks, Wastes, and Algae: Science, Technology, Health, and Environment*, ed. O. Konur. Boca Raton, FL: CRC Press.

Konur, O. 2021t. Algal biodiesel fuels: A scientometric review of the research. In *Handbook of Biodiesel and Petrodiesel Fuels: Science, Technology, Health, and Environment. Volume 2. Biodiesel Fuels based on the Edible and Nonedible Feedstocks, Wastes, and Algae: Science, Technology, Health, and Environment*, ed. O. Konur. Boca Raton, FL: CRC Press.

Konur, O. 2021u. Algal biomass production for biodiesel production: A review of the research. In *Handbook of Biodiesel and Petrodiesel Fuels: Science, Technology, Health, and Environment. Volume 2. Biodiesel Fuels based on the Edible and Nonedible Feedstocks, Wastes, and Algae*, Ed. O. Konur. Boca Raton, FL: CRC Press.

Konur, O. 2021v. Algal biomass production in wastewaters for biodiesel production: A review of the research. In *Handbook of Biodiesel and Petrodiesel Fuels: Science, Technology, Health, and Environment. Volume 2. Biodiesel Fuels based on the Edible and Nonedible Feedstocks, Wastes, and Algae*, ed. O. Konur. Boca Raton, FL: CRC Press.

Konur, O. 2021x. Algal lipid production for biodiesel production: A review of the research. In *Handbook of Biodiesel and Petrodiesel Fuels: Science, Technology, Health, and Environment. Volume 2. Biodiesel Fuels based on the Edible and Nonedible Feedstocks, Wastes, and Algae*, Ed. O. Konur. Boca Raton, FL: CRC Press.

Konur, O. 2021y. Crude oils: A scientometric review of the research. In *Handbook of Biodiesel and Petrodiesel Fuels: Science, Technology, Health, and Environment. Volume 3. Petrodiesel Fuels: Science, Technology, Health, and Environment*, ed. O. Konur. Boca Raton, FL: CRC Press.

Konur, O. 2021z. Petrodiesel fuels: A scientometric review of the research. In *Handbook of Biodiesel and Petrodiesel Fuels: Science, Technology, Health, and Environment. Volume 3. Petrodiesel Fuels: Science, Technology, Health, and Environment*, ed. O. Konur. Boca Raton, FL: CRC Press.

Konur, O. 2021aa. Bioremediation of petroleum hydrocarbons in the contaminated soils: A review of the research. In *Handbook of Biodiesel and Petrodiesel Fuels: Science, Technology, Health, and Environment. Volume 3. Petrodiesel Fuels: Science, Technology, Health, and Environment*, ed. O. Konur. Boca Raton, FL: CRC Press.

Konur, O. 2021ab. Desulfurization of diesel fuels: A review of the research. In *Handbook of Biodiesel and Petrodiesel Fuels: Science, Technology, Health, and Environment. Volume 3. Petrodiesel Fuels: Science, Technology, Health, and Environment*, ed. O. Konur. Boca Raton, FL: CRC Press.

Konur, O. 2021ac. Diesel fuel exhaust emissions: A scientometric review of the research. In *Handbook of Biodiesel and Petrodiesel Fuels: Science, Technology, Health, and Environment. Volume 3. Petrodiesel Fuels: Science, Technology, Health, and Environment*, ed. O. Konur. Boca Raton, FL: CRC Press.

Konur, O. 2021ad. The adverse health and safety impact of diesel fuels: A scientometric review of the research. In *Handbook of Biodiesel and Petrodiesel Fuels: Science, Technology, Health, and Environment. Volume 3. Petrodiesel Fuels: Science, Technology, Health, and Environment*, ed. O. Konur. Boca Raton, FL: CRC Press.

Konur, O. 2021ae. Respiratory illnesses caused by the diesel fuel exhaust emissions: A review of the research. In *Handbook of Biodiesel and Petrodiesel Fuels: Science, Technology, Health, and Environment. Volume 3. Petrodiesel Fuels: Science, Technology, Health, and Environment*, ed. O. Konur. Boca Raton, FL: CRC Press.

Konur, O. 2021af. Cancer caused by the diesel fuel exhaust emissions: A review of the research. In *Handbook of Biodiesel and Petrodiesel Fuels: Science, Technology, Health, and Environment. Volume 3. Petrodiesel Fuels: Science, Technology, Health, and Environment*, ed. O. Konur. Boca Raton, FL: CRC Press.

Konur, O. 2021ag. Cardiovascular and other illnesses caused by the diesel fuel exhaust emissions: A review of the research. In *Handbook of Biodiesel and Petrodiesel Fuels: Science, Technology, Health, and Environment. Volume 3. Petrodiesel Fuels: Science, Technology, Health, and Environment*, ed. O. Konur. Boca Raton, FL: CRC Press.

Koski, A., S. Pekkarinen, A. Hopia, K. Wahala, and M. Heinonen. 2003. Processing of rapeseed oil: effects on sinapic acid derivative content and oxidative stability. *European Food Research and Technology* 217: 110–114.

Kulkarni, M. G., A. K. Dalai, and N. N. Bakhshi. 2007. Transesterification of canola oil in mixed methanol/ethanol system and use of esters as lubricity additive. *Bioresource Technology* 98: 2027–2033.

Kusdiana, D. and S. Saka. 2001a. Kinetics of transesterification in rapeseed oil to biodiesel fuel as treated in supercritical methanol. *Fuel* 80: 693–698.

Kusdiana, D. and S. Saka. 2001b. Methyl esterification of free fatty acids of rapeseed oil as treated in supercritical methanol. *Journal of Chemical Engineering of Japan* 34: 383–387.

Labeckas, G. and S. Slavinskas. 2006. The effect of rapeseed oil methyl ester on direct injection diesel engine performance and exhaust emissions. *Energy Conversion and Management* 47: 1954–1967.

Lam, M. K., K. T. Tan, K. T. Lee, and A. R. Mohamed. 2009. Malaysian palm oil: Surviving the food versus fuel dispute for a sustainable future. *Renewable and Sustainable Energy Reviews* 13:1456–1464.

Leclercq, E., A. Finiels, and C. Moreau. 2001. Transesterification of rapeseed oil in the presence of basic zeolites and related solid catalysts. *Journal of the American Oil Chemists Society* 78: 1161–1165.

Li, L. L., W. Du, D. H. Liu, L. Wang, and Z. B. Li. 2006. Lipase-catalyzed transesterification of rapeseed oils for biodiesel production with a novel organic solvent as the reaction medium. *Journal of Molecular Catalysis B-Enzymatic* 43: 58–62.

Lobell, D. B., M. B. Burke, and C. Tebaldi, et al. 2008. Prioritizing climate change adaptation needs for food security in 2030. *Science* 319: 607–610.

Milazzo, M. F., F. Spina, A. Vinci, C. Espro, and J. C. J. Bart. 2013. Brassica biodiesels: Past, present and future. *Renewable & Sustainable Energy Reviews* 18: 350–389.

Mittelbach, M. and S. Gangl. 2001. Long storage stability of biodiesel made from rapeseed and used frying oil. *Journal of the American Oil Chemists Society* 78: 573–577.

Mohan, D., C. U. Pittman Jr, and P. H. Steele. 2006. Pyrolysis of wood/biomass for bio oil: A critical review. *Energy & Fuels* 20: 848–889.

Moser, B. R. 2008. Influence of blending canola, palm, soybean, and sunflower oil methyl esters on fuel properties of biodiesel. *Energy & Fuels* 22: 4301–4306.

Naczk, M. and F. Shahidi. 1989. The effect of methanol ammonia water-treatment on the content of phenolic-acids of canola. *Food Chemistry* 31:159–164.

Naczk, M., R. Amarowicz, A. Sullivan, and F. Shahidi. 1998. Current research developments on polyphenolics of rapeseed/canola: A review. *Food Chemistry* 62: 489–502.

Naylor, R. L., A. J. Liska, and M. B. Burke, et al. 2007. The ripple effect: Biofuels, food security, and the environment. *Environment: Science and Policy for Sustainable Development* 49: 30–43.

North, D. C. 1991a. *Institutions, Institutional Change and Economic Performance*. Cambridge, Mass.: Cambridge University Press.

North, D.C. 1991b. Institutions. *Journal of Economic Perspectives* 5: 97–112.

Noureddini, H. and D. Zhu. 1997. Kinetics of transesterification of soybean oil. *Journal of the American Oil Chemists Society* 74.1457–1463.

Ozsezen, A. N. and M. Canakci. 2011. Determination of performance and combustion characteristics of a diesel engine fueled with canola and waste palm oil methyl esters. *Energy Conversion and Management* 52: 108–116.

Ozsezen, A. N., M. Canakci, A. Turkcan, and C. Sayin. 2009. Performance and combustion characteristics of a DI diesel engine fueled with waste palm oil and canola oil methyl esters. *Fuel* 88: 629–636.

Perron, P. 1989. The great crash, the oil price shock, and the unit root hypothesis. *Econometrica: Journal of the Econometric Society* 57: 1361–1401.

Peterson, G. R. and W. P. Scarrah. 1984. Rapeseed oil transesterification by heterogeneous catalysis. *Journal of the American Oil Chemists Society* 61:1593–1597.

Rashid, U. and F. Anwar. 2008. Production of biodiesel through optimized alkaline-catalyzed transesterification of rapeseed oil. *Fuel* 87: 265–273.

Rosegrant, M. W. and S. A. Cline. 2003. Global food security: Challenges and policies. *Science* 302: 1917–1919.

Saka, S. and D. Kusdiana. 2001. Biodiesel fuel from rapeseed oil as prepared in supercritical methanol. *Fuel* 80: 225–231.

Schmidhuber, J. and F. N. Tubiello. 2007. Global food security under climate change. *Proceedings of the National Academy of Sciences* 104:19703–19708.

Sotelo-Boyas, R., Y. Y. Liu, and T. Minowa. 2011. Renewable diesel production from the hydrotreating of rapeseed oil with Pt/zeolite and NiMo/Al$_2$O$_3$ catalysts. *Industrial & Engineering Chemistry Research* 50: 2791–2799.

Tenenbaum, D. J. 2008. Food vs. fuel: Diversion of crops could cause more hunger. *Environmental Health Perspectives* 116: A254–A257.

Tsolakis, A. 2006. Effects on particle size distribution from the diesel engine operating on RME-biodiesel with EGR. *Source: Energy & Fuels* 20: 1418–1424. 135.

Tsolakis, A., A. Megaritis, M. L. Wyszynski, and K. Theinnoi. 2007. Engine performance and emissions of a diesel engine operating on diesel-RME (rapeseed methyl ester) blends with EGR (exhaust gas recirculation). *Energy* 32: 2072–2080.

Verziu, M., B. Cojocaru, and J. C. Hu, et al. 2008. Sunflower and rapeseed oil transesterification to biodiesel over different nanocrystalline MgO catalysts. *Green Chemistry* 10: 373–381.

Warabi, Y., D. Kusdiana, and S. Saka. 2004. Reactivity of triglycerides and fatty acids of rapeseed oil in supercritical alcohols. *Bioresource Technology* 91: 283–287.

Westbrook, C. K., C. V. Naik, and O. Herbinet, et al. 2011. Detailed chemical kinetic reaction mechanisms for soy and rapeseed biodiesel fuels. *Combustion and Flame* 158: 742–755.

Yan, S. L., H. F. Lu, and B. Liang. 2008. Supported CaO catalysts used in the transesterification of rapeseed oil for the purpose of biodiesel production. *Energy & Fuels* 22: 646–651.

Yuan, X., J. Liu, and G. Zeng, et al. 2008. Optimization of conversion of waste rapeseed oil with high FFA to biodiesel using response surface methodology. *Renewable Energy* 33:1678–1684.

Zeng, H. Y., Z. Feng, X. Deng, and Y. Q. Li. 2008. Activation of Mg-Al hydrotalcite catalysts for transesterification of rape oil. *Fuel* 87: 3071–3076.

Zhang, Q., J. Chang, T. J. Wang, and Y. Xu. 2007. Review of biomass pyrolysis oil properties and upgrading research. *Energy Conversion and Management* 48: 87–92.

Part VI

Nonedible Oil-based Biodiesel Fuels

25 Nonedible Oil-based Biodiesel Fuels

A Scientometric Review of the Research

Ozcan Konur

CONTENTS

25.1 INTRODUCTION

Crude oils have been primary sources of energy and fuels, such as petrodiesel (Busca et al., 1998; Khalili et al., 1995; Rogge et al., 1993; Schauer et al., 1999). However, significant public concern about the sustainability, price fluctuations, and adverse environmental impact of crude oils have emerged since the 1970s (Ahmadun et al.,

2009; Atlas, 1981; Babich and Moulijn, 2003; Kilian, 2009; Perron, 1989). Thus, biooils have emerged as an alternative to crude oils in recent decades (Bridgwater and Peacocke, 2000; Czernik and Bridgwater, 2004; Gallezot, 2012; Mohan et al., 2006). In this context, NOBD fuels (Achten et al., 2008; Atabani et al., 2013; Berchmans and Hirata, 2008; Gui et al., 2008; Pramanik, 2003; Raheman and Phadatare, 2004; Ramadhas et al., 2005a–b; Sahoo et al., 2007; Tiwari et al., 2007) have emerged as a viable alternative to crude oil-based petrodiesel fuels (Busca et al., 1998; Khalili et al., 1995; Rogge et al., 1993; Schauer et al., 1999). Nowadays, both petrodiesel and biodiesel fuels are being used on a global scale (Konur, 2021a–ag).

However, for the efficient progression of the research in this field, it is necessary to develop efficient incentive structures for the primary stakeholders and to inform these stakeholders about the research (Konur, 2000, 2002a–c, 2006a–b, 2007a–b; North, 1991a–b).

Scientometric analysis offers ways to evaluate the research in a respective field (Garfield, 1955, 1972; Konur, 2011, 2012a–n, 2015, 2016a–f, 2017a–f, 2018a–b, 2019a–b). However, there has been no current scientometric study of this field.

This chapter presents a study on the scientometric evaluation of the research in this field using two datasets. The first dataset includes the 100-most-cited papers ($n = 100$ sample papers) whilst the second set includes population papers ($n = $ over 2,150 population papers) published between 1980 and 2019. This complements the chapters on the crude oil-based petrodiesel fuels and other biooil-based biodiesel fuels.

The data on the indices, document types, authors, institutions, funding bodies, source titles, 'Web of Science' subject categories, keywords, research fronts, and citation impact are presented and discussed.

25.2 MATERIALS AND METHODOLOGY

The search for the literature was carried out in the 'Web of Science' (WOS) database in January 2020. It contains the 'Science Citation Index–Expanded' (SCI-E), the Social Sciences Citation Index' (SSCI), the 'Book Citation Index–Science' (BCI-S), the 'Conference Proceedings Citation Index–Science' (CPCI-S), the 'Emerging Sources Citation Index' (ESCI), the 'Book Citation Index–Social Sciences and Humanities' (BCI-SSH), the 'Conference Proceedings Citation Index–Social Sciences and Humanities' (CPCI-SSH), and the 'Arts and Humanities Citation Index' (A&HCI).

The keywords for the search of the literature are collated from the screening of abstract pages for the first 500 highly cited papers. This keyword set is provided in the Appendix.

Two datasets are used for this study. The highly cited 100 papers comprise the first dataset (sample data set, $n = 100$ papers) whilst all the papers form the second dataset (population data set, $n = $ over 2,150 papers).

The data on the indices, document types, publication years, institutions, funding bodies, source titles, countries, 'Web of Science' subject categories, citation impact, keywords, and research fronts are collated from these datasets. The key findings are provided in the relevant tables and figure, supplemented with explanatory notes in

the text. The findings are discussed and a number of conclusions are drawn and a number of recommendations for further study are made.

25.3   RESULTS

25.3.1   INDICES AND DOCUMENTS

There are over 2,500 papers in this field in the 'Web of Science' as of January 2020. This original population dataset is refined for the document type (article, review, book chapter, book, editorial material, note, and letter) and language (English), resulting in over 2,150 papers comprising over 85.3% of the original population dataset.

The primary index is the SCI-E for both the sample and population papers. About 90.3% of the population papers are indexed by the SCI-E database. Additionally 2.2, 8.7, and 0.5% of these papers are indexed by the CPCI-S, ESCI, and BCI-S databases, respectively. The papers on the social and humanitarian aspects of this field are relatively negligible with 1.4 and 0.0% of the population papers indexed by the SSCI and A&HCI, respectively.

Brief information on the document types for both datasets is provided in Table 25.1. The key finding is that article types of documents are the primary documents for both the sample and population papers whilst reviews form 19% of the sample papers.

25.3.2   AUTHORS

Brief information about the 20-most-prolific authors with at least three sample papers each is provided in Table 25.2. Around 360 and 5,050 authors contribute to the sample and population papers, respectively.

The most-prolific author is 'Hassan H. J. Masjuki' with ten sample papers working primarily on 'biodiesel fuel production' and 'biodiesel fuel properties' using the oils of *Jatropha*, *Calophyllum*, *Sterculia*, and *Ceiba*. The other top four authors are 'Lalit Mohan Das', 'T. M. Indra Mahlia', 'Hwai Chyuan Ong', and 'Hiflur Raheman' with five sample papers each.

TABLE 25.1
Document Types

	Document Type	Sample Dataset (%)	Population Dataset (%)	Difference (%)
1	Article	81	96.0	−15.0
2	Review	19	3.4	15.6
3	Book chapter	0	0.5	−0.5
4	Proceeding paper	2	2.2	−0.2
5	Editorial material	0	0.2	−0.2
6	Letter	0	0.2	−0.2
7	Book	0	0.0	0.0
8	Note	0	0.1	−0.1

TABLE 25.2
Authors

	Author	Sample Papers (%)	Population Papers (%)	Surplus (%)	Institution	Country	Nonedible Oils	Research Front
1	Masjuki, H. Hassan*	10	2.7	7.3	Univ. Malaya	Malaysia	*Jatropha, Calophyllum, Sterculia, Ceiba,* nonedible	Production, properties
2	Das, Lalit Mohan	5	0.7	4.3	Indian Inst. Technol.	India	Polanga, *Jatropha,* karanja	Production, properties
3	Mahlia, T. M. Indra*	5	1.2	3.8	Univ. Malaya	Malaysia	*Jatropha, Calophyllum, Sterculia, Ceiba,* nonedible	Production, properties
4	Ong, Hwai Chyuan*	5	1.5	3.5	Univ. Malaya	Malaysia	*Jatropha, Calophyllum, Sterculia, Ceiba,* nonedible	Production, properties
5	Raheman, Hiflur	5	0.4	4.6	Indian Inst. Technol.	India	*Jatropha,* mahua, karanja	Production, properties
6	Atabani, Abdulaziz E.	4	0.8	3.2	Univ. Malaya	Malaysia	*Jatropha,* nonedible	Production, properties
7	Kalam, M. Abul*	4	1.7	2.3	Univ. Malaya	Malaysia	*Jatropha,* nonedible	Properties
8	Kumar, Naveen	4	0.2	2.8	Delhi Technol. Univ.	India	*Jatropha,* karanja, polanga	Properties
9	Naik, Satya Narayan	4	0.8	3.2	Indian Inst. Technol.	India	Karanja	Production, properties
10	Sahoo, Prasanta K.	4	0.5	3.5	Indian Inst. Technol.	India	Karanja	Production, properties
11	Silitonga, Arridina Susan	4	1.2	2.8	Univ. Malaya	Malaysia	*Jatropha, Calophyllum, Sterculia, Ceiba,* nonedible	Production, properties
12	Agarwal, Avinash K.	3	1.2	1.8	Indian Inst. Technol.	India	Karanja	Properties
13	Chauhan, Bhupendra Singh	3	0.1	2.9	Delhi Technol. Univ.	India	*Jatropha,* Karanja	Properties
14	Ghadge, Shahikant Vilas	3	0.1	2.9	Indian Inst. Technol.	India	Mahua	Production, properties
15	Jayaraj, Sivasubramanian	3	0.2	2.8	Natl. Inst. Technol.	India	Rubber	Production, properties
16	Meher, Lekha Charan	3	0.3	2.7	Indian Inst. Technol.	India	Karanja	Production
17	Mofijur, M.	3	0.5	2.5	Univ. Malaya	Malaysia	*Jatropha,* nonedible	Properties
18	Muraleedharan, Chandrasekharan	3	0.4	2.6	Natl. Inst. Technol.	India	Rubber	Properties
19	Ramadhas, A. S.	3	0.2	2.8	Natl. Inst. Technol.	India	Rubber	Properties
20	Singh, Virendra P.	3	0.2	2.8	World Agrofore. Ctr.	India	*Jatropha,* karanja, polanga	Production, properties

*'Highly Cited Researchers' in 2019 (Clarivate Analytics, 2019).

On the other hand, a number of authors have a significant presence in the population papers: 'Keat Teong Lee', 'B. Ashok', 'M. P. Sharma', 'K. Nanthagopal', 'Suzana Yusup', 'Nagaraj R. Banapurmath', 'Yun Hin Taufiq-Yap', 'Wen Tong Chong, 'V. Edwin Geo', 'Siddharth Jain', 'Mushtaq Ahmad', 'Swanup Kumar Nayak', and 'Bhaskar Singh' with at least 0.7% of the population papers each.

The most-prolific institutions for these top authors are the 'University of Malaya' and the 'Indian Institute of Technology' with seven authors each. The other prolific institutions are the 'National Institute of Technology' and 'Delhi Technological University' of India with three and two authors, respectively. Thus, in total, four institutions house these top authors.

It is notable that four of these top researchers are listed in the 'Highly Cited Researchers' (HCR) in 2019 (Clarivate Analytics, 2019; Docampo and Cram, 2019).

The most-prolific country for these top authors is India with 13. The other prolific country is Malaysia with seven authors. Thus, in total, two countries contribute to these top papers.

There are two key research fronts for these top researchers: 'biodiesel fuel production' and 'biodiesel fuel properties' with 9 and 13 authors, respectively. On the other hand, these top authors focus on nine nonedible oils. The most-prolific oil is 'Jatropha oil' with 12 authors. The other prolific oils are 'karanja' and 'nonedible oils in general' with nine and eight authors, respectively. The other prolific oils are 'Calophyllum', 'polanga', 'rubber', 'Sterculia', 'Ceiba', and 'mahua' with at least two authors each.

It is further notable that there is a significant gender deficit among these top authors as only one of them is female (Lariviere et al., 2013; Xie and Shauman, 1998).

The authors with the most impact are 'Hassan H. Masjuki', 'Hiflur Raheman', 'Lalit Mohan Das', 'T. M. Indra Mahlia', 'Hwai Chyuan Ong', and 'Prasanta K. Sahoo' with at least a 3.5% publication surplus each. On the other hand, the authors with the least impact are 'Avinash K. Agarwal', 'M. Abul Kalam', and 'M. Mofijur' with at least a 1.5% publication surplus each.

25.3.3 Publication Years

Information about the publication years for both datasets is provided in Figure 25.1. This figure shows that 0, 0, 63, and 37% of the sample papers and 0.8, 1.2, 10.9, and 86.9% of the population papers were published in the 1980s, 1990s, 2000s, and 2010s, respectively.

Similarly, the most-prolific publication years for the sample dataset are 2009, 2011, 2008, and 2007 with 16, 14, 12, and 11 papers, respectively. On the other hand, the most-prolific publication years for the population dataset are 2017, 2016, 2019, and 2015 with 11.6, 11.0, 10.7, and 9.9% of the population papers, respectively. It is notable that there is a sharply rising trend for the population papers in the 2000s and more strikingly in the 2010s.

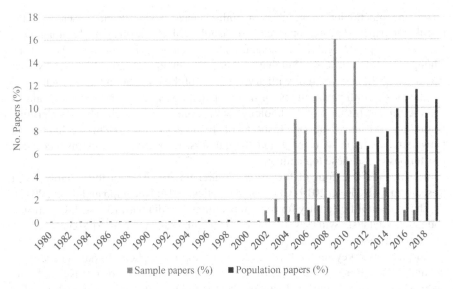

FIGURE 25.1   The research output between 1980 and 2019.

25.3.4 INSTITUTIONS

Brief information on the top nine institutions with at least 3% of the sample papers each is provided in Table 25.3. In total, around 100 and 1,550 institutions contribute to the sample and population papers, respectively.

These top institutions publish 63.0 and 23.4% of the sample and population papers, respectively. The top institution is the 'Indian Institute of Technology' with 27 sample papers and a 17.9% publication surplus. This top institution is followed by the 'University of Malaya' of Malaysia with 11 sample papers and a 6.5% publication surplus. The other top institutions are 'Anna University' and the 'National Institute of Technology' of India with six and four sample papers, respectively.

TABLE 25.3
Institutions

	Institution	Country	No. of Sample Papers (%)	No. of Population Papers (%)	Difference (%)
1	Indian Inst. Technol.	India	27	9.1	17.9
2	Univ. Malaya	Malaysia	11	4.5	6.5
3	Anna Univ.	India	6	2.7	3.3
4	Natl. Inst. Technol.	India	4	1.8	2.2
5	Delhi Univ. Technol.	India	3	1.3	1.7
6	Kongju Natl. Univ.	S. Korea	3	0.3	2.7
7	State Polytech. Medan	Indonesia	3	1.0	2.0
8	Univ. Putra Malaysia	Malaysia	3	1.5	1.5
9	Univ. Tenaga Natl.	Malaysia	3	1.2	1.8

The most-prolific countries for these top institutions are India and Malaysia with four and three sample papers, respectively. The other countries are Indonesia and South Korea. In total, four countries house these institutions.

The institutions with the most impact are the 'Indian Institute of Technology', the 'University of Malaysia', and 'Anna University' with 17.9, 6.5, and 3.3% publication surpluses, respectively. On the other hand, the institutions with the least impact are the 'University of Putra Malaysia', 'Delhi University of Technology', and the 'University of Tenaga National' with 1.5, 1.7, and 1.8% publication surpluses, respectively.

It is notable that some institutions have a heavy presence in the population papers: the 'Council of Scientific Industrial Research', 'Vellore Institute of Technology', the 'Chinese Academy of Sciences', 'Petronas University of Technology', the US 'Department of Agriculture', the 'SRM Institute of Science Technology', and the 'University of Petroleum Energy Studies' of India have at least a 1% presence in the population papers each.

25.3.5 FUNDING BODIES

Brief information about the top 12 funding bodies with at least 2% of the sample papers each is provided in Table 25.4. It is significant that only 31.0 and 50.1% of the sample and population papers declare any funding, respectively. Around 45 and 1,100 bodies fund the research for the sample and population papers, respectively.

The top funding body is the 'University of Malaya' of Malaysia, funding 10% of the sample papers and having a 7.6% publication surplus. This top funding body is followed by the 'Ministry of Education' of Malaysia with four sample papers.

TABLE 25.4
Funding Bodies

	Institution	Country	No. of Sample Papers (%)	No. of Population Papers (%)	Difference (%)
1	Univ. Malaya	Malaysia	10	2.4	7.6
2	Ministry Educ.	Malaysia	4	1.5	2.5
3	Chinese Acad. Sci.	China	2	0.7	1.3
4	Flemish InterUniv. Counc.	Netherlands	2	0.1	1.9
5	Cath. Univ. Leuven	Belgium	2	0.2	1.8
6	Ministry Ener. Sci. Technol.	Malaysia	2	0.4	1.6
7	Natl. High Technol. Res. Devnt. Prog.	China	2	0.3	1.7
8	Natl. Natr. Sci. Found.	China	2	3.0	−1.0
9	Nirma Univ.	India	2	0.1	1.9
10	Shell	Netherlands	2	0.2	1.8
11	Univ. Sains Malaysia	Malaysia	2	1.0	1.0
12	US Dept. Energy	USA	2	0.3	1.7

The most-prolific countries for these funding bodies are Malaysia and China with four and three funding bodies respectively. In total, these funding bodies are from four countries only.

It is notable that some top funding agencies have a heavy presence in the population studies. Some of them are the 'National Council for Scientific and Technological Development' of Brazil, the 'Department of Science Technology' of India, the 'Council of Scientific Industrial Research' of India, the 'University Grants Commission' of India, the 'National Council for Science and Technology' of Mexico, and 'Science Technology and Innovation' of Brazil with 4.0, 2.6, 1.8, 1.4, 0.9, and 0.8% of the population papers, respectively. These funding bodies are from Brazil, India, and Mexico.

25.3.6 SOURCE TITLES

Brief information about the top 11 source titles with at least two sample papers each is provided in Table 25.5. In total, 19 and over 480 source titles publish the sample and population papers, respectively. On the other hand, these top 11 journals publish 92 and 30.1% of the sample and population papers, respectively.

The top journal is 'Fuel' publishing 20 sample papers with an 11.9% publication surplus. This top journal is followed by 'Renewable Sustainable Energy Reviews', 'Biomass Bioenergy', 'Energy', and 'Renewable Energy' with 14, 13, 10, and 9 sample papers, respectively.

TABLE 25.5
Source Titles

	Source Title	Wos Subject Category	No. of Sample Papers (%)	No. of Population Papers (%)	Difference (%)
1	Fuel	Ener. Fuels, Eng. Chem.	20	8.1	11.9
2	Renewable Sustainable Energy Reviews	Green Sust. Sci. Technol., Ener. Fuels	14	2.1	11.9
3	Biomass Bioenergy	Agr. Eng., Biot. Appl. Microb., Ener. Fuels	13	2.1	10.9
4	Energy	Therm., Ener. Fuels	10	3.1	6.9
5	Renewable Energy	Green Sust. Sci. Technol., Ener. Fuels	9	4.3	4.7
6	Bioresource Technology	Agr. Eng., Biot. Appl. Microb., Ener. Fuels	6	2.4	3.6
7	Energy Conversion and Management	Therm., Ener. Fuels, Mechs.	6	3.3	2.7
8	Energy Fuels	Ener. Fuels, Eng. Chem.	5	2.5	2.5
9	Applied Energy	Ener. Fuels, Eng. Chem.	4	1.2	2.8
10	Applied Thermal Engineering	Therm., Ener. Fuels, Eng. Mech., Mechs.	3	0.6	2.4
11	European Journal of Lipid Science and Technology	Food Sci. Technol., Nutr. Diet.	2	0.4	1.6

Although these journals are indexed by nine subject categories, the top category is 'Energy Fuels' with nine journals. The other prolific subject categories are 'Engineering Chemical' and 'Thermodynamics' with three journals each.

The journals with the most impact are 'Fuel', 'Renewable Sustainable Energy Reviews', and 'Biomass Bioenergy' with 11.9, 11.9, and 10.9% publication surpluses, respectively. On the other hand, the journals with the least impact are the 'European Journal of Lipid Science and Technology', 'Applied Thermal Engineering', and 'Energy Fuels' with 1.6, 2.4, and 2.5% publication surpluses each.

It is notable that some journals have a heavy presence in the population papers. Some of them are 'Energy Sources Part A Recovery Utilization and Environmental Effects', 'Industrial Crops and Products', 'Fuel Processing Technology', the 'International Journal of Green Energy', 'Biofuels UK', the 'International Journal of Ambient Energy', the 'Journal of Cleaner Production', the 'Journal of Scientific Industrial Research', the 'Journal of the American Oil Chemists Society', 'Environmental Science and Pollution Research', and 'RSC Advances' with at least a 1.0% presence in the population papers each.

25.3.7 COUNTRIES

Brief information about the top 14 countries with at least two sample papers each is provided in Table 25.6. In total, around 30 and 100 countries contribute to the sample and population papers, respectively.

The top country is India publishing 52.0 and 44.3% of the sample and population papers, respectively. Malaysia follows India with 16.0 and 10.9% of the sample and population papers, respectively. The other prolific countries are China, Indonesia, Brazil, South Korea, and the USA, publishing at least four sample papers each.

TABLE 25.6
Countries

Country	No. of Sample Papers (%)	No. of Population Papers (%)	Difference (%)
1 India	52	44.3	7.7
2 Malaysia	16	10.9	5.1
3 China	7	8.9	−1.9
4 Indonesia	5	4.6	0.4
5 Brazil	4	7.3	−3.3
6 South Korea	4	1.4	2.6
7 USA	4	4.8	−0.8
8 Japan	3	2.9	0.1
9 Belgium	2	0.5	1.5
10 Canada	2	1.0	1.0
11 Egypt	2	1.8	0.2
12 Kenya	2	0.4	1.6
13 Serbia	2	0.1	1.9
14 Turkey	2	1.5	0.5
Europe-2	4	0.6	3.4
Asia-6	87	73	14

On the other hand, the European and Asian countries represented in this table publish altogether 4 and 87% of the sample papers and 0.6 and 73.0% of the population papers, respectively.

It is notable that the publication surplus for the USA and these European and Asian countries is 7.7, 3.4, and 14.4%, respectively. On the other hand, the countries with the most impact are India, Malaysia, and South Korea with 7.7, 5.1, and 2.6% publication surpluses, respectively. Furthermore, the countries with the least impact are Brazil, China, and the USA with −3.3, −1.9, and −0.8% publication deficits, respectively.

It is also notable that some countries have a heavy presence in the population papers. The major producers of these papers are Nigeria, Spain, Australia, Pakistan, Taiwan, Mexico, Thailand, France, South Africa, Saudi Arabia, the UK, and Italy, with at least 1.0% of the population papers each.

25.3.8 'WEB OF SCIENCE' SUBJECT CATEGORIES

Brief information about the top 12 'Web of Science' subject categories with at least two sample papers each is provided in Table 25.7. The sample and population papers are indexed by 14 and 87 subject categories, respectively.

For the sample papers, the top subject is 'Energy Fuels' with 91.0 and 51.1% of the sample and population papers, respectively. This top subject category is followed by 'Engineering Chemical', 'Green Sustainable Science Technology', 'Biotechnology Applied Microbiology', 'Agricultural Engineering', and 'Thermodynamics' with 34, 23, 22, 20, and 20% of the sample papers, respectively.

It is notable that the publication surplus is most significant for 'Energy Fuels', 'Biotechnology Applied Microbiology', 'Agricultural Engineering', and 'Green Sustainable Science Technology' with 39.9, 14.3, 12.4, and 10.4% publication surpluses, respectively. On the other hand, the subjects with least impact are 'Chemistry

TABLE 25.7
Web of Science Subject Categories

	Subject	No. of Sample Papers (%)	No. of Population Papers (%)	Difference (%)
1	Energy Fuels	91	51.1	39.9
2	Engineering Chemical	34	29.7	4.3
3	Green Sustainable Science Technology	23	12.6	10.4
4	Biotechnology Applied Microbiology	22	7.7	14.3
5	Agricultural Engineering	20	7.6	12.4
6	Thermodynamics	20	12.3	7.7
7	Mechanics	9	4.7	4.3
8	Engineering Mechanical	3	3.5	−0.5
9	Biochemistry Molecular Biology	2	1.6	0.4
10	Chemistry Multidisciplinary	2	6.3	−4.3
11	Food Science Technology	2	2.9	−0.9
12	Nutrition Dietetics	2	0.5	1.5

Multidisciplinary', 'Food Science Technology', 'Engineering Mechanical', and 'Biochemistry Molecular Biology' with –4.3, –0.9, –0.5, and 0.4% publication deficits, respectively. This latter group of subject categories are under-represented in the sample papers.

Additionally, some subject categories have a heavy presence in the population papers: 'Environmental Sciences', 'Chemistry Applied', 'Engineering Environmental', 'Chemistry Physical', 'Agronomy', 'Engineering Multidisciplinary', 'Multidisciplinary Sciences', 'Plant Sciences', and 'Materials Science Multidisciplinary' have at least a 1.7% presence in the population papers each.

25.3.9 CITATION IMPACT

These sample and population papers receive about 20,300 and 55,000 citations, respectively as of January 2020. Thus, the average number of citations per paper are about 203 and 25, respectively.

25.3.10 KEYWORDS

Although a number of keywords are listed in the Appendix for the datasets related to this field, some of them are more significant for the sample papers.

The most-prolific keyword related to nonedible oils is 'Jatropha' with 49 papers. This top keyword is followed by 'non-edible*' and 'karanj*' with 14 and 13 sample papers, respectively. The other prolific keywords are 'pongamia' (5), 'castor' (5), 'calophyllum' (2), 'polanga' (4), 'mahua' (5), 'Madhuca' (5), 'rubber' (3), 'tobacco' (4), and 'jojoba' (3).

On the other hand, the most-prolific keyword related to diesel fuels is '*diesel' with 100 sample papers. The other prolific keywords are 'methyl-ester*' (12), 'transester*' (9), 'compression ignition engine' (9), 'methanolysis' (2), and 'CI engine' (4).

25.3.11 RESEARCH FRONTS

Brief information about the key research fronts is provided in Table 25.8. There are two major topical research fronts for these sample papers: 'NOBD production' (Achten et al., 2008; Berchmans and Hirata, 2008; Gui et al., 2008; Ramadhas et al., 2005a; Tiwari et al., 2007) and 'properties and characterization of NOBD fuels' (Atabani et al., 2013; Pramanik, 2003; Raheman and Phadatare, 2004; Ramadhas et al., 2005b, Sahoo et al., 2007) with 59 and 54 sample papers, respectively.

On the other hand, there are nine research fronts for the nonedible oils used for biodiesel production. The most-prolific research front is 'Jatropha oil-based biodiesel fuels' with 49 sample papers (Achten et al., 2008; Berchmans and Hirata, 2008; Pramanik, 2003; Tiwari et al., 2007). This top research front is followed by 'karanja oil-based biodiesel fuels' (Agarwal and Rajamanoharan, 2009; Raheman and Phadatare, 2004) and 'NOBD fuels' in general (Atabani et al., 2013; Gui et al., 2008) with 16 sample papers each. These top three research fronts form 91% of the sample papers in total.

TABLE 25.8

Research Fronts

Research Front	No. of Sample Papers (%)
Topical research fronts	
Biodiesel fuel production	59
Biodiesel fuel characterization and properties	54
Biomass-based research fronts	
Jatropha oil-based biodiesel fuels	49
Karanja oil-based biodiesel fuels	16
NOBD fuels in general	16
Castor oil-based biodiesel fuels	5
Mahua oil-based biodiesel fuels	5
Polanga oil-based biodiesel fuels	4
Tobacco oil-based biodiesel fuels	4
Rubber oil-based biodiesel fuels	3
Other edible oil-based biodiesel fuels	14

The other prolific research fronts are 'castor oil-based biodiesel fuels' (Conceicao et al., 2007; Varma and Madras, 2007), 'mahua oil-based biodiesel fuels' (Ghadge and Raheman, 2005, 2006), 'polanga oil-based biodiesel fuels' (Sahoo et al., 2007; Saho and Das, 2009), 'tobacco oil-based biodiesel fuels' (Usta, 2005; Veljkovic et al., 2006), 'rubber oil-based biodiesel fuels' (Ramadhas et al., 2005a–b) and 'other NOBD fuels' (Huzayyiln et al., 2004; Lujaji et al., 2011; Ong et al., 2011, 2013) with 5, 5, 4, 4, 3, and 14 sample papers, respectively.

25.4 DISCUSSION

The size of the research in this field has increased to over 2,150 papers as of January 2020. It is expected that the number of the population papers in this field will exceed 6,000 papers by the end of the 2020s.

The research has developed more in the technological aspects of this field, rather than in the social and humanitarian pathways, as evidenced by the negligible number of population papers in the indices of the 'Web of Science', SSCI, and A&HCI.

The article types of documents are the primary documents for both datasets; reviews are over-represented by 15.4% in the sample papers whilst articles are under-represented by 15% (Table 25.1). Thus, the contribution of reviews by 19% of the sample papers in this field is highly exceptional (cf. Konur, 2011, 2012a–n, 2015, 2016a–f, 2017a–f, 2018a–b, 2019a–b).

Twenty authors from four institutions have at least three sample papers each (Table 25.2). Thirteen and seven of these authors are from India and Malaysia, respectively.

These authors focus on 'biodiesel fuel production' and 'biodiesel fuel properties'. There is significant 'gender deficit' among these top authors as only one of them is female (Lariviere et al., 2013; Xie and Shauman, 1998).

The population papers have built on the sample papers, primarily published in the 2000s and 2010s (Figure 25.1). Following this rising trend, particularly in the 2000s and 2010s, it is expected that the number of papers will reach 6,000 by the end of the 2020s, nearly tripling the current size.

The engagement of the institutions in this field at the global scale is significant as around 100 and 1,550 institutions contribute to the sample and population papers, respectively.

Nine top institutions publish 63.0 and 23.4% of the sample and population papers, respectively (Table 25.3). The top institution is the 'Indian Institute of Technology' of India and the 'University of Malaya' of Malaysia with 27 and 11 sample papers, respectively. The other top institutions are 'Anna University' and the 'National Institute of Technology' of India with six and four sample papers, respectively.

The most-prolific countries for these top institutions are India and Malaysia. It is notable that some institutions with a heavy presence in the population papers are under-represented in the sample papers.

It is significant that only 31.0% and about 50.1% of the sample and population papers declare any funding, respectively. It is notable that the prolific countries for these funding bodies are Malaysia and China (Table 25.4). It is further notable that some top funding agencies for the population studies do not enter this top funding body list.

However, the lack of substantial Chinese funding bodies is notable. This finding is in contrast with the studies showing the heavy research funding in China; the NSFC is the primary funding agency in China (Wang et al., 2012).

The sample and population papers are published by 19 and over 480 journals, respectively. It is significant that the top 11 journals publish 92.0 and 30.1% of the sample and population papers, respectively (Table 25.5).

The top journal is 'Fuel' publishing 20 sample papers with an 11.9% publication surplus. This top journal is followed by 'Renewable Sustainable Energy Reviews', 'Biomass Bioenergy', 'Energy', and 'Renewable Energy' with 14, 13, 10, and 9 sample papers, respectively.

The top subject categories for these journals are 'Energy Fuels', 'Engineering Chemical', and 'Thermodynamics' with nine, three, and three journals, respectively. It is notable that some journals with a heavy presence in the population papers are relatively under-represented in the sample papers.

In total, around 30 and 100 countries contribute to the sample and population papers, respectively. The top country is India publishing 52.0 and 44.3% of the sample and population papers, respectively (Table 25.6). Malaysia follows India with 16.0 and 10.9% of the sample and population papers, respectively. These findings are in line with the studies showing heavy research activity in India and Malaysia in recent decades (Bhattacharya et al., 2012; Kumar and Jan, 2014).

The USA publishes only 4.0 and 4.8% of the sample and population papers, respectively. This finding is in contrast with the studies arguing that the USA is not losing ground in science and technology (Leydesdorff and Wagner, 2009).

Indonesia, Brazil, and South Korea publish at least four sample papers each. These findings are in line with the studies showing huge research incentives in these countries (Glanzel et al., 2006, Hassan et al., 2012; Leydesdorff and Zhou, 2005).

On the other hand, the European and Asian countries represented in this table publish together 4.0 and 87% of the sample papers and 0.6 and 73.0% of the population papers, respectively. These findings are in contrast with the studies showing that European countries have superior publication performance in science and technology (Bordons et al., 2015; Youtie et al., 2008).

It is notable that the publication surplus for the USA and these European and Asian countries is 7.7, 3.4, and 14.4%, respectively. On the other hand, the countries with the most impact are India, Malaysia, and South Korea. Furthermore, the countries with the least impact are Brazil, China, and the USA.

It is further notable that China follows India and Malaysia with 7.0 and 8.9% of the sample and population papers, respectively. This finding is in line with China's efforts to be a leading nation in science and technology (Guan and Ma, 2007; Youtie et al., 2008; Zhou and Leydesdorff, 2006).

It is also notable that some countries have a heavy presence in the population papers. The major producers of the population papers are Nigeria, Spain, Australia, Pakistan, Taiwan, Mexico, Thailand, France, South Africa, Saudi Arabia, the UK, and Italy with at least 1.0% of the population papers each (Braun et al., 1991; de Bruin et al., 1991; Farooq et al., 2018; Huang et al. 2006).

The sample and population papers are indexed by 14 and 87 subject categories, respectively. For the sample papers, the top subject is 'Energy Fuels' with 91.0 and 51.1% of the sample and population papers, respectively (Table 25.7). This top subject category is followed by 'Engineering Chemical', 'Green Sustainable Science Technology', 'Biotechnology Applied Microbiology', 'Agricultural Engineering', and 'Thermodynamics' with 34, 23, 22, 20, and 20% of the sample papers, respectively.

It is notable that the publication surplus is most significant for 'Energy Fuels', 'Biotechnology Applied Microbiology', 'Agricultural Engineering', and 'Green Sustainable Science Technology' with 39.9, 14.3, 12.4, and 10.4% publication surpluses, respectively. On the other hand, the subjects with least impact are 'Chemistry Multidisciplinary', 'Food Science Technology', 'Engineering Mechanical', and 'Biochemistry Molecular Biology' with −4.3, −0.9, −0.5, and 0.4% publication deficits/surpluses, respectively. This latter group of subject categories are under-represented in the sample papers.

These sample and population papers received about 20,300 and 55,000 citations, respectively as of January 2020. Thus, the average numbers of citations per paper are about 203 and 25, respectively. Hence, the citation impact of these top 100 papers in this field has been significant.

Although a number of keywords are listed in the Appendix for the datasets related to this field, some of them are more significant for the sample papers.

The most-prolific keyword related to nonedible oils is 'Jatropha' with 49 papers. This top keyword is followed by 'non-edible*' and 'karanj*' with 14 and 13 sample papers, respectively. The other prolific keywords are 'pongamia' (5), 'castor' (5), 'Calophyllum' (2), 'polanga' (4), 'mahua' (5), 'Madhuca' (5), 'rubber' (3), 'tobacco' (4), and 'jojoba' (3).

On the other hand, the most-prolific keyword related to diesel fuels is '*diesel' with 100 sample papers. The other prolific keywords are 'methyl-ester*' (12), 'trans-ester*' (9), 'compression ignition engine' (9), 'methanolysis' (2), and 'CI engine' (4). As expected, these keywords provide valuable information about the pathways of the research in this field.

There are two major topical research fronts for these sample papers: 'NOBD production' and 'properties and characterization of NOBD fuels' with 59 and 54 sample papers, respectively (Table 25.8).

On the other hand, there are nine research fronts for the nonedible oils used for biodiesel production. The most-prolific research front is 'Jatropha oil-based biodiesel fuels' with 49 sample papers. This top research front is followed by 'karanja oil-based biodiesel fuels' and 'NOBD fuels' in general with 16 sample papers each. These top three research fronts form 91% of the sample papers in total.

The other prolific research fronts are 'castor oil-based biodiesel fuels', 'mahua oil-based biodiesel fuels', 'polanga oil-based biodiesel fuels', 'tobacco oil-based biodiesel fuels', 'rubber oil-based biodiesel fuels', and 'other NOBD fuels' with 5, 5, 4, 4, 3, and 14 sample papers, respectively.

The key emphasis in these research fronts is the exploration of the structure–processing–property relationships of biomass, biooils, and biodiesel (Cheng and Ma, 2011; Konur and Matthews, 1989; Rogers and Hopfinger, 1994; Scherf and List, 2002).

25.5 CONCLUSION

This chapter has mapped the research on the nonedible oil-based biodiesel fuels using a scientometric method.

The size of over 2,150 population papers shows the public importance of this interdisciplinary research field. However, it is significant that the research has developed more in the technological aspects in this field, rather than the social and humanitarian pathways.

Articles dominate both the sample and population papers. The population papers, primarily published in the 2010s, build on these sample papers, primarily published in the 2000s and 2010s.

The data presented in the tables and figure show that a small number of authors, institutions, funding bodies, journals, keywords, research fronts, subject categories, and countries have shaped the research in this field.

It is notable that the authors, institutions, and funding bodies from India, Malaysia, and to a lesser extent China dominate the research. Furthermore, it is also notable that some countries have a heavy presence in the population papers. The major producers of the population papers are Nigeria, Spain, Australia, Pakistan, Taiwan, Mexico, Thailand, France, South Africa, Saudi Arabia, the UK, and Italy with at least 1.0% of the population papers. Additionally, Brazil, China, and the USA are underrepresented significantly in the sample papers.

These findings show the importance of the development of efficient incentive structures for the development of the research in this field as in other fields. It seems that the Asian countries (such as India, Malaysia, China, and South Korea) have efficient incentive structures for the development of the research in this field, contrary to Nigeria, Spain, Australia, Pakistan, Taiwan, Mexico, Thailand, France, South Africa, Saudi Arabia, the UK, and Italy.

It further seems that, although the research funding is a significant element of these incentive structures, it might not be a sole solution for increasing the incentives for the research in this field as in the case of Brazil, China, Nigeria, Spain, Australia, Pakistan, Taiwan, Mexico, Thailand, France, South Africa, Saudi Arabia, the UK, and Italy in this field.

On the other hand, it seems there is more to do to reduce the significant gender deficit in this field, as in other fields of science and technology (Lariviere et al., 2013; Xie and Shauman, 1998).

The data on the research fronts, keywords, source titles, and subject categories provide valuable evidence for the interdisciplinary (Lariviere and Gingras, 2010; Morillo et al., 2001) nature of the research in this field.

There is ample justification for the broad search strategy employed in this study due to the interdisciplinary nature of this research field as evidenced by the top subject categories. The search strategy employed in this study is in line with those employed in related and other research fields (Konur, 2011, 2012a–n, 2015, 2016a–f, 2017a–f, 2018a–b, 2019a–b). It is particularly noted that 91.0 and 51.1% of the sample and population papers are indexed by the 'Energy Fuels' subject category, respectively.

There are two major topical research fronts for these sample papers: 'NOBD fuel production' and 'properties and characterization of NOBD fuels' (Table 25.8). On the other hand, there are nine research fronts for the nonedible oils used for biodiesel production. The most-prolific research front is 'Jatropha oil-based biodiesel fuels' with 49 sample papers. This top research front is followed by 'karanja oil-based biodiesel fuels' and 'NOBD fuels' in general with 16 sample papers each. These top three research fronts form 91% of the sample papers in total.

The other prolific research fronts are 'castor oil-based biodiesel fuels', 'mahua oil-based biodiesel fuels', 'polanga oil-based biodiesel fuels', 'tobacco oil-based biodiesel fuels', 'rubber oil-based biodiesel fuels', and 'other NOBD fuels' with 5, 5, 4, 4, 3, and 14 sample papers, respectively.

It is recommended that further scientometric studies are carried out for each of these research fronts building on the pioneering studies in these fields.

ACKNOWLEDGMENTS

The contribution of the highly cited researchers in the fields of NOBD fuels is greatly acknowledged.

25.A APPENDIX

The keyword set for the NOBD fuels
Syntax: 1 AND 2

25.A.1 KEYWORDS: RELATED TO NON-EDIBLE OILS

TI=(castor* or "non-edible" or nonedible or inedible or jatropha or calophyllum or neem or trisperma or honge or starflower* or borage or eucalyptus or Sterculia or croton or patchouli or pangim or karanj* or polanga or millettia or pongamia or Ceiba or euphorbia or sapium or mahua or Madhuca or simarouba or poon or tamanu or nagchampa or honne or rubber or hevea or azadirachta or tobacco or nicotiana or jojoba or pennycress or thlaspi or zanthoxylum or tung or vernicia or cuphea or milkweed or thumba or citrullus or pine or silybum or macauba or acrocomia or coriander or deccan or *melon or guizotia or thevetia or oleander or kusum or schleichera or cerbera or datura or putranjiva or cassia or "stone fruit*" or salvadora or rhus or styrax or aleurite* or terminalia or sandbox or hura or slug or prunus or "tallow seed*") NOT TI=(animal*).

25.A.2 KEYWORDS: RELATED TO DIESEL FUELS

TI=(*diesel or transester* or "trans-ester*" or "methyl-ester*" or "compression ignition engine*" or methanolysis or "CI engine*" or fame* or hydrotreat* or hydroprocessing or alcoholysis or "ethyl-ester*" or ethanolysis or hydrodeoxygenation or hydrocracking or "thermal cracking" or *cracking).

REFERENCES

Achten, W. M. J., L. Verchot, and Y. J. Franken, et al. 2008. *Jatropha* bio-diesel production and use. *Biomass & Bioenergy* 32: 1063–1084.

Agarwal, A. K. and K. Rajamanoharan. 2009. Experimental investigations of performance and emissions of karanja oil and its blends in a single cylinder agricultural diesel engine. *Applied Energy* 86: 106–112.

Ahmadun, F. R., A. Pendashteh, and L. C. Abdullah, et al. 2009. Review of technologies for oil and gas produced water treatment. *Journal of Hazardous Materials* 170: 530–551.

Atabani, A. E., A. S. Silitonga, and H. C. Ong, et al. 2013. Non-edible vegetable oils: A critical evaluation of oil extraction, fatty acid compositions, biodiesel production, characteristics, engine performance and emissions production. *Renewable & Sustainable Energy Reviews* 18: 211–245.

Atlas, R. M. 1981. Microbial degradation of petroleum hydrocarbons: An environmental perspective. *Microbiological Reviews* 45: 180–209.

Babich, I. V. and J. A. Moulijn. 2003. Science and technology of novel processes for deep desulfurization of oil refinery streams: A review. *Fuel* 82: 607–631.

Berchmans, H. J. and S. Hirata. 2008. Biodiesel production from crude *Jatropha curcas* L. seed oil with a high content of free fatty acids. *Bioresource Technology* 99: 1716–1721.

Bhattacharya, S., Shilpa, and M. Bhati. 2012. China and India: The two new players in the nanotechnology race. *Scientometrics* 93: 59–87.

Bordons, M., B. Gonzalez-Albo, J. Aparicio, and L. Moreno. 2015. The influence of R & D intensity of countries on the impact of international collaborative research: Evidence from Spain. *Scientometrics* 102: 1385–1400.

Braun, T., W. Glanzel, and A. Schubert. 1991. The bibliometric assessment of UK scientific performance: Some comments on Martin's "reply". *Scientometrics* 20: 359–362.

Bridgwater, A. V. and G. V. C. Peacocke. 2000. Fast pyrolysis processes for biomass. *Renewable & Sustainable Energy Reviews* 4: 1–73.

Busca, G., L. Lietti, G. Ramis, and F. Berti. 1998. Chemical and mechanistic aspects of the selective catalytic reduction of NO_x by ammonia over oxide catalysts: A review. *Applied Catalysis B-Environmental* 18: 1–36.

Cheng, Y. Q. and E. Ma. 2011. Atomic-level structure and structure–property relationship in metallic glasses. *Progress in Materials Science* 56: 379–473.

Clarivate Analytics. 2019. *Highly cited researchers: 2019 Recipients*. Philadelphia, PA: Clarivate Analytics. https://recognition.webofsciencegroup.com/awards/highly-cited/2019/ (accessed January, 3, 2020).

Conceicao, M. M., R. A. Candeia, and F. C. Silva, et al. 2007. Thermo analytical characterization of castor oil biodiesel. *Renewable & Sustainable Energy Reviews* 11: 964–975.

Czernik, S. and A. V. Bridgwater. 2004. Overview of applications of biomass fast pyrolysis oil.
Energy & Fuels 18: 590–598.

de Bruin, R. E., R. R. Braam, and H. F. Moed. 1991. Bibliometric lines in the sand. *Nature*
349: 559–562.

Docampo, D. and L. Cram. 2019. Highly cited researchers: A moving target. *Scientometrics*
118: 1011–1025.

Farooq, M., M. Asim, and M. Imran, et al. 2018. Mapping past, current and future energy
research trend in Pakistan: A scientometric assessment. *Scientometrics* 117:
1733–1753.

Gallezot, P. 2012. Conversion of biomass to selected chemical products. *Chemical Society
Reviews* 41: 1538–1558.

Garfield, E. 1955. Citation indexes for science. *Science* 122: 108–111.

Garfield, E. 1972. Citation analysis as a tool in journal evaluation. *Science* 178: 471–479.

Ghadge, S. V. and H. Raheman. 2005. Biodiesel production from mahua (*Madhuca indica*) oil
having high free fatty acids. *Biomass & Bioenergy* 28: 601–605.

Ghadge, S. V. and H. Raheman. 2006. Process optimization for biodiesel production from
mahua (*Madhuca indica*) oil using response surface methodology. *Bioresource
Technology* 97: 379–384.

Glanzel, W., J. Leta, and B. Thijs. 2006. Science in Brazil. Part 1: A macro-level comparative
study. *Scientometrics* 67: 67–86.

Guan, J. C. and N. Ma. 2007. China's emerging presence in nanoscience and nanotechnology:
A comparative bibliometric study of several nanoscience 'giants'. *Research Policy* 36:
880–886.

Gui, M. M., K. T. Lee, and S. Bhatia. 2008. Feasibility of edible oil vs. non-edible oil vs. waste
edible oil as biodiesel feedstock. *Energy* 33: 1646–1653.

Hassan, S. U., P. Haddawy, P. Kuinkel, A. Degelsegger, and C. Blasy. 2012. A bibliometric
study of research activity in ASEAN related to the EU in FP7 priority areas.
Scientometrics 91: 1035–1051.

Huang, M. H., H. W. Chang, and D. Z. Chen. 2006. Research evaluation of research-oriented
universities in Taiwan from 1993 to 2003. *Scientometrics* 67: 419–435.

Huzayyiln, A. S., A. H. Bawady, M. A. Rady, and A. Dawood. 2004. Experimental evaluation
of Diesel engine performance and emission using blends of jojoba oil and diesel fuel.
Energy Conversion and Management 45: 2093–2112.

Khalili, N. R., P. A. Scheff, and T. M. Holsen. 1995. PAH source fingerprints for coke ovens,
diesel and gasoline-engines, highway tunnels, and wood combustion emissions.
Atmospheric Environment 29: 533–542.

Kilian, L. 2009. Not all oil price shocks are alike: Disentangling demand and supply shocks in
the crude oil market. *American Economic Review* 99: 1053–1069.

Konur, O. 2000. Creating enforceable civil rights for disabled students in higher education: An
institutional theory perspective. *Disability & Society* 15: 1041–1063.

Konur, O. 2002a. Access to Nursing Education by disabled students: Rights and duties of nursing programs. *Nurse Education Today* 22: 364–374.

Konur, O. 2002b. Assessment of disabled students in higher education: Current public policy
issues. *Assessment and Evaluation in Higher Education* 27: 131–152.

Konur, O. 2002c. Access to employment by disabled people in the UK: Is the Disability
Discrimination Act working? *International Journal of Discrimination and the Law* 5:
247–279.

Konur, O. 2006a. Participation of children with dyslexia in compulsory education: Current
public policy issues. *Dyslexia* 12: 51–67.

Konur, O. 2006b. Teaching disabled students in Higher Education. *Teaching in Higher
Education* 11: 351–363.

Konur, O. 2007a. A judicial outcome analysis of the Disability Discrimination Act: A windfall for the employers? *Disability & Society* 22: 187–204.

Konur, O. 2007b. Computer-assisted teaching and assessment of disabled students in higher education: The interface between academic standards and disability rights. *Journal of Computer Assisted Learning* 23: 207–219.

Konur, O. 2011. The scientometric evaluation of the research on the algae and bio-energy. *Applied Energy* 88: 3532–3540.

Konur, O. 2012a. Evaluation of the research on the social sciences in Turkey: A scientometric approach. *Energy Education Science and Technology Part B: Social and Educational Studies* 4: 1893–1908.

Konur, O. 2012b. Prof. Dr. Ayhan Demirbas' scientometric biography. *Energy Education Science and Technology Part A: Energy Science and Research* 28: 727–738.

Konur, O. 2012c. The evaluation of the biogas research: A scientometric approach. *Energy Education Science and Technology Part A: Energy Science and Research* 29: 1277–1292.

Konur, O. 2012d. The evaluation of the educational research: A scientometric approach. *Energy Education Science and Technology Part B: Social and Educational Studies* 4: 1935–1948.

Konur, O. 2012e. The evaluation of the global energy and fuels research: A scientometric approach. *Energy Education Science and Technology Part A: Energy Science and Research* 30: 613–628.

Konur, O. 2012f. The evaluation of the research on the Arts and Humanities in Turkey: A scientometric approach. *Energy Education Science and Technology Part B: Social and Educational Studies* 4: 1603–1618.

Konur, O. 2012g. The evaluation of the research on the biodiesel: A scientometric approach. *Energy Education Science and Technology Part A: Energy Science and Research* 28: 1003–1014.

Konur, O. 2012h. The evaluation of the research on the bioethanol: A scientometric approach. *Energy Education Science and Technology Part A: Energy Science and Research* 28: 1051–1064.

Konur, O. 2012i. The evaluation of the research on the biofuels: A scientometric approach. *Energy Education Science and Technology Part A: Energy Science and Research* 28: 903–916.

Konur, O. 2012j. The evaluation of the research on the biohydrogen: A scientometric approach. *Energy Education Science and Technology Part A: Energy Science and Research* 29: 323–338.

Konur, O. 2012k. The evaluation of the research on the microbial fuel cells: A scientometric approach. *Energy Education Science and Technology Part A: Energy Science and Research* 29: 309–322.

Konur, O. 2012l. The scientometric evaluation of the research on the production of bioenergy from biomass. *Biomass and Bioenergy* 47: 504–515.

Konur, O. 2012m. The scientometric evaluation of the research on the deaf students in higher education. *Energy Education Science and Technology Part B: Social and Educational Studies* 4: 1573–1588.

Konur, O. 2012n. The scientometric evaluation of the research on the students with ADHD in higher education. *Energy Education Science and Technology Part B: Social and Educational Studies* 4: 1547–1562.

Konur, O. 2015. Current state of research on algal biodiesel. In *Marine Bioenergy: Trends and Developments*, S. K. Kim, and C. G. Lee, ed., 487-512. Boca Raton, FL: CRC Press.

Konur, O. 2016a. Scientometric overview in nanobiodrugs. In *Nanoarchitectonics for Smart Delivery and Drug Targeting*, A. M. Holban and A.M. Grumezescu, ed., 405–428. Amsterdam: Elsevier.

Konur, O. 2016b. Scientometric overview regarding nanoemulsions used in the food industry. In *Emulsions: Nanotechnology in the Agri-Food Industry*, A. M. Grumezescu, ed., 689–711. Amsterdam: Elsevier.

Konur, O. 2016c. Scientometric overview regarding the nanobiomaterials in antimicrobial therapy. In *Nanobiomaterials in Antimicrobial Therapy*, A. M. Grumezescu, ed., 511–535. Amsterdam: Elsevier.

Konur, O. 2016d. Scientometric overview regarding the nanobiomaterials in dentistry. In *Nanobiomaterials in Dentistry*, A. M. Grumezescu, ed., 425–453. Amsterdam: Elsevier.

Konur, O. 2016e. Scientometric overview regarding the surface chemistry of nanobiomaterials. In *Surface Chemistry of Nanobiomaterials*, A. M. Grumezescu, ed., 463–486. Amsterdam: Elsevier.

Konur, O. 2016f. The scientometric overview in cancer targeting. In *Nanoarchitectonics for Smart Delivery and Drug Targeting*, A. M. Holban and A. Grumezescu, ed., 871–895. Amsterdam: Elsevier.

Konur, O. 2017a. Recent citation classics in antimicrobial nanobiomaterials. In *Nanostructures for Antimicrobial Therapy*, A. Ficai and A. M. Grumezescu, ed., 669–685. Amsterdam: Elsevier.

Konur, O. 2017b. Scientometric overview in nanopesticides. In *New Pesticides and Soil Sensors*, A. M. Grumezescu, ed. 719–744. Amsterdam: Elsevier.

Konur, O. 2017c. Scientometric overview regarding oral cancer nanomedicine. In *Nanostructures for Oral Medicine*, E. Andronescu, A. M. Grumezescu, ed., 939–962. Amsterdam: Elsevier.

Konur, O. 2017d. Scientometric overview regarding water nanopurification. In *Water Purification*, A. M. Grumezescu, ed., 693–716. Amsterdam: Elsevier.

Konur, O. 2017e. Scientometric overview in food nanopreservation. In *Food Preservation*, A. M. Grumezescu, ed., 703–729. Amsterdam: Elsevier.

Konur, O. 2017f. The top citation classics in alginates for biomedicine. In *Seaweed Polysaccharides: Isolation, Biological and Biomedical Applications*, J. Venkatesan, S. Anil, S. K. Kim, ed., 223–249. Amsterdam: Elsevier.

Konur, O. 2018a. Scientometric evaluation of the global research in spine: An update on the pioneering study by Wei et al. *European Spine Journal* 27: 525–529.

Konur, O. 2018b. Bioenergy and biofuels science and technology: Scientometric overview and citation classics. In *Bioenergy and Biofuels*, O. Konur, ed., 3–63. Boca Raton: CRC Press.

Konur, O. 2019a. Cyanobacterial bioenergy and biofuels science and technology: A scientometric overview. In *Cyanobacteria: From Basic Science to Applications*, ed. A. K. Mishra, D. N. Tiwari and A. N. Rai, 419–442. Amsterdam: Elsevier.

Konur, O. 2019b. Nanotechnology applications in food: A scientometric overview. In *Nanoscience for Sustainable Agriculture*, R. N., Pudake, N. Chauhan, and C. Kole, ed., 683–711. Cham: Springer.

Konur, O., ed. 2021a. *Handbook of Biodiesel and Petrodiesel Fuels: Science, Technology, Health, and Environment*. Boca Raton, FL: CRC Press.

Konur, O., ed. 2021b. *Handbook of Biodiesel and Petrodiesel Fuels: Science, Technology, Health, and Environment. Volume 1. Biodiesel Fuels: Science, Technology, Health, and Environment*. Boca Raton, FL: CRC Press.

Konur, O., ed. 2021c. *Handbook of Biodiesel and Petrodiesel Fuels: Science, Technology, Health, and Environment. Volume 2. Biodiesel Fuels based on the Edible and Nonedible Feedstocks, Wastes, and Algae: Science, Technology, Health, and Environment*. Boca Raton, FL: CRC Press.

Konur, O., ed. 2021d. *Handbook of Biodiesel and Petrodiesel Fuels: Science, Technology, Health, and Environment. Volume 3. Petrodiesel Fuels: Science, Technology, Health, and Environment*. Boca Raton, FL: CRC Press.

Konur, O. 2021e. Biodiesel and petrodiesel fuels: Science, technology, health, and environment. In *Handbook of Biodiesel and Petrodiesel Fuels: Science, Technology, Health, and Environment. Volume 1. Biodiesel Fuels: Science, Technology, Health, and Environment*, ed. O. Konur. Boca Raton, FL: CRC Press.

Konur, O. 2021f. Biodiesel and petrodiesel fuels: A scientometric review of the research. In *Handbook of Biodiesel and Petrodiesel Fuels: Science, Technology, Health, and Environment. Volume 1. Biodiesel Fuels: Science, Technology, Health, and Environment*, ed. O. Konur. Boca Raton, FL: CRC Press.

Konur, O. 2021g. Biodiesel and petrodiesel fuels: A review of the research. In *Handbook of Biodiesel and Petrodiesel Fuels: Science, Technology, Health, and Environment. Volume 1. Biodiesel Fuels: Science, Technology, Health, and Environment*, ed. O. Konur. Boca Raton, FL: CRC Press.

Konur, O. 2021h Nanotechnology applications in the diesel fuels and the related research fields: A review of the research. In *Handbook of Biodiesel and Petrodiesel Fuels: Science, Technology, Health, and Environment. Volume 1. Biodiesel Fuels: Science, Technology, Health, and Environment*, ed. O. Konur. Boca Raton, FL: CRC Press.

Konur, O. 2021i. Biooils: A scientometric review of the research. In *Handbook of Biodiesel and Petrodiesel Fuels: Science, Technology, Health, and Environment. Volume 1. Biodiesel Fuels: Science, Technology, Health, and Environment*, ed. O. Konur. Boca Raton, FL: CRC Press.

Konur, O. 2021j. Characterization and properties of biooils: A review of the research. In *Handbook of Biodiesel and Petrodiesel Fuels: Science, Technology, Health, and Environment. Volume 1. Biodiesel Fuels: Science, Technology, Health, and Environment*, ed. O. Konur. Boca Raton, FL: CRC Press.

Konur, O. 2021k. Biomass pyrolysis and pyrolysis oils: A review of the research. In *Handbook of Biodiesel and Petrodiesel Fuels: Science, Technology, Health, and Environment. Volume 1. Biodiesel Fuels: Science, Technology, Health, and Environment*, ed. O. Konur. Boca Raton, FL: CRC Press.

Konur, O. 2021l. Biodiesel fuels: A scientometric review of the research. In *Handbook of Biodiesel and Petrodiesel Fuels: Science, Technology, Health, and Environment. Volume 1. Biodiesel Fuels: Science, Technology, Health, and Environment*, ed. O. Konur. Boca Raton, FL: CRC Press.

Konur, O. 2021m. Glycerol: A scientometric review of the research. In *Handbook of Biodiesel and Petrodiesel Fuels: Science, Technology, Health, and Environment. Volume 1. Biodiesel Fuels: Science, Technology, Health, and Environment*, ed. O. Konur. Boca Raton, FL: CRC Press.

Konur, O. 2021n. Propanediol production from glycerol: A review of the research. In *Handbook of Biodiesel and Petrodiesel Fuels: Science, Technology, Health, and Environment. Volume 1. Biodiesel Fuels: Science, Technology, Health, and Environment*, ed. O. Konur. Boca Raton, FL: CRC Press.

Konur, O. 2021o. Edible oil-based biodiesel fuels: A scientometric review of the research. In *Handbook of Biodiesel and Petrodiesel Fuels: Science, Technology, Health, and Environment. Volume 2. Biodiesel Fuels based on the Edible and Nonedible Feedstocks, Wastes, and Algae: Science, Technology, Health, and Environment*, ed. O. Konur. Boca Raton, FL: CRC Press.

Konur, O. 2021p. Palm oil-based biodiesel fuels: A review of the research. In *Handbook of Biodiesel and Petrodiesel Fuels: Science, Technology, Health, and Environment. Volume 2. Biodiesel Fuels based on the Edible and Nonedible Feedstocks, Wastes, and Algae*, ed. O. Konur. Boca Raton, FL: CRC Press.

Konur, O. 2021q. Rapeseed oil-based biodiesel fuels: A review of the research. In *Handbook of Biodiesel and Petrodiesel Fuels: Science, Technology, Health, and Environment. Volume 2. Biodiesel Fuels based on the Edible and Nonedible Feedstocks, Wastes, and Algae*, ed. O. Konur. Boca Raton, FL: CRC Press.

Konur, O. 2021r. Nonedible oil-based biodiesel fuels: A scientometric review of the research. In *Handbook of Biodiesel and Petrodiesel Fuels: Science, Technology, Health, and*

Environment. Volume 2. Biodiesel Fuels based on the Edible and Nonedible Feedstocks, Wastes, and Algae: Science, Technology, Health, and Environment, ed. O. Konur. Boca Raton, FL: CRC Press.

Konur, O. 2021s. Waste oil-based biodiesel fuels: A scientometric review of the research. In *Handbook of Biodiesel and Petrodiesel Fuels: Science, Technology, Health, and Environment. Volume 2. Biodiesel Fuels based on the Edible and Nonedible Feedstocks, Wastes, and Algae: Science, Technology, Health, and Environment*, ed. O. Konur. Boca Raton, FL: CRC Press.

Konur, O. 2021t. Algal biodiesel fuels: A scientometric review of the research. In *Handbook of Biodiesel and Petrodiesel Fuels: Science, Technology, Health, and Environment. Volume 2. Biodiesel Fuels based on the Edible and Nonedible Feedstocks, Wastes, and Algae: Science, Technology, Health, and Environment*, ed. O. Konur. Boca Raton, FL: CRC Press.

Konur, O. 2021u. Algal biomass production for biodiesel production: A review of the research. In *Handbook of Biodiesel and Petrodiesel Fuels: Science, Technology, Health, and Environment. Volume 2. Biodiesel Fuels based on the Edible and Nonedible Feedstocks, Wastes, and Algae*, Ed. O. Konur. Boca Raton, FL: CRC Press.

Konur, O. 2021v. Algal biomass production in wastewaters for biodiesel production: A review of the research. In *Handbook of Biodiesel and Petrodiesel Fuels: Science, Technology, Health, and Environment. Volume 2. Biodiesel Fuels based on the Edible and Nonedible Feedstocks, Wastes, and Algae*, ed. O. Konur. Boca Raton, FL: CRC Press.

Konur, O. 2021x. Algal lipid production for biodiesel production: A review of the research. In *Handbook of Biodiesel and Petrodiesel Fuels: Science, Technology, Health, and Environment. Volume 2. Biodiesel Fuels based on the Edible and Nonedible Feedstocks, Wastes, and Algae*, Ed. O. Konur. Boca Raton, FL: CRC Press.

Konur, O. 2021y. Crude oils: A scientometric review of the research. In *Handbook of Biodiesel and Petrodiesel Fuels: Science, Technology, Health, and Environment. Volume 3. Petrodiesel Fuels: Science, Technology, Health, and Environment*, ed. O. Konur. Boca Raton, FL: CRC Press.

Konur, O. 2021z. Petrodiesel fuels: A scientometric review of the research. In *Handbook of Biodiesel and Petrodiesel Fuels: Science, Technology, Health, and Environment. Volume 3. Petrodiesel Fuels: Science, Technology, Health, and Environment*, ed. O. Konur. Boca Raton, FL: CRC Press.

Konur, O. 2021aa. Bioremediation of petroleum hydrocarbons in the contaminated soils: A review of the research. In *Handbook of Biodiesel and Petrodiesel Fuels: Science, Technology, Health, and Environment. Volume 3. Petrodiesel Fuels: Science, Technology, Health, and Environment*, ed. O. Konur. Boca Raton, FL: CRC Press.

Konur, O. 2021ab. Desulfurization of diesel fuels: A review of the research. In *Handbook of Biodiesel and Petrodiesel Fuels: Science, Technology, Health, and Environment. Volume 3. Petrodiesel Fuels: Science, Technology, Health, and Environment*, ed. O. Konur. Boca Raton, FL: CRC Press.

Konur, O. 2021ac. Diesel fuel exhaust emissions: A scientometric review of the research. In *Handbook of Biodiesel and Petrodiesel Fuels: Science, Technology, Health, and Environment. Volume 3. Petrodiesel Fuels: Science, Technology, Health, and Environment*, ed. O. Konur. Boca Raton, FL: CRC Press.

Konur, O. 2021ad. The adverse health and safety impact of diesel fuels: A scientometric review of the research. In *Handbook of Biodiesel and Petrodiesel Fuels: Science, Technology, Health, and Environment. Volume 3. Petrodiesel Fuels: Science, Technology, Health, and Environment*, ed. O. Konur. Boca Raton, FL: CRC Press.

Konur, O. 2021ae. Respiratory illnesses caused by the diesel fuel exhaust emissions: A review of the research. In *Handbook of Biodiesel and Petrodiesel Fuels: Science, Technology,*

Health, and Environment. Volume 3. Petrodiesel Fuels: Science, Technology, Health, and Environment, ed. O. Konur. Boca Raton, FL: CRC Press.

Konur, O. 2021af. Cancer caused by the diesel fuel exhaust emissions: A review of the research. In *Handbook of Biodiesel and Petrodiesel Fuels: Science, Technology, Health, and Environment. Volume 3. Petrodiesel Fuels: Science, Technology, Health, and Environment*, ed. O. Konur. Boca Raton, FL: CRC Press.

Konur, O. 2021ag. Cardiovascular and other illnesses caused by the diesel fuel exhaust emissions: A review of the research. In *Handbook of Biodiesel and Petrodiesel Fuels: Science, Technology, Health, and Environment. Volume 3. Petrodiesel Fuels: Science, Technology, Health, and Environment*, ed. O. Konur. Boca Raton, FL: CRC Press.

Konur, O. and F. L. Matthews. 1989. Effect of the properties of the constituents on the fatigue performance of composites: A review. *Composites* 20: 317–328.

Kumar, S. and J. M. Jan. 2014. Research collaboration networks of two OIC nations: Comparative study between Turkey and Malaysia in the field of 'Energy Fuels', 2009–2011. *Scientometrics* 98: 387–414.

Lariviere, V. and Y. Gingras. 2010. On the relationship between interdisciplinarity and scientific impact. *Journal of the American Society for Information Science and Technology* 61: 126–131.

Lariviere, V., C. Ni, Y. Gingras, B. Cronin, B., and C. R. Sugimoto. 2013. Bibliometrics: Global gender disparities in science. *Nature News* 504: 211–213.

Leydesdorff, L. and C. Wagner. 2009. Is the United States losing ground in science? A global perspective on the world science system. *Scientometrics* 78: 23–36.

Leydesdorff, L. and P. Zhou. 2005. Are the contributions of China and Korea upsetting the world system of science? *Scientometrics* 63: 617–630.

Lujaji, F., L. Kristof, A. Bereczky, and M. Mbarawa. 2011. Experimental investigation of fuel properties, engine performance, combustion and emissions of blends containing croton oil, butanol, and diesel on a CI engine. *Fuel* 90: 505–510.

Mohan, D., C. U. Pittman, and P. H. Steele. 2006. Pyrolysis of wood/biomass for bio-oil: A critical review. *Energy & Fuels* 20: 848–889.

Morillo, F., M. Bordons, and I. Gomez. 2001. An approach to interdisciplinarity through bibliometric indicators. *Scientometrics* 51: 203–222.

North, D. C. 1991a. *Institutions, Institutional Change and Economic Performance*. Cambridge, Mass.: Cambridge University Press.

North, D.C. 1991b. Institutions. *Journal of Economic Perspectives* 5: 97–112.

Ong, H. C., T. M. I. Mahlia, H. H. Masjuki, and R. S. Norhasyima. 2011. Comparison of palm oil, *Jatropha curcas* and *Calophyllum inophyllum* for biodiesel: A review. *Renewable & Sustainable Energy Reviews* 15: 3501–3515.

Ong, H. C., A. S. Silitonga, and H. H. Masjuki, et al. 2013. Production and comparative fuel properties of biodiesel from non-edible oils: *Jatropha curcas, Sterculia foetida* and *Ceiba pentandra. Energy Conversion and Management* 73: 245–255.

Perron, P. 1989. The great crash, the oil price shock, and the unit root hypothesis. *Econometrica: Journal of the Econometric Society* 57: 1361–1401.

Pramanik, K. 2003. Properties and use of *Jatropha curcas* oil and diesel fuel blends in compression ignition engine. *Renewable Energy* 28: 239–248.

Raheman, H. and A. G. Phadatare. 2004. Diesel engine emissions and performance from blends of karanja methyl ester and diesel. *Biomass & Bioenergy* 27: 393–397.

Ramadhas, A. S., C. Muraleedharan, and S. Jayaraj. 2005a. Performance and emission evaluation of a diesel engine fueled with methyl esters of rubber seed oil. *Renewable Energy* 30: 1789–1800.

Ramadhas, A. S., S. Jayaraj, and C. Muraleedharan. 2005b. Biodiesel production from high FFA rubber seed oil. *Fuel* 84: 335–340.

Rogers, D. and A. J. Hopfinger. 1994. Application of genetic function approximation to quantitative structure-activity relationships and quantitative structure-property relationships. *Journal of Chemical Information and Computer Sciences* 34: 854–866.

Rogge, W. F., L. M. Hildemann, M. A. Mazurek, G. R. Cass, and B. R. T. Simoneit. 1993. Sources of fine organic aerosol. 2. Noncatalyst and catalyst-equipped automobiles and heavy-duty diesel trucks. *Environmental Science & Technology* 27: 636–651.

Sahoo, P. K. and L. M. Das. 2009. Combustion analysis of *Jatropha*, karanja and polanga based biodiesel as fuel in a diesel engine. *Fuel* 88: 994–999.

Sahoo, P. K., L. M. Das, M. K. G. Babu, and S. N. Naik. 2007. Biodiesel development from high acid value polanga seed oil and performance evaluation in a CI engine. *Fuel* 86: 448–454.

Schauer, J. J., M. J. Kleeman, G. R. Cass, and B. R. T. Simoneit. 1999. Measurement of emissions from air pollution sources. 2. C_1 through C_{30} organic compounds from medium duty diesel trucks. *Environmental Science & Technology* 33: 1578–1587.

Scherf, U. and E. J. List. 2002. Semiconducting polyfluorenes-towards reliable structure–property relationships. *Advanced Materials* 14: 477–487.

Tiwari, A. K., A. Kumar, and H. Raheman. 2007. Biodiesel production from *Jatropha* (*Jatropha curcas*) with high free fatty acids: An optimized process. *Biomass & Bioenergy* 31: 569–575.

Usta, N. 2005. An experimental study on performance and exhaust emissions of a diesel engine fuelled with tobacco seed oil methyl ester. *Energy Conversion and Management* 46: 2373–2386.

Varma, M. N. and G. Madras. 2007. Synthesis of biodiesel from castor oil and linseed oil in supercritical fluids. *Industrial & Engineering Chemistry Research* 46: 1–6.

Veljkovic, V. B., S. H. Lakicevic, and O. S. Stamenkovic, et al., 2006. Biodiesel production from tobacco (*Nicotiana tabacum* L.) seed oil with a high content of free fatty acids. *Fuel* 85: 2671–2675.

Wang, X., D. Liu, K. Ding, K., and X. Wang. 2012. Science funding and research output: A study on 10 countries. *Scientometrics* 91: 591–599.

Xie, Y. and K. A. Shauman. 1998. Sex differences in research productivity: New evidence about an old puzzle. *American Sociological Review* 63: 847–870.

Youtie, J, P. Shapira, and A. L. Porter. 2008. Nanotechnology publications and citations by leading countries and blocs. *Journal of Nanoparticle Research* 10: 981–986.

Zhou, P. and L. Leydesdorff. 2006. The emergence of China as a leading nation in science. *Research Policy* 35: 83–104.

26 An Exhaustive Study on the Use of Jatropha Based Biodiesel for Modern Diesel Engine Applications

Nagaraj R. Banapurmath

Manzoore Elahi Soudagar

Sharanabasava V. Ganachari

Pratap S. Kulkarni

N. Keerthi Kumar

V. S. Yaliwal

CONTENTS

26.1 INTRODUCTION

The ever increasing global energy requirement for fossil fuel has led to extreme environmental effects regarding climate change, air pollution, and oil spills (IEA, 2018; Soudagar et al., 2018). Fuel dependent Asian countries are in economic disparity due to turbulent fluctuations of fuel prices (Agarwal, 2007). Emissions from the combustion of fossil fuels affect both the ecosystem and human health. Diesel engine emissions have higher levels of NO_x, smoke, and 'polycyclic aromatic hydrocarbons' (PAHs) which are highly carcinogenic to humans (Pereira et al., 2002). Engine emissions is a major research field due to environmental protection concerns and stringent emission regulations.

In order to address these problems, it is necessary to produce alternative fuels that are locally available, globally acceptable, and technically feasible in terms of cost and ease of production. In this direction, vegetable oils provide an alternative to diesel oil as they are renewable and have comparable properties. Renewable energy sources have greater potential to meet world energy demand to a greater extent. Biodiesel has become a popular alternative fuel source to petrodiesel (Demirbas, 2009). It is a renewable fuel source and can be obtained from different places (Mofijur et al., 2013b). It addresses fuel economy with only a marginal reduction in the energy basis (about 10%), has a higher density, lower cloud and pour points, and cold starting problems (Mofijur et al., 2013a–b; Palash et al., 2013). Use of fuel additives results in the enhancement of engine performance and acceptable exhaust emissions (Kadarohman et al., 2010). Fuel modification, by way of blending diesel and biodiesel and additives of volatile fuels such as alcohols in biodiesels, plays a significant role in addressing the emission issues as well as delivering enhanced diesel engine performance. NPs have emerged as a novel and potential additive in liquid fuels for the reduction of engine tailpipe emissions and increased engine performance. Many researchers have focused their attention on higher fuel yield as well as fuel modification methods by using nanoadditives for achieving improved performance and emission characteristics (Aalam and Saravana, 2017; Arockiasamy and Anand, 2015; Balaji and Cheralathan, 2015; Basha, 2016; Bazooyar et al., 2018; Bet-Moushoul et al., 2016; Shaafi and Velraj, 2015). Adding NPs into liquid fuels facilitates improved combustion characteristics due to their higher surface energy and enhanced catalytic activity.

In this direction, investigators have reported on the modified fuels in diesel engine applications. Metal oxides of Cu, Fe, Ce, Pt, B, Al, and Co have been proposed as additives in diesel and biodiesel fuel blends (Balaji and Cheralathan, 2015; Basha, 2016; Debbarma and Misra, 2018; El-Seesy et al., 2018; Hosseini et al., 2017; Keskin et al., 2007, 2018; Pang et al., 2012; Ranjan et al., 2018). MPs as a fuel additive and subsequent preparations of stable blends are the major issues to be resolved before they can be used as potential fuels for diesel engine applications.

This is because NPs aggregate due to their large surface area and surface activity. Clogging of the fuel injection system occurs as NPs coagulate. A certain quantity of surfactant will reduce the coagulation of NPs in the fluid (Christian and Scamehorn, 2019; Farn, 2006; Gardy et al., 2017; Ghadimi and Metselaar, 2013; Paramashivaiah and Rajashekhar, 2016; Salager, 2002; Venu and Madhavan, 2016). The selection of surfactant for NP dispersion in the base fluid is a very critical issue in the preparation of nanofuels. Only a very limited number of articles pertaining to the stability of petrodiesel and biodiesel fuel blends with NPs has been published.

Nanofluids are more susceptible to agglomeration which causes a decrease in zeta potential, thereby producing unstable nanosuspensions (Saxena et al., 2017). The mechanism of aggregation and sedimentation of nanosuspensions have been explained using a Brownian dynamics sedimentation model (Ansell and Dickinson, 1986), fractal models (Schaefer et al., 1984), and a diffusion-limited aggregation model (Witten and Sander, 1981).

The combustion and emission characteristics of a high pressure CRDI engine fueled with Jatropha biodiesel and its blends with graphene NPs at different dosages were studied.

26.2 CHARACTERIZATION OF JATROPHA OIL BLENDS WITH DIESEL AND THEIR NANOADDITIVE BLENDS

Jatropha belongs to the family *Euphorbiaceae* and grows throughout India and the tropics. It has a life of around 50 years and could provide economic returns annually. The oil content of Jatropha seed varies from 30 to 40% (w/w). Jatropha oil is slow drying, odorless, and colorless, and it turns yellow with aging. The cost of biodiesel production can be reduced if nonedible oils are used instead of edible oils (Agarwal, 2007). Figure 26.1 illustrates the Jatropha seed.

The properties of petrodiesel, Jatropha oil, JOME, a blend of 80% diesel with 20% JOME (B20), and blends of B20 with graphene NPs in varied percentages have been determined and are summarized in Table 26.1.

26.3 PREPARATION OF NANOADDITIVE–BIODIESEL BLENDS

The nanofluids produced should be stable without any NP agglomeration. An ultrasonication technique has been proposed for nanofluid preparation (Saxena et al., 2017; Soudagar et al., 2018). The methods suggest a suitable procedure for addressing the agglomeration of NPs. An ultrasound frequency of waves evaluates the size of the bubbles in the liquid. Rapidly expanding nucleating bubbles at the solid surfaces can force particle separation where large bubbles are produced due to low-frequency waves and high energy forces arising as they collapse (Gumus et al., 2016; Shaafii and Velraj, 2015).

(a)   (b)   (c)

FIGURE 26.1   Jatropha feedstock: (a) Jatropha tree, (b) flowers and kernels, (c) dried seeds and shell.

TABLE 26.1
Properties of Diesel, Jatropha Oil, JOME, and JOMEB20

Property	Diesel	Jatropha Oil	JOME	JOMEB20
Density, kg/m³	840	917	870	855
Specific gravity	0.840	0.917	0.870	0.855
Kinematic viscosity at 40°C, centistokes	3.5	44.5	5.65	3.99
Flash point, °C	56	280	170	75
Calorific value, kJ/kg	43,000	35,600	38,450	39,800
Cetane number	40–48	40	44	—

JOME: Jatropha oil methyl esters; JOMEB20: Jatropha oil methyl esters (20%) and petrodiesel (80%) blends.

Graphene NPs were mixed with JOMEB20 in an ultrasonication bath for 120 min. A probe sonicator using a 12 mm diameter probe and a frequency of 40 Hz was used in blending the solution.

Figure 26.2 shows the schematic pathway to the preparation of nanoadditive Jatropha-based biodiesel. Figure 26.3 is a graphical representation of graphene NP incorporated in a biodiesel preparation. Ultrasonic waves were passed through each fuel blend for a duration of 20–30 min.

For a fixed graphene dosage, an aqueous solution of deionized (DI) water with varied mass fractions of surfactants in the ratios of 1:2, 1:3, 1:4, and 1:5 was prepared. Figure 26.4 shows blends of JOME (B20) with different concentrations of nanofluid blends. Stable dispersion was obtained for a surfactant to graphene ratio of 1:4. The obtained zeta potential was at the value of 74.6 mV, well above 60 mV. Hence, the dispersion could be considered as that which results in a homogeneous and stable blend.

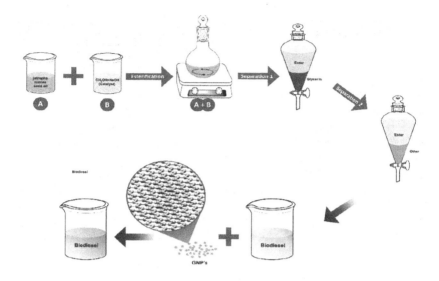

FIGURE 26.2 Schematic pathway to the preparation of nano additive Jatropha-based biodiesel.

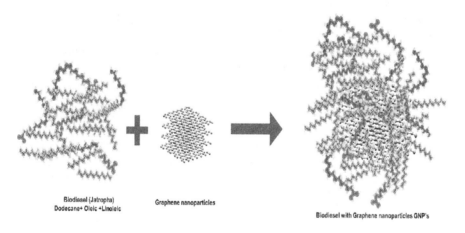

Biodiesel (Jatropha)
Dodecane+ Oleic +Linoleic

Graphene nanoparticles

Biodiesel with Graphene nanoparticles GNP's

FIGURE 26.3 Graphical representation of graphene NPs incorporated in a biodiesel preparation.

26.3.1 EXPERIMENTAL HEAT-RELEASE RATE ESTIMATION

The 'heat release rate' (HRR) was calculated by first law analysis of the pressure crank angle (^{0}CA) data. A program was developed to obtain the ensemble averaged pressure (^{0}CA) data of 100 cycles. For a single-zone model, cylinder contents were assumed to be homogeneous. The net heat-release rate using a single-zone heat-release model was calculated (Heywood, 1988).

FIGURE 26.4 Palm oil methyl ester (POME)(B20) fuel blend samples of nanofluids with varied concentrations of graphene NPs.

26.3.2 EXPERIMENTAL TEST RIG

Experiments were conducted on a 4-stroke 1-cylinder direct-injection water cooled 'compression ignition' (CI) engine modified to operate in CRDI mode as shown in Figure 26.5 and whose specifications are given in Table 26.2.

The engine was operated at a rated speed of 1,500 rpm and fueled with Jatropha oil and JOME at optimized injection timings (ITs) (10°BTDC (before top dead center)) and injection pressures (IPs) (900 bars). Optimum conditions were achieved

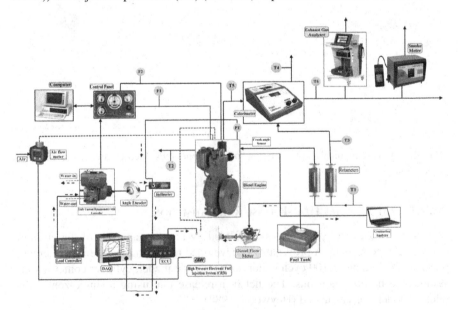

FIGURE 26.5 Single-cylinder CRDI engine test rig.

TABLE 26.2
Test Engine Specifications

CRDI Engine Specifications

Engine	TV1 Kirloskar Engine
Software used	Engine soft
No. of cylinders	1
No. of strokes	4
Bore × stroke (mm)	87.5 × 110
Compression ratio	17.5:1
Dynamometer	Eddy current
Combustion chamber	Toroidal
Governor type	Mechanical centrifugal type
Rated power	5.2 kW

Air Measurement Manometer

Made	MX 201
Type	U-Type
Range	100-0-100 mm

Eddy Current Dynamometer

Model	AG-10
Type	Eddy current
Flow	Water must flow through dynamometer during use
Dynamometer arm length	0.180 m

Port Fuel Injector

Injector opening pressure	5 bar
Injector	Solenoid operated

Diesel Injector Specifications

No. of nozzle holes	6
Injector opening pressure	900 bar
Nozzle hole diameter	0.1 mm
Injector	Solenoid operated

with exhaustive engine tests in which three ITs of 25°, 10°BTDC, and 5°ATDC (after top dead center) and IPs from 600 to 1,000 bars were used (Figure 26.6). The fuel flow rate was measured on a mass basis using a sensitive weighing machine and stopwatch. The engine and dynamometer were coupled. Emission characteristics were recorded using a Hartridge Smokemeter (London, UK) and equipment from AVL (New Delhi, India) during the steady state operation. The tests were conducted with neat B20 (D80% + JOME 20%), B20Gr15 (B20 + 15 mg graphene), B20Gr20 (B20 + 20 mg graphene), and B20Gr30 (B20 + 30 mg graphene) respectively.

26.4 RESULTS AND DISCUSSION

Experimental investigations were done to determine the performance, emission, and combustion characteristics of a CRDI engine fueled with petrodiesel, JOMEB20, and nanoadditive blends of JOMEB20 with varied dosages of graphene, i.e. JOMEB20Gr15, JOMEB20Gr20, and JOMEB20Gr30, respectively.

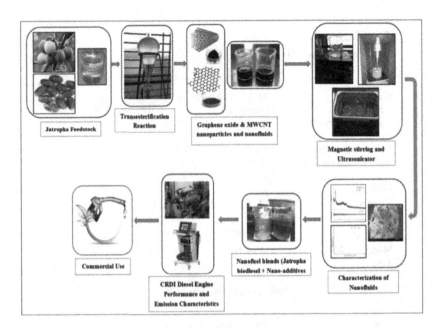

FIGURE 26.6 The comprehensive steps involved in the investigation of nanofuel blends in diesel engines.

26.4.1 The Effect of Graphene Nanoadditives on for Different Fuel Blends

Figure 26.7 shows the variation of BTE with 'brake power' (BP) for petrodiesel and JOME blended with various dosage levels of graphene NPs. JOMEB20 showed a lower BTE compared to nanoadditive blends. The lower calorific value associated with poorer atomization and marginally higher viscosity were the reasons for this behavior. Further, the fuel-spray pattern of JOMEB20 with an extended 'ignition delay' (ID) period promotes poorer atomization and increased dilution in the pre-flame region (Banapurmath et al., 2008).

The nanoadditive blended JOMEB20 blends showed a higher BTE. As the dosing level of graphene NPs increased in the B20 blends, BTE improved. This could be due to higher rates of pressure rise and HRRs observed at all BPs. At 80% load conditions, the BTE found for JOMEB20Gr15, JOMEB20Gr20, and JOMEB20Gr30 was 28.54%, 29.50%, and 29.00% respectively. Higher convective heat transfer coefficients and improved combustion characteristics of nanoadditive fuel blends resulted in these trends (El-Seesy et al., 2017). A reduced ID period and 'combustion duration' (CD) for the nanoadditive fuel blends resulted in the improved combustion of the fuel, which increased the BTE using maximum fuel energy.

Decreased BTE was observed when the dosage level of graphene NPs was increased to 30 ppm. The reasons could be attributed to the non-uniform dispersion of the graphene NPs at higher dosages which further increased the viscosity.

The fuel amount required by an engine to generate a unit power output refers to the BSFC. The fuel consumption for diesel, JOMEB20, JOMEB20Gr15,

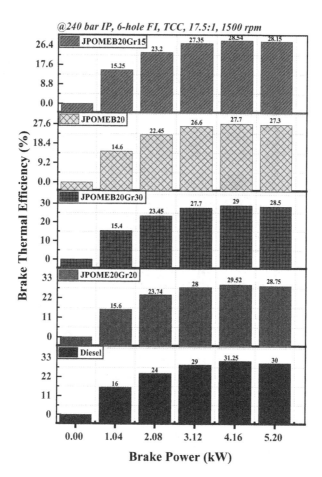

FIGURE 26.7 The variation of BTE with BP for various fuel blends.

JOMEB20Gr20, and JOMEB20Gr30 blends at varied BP is shown in Figure 26.8. An increased BTE for graphene NP dispersed JOMEB20 fuels exhibited a higher amount of fuel burning in premixed combustion and hence resulted in a lower BSFC when compared to that of a JOMEB20 blend. Among the nanoadditive fuel blends, JOMEB20Gr20 showed the lowest BSFC, followed by JOMEB20Gr30, at all BPs.

26.4.2 THE EMISSION CHARACTERISTICS OF FUEL BLENDS WITH GRAPHENE NPS

26.4.2.1 Effect of Graphene NP-Based Fuels on Smoke Emission

Smoke primarily occurs in the fuel rich zones with non-homogeneous air–fuel mixture formation and increases at maximum loading conditions. It refers to the concentration of soot particles in exhaust emissions.

Figure 26.9 shows the variations of the smoke emission patterns at varied BPs for different fuel combinations. The higher smoke emission result for the JOMEB20 (63 HSU at 4.16 kW) relates to the partial combustion of fuel at all BPs. A

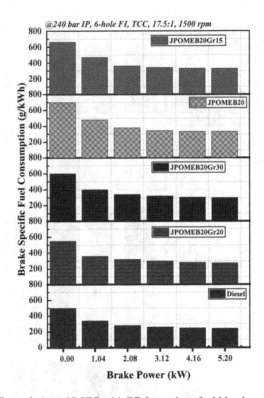

FIGURE 26.8   The variation of BSFC with BP for various fuel blends.

subsequent reduction in smoke emissions with the addition of graphene NPs in JOMEB20 was observed. Smoke emissions reported for JOMEB20Gr15, JOMEB20Gr20, and JOMEB20Gr30 fuel blends were 57, 52 HSU, and 55 respectively at maximum loading conditions. As the dosage of graphene NPs in JOMEB20 increased, the smoke emissions reduced to a 20 mg dosage, beyond which the same increased and this could be due to the increased viscosity of the fuel blends and non-uniform dispersion of the NPs with the higher dosage used. However, the addition of a 'sodium dodecyl sulfate' (SDS) surfactant to enable blending reduced viscosity and facilitated improved atomization of the fuel droplets, thereby increasing the BTE.

26.4.2.2   Effect of Graphene NP-Based Fuels on UHC Emissions

Figure 26.10 depicts the UHC emission pattern vs. the BP for different fuel blends. UHC refers to the partial combustion of fuel in the combustion chamber due to wall wetting by the fuel blends used, thereby reducing the BTE of the CRDI engine. Decreased UHC emissions were found with the increased dosage of graphene NPs in JOMEB20, which showed a higher UHC due to its poor combustion and increased wall wetting. UHC emissions moderately reduced with the dispersion of graphene NPs in JOMEB20 up to 20 ppm, beyond which it increased due to the higher viscosity of fuel blending and associated wall wetting. The further lowered dispersion of

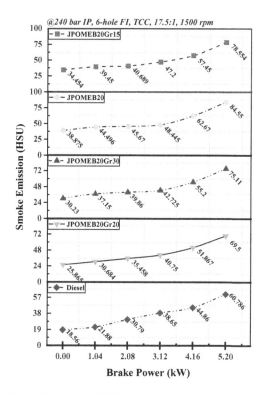

FIGURE 26.9 Variation of smoke with BP for various fuel blends.

NPs with the higher dosage of 30 ppm added to this behavior. The JOMEB20Gr20 nanofuel blend emitted 40 ppm of UHC at 4.16 kW (around a 40% reduction) compared with JOMEB20. The UHC emissions of nanofuel blends were comparable with diesel fuel.

26.4.2.3 Effect of Graphene NP-Based Fuels on CO Emissions

CO primarily occurs due to non-homogeneous air–fuel mixtures used in the combustion process. Thus, partial combustion occurs according to IP, engine temperature, the A:F ratio, the dwell time for oxidation, and IT. Figure 26.11 shows the effect of CO with BP for different fuel blends. Poorer performance of the JOMEB20 fuel blend in CRDI engines with reduced BTE increases CO formation in the engine cylinder. CO emission is reduced with an increased dosage of graphene NPs in JOMEB20 up to 20 ppm, beyond which it increased for the JOMEB20Gr30 fuel blend. Increased viscosity, density, and non-uniform dispersion of NPs with higher dosage could be responsible for this trend. The fuel blend JOMEB20Gr20 displayed the lowest CO emission (around a 40% reduction), compared with the JOMEB20 blend.

26.4.2.4 Effect of Graphene NP-Based Fuels on NO_x Emissions

Variations of NO_x emission with BP is illustrated in Figure 26.12. JOMEB20 showed lowered NO_x emissions while nanoadditive fuel blends showed higher NO_x

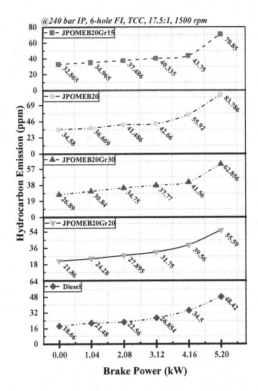

FIGURE 26.10 Variation of UHC with BP for various fuel blends.

emissions. Increased NO_x is mainly associated with higher peak flame temperatures with more premixed combustion and oxygen availability in the combustion chamber and residence time. NO_x emissions increased up to 20 mg graphene as this blend exhibited higher BTE while JOMEB20Gr30 showed lower NO_x. Lowered NO_x emissions for JOMEB20Gr30 could be attributed to lowered premixed combustion as higher NP dosages resulted in non-uniform fuel blends.

26.4.3 THE COMBUSTION CHARACTERISTICS OF FUEL BLENDS WITH A GRAPHENE NP NANOADDITIVE

26.4.3.1 Effect of Graphene NP-Based Fuels on ID, CD, and PP for Different Fuel Blends

The variations of ID, CD, and 'peak pressure' (PP) with respect to BP are shown in Figures 26.13 through 26.15 respectively. Added graphene NPs provide higher thermal conductivity which further enhances combustion activity with an increased convective heat transfer coefficient. JOMEB20 showed higher ID and CD as this blend had a lower BTE and PPs as well as a lower HRR. The addition of graphene NP dosages in JOMEB20 increased combustion activity and hence resulted in decreased ID and CD respectively up to a 20 mg dosage, beyond which both increased. The reasons could be mainly attributed to higher dosages resulting in unstable fuel blends.

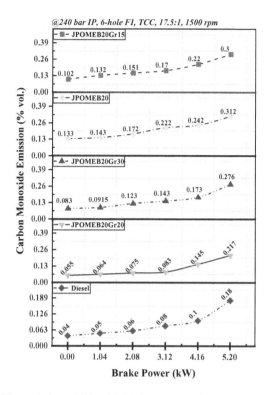

FIGURE 26.11 The variation of CO with BP for various fuel blends.

Higher dosages of NPs in liquid fuels is still a research issue and many investigations are focused on developing stable nanoadditive fuel blends.

PPs as depicted in Figure 26.12 followed a similar pattern with BP as indicated by both ID and CD variation. Higher BTE associated with JOMEB20Gr20 resulting in higher PP as shown in both Figure 26.12 ensures less fuel burning in the diffusion combustion phase. As the dosage of graphene NPs increase, the PP is reduced. The addition of NPs to JOMEB20 led to improved fuel droplet evaporation, resulting in a pressure rise (El-Seesy et al., 2017). The dosing of NPs enhances in-cylinder combustion characteristics which increase PP and reduces CD.

26.4.3.2 Effect of a Graphene Nanoadditive on the Cylinder Pressure and HRR

Figures 26.16 and 26.17 show variation of both in-cylinder pressure and HRR with ^{0}CA for various fuel combinations used. The amount of fuel participating during the uncontrolled combustion phase is proportional to the enhancement of pressure in the engine cylinder.

The addition of graphene NPs in JOMEB20 leads to an increase in cylinder pressure and HRR respectively. Nanoadditives increase cylinder pressure as they facilitate a higher 'cetane number' (CN), due to their high surface to volume ratio and increased oxygen supply to the fuel. JOMEB20Gr20 showed a maximum cylinder pressure (Figure 26.16) and HRR, followed by JOMEB20Gr30 and JOMEB20Gr15.

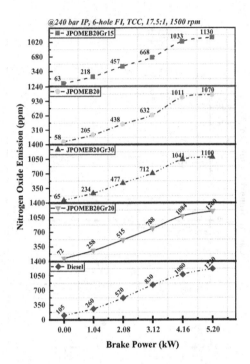

FIGURE 26.12   The variation of NO$_x$ with BP for various fuel blends.

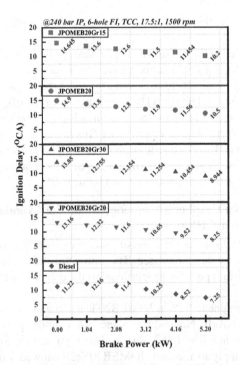

FIGURE 26.13   The variation of ID with BP for various fuel blends.

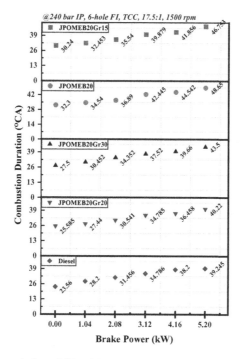

FIGURE 26.14 The variation of CD with BP for various fuel blends.

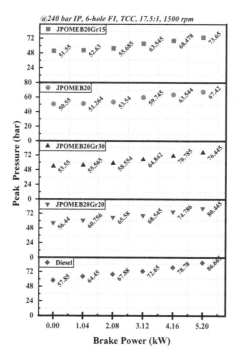

FIGURE 26.15 The variation of PP with BP for various fuel blends.

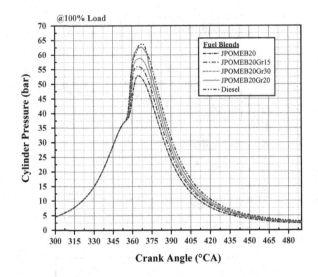

FIGURE 26.16 The variation of PP with (^{0}CA) for various fuel blends.

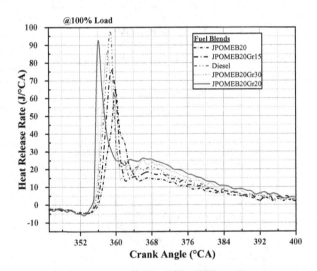

FIGURE 26.17 The variation of HRR with (^{0}CA) for various fuel blends.

Figure 26.17 illustrates the variation of HRR with ^{0}CA at maximum BP. The addition of graphene NPs for JOMEB20Gr20 resulted in an HRR comparable to diesel due to enhancement in the calorific value of the fuel, a lower ignition delay period, and an improved catalytic effect. The maximum values of HRR observed for diesel, JOMEB20, JOMEB20Gr15, JOMEB20Gr20, and JOMEB20Gr30 were 65, 75, 92, and 82 J/^{0}CA respectively. For a dosage level of 30 ppm, the HRR decreases. This may be due to the increase in viscosity and unstable fuel blends obtained at that dosage level.

26.5 CONCLUSION

The combustion characteristics of a 4-stroke CRDI engine powered with JOMEB20, JOMEB20Gr15, JOMEB20Gr20, and JOMEB20Gr30 blends at a constant speed of 1,500 rpm have been determined and compared with diesel fuel operation. Based on the above tests the following conclusions may be drawn.

1. The stability of nanoadditive blends plays an important role in their use as alternative fuels for 'internal combustion' (IC) engines. The use of appropriate surfactants compatible with the chosen biodiesel and NP to surfactant ratio is essential for the feasibility of realizing such future fuels. The SDS surfactant resulted in a stable nanofuel blend. Fuel blends with an NP to surfactant ratio of 1:4 exhibited the highest stability of all the fuel blends.
2. The use of a NP concentration of 20 ppm in the fuel blends reduced engine parameters; above 30 ppm leads to fuel stability issues.
3. The JOMEB20Gr20 fuel blend exhibited overall engine performance, combustion, and emission characteristics. This combination led to significant reductions in CO, UHCs, and smoke emissions by 40, 40.44, and 21.15% respectively, compared to JOMEB20 fuel.
4. The BTE for the fuel blend JOMEB20Gr20 increased by 7% and reduced the BSFC by 9.44%, compared to JOME(B20) fuel.
5. The NO_x emission for the nanofuel blend of JOMEB20Gr20 increased due to an increase in the HRR and PP.
6. The nanofuel JOMEB20Gr20 enhanced combustion phenomena, which led to increased PP and HRRs. A lower ID period and CD were achieved for JOMEB20Gr20. Graphene NPs are suitable as nanoadditives to improve the overall performance of CRDI diesel engines. Graphene NPs can be used as a suitable additive to further improve the existing properties of JOMEB20 fuel.

REFERENCES

Aalam, C. S. and C. G. Saravanan. 2017. Effects of nano metal oxide blended Mahua biodiesel on CRDI diesel engine. *Ain Shams Engineering Journal* 8:689–696.

Agarwal, A. K. 2007. Biofuels (alcohols and biodiesel) applications as fuels for internal combustion engines. *Progress in Energy and Combustion Science* 33:233–271.

Ansell, G. C. and E. Dickinson. 1986. Sediment formation by Brownian dynamics simulation: Effect of colloidal and hydrodynamic interactions on the sediment structure. *Journal of Chemical Physics* 85:4079–4086.

Arockiasamy, P. and R. B. Anand. 2015. Performance, combustion and emission characteristics of a DI diesel engine fuelled with nanoparticle blended jatropha biodiesel. *Periodica Polytechnica Mechanical Engineering* 59:88–93.

Balaji, G. and M. Cheralathan. 2015. Effect of CNT as additive with biodiesel on the performance and emission characteristics of a DI diesel engine. *International Journal of ChemTech Research* 7:1230–1236.

Banapurmath, N. R., P. G. Tewari, and R.S. Hosmath. 2008. Performance and emission characteristics of a DI compression ignition engine operated on Honge, Jatropha and sesame oil methyl esters. *Renewable Energy* 33:1982–1988.

Basha, J.S. 2016. Impact of carbon nanotubes and di-ethyl ether as additives with biodiesel emulsion fuels in a diesel engine: An experimental investigation. *Journal of the Energy Institute* 91:289–303.

Bazooyar, B., S. Y. Hosseini, and S. M. G. Begloo, et al. 2018. Mixed modified Fe_2O_3-WO_3 as new fuel borne catalyst (FBC) for biodiesel fuel. *Energy* 149:438–453.

Bet-Moushoul, E., K. Farhadi, and Y. Mansourpanah, et al. 2016. Application of CaO-based/ Au nanoparticles as heterogeneous nanocatalysts in biodiesel production. *Fuel* 164:119–127.

Christian, S. D. and J. F. Scamehorn. 2019. *Solubilization in Surfactant Aggregates*. Boca Raton: CRC Press.

Debbarma, S. and R. D. Misra. 2018. Effects of iron nanoparticle fuel additive on the performance and exhaust emissions of a compression ignition engine fueled with diesel and biodiesel. *Journal of Thermal Science and Engineering Applications* 10:041002.

Demirbas, A. 2009. Progress and recent trends in biodiesel fuels. *Energy Conversion and Management* 50:14–34.

El-Seesy A. I., H. Hassan, and S. Ookawara. 2018. Effects of graphene nanoplatelet addition to jatropha biodiesel-diesel mixture on the performance and emission characteristics of a diesel engine. *Energy* 147:1129–1152.

El-Seesy, A. I., A. K. Abdel-Rahman, M. Bady, and S. Ookawara. 2017. Performance, combustion, and emission characteristics of a diesel engine fueled by biodiesel-diesel mixtures with multi-walled carbon nanotubes additives. *Energy Conversion and Management* 135:373–393.

Farn, R. J. 2006. *Chemistry and Technology of Surfactants*. Chichester: John Wiley & Sons.

Gardy, J., A. Hassanpour, X. Lai, M. H. Ahmed, and M. Rehan. 2017. Biodiesel production from used cooking oil using a novel surface functionalised TiO_2 nano-catalyst. *Applied Catalysis B: Environmental* 207:297–310.

Ghadimi, A. and I. H. Metselaar. 2013. The influence of surfactant and ultrasonic processing on improvement of stability, thermal conductivity and viscosity of titania nanofluid. *Experimental Thermal and Fluid Science* 51:1–9.

Gumus, S., H. Ozcan, M. Ozbey, and B. Topaloglu, 2016. Aluminum oxide and copper oxide nanodiesel fuel properties and usage in a compression ignition engine, *Fuel* 163:80–87.

Heywood, J. B. 1988. *Internal Combustion Engine Fundamentals*. New York: McGraw-Hill Book Co.

Hosseini, S. H., A. Taghizadeh-Alisaraei, B. Ghobadian, and A. Abbaszadeh-Mayvan. 2017. Performance and emission characteristics of a CI engine fuelled with carbon nanotubes and diesel-biodiesel blends. *Renewable Energy* 111:201–213.

IEA. 2018. *World Energy Outlook*. Paris, Cedex: Intermational Energy Agency.

Kadarohman, A., Hernani, F. Khoerunisa, and R. M. Astuti. 2010. A potential study on clove oil, eugenol and eugenyl acetate as diesel fuel bio-additives and their performance on one cylinder engine. *Transport* 25:66–76.

Keskin, A., A. Yasar, and S. Yildizhan, et al. 2018. Evaluation of diesel fuel-biodiesel blends with palladium and acetylferrocene based additives in a diesel engine. *Fuel* 216:349–355.

Keskin, A., M. Guru, and D. Altiparmak. 2007. Biodiesel production from tall oil with synthesized Mn and Ni based additives: Effects of the additives on fuel consumption and emissions. *Fuel* 86:1139–1143.

Mofijur, M., Atabani, A. E., Masjuki, H. H., M. A. Kalam, and B. M. Masum. 2013a. A study on the effects of promising edible and non-edible biodiesel feedstocks on engine performance and emissions production: a comparative evaluation. *Renewable and Sustainable Energy Reviews* 23:391–404.

Mofijur, M., H. H. Masjuki, and M. A. Kalam, et al. 2013b. Effect of biodiesel from various feedstocks on combustion characteristics, engine durability and materials compatibility: a review. *Renewable and Sustainable Energy Reviews* 28:441–455.

Palash, S. M., M. A. Kalam, and H. H. Masjuki, et al. 2013. Impacts of biodiesel combustion on NO_x emissions and their reduction approaches. *Renewable and Sustainable Energy Reviews* 23:473–490.

Pang, C., J.-Y. Jung, J. W. Lee, and Y. T. Kang. 2012. Thermal conductivity measurement of methanol-based nanofluids with Al_2O_3 and SiO_2 nanoparticles. *International Journal of Heat and Mass Transfer* 55:5597–5602.

Paramashivaiah, B. M. and C. R. Rajashekhar. 2016. Studies on effect of various surfactants on stable dispersion of graphene nano particles in simarouba biodiesel. *IOP Conference Series: Materials Science and Engineering* 149:012083.

Pereira, P. A. P., de Andrade, J. B., and A. H. Miguel. 2002. Measurements of semivolatile and particulate polycyclic aromatic hydrocarbons in a bus station and an urban tunnel in Salvador, Brazil. *Journal of Environmental Monitoring* 4:558–561.

Ranjan, A., S. S. Dawn, and J. Jayaprabakar, et al. 2018. Experimental investigation on effect of MgO nanoparticles on cold flow properties, performance, emission and combustion characteristics of waste cooking oil biodiesel. *Fuel* 220:780–791.

Salager, J.-L. 2002. *Surfactants Types and Uses*. Bogota: University of Los Andes.

Saxena, V., N. Kumar, and V. K. Saxena. 2017. A comprehensive review on combustion and stability aspects of metal nanoparticles and its additive effect on diesel and biodiesel fuelled CI engine. *Renewable and Sustainable Energy Reviews* 70:563–588.

Schaefer, D. W., J. E. Martin, P. Wiltzius, and D. S. Cannell. 1984. Fractal geometry of colloidal aggregates. *Physical Review Letters* 52:2371–2374.

Shaafi, T. and R. Velraj. 2015. Influence of alumina nanoparticles, ethanol and isopropanol blend as additive with diesel-soybean biodiesel blend fuel: Combustion, engine performance and emissions. *Renewable Energy* 80:655–663.

Soudagar, M. E. M., N.-N. Nik-Ghazali, and M. A. Kalam, et al. 2018. The effect of nano-additives in diesel-biodiesel fuel blends: A comprehensive review on stability, engine performance and emission characteristics. *Energy Conversion and Management* 178:146–177.

Venu, H. and V. Madhavan. 2016. Effect of nano additives (titanium and zirconium oxides) and diethyl ether on biodiesel-ethanol fuelled CI engine. *Journal of Mechanical Science and Technology* 30:2361–2368.

Witten, T. A. and L. M. Sander. 1981. Diffusion-limited aggregation, a kinetic critical phenomenon. *Physical Review Letters* 47:1400–1403.

27 The Effects of Additives with *Calophyllum inophyllum* Methyl Ester in CI Engine Applications

Madhu Sudan Reddy Dandu
K. Nanthagopal
B. Ashok

CONTENTS

27.1 INTRODUCTION

The seeds from the *Calophyllum inophyllum* tree have been chosen as a raw material in order to produce biodiesel; the selected feedstock belongs to the family of

'Clusiaceae'. The literature reflects the effective utilization of '*Calophyllum inophyllum* methyl ester' (CIME) biodiesel as a fuel in CI engine applications with respect to engine characteristics when compared with diesel fuel. In addition, additives play a key role with CIME biodiesel in enhancing engine characteristics. These additives include NPs, ethers, antioxidants, and alcohols. These can be blended with CIME fuel in order to reduce exhaust emissions and also the scarcity of fuel. *Calophyllum inophyllum* is extensively used in several areas around the world and hence it can be utilized as an alternative source of energy for applications with respect to CI engines. The present work focusses on the extensive literature carried out by many researchers in evaluating the usage of CIME blends in diesel engine applications. This work presents the performance and emission characteristics of diesel engines by adopting the CIBD with various blends and also with additives. In addition, this chapter presents a detailed explanation of diesel engine characteristics in varying their different parameters by using CIME biodiesel with various blends.

27.2 EFFECT OF CIME–DIESEL BLENDS ON ENGINE CHARACTERISTICS

The concentration of CIME biodiesel percentage in a blend has a foremost influence with respect to the fuel properties in diesel engine applications. The addition of CIME in diesel increases the properties of viscosity and density but reduces the calorific value of the blend. These parameters may affect engine characteristics when compared to base diesel fuel (Ashok et al., 2018). Table 27.1 represents the summarized data presented by researchers.

27.2.1 PERFORMANCE AND COMBUSTION CHARACTERISTICS

Many research studies revealed that by increasing the percentage of biodiesel in the diesel blend, there will be a decreasing trend in the 'brake thermal efficiency' (BTE) of the engine (Nanthagopal et al., 2016). This can be attributed to the lower calorific value of the biodiesel which also leads to more consumption of fuel, leading to an increase in the 'brake specific fuel consumption' (BSFC). The decreasing trend of BSFC could be attributed to the varying properties of the blend by adding the biodiesel; the major contributing factors are density and viscosity of the blend (Atabani et al., 2013). On the other hand, the higher viscosity and boiling point of the diesel and biodiesel affect the values of 'brake specific energy consumption' (BSEC). Further, the temperature of the exhaust gases also rises with intensification in the concentration of CIME in the blend. This is because of the viscosity of the blend which in turn affects the characterization that includes atomization of fuel blends in the engine cylinder during the combustion process (Vedharaj et al., 2013). While addressing the combustion characteristics of the engine using fuel blends, it was perceived that 'methyl ester' (ME) had peak pressures in the engine cylinder along with a high 'heat release rate' (HRR) affecting the ignition delay when compared to neat diesel fuel. The variation of the CIME volume in the blend affects the pressure fluctuations because of the lower calorific value of the CIME blends, with the diesel ending in poor fuel burning during the phase of premixed combustion. The peak HRR

TABLE 27.1

Blend Variation of CIME with Diesel and its Influence on Engine Characteristics

Engine	Reference Fuel	Blend	Combustion	Performance	Emission	Reference
Single cylinder constant speed engine	Diesel	B30	↓: Cylinder Pressure, HRR, Ignition Delay	↓: BTE ↑: BSFC, BSEC	↓: UBHC, CO, Smoke; ↑: NOx	Ashok et al., 2018
		B60	↓: Cylinder Pressure, HRR, Ignition Delay	↓: BTE ↑: BSFC, BSEC	↓: UBHC, CO, Smoke ↑: NOx	
		B100	↑: Cylinder Pressure; ↓: HRR, Ignition Delay	↓: BTE ↑: BSFC, BSEC	↓: UBHC, CO ↑: NOx, Smoke	
		B25	↔: HRR, Cylinder Pressure	↓: BTE	↓: NOx ↑: UBHC, Smoke, CO	
Single cylinder constant speed engine	Diesel	B50	↓: HRR, Cylinder Pressure	↓: BTE	↓: NOx ↑: UBHC, Smoke, CO	Al-Dawody and Bhatti, 2013
		B75	↓: HRR, Cylinder Pressure	↓: BTE	↓: NOx ↑: UBHC, Smoke, CO	
		B100	↓: HRR, Cylinder Pressure	↓: BTE	↓: NOx ↑: UBHC, Smoke, CO	
Single cylinder variable speed engine	Diesel	B10		↓: BSFC, EGT ↑: BTE	↑: CO, Smoke ↓: NOx	Ong et al., 2014
		B20		↓: BTE ↑: BSFC, EGT	↓: Smoke ↑: NOx, CO	
		B30	–	↓: BTE ↑: BSFC, EGT	↓: Smoke ↑: NOx, CO	
		B50		↓: BTE ↑: BSFC, EGT	↓: Smoke ↑: NOx, CO	

observed for the blends with more concentration of CIME, due to higher viscosity, hinders the reaction of fuel with air for combustion process. In addition, it was projected that shortening the delay period of ignition would be observed by increasing the percentage of biodiesel in the blend. This was due to the higher 'cetane number' (CN) of the fuel blends and abridged exhaust gas dilution (Rahman et al., 2013).

27.2.2 EMISSION CHARACTERISTICS

The increment in the content of CIME in the blends resulted in a declining trend of the emissions, particularly 'unburned hydrocarbons' (UHCs), smoke, and carbon monoxide (CO). At the same time, there was an increasing trend for 'nitrogen oxide' (NO_x) emissions. This can be attributed to the CN which will be higher for the blends affecting the temperature in the engine cylinder, resulting in complete combustion when compared to diesel fuel (Silitonga et al., 2013). These factors increase NO_x emissions and at the same time the inherent content of oxygen available in the blends to promote the oxidation of 'carbon monoxide' (CO) into 'carbon dioxide' (CO_2) completely, thus reducing the CO emissions (Vallinayagam et al., 2013). It was further concluded that, by increasing the volume of CIME in the blends, the emissions continued to be reduced further. The greater temperatures in the engine cylinder along with the increased content of oxygen in the biodiesel blends acts as the causative factor for an upsurge in NO_x emissions (Qi et al., 2014), which intensifies with the greater concentration of CIME in the blends. In addition, emissions of smoke were detected with higher values with diesel fuel when compared with biodiesel blends. This can be attributed to the existence of aromatic compounds in the MEs when compared to diesel and also the ratio of carbon to hydrogen being less (Usta, 2005). Thus, the emissions of smoke deteriorate further with an increase in the CIME volume in the blend. Such an incongruity was detected as the viscosity of the greater blends disturbs the spray characteristics and also the fuel diffusion in the combustion chamber (Vedharaj et al., 2015). In this situation, soot precursor oxidation is infrequently suitable, which leads to increased smoke during the phase of diffusion combustion.

27.3 EFFECT OF OXYGENATED ADDITIVES WITH CIME ON ENGINE CHARACTERISTICS

The literature studies with respect to the influence of OAs added to CIBD and their blends is briefly discussed here. Past research work focused on the biodiesel blends with OAs which showed better results in certain conditions. Various additives like antioxidants, ethers, higher alcohols, and NPs were added to biodiesel and diesel blends in order to achieve enhanced engine characteristics in all respects. The most suitable additives used in the CIME are represented in Figure 27.1. Antioxidants or OAs, namely ethanox (Ashok et al., 2017b), BHT, and butylated hydroxyanisole (BHA), help in reducing emissions, particularly NOx, by reducing the inherent quantity of oxygen present in the biodiesel such that it enhances the oxidation stability of the biodiesel and its various blends (Rashed et al., 2016). Table 27.2 shows the summarized data which has been produced by researchers.

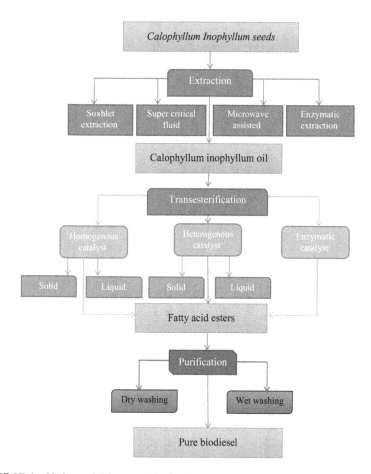

FIGURE 27.1   Various additives used in CMIE biodiesel.

27.3.1 PERFORMANCE AND COMBUSTION CHARACTERISTICS

The influence of OAs in the form of antioxidants were considered when added to the CIME blends: the results were comparable to that of neat biodiesel and their blends, without these additives, across all criteria (Ashok et al., 2017a). By considering engine characteristics with respect to the performance parameters, BTE reduces when antioxidants increase in concentration in the blends. This can be attributed to the substantial reduction of the available oxygen content in the additive blend, which tends to decrease the efficiency to a significant level. Furthermore, the BSFC increases with the insertion of an OA or antioxidant. This is because the quality of ignition tends to depreciate with respect to many operational parameters, such as the level of oxygen content existing in the fuel and lower values of viscosity and density for the test fuel blends (Palash et al., 2014). Addressing the combustion aspects of biodiesel fuel blends with additives unveiled that the peak pressure in the engine cylinder was reduced when compared with diesel fuel. The value of peak pressures does not have a direct relation with the concentration of OAs that have been added to

TABLE 27.2

Effect of Oxygenated Additives (OAs) with CIME Fuel on Engine Characteristics

Test Engine	Blend	Antioxidant	Combustion	Performance	Emission	Reference
Single cylinder variable speed engine	B20	B20 without additive	–	↓: BTE ↑: BSFC	↓: HC, CO, Smoke ↑: NOx	Rashed et al., 2016
		NPPD @1000 ppm		↓: BTE ↑: BSFC	↓: HC, CO, Smoke ↑: NOx	
		DPPD @1000 ppm		↓: BTE ↑: BSFC	↓: HC, CO, Smoke ↑: NOx	
		EHN @1000 ppm		↓: BTE ↑: BSFC	↓: HC, CO, Smoke ↑: NOx	
		BHT @200 ppm	↓: Cylinder Pressure, HRR	↓: BTE ↑: BSFC	↓: UBHC, Smoke ↑: NOx, CO	
		BHT @500 ppm	↓: Cylinder Pressure, HRR	↓: BTE ↑: BSFC	↑: NOx, CO, UBHC, Smoke	
Twin cylinder constant speed engine	B100	BHT @1000 ppm	↓: Cylinder Pressure, HRR	↓: BTE ↑: BSFC	↑: NOx, CO, UBHC, Smoke	Ashok et al., 2017a
		Ethanox @200ppm	↓: Cylinder Pressure, HRR	↓: BTE ↑: BSFC	↓: UBHC ↑: NOx, CO, Smoke	
		Ethanox @500ppm	↓: Cylinder Pressure, HRR	↓: BTE ↑: BSFC	↓: UBHC, Smoke ↑: NOx, CO	
		Ethanox @1000 ppm	↓: Cylinder Pressure, HRR	↓: BTE ↑: BSFC	↓: UBHC, Smoke ↑: NOx, CO	
		BHA @100 ppm		↓: Power ↑: BSFC, BSEC	↓: CO, HC ↑: NOx	
4 cylinder variable speed engine	B20	BHT @100 ppm	–	↓: Power ↑: BSFC, BSEC	↓: CO, HC ↑: NOx	Fattah et al., 2014
		BHQ @100 ppm		↓: Power ↑: BSFC, BSEC	↓: CO, HC ↑: NOx	

the biodiesel, though the addition of greater concentrations of antioxidant additives reflected a pressure drop in the engine cylinder. This can be attributed to the influence of methyl groups with their positive inductive effect on effective oxidation (Hess et al., 2001). A similar trend was also observed with HRR because of incomplete combustion.

27.3.2 EMISSION CHARACTERISTICS

OA inclusion to biodiesel and its blends diminishes NO_x emissions in diesel engines with respect to various operating environments, by impeding the formation of free radicals that enables NO development (Varatharajan and Cheralathan, 2013). The OAs added to CIME blends compel the formation of hydroxyl radicals throughout the combustion process. The oxidation of CO and UHC is inhibited by the deterioration of OH radical development which tends to decrease the temperature in the engine cylinder and henceforth lower the emissions of NOx. A similar trend reflects the rising emissions of CO and UHC due to the oxidation effect by adding the OA, resulting in incomplete combustion (Palash et al., 2014). On the other hand, the inclusion of OA exhibited increased emission with respect to smoke. This can be attributed to the intensification in C-C bonds, which tends to diminish the accessibility of oxygen content and augmentation in aromatic substances.

27.4 EFFECT OF ETHERS WITH CIME ON ENGINE CHARACTERISTICS

The present section is focused on previous research addressing the impact of ether additives with CIBD and its blends with diesel. The ether additives generally used were 'methyl tert-butyl ether' (MTBE) and 'diethyl ether' (DEE). Table 27.3 shows the summarized data presented by researchers.

27.4.1 PERFORMANCE AND COMBUSTION CHARACTERISTICS

The ether additions to CIME support the reducing emissions from engines but at the cost of performance characteristics. The reducing nature of BTE was observed with ether addition to the biodiesel and continues increasing with the concentration of the content of ethers in the fuel blends. This can be attributed to the collective effect of cooling with ethers and the decline in the calorific value of the blends, signifying that they play a prevailing role in reducing BTE. Furthermore, the ether additions to the biodiesel were reflected in decreasing the calorific values, an inclining trend in the values of density, along with the viscosity of the test biodiesel blends, thereby maintaining similar trends of power output by supplying greater quantities of fuel blends and hence increasing the BSFC (Rakopoulos et al., 2013). Coming to the analysis of combustion parameters of the diesel engine with ether–biodiesel blends, ethers play a critical role, especially in the development of peak pressures and ignition delay as well. However, by adding ether content to the biodiesel marginally delays ignition, but reduces the development of peak pressures in the engine cylinder. In addition, ethers in the blends tend to decrease the value of the CN, resulting in the decelerating

TABLE 27.3

Effect of Ethers with CIME Fuel on Engine Characteristics

Test Engine	Blend	Antioxidant/Ether	Combustion	Performance	Emission	Reference
Single cylinder constant speed engine	B45	5%DEE	↓: Cylinder Pressure, HRR ↑: Ignition Delay	↓: BTE ↑: BSFC, BSEC	↓: UBHC, Smoke, CO, NOx	Bragadeshwaran et al., 2018
	B42.5	7.5%DEE	↓: Cylinder Pressure, HRR ↑: Ignition Delay	↑: BTE ↑: BSFC, BSEC	↓: UBHC, Smoke, CO, NOx	
	B40	10%DEE	↓: Cylinder Pressure, HRR ↑: Ignition Delay	↑: BTE ↑: BSFC, BSEC	↓: UBHC, CO, NOx ↑: Smoke	
	B37.5	12.5%DEE	↓: Cylinder Pressure, HRR ↑: Ignition Delay	↑: BTE ↑: BSFC, BSEC	↓: UBHC, CO, NOx ↑: Smoke	
Single cylinder constant speed engine	B45	5%MTBE	↓: Cylinder Pressure, HRR ↑: Ignition Delay	↓: BTE ↑: BSFC, BSEC	↓: UBHC, CO, NOx, CO_2 ↑: Smoke	Nanthagopal et al., 2019b
	B40	10%MTBE	↓: Cylinder Pressure, HRR ↑: Ignition Delay	↓: BTE ↑: BSFC, BSEC	↓: UBHC, CO, NOx, CO_2 ↑: Smoke	
	B35	15%MTBE	↓: Cylinder Pressure, HRR ↑: Ignition Delay	↓: BTE ↑: BSFC, BSEC	↓: UBHC, CO, NOx Smoke, CO_2	
	B30	20%MTBE	↓: Cylinder Pressure, HRR ↑: Ignition Delay	↓: BTE 3↑: BSFC, BSEC	↓: UBHC, CO, NOx, CO_2 ↑: Smoke	

of the process of combustion and its influence on diesel aromatics, thus depressing the rate of heat release and pressures inside the engine cylinder (Varatharajan et al., 2011).

27.4.2 EMISSION CHARACTERISTICS

Ether addition to the MEs tends to diminish the emissions of NO_x. This can be attributed to the inferior heat of vaporization possessed by ethers, thus absorbing more heat reflected in suppressing the process of combustion in the engine cylinder. On the other hand, ether addition to biodiesel results in shoddier fraternization of fuel with air and lower dissipation because of higher values of latent heat – all these factors tend to intensify the emissions of CO and UHC. This increased trend was proportional to the amount of ether content in the blends (Sivakumar et al., 2010). The results projected that ether addition to CIME increased smoke emissions because of deprived atomization characteristics and also a deficiency in the time accessible in order to oxidize the soot nuclei.

27.5   EFFECT OF HIGHER ALCOHOLS WITH CIME ON ENGINE CHARACTERISTICS

The recent development of carbon content with longer chains seems to be one of the partial substitutes for biodiesel blends, namely 'higher alcohols' at a certain level of concentration. In particular, CIME with additives of higher alcohols reflects better results in diesel. This can be attributed to many factors, such as the desirable property of miscibility, the ability for auto-ignition, and higher flame speed. These properties of higher alcohols help in achieving the enhanced characteristics of the CI engine. Some of the literature review is presented in Table 27.4 with *Calophyllum inophyllum* blends of various higher alcohols. The various alcohols generally used were decanol, octanol, hexanol (Ashok et al., 2019c), and pentanol.

27.5.1   PERFORMANCE AND COMBUSTION CHARACTERISTICS

The BTE of the alcohol additive blend with CIME was observed with inferior results when compared to base diesel fuel – this happened because of the calorific value for the respective blends being lower than that of diesel fuel (Wang et al., 2013). This trend of increasing the BTE value depends on the concentration of alcohol present in the ME blend and is observed with a proportional rise. This may occur due to the existence of a superfluous content of oxygen in alcohol and the CN being higher for alcohol fuels generally, so as to have a greater heat resonant capacity, etc. (Ashok et al., 2019b). In addition, the calorific value of the alcohol blend and better atomization with alcohol fuel also tends to reciprocate with an increasing trend of BTE for biodiesel and alcohol fuel blends.

Neat biodiesel fuel generally is observed with a higher fuel consumption rate or BSFC, and this trend may be because of the various properties of biodiesel, such as lower calorific value, and higher viscosity and density values. But, by adding the alcohol additive to CIME, this trend reverses, such that the BSFC value reduces as

TABLE 27.4

Effect of Alcohol as an Additive with CIME on Engine Characteristics

Engine	Blend	Alcohol Additive	Combustion	Performance	Emission	Reference
Single cylinder variable speed engine	B20	Without additive	↓: Cylinder Pressure, HRR	↓: BTE ↑: BSFC	↑: NO, CO_2 ↓: CO, HC, Smoke	Imdadul et al., 2016
	B10	Pentanol@10%	↓: Cylinder Pressure, HRR	↑: BTE, BSFC	↑: NO, CO_2 ↓: CO, HC, Smoke	
	B15	Pentanol@15%	↓: Cylinder Pressure, HRR	↑: BTE, BSFC	↑:NO, CO_2 ↓: CO, HC, Smoke	
	B20	Pentanol@20%	↓: Cylinder Pressure, HRR	↑: BTE, BSFC	↑: NO, CO_2 ↓: CO, HC, Smoke	
Single cylinder constant speed engine	B60	1-Pentanol@40%	↓: Cylinder Pressure, HRR ↑: Ignition Delay	↑: BSFC, BSEC ↓: BTE	↑: NOx, CO_2 ↓: CO, HC, Smoke	Nanthagopal et al., 2018a
	B50	1-Pentanol@50%	↓: Cylinder Pressure, HRR ↑: Ignition Delay	↑: BSFC, BSEC ↓: BTE	↑: NOx, CO_2 ↓: CO, HC, Smoke	
	B40	1-Pentanol@60%	↓: Cylinder Pressure, HRR ↑: Ignition Delay	↑: BSFC, BSEC ↓: BTE	↑: NOx, CO, CO_2 ↓: HC, Smoke	
	B60	1-Butanol@40%	↓: Cylinder Pressure, HRR ↑: Ignition Delay	↑: BSFC, BSEC ↓: BTE	↑: NOx, CO_2 ↓: CO, HC, Smoke	
	B50	1-Butanol@50%	↓: Cylinder Pressure, HRR ↑: Ignition Delay	↑: BSFC, BSEC ↓: BTE	↑: NOx, CO_2 ↓: CO, HC, Smoke	
	B40	1-Butanol@60%	↓: Cylinder Pressure, HRR ↑: Ignition Delay	↑: BSFC, BSEC ↓: BTE	↑: NOx, CO_2 ↓: CO, HC, Smoke	
Single cylinder constant speed engine	B90	n-Octanol@10%	↓: Cylinder Pressure, HRR, Ignition Delay	↓: BTE	↑: NO, ↓: CO, HC, Smoke	Ashok et al., 2019d
	B80	n-Octanol@20%	↓: Cylinder Pressure, HRR, Ignition Delay	↓: BTE	↑: NO, ↓: CO, HC, Smoke	
	B70	n-Octanol@30%	↓: Cylinder Pressure, HRR, Ignition Delay	↑: BTE	↑: NO, ↓: CO, HC, Smoke	
	B60	n-Octanol@40%	↓: Cylinder Pressure, HRR, Ignition Delay	↓: BTE	↑: NO, ↓: CO, HC, Smoke	
	B50	n-Octanol@50%	↓: Ignition Delay ↑: HRR, Cylinder Pressure	↓: BTE	↑: NO, ↓: CO, HC, Smoke	
	Diesel @80%	Isobutanol@20%	↓: Cylinder Pressure, HRR ↑: Ignition Delay	↑: BSFC, BSEC ↓: BTE	↑: NOx, HC ↓: CO, Smoke	
Single cylinder constant speed engine	B80	Isobutanol@20%	↓: Cylinder Pressure, HRR ↑: Ignition Delay	↑: BSFC, BSEC ↓: BTE	↑: NOx, HC ↓: CO, Smoke	Ashok et al., 2019b
	B20	Isobutanol@10%	↓: Cylinder Pressure, HRR ↑: Ignition Delay	↑: BSFC, BSEC ↓: BTE	↑: NOx, HC ↓: CO, Smoke	
	B15	Isobutanol@15%	↓: Cylinder Pressure, HRR ↑: Ignition Delay	↑: BSFC, BSEC ↓: BTE	↑: NOx, HC ↓: CO, Smoke	
	B10	Isobutanol@20%	↓: Cylinder Pressure, HRR ↑: Ignition Delay	↑: BSFC, BSEC ↓: BTE	↑: NOx, HC ↓: CO, Smoke	

Engine	Blend	Additive	Combustion	BSFC/BTE	Emissions	Reference
Single cylinder constant speed engine	B90	n-Pentanol@10%	↓: Cylinder Pressure, HRR ↑: Ignition Delay	↑: BSFC, BSEC ↓: BTE	↑: NO_x ↓: CO, HC, Smoke	Nanthagopal et al., 2018b
	B80	n-Pentanol@20%	↓: Cylinder Pressure, HRR ↑: Ignition Delay	↑: BSFC, BSEC ↓: BTE	↑: NO_x ↓: CO, HC, Smoke	
	B70	n-Pentanol@30%	↓: Cylinder Pressure, HRR ↑: Ignition Delay	↑: BSFC, BSEC ↓: BTE	↑: NO_x ↓: CO, HC, Smoke	
	B90	n-Octanol@10%	↓: Cylinder Pressure, HRR, Ignition Delay	↑: BSFC, BSEC ↓: BTE	↑: NO_x ↓: CO, HC, Smoke	
	B80	n-Octanol@20%	↓: Cylinder Pressure, HRR, Ignition Delay	↑: BSFC, BSEC ↓: BTE	↑: NO_x ↓: CO, HC, Smoke	
	B70	n-Octanol@30%	↓: Cylinder Pressure, HRR, Ignition Delay	↑: BSFC, BSEC ↓: BTE	↑: NO_x ↓: CO, HC, Smoke	
Single cylinder constant speed engine	B50	Without additive	↓: Cylinder Pressure, HRR ↑: Ignition Delay	↑: BSFC, BSEC ↓: BTE	↑: NO_x ↓: CO, HC, Smoke	Nanthagopal et al., 2019a
	B40	Decanol@10%	↓: Cylinder Pressure, HRR ↑: Ignition Delay	↑: BSFC, BSEC ↓: BTE	↑: NO_x ↓: CO, HC, Smoke	
	B35	Decanol@15%	↓: Cylinder Pressure, HRR ↑: Ignition Delay	↑: BSFC, BSEC ↓: BTE	↑: NO_x ↓: CO, HC, Smoke	
	B30	Decanol@20%	↓: Cylinder Pressure, HRR ↑: Ignition Delay	↑: BSFC, BSEC ↓: BTE	↑: NO_x ↓: CO, HC, Smoke	
	B20	Decanol@30%	↓: Cylinder Pressure, HRR ↑: Ignition Delay	↑: BSFC, BSEC ↓: BTE	↑: NO_x ↓: CO, HC, Smoke	
	B10	Decanol@40%	↓: Cylinder Pressure, HRR ↑: Ignition Delay	↑: BSFC, BSEC ↓: BTE	↑: NO_x ↓: CO, HC, Smoke	
Single Cylinder	100% CIME	Without additive	↓: Cylinder Pressure, HRR ↑: Ignition Delay	↓: BTE ↑: BSFC	↓: NO_x ↓: CO, HC, Smoke	Ashok et al., 2019a
	B90	Pentanol@10%	↓: Cylinder Pressure, HRR ↑: Ignition Delay	↓: BTE ↑: BSFC	↓: NO_x ↓: CO, HC, Smoke	
	B80	Pentanol@15%	↓: Cylinder Pressure, HRR ↑: Ignition Delay	↓: BTE ↑: BSFC	↓: NO_x ↓: CO, HC, Smoke	
	B70	Pentanol@20%	↓: Cylinder Pressure, HRR ↑: Ignition Delay	↓: BTE ↑: BSFC	↓: NO_x ↓: CO, HC, Smoke	
	B60	Pentanol@40%	↓: Cylinder Pressure, HRR ↑: Ignition Delay	↓: BTE ↓: BSFC	↑: NO_x ↓: CO, HC, Smoke	
	B50	Pentanol@50%	↓: Cylinder Pressure, HRR ↑: Ignition Delay	↓: BTE ↑: BSFC	↑: NO_x ↓: CO, HC, Smoke	

the increased amount of oxygen present in the alcohol catalyzes the combustion reaction the same as that of ME fuels. The superfluous accessibility of free molecules of oxygen existing in the fuel operates the diesel engine with leaner mixtures of fuel and air, leading to an amended combustion rate (Babu and Anand, 2017). With an increase in the concentration of the alcohol percentage in the ME fuel blends, the BSFC value tends to diminish further, almost approaching the values related to base diesel fuel. The addition of higher alcohols to CIME tends to increase the content carbon further in the blend, such that the overall content of energy production increases, which enhances the rate of combustion, in order to have similar power output with the consumption of less fuel. A similar trend was repeated with the BSEC and for CIME fuels blended with higher alcohols when compared to base diesel fuel; this happened because of the superior calorific value of the alcohol blends (Nanthagopal et al., 2018a). An increase in the percentage of higher alcohol in the ME blends, resulting in the enhanced efficiency of the combustion process due to improved atomization of the fuel blend, leads to a reduction of the overall BSEC.

In general, the calorific value of the base diesel fuel was higher when compared to ME fuels and their blends, and this tends to raise the pressure values in the engine cylinder. Neat CIBD normally shows lower values with respect to pressures inside the engine combustion chamber, which arises because of the collective effect of the calorific value being lower, along with that of the latent heat of vaporization. The addition of higher alcohols to the ME blends leads to the raising of the overall calorific value, along with a substantial decline in the value of surface tension and viscosity. The stated properties with respect to the alcohol fuel blends tend to diverge further, empowering a superior process of atomization of the fuel blend emerging from the nozzle, enhancing the process of self-ignition, and increasing the in-cylinder pressures more than for alcohols with lower chains (Nanthagopal et al., 2018b). Increasing the percentage content of higher alcohols in CIME blends results in a surplus amount of carbon atoms being added from the chain, providing a greater amount of combustible mixture and accordingly revealing greater pressures inside the engine cylinder. Correspondingly, the values of HRR for biodiesel fuels blended with higher alcohols are found to be inferior to that of base diesel. The CN of the CIME being higher tends to reduce the ignition delay, leading to a decline in the duration of the combustion process. But, higher alcohols generally have a lower value with respect to the CN than pure biodiesel, thereby reducing the overall CN of the alcohol fuel blends with CIME. This results in a delay in the starting of the combustion process (Imdadul et al., 2016).

27.5.2 EMISSION CHARACTERISTICS

The well-known fact that biodiesel fuel emits the maximum amount of NO_x emissions in diesel engine applications has been established in many previous studies. This is because the excess amount of oxygen availability in biodiesel and its greater atomic weight, through which increasing the temperature of the flame in the combustion zones results in surplus amount of producing NOx. But, reverse results were experienced with higher alcohol blends. With an increase in the concentration of alcohols in ME blends there is substantial diminution in NO_x emissions. This can be attributed to the effect of the cooling nature of alcohol and also the latent heat of vaporization being higher for alcohol blends, which tends to decrease the temperature in the combustion chamber of the cylinder, thus reducing the development of NO_x (Oliveira and da Silva, 2013). In a similar way, the CO emissions for biodiesel blended fuels were found to be lower than that of diesel because of an excess amount of oxygen content existing in biodiesel fuels. This interpretation results in thorough combustion leading to the complete conversion of CO to CO_2 and hence reducing the emissions of CO (Kumar and Saravanan, 2015). The addition of higher alcohols as additives to CIME marginally promotes the production of CO emissions, and this may be due to the deprived combustion rate and ignition characteristics with respect to alcohol fuels. The primitive properties of alcohol fuel that have the quality of the CN being lower and the latent heat of vaporization being higher for alcohols decelerates the fuel combustion and tends to partial combustion. With an increase in the concentration of the alcohol percentage in the test fuel blends, there is an increase in the emissions of CO, and this occurs due to the cooling nature of the higher alcohols. In addition, the length of carbon chain increases when alcohols are added to CIME blends, leading to an increment in CO emissions owing to the presence of more carbon atoms. In general, the emissions of UHC tend to decrease particularly at higher loads through enhanced efficiency in the combustion rate. Clean CIMEs with their blends of alcohols develop fewer emissions with respect to UHC as there is a superfluous amount of oxygen content available in the ME and also the CN is higher for alcohols. In addition, the latent heat of vaporization also plays a vigorous role in suppressing UHC emissions. The emissions of UHC tend to increase further when there is a rise in the concentration of the alcohol percentage in biodiesel blends. Addressing smoke emissions, the ME fuel itself produces lower emissions, and alcohol blended CIBD fuels also produce lower smoke emissions. This can be attributed to the existence of more oxygen in ME blends, with alcohol additives leading to complete combustion and tending to reduce the formation of soot. The studies revealed that the smoke emissions increase with an increase in the percentage of alcohol content in the blends. This can be attributed to the reduced auto-ignition and also the higher latent heat with respect to the higher alcohols (Nanthagopal et al., 2019a). The nature of the cooling effect possessed by the alcohols tends to intensify the flame quenching regions leading to amplified formation of soot.

27.6   EFFECT OF NPs WITH CIME ON ENGINE CHARACTERISTICS

In order to enhance the properties of CIME blends, a variety of additives is used, among which NPs tend to intensify the properties of the biodiesel blends and in turn help to enhance engine characteristics in all respects. The engine characteristics, including performance, combustion, and emissions, using CIME blends with NPs as additives are presented in Table 27.5.

27.6.1   PERFORMANCE AND COMBUSTION CHARACTERISTICS

The BTE of CIME blends increases when NPs are added as they improve the oxidation rate of test fuel blends in the reaction and also the HRR (Tamilvanan et al.,

TABLE 27.5

Effect of NPs as Additives with CIME on Engine Characteristics

Test engine	Blend	Nanoparticles	Combustion	Performance	Emission	Reference
Twin Cylinder Constant Speed Engine	B100	ZnO@50 ppm	↓: HRR, Cylinder Pressure, CHRR	↓: BTE, BSFC, BSEC	↓: CO, HC, Smoke ↑: NOx, CO_2	Nanthagopal et al., 2017
		ZnO@100 ppm	↓: HRR, Cylinder Pressure, CHRR	↓: BTE, BSFC, BSEC	↓: CO, HC, Smoke ↑: NOx, CO_2	
		TiO_2@50 ppm	↓: HRR, Cylinder Pressure, CHRR	↓: BTE, BSFC, BSEC	↓: CO, HC, Smoke ↑: NOx, CO_2	
		TiO_2@100 ppm	↓: HRR, Cylinder pressure, CHRR CHRR	↓: BTE, BSFC, BSEC	↓: CO, HC, Smoke ↑: NOx, CO_2	
Single cylinder constant speed engine	B100	CoO@100 ppm		↓: BTE, BSFC	↑: NOx, CO, HC	Jeryrajkumar et al., 2016
		CoO@150 ppm	-	↓: BTE, BSFC	↑: NOx, CO, HC	
Single cylinder constant speed engine	B50	CeO_2@20 ppm	↑: HRR, Cylinder Pressure	↓: BTE ↑: BSFC	↓: NOx, CO_2 ↑: CO, HC, Smoke	Vairamuthu et al., 2016
		CeO_2@40 ppm	↑: HRR, Cylinder Pressure	↑: BTE, BSFC	↓: NOx, CO_2 ↑: CO, HC, Smoke	
		CeO_2@60 ppm	↑: HRR, Cylinder Pressure	↓: BTE ↑: BSFC	↓: NOx, CO_2 ↑: CO, HC, Smoke	
	B100	CeO_2@20 ppm	↑: HRR, Cylinder Pressure	↓: BTE ↑: BSFC	↑: CO, HC, Smoke, NOx, CO_2	
		CeO_2@40 ppm	↑: HRR, Cylinder Pressure	↓: BTE ↑: BSFC	↑: CO, HC, Smoke, NOx, CO_2	
		CeO_2@60 ppm	↑: HRR, Cylinder Pressure	↓: BTE ↑: BSFC	↑: CO, HC, Smoke, NOx, CO_2	
Single cylinder constant speed engine	B10	Cu@30 mgL^{-1}	↑: Cylinder Pressure ↓: HRR	↓: BTE ↑: BSFC	↓: CO, HC, Smoke, NO_x ↑: CO_2	Tamilvanan et al., 2019
	B20	Cu@30 mgL^{-1}	↓: HRR, Cylinder Pressure	↓: BTE ↑: BSFC	↓: CO, HC, Smoke, NO_x ↑: CO_2	
	B100	Cu@30 mgL^{-1}	↓: HRR, Cylinder Pressure	↓: BTE ↑: BSFC	↓: CO, HC, Smoke, NO_x ↑: CO_2	

2019). The NPs tend to catalyze the reaction at a faster response; the microeruption of water molecules in the nanoemulsion benefits the evaporation of fuel blends at a rapid rate and with better formation of the air–fuel mixture (Tamilvanan et al., 2019). It can be seen from a few studies that the BSFC value is lower for the base diesel fuel when compared with pure ME blends of *Calophyllum inophyllum* fuel. This is due to the calorific value being higher for diesel fuel. As the calorific value for neat ME fuel was found to be less, so the more the fuel–air mixture was required to produce the same power output in the engine cylinder (Shaafi and Velraj, 2015). The BSFC tends to reduce when NPs are added to CIME fuel as additives, and this happens owing to the micro-exploding nature of NPs. In addition, the water existing in the emulsion attenuates the size of the fuel droplet due to the superior atomization of the test fuel blends (Annamalai et al., 2016). The role of NPs in CIME biodiesel blends seems to be to act as an oxygen buffer through complete combustion. The additional amount of oxygen supply increases the rate of heat transfer between unburned and burned atoms. This effect governs the calorific value disturbing the energy as well as fuel consumption.

With respect to the combustion characteristics of the diesel engine, CIME fuel reciprocates with a lower value of peak pressures and also HRR compared to base diesel fuel. It can be seen that, due to the surplus content of oxygen present in the biodiesel and its CN being higher, there is a rapid rate of fuel atomization which instigates a diminution in the ignition delay period, triggering a reduced peak pressure value inside the engine cylinder (Nanthagopal et al., 2017). In addition, the inclusion of the NP additive results in an upsurge in peak pressure values in the engine cylinder. This can be attributed to the trend of NPs as they contribute to enhancing the oxidation process, leading to an improvement in the combustion process. Furthermore, combustion instigated early for ME blends with NPs owing to several factors, such as ignition delay being diminutive and a higher CN leading to the production of a greater HRR than that of base diesel fuel. Therefore, inclusion of NPS in biodiesel increases the HRR owing to an amplified oxidation process, promoting enhanced combustion (Tamilvanan et al., 2019).

27.6.2 Emission Characteristics

The existing literature reveals that the emissions of UHC tend to fall by adding the NPs to the CIME fuel. This can be attributed to the supply of the required quantity of oxygen in order to oxidize the exhaust emissions by NPs (Jeryrajkumar et al., 2016). In addition, the NPs retaining the activation energy tend to evade the non-polar elements in order to deposit on the engine cylinder wall, thus reducing the emission of UHC (Vairamthu et al., 2016). Furthermore, it was detected that adding the NPs to the ME leads to a reduction of CO emissions, since more oxygen existing in the nanoemulsion converts the CO molecules into CO_2 molecules (Karthikeyan et al., 2014). Similarly, a reducing trend of NO_x emissions is observed with the NPs addition to CIME since the NPs behave as catalytic convertors by flouting down NOx into nitrogen and oxygen (Borhanipour et al., 2014). Coming to the emissions of smoke opacity, the addition of NPs to CIME tends to decline further the occurrence of metal oxides in the form of nanoemulsion, bounces thermal stability through the combustion period and augments the combustion rate owing to rapid evaporation, thus reducing the emissions of smoke.

27.7 CONCLUSIONS

The foremost objective of the present review has been to explore the main characteristics of CIME biodiesel as a fuel in diesel engine applications in all respects, including the properties as a capable substitute for diesel fuel. As per the existing literature studies, various types of feedstocks are available for the production of biodiesel fuels, among which *Calophyllum inophyllum* is one feedstock having several advantages, including abundant availability in nature and with superior properties with a modest extraction process. The application of CIME and their blends in diesel engines are reported to have better characteristics in all respects of the engine as compared to that of base diesel fuel. In order to enhance the properties of the CIME fuel, various additives have been identified to achieve improved characteristics of the engine. The additives generally suitable for CIME blends are OAs like ethers and alcohols and NPs. All the experimental results projected a slightly poor performance and better emission behavior, due to the reduction in net heat content. In particular, the earlier studies on the impact of higher alcohol addition with CIME, such as pentanol and hexanol, have been discussed in a detailed way and all the results revealed that the CIME fuel performance was highly improved and on a par with the BTE of diesel fuel. On the other hand, the general drawback of CIME fuel, such as NO_x emissions, is easily eradicated by the addition of higher alcohols.

REFERENCES

Al-Dawody, M. F. and S. K. Bhatti. 2013. Optimization strategies to reduce the biodiesel NO_x effect in diesel engine with experimental verification. *Energy Conversion and Management* 68: 96–104.

Annamalai, M., B. Dhinesh, and K. Nanthagopal, et al. 2016. An assessment on performance, combustion and emission behavior of a diesel engine powered by ceria nanoparticle blended emulsified biofuel. *Energy Conversion and Management* 123: 372–380.

Ashok, B., K. Nanthagopal, and A. K. Jeevanantham, et al. 2017a. An assessment of *Calophyllum inophyllum* biodiesel fuelled diesel engine characteristics using novel antioxidant additives. *Energy Conversion and Management* 148: 935–943.

Ashok, B., K. Nanthgopal, A. Mohan, A. Johny, and A. Tamilarasu. 2017b. Comparative analysis on the effect of zinc oxide and ethanox as additives with biodiesel in CI engine. *Energy* 140: 352–364.

Ashok, B., K. Nanthagopal, and D. S. Vignesh. 2018. *Calophyllum inophyllum* methyl ester biodiesel blend as an alternate fuel for diesel engine applications. *Alexandria Engineering Journal* 57: 1239–1247.

Ashok, B., A. K. Jeevanantham, and K. Nanthagopal, et al. 2019a. An experimental analysis on the effect of n-pentanol-*Calophyllum inophyllum* biodiesel binary blends in CI engine characteristics. *Energy* 173: 290–305.

Ashok, B., K. Nanthagopal, and B. Saravanan, et al. 2019b. Study on isobutanol and *Calophyllum inophyllum* biodiesel as a partial replacement in CI engine applications. *Fuel* 235: 984–994.

Ashok, B., K. Nanthagopal, and S. Darla, et al. 2019c. Comparative assessment of hexanol and decanol as oxygenated additives with *Calophyllum Inophyllum* biodiesel. *Energy* 173: 494–510.

Ashok, B., K. Nanthagopal, and V. Anand, et al. 2019d. Effects of n-octanol as a fuel blend with biodiesel on diesel engine characteristics. *Fuel* 235: 363–373.

Atabani, A. E., T. M. I. Mahlia, and H. H. Masjuki, et al. 2013. A comparative evaluation of physical and chemical properties of biodiesel synthesized from edible and non-edible oils and study on the effect of biodiesel blending. *Energy* 58: 296–304.

Babu, D. and R. Anand. 2017. Effect of biodiesel-diesel-n-pentanol and biodiesel-diesel-nhexanol blends on diesel engine emission and combustion characteristics. *Energy* 133: 761–776.

Borhanipour, M., P. Karin, M. Tongroon, N. Chollacoop, K. Hanamura. 2014. Comparison study on fuel properties of biodiesel from Jatropha, palm and petroleum based diesel fuel. SAE Technical Paper 2014-01-2017.

Bragadeshwaran, A., N. Kasianantham, and S. Ballusamy, et al. 2018. Experimental study of methyl tert-butyl ether as an oxygenated additive in diesel and *Calophyllum inophyllum* methyl ester blended fuel in CI engine. *Environmental Science and Pollution Research* 25: 33573–33590.

Fattah, I. M. R., H. H. Masjuki, and M. A. Kalam, et al. 2014. Experimental investigation of performance and regulated emissions of a diesel engine with *Calophyllum inophyllum* biodiesel blends accompanied by oxidation inhibitors. *Energy Conversion and Management* 83: 232–240.

Hess, H. S., J. Szybist, and A. L. Boehman, et al. 2001. Impact of oxygenated fuel on diesel engine performance and emissions. *Proceedings of the National Heat Transfer Conference*, 1: 931–941.

Imdadul, H. K., H. H. Masjuki, and M. A. Kalam, et al. 2016. Higher alcohol-biodiesel-diesel blends: An approach for improving the performance, emission, and combustion of a light-duty diesel engine. *Energy Conversion and Management* 111: 174–185.

Jeryrajkumar, L., G. Anbarasu, and T. Elangovan. 2016. Effects on nano additives on performance and emission characteristics of *Calophyllum inophyllum* biodiesel. *International Journal of Chemtech Research* 9: 210–219.

Karthikeyan, S., A. Elango, and A. Prathima. 2014. Diesel engine performance and emission analysis using canola oil methyl ester with the nano sized zinc oxide particles. *IJEMS* 21: 83–87.

Kumar, B. R. and S. Saravanan. 2015. Effect of exhaust gas recirculation (EGR) on performance and emissions of a constant speed DI diesel engine fueled with pentanol/diesel blends. *Fuel* 160: 217

Nanthagopal, K., B. Ashok, and R. T. Raj. 2016. Influence of fuel injection pressures on *Calophyllum inophyllum* methyl ester fuelled direct injection diesel engine. *Energy Conversion and Management* 116: 165–173.

Nanthagopal, K., B. Ashok, and A. Tamilarasu, A. Johny, and A. Mohan. 2017. Influence on the effect of zinc oxide and titanium dioxide nanoparticles as an additive with *Calophyllum inophyllum* methyl ester in a CI engine. *Energy Conversion and Management* 146: 8–19.

Nanthagopal, K., B. Ashok, and B. Saravanan, et al. 2018a. An assessment on the effects of 1-pentanol and 1-butanol as additives with *Calophyllum inophyllum* biodiesel. *Energy Conversion and Management* 158: 70–80.

Nanthagopal, K., B. Ashok, B. Saravanan, S. M. Korah, and S. Chandra. 2018b. Effect of next generation higher alcohols and *Calophyllum inophyllum* methyl ester blends in diesel engine. *Journal of Cleaner Production* 180: 50–63.

Nanthagopal, K., B. Ashok, and B. Saravanan, et al. 2019a. Study on decanol and *Calophyllum inophyllum* biodiesel as ternary blends in CI engine. *Fuel* 239: 862–873.

Nanthagopal, K., B. Ashok, R. S. Garnepudi, K. R. Tarun, and B. Dhinesh 2019b. Investigation on diethyl ether as an additive with *Calophyllum inophyllum* biodiesel for CI engine application. *Energy Conversion and Management* 179: 104–113.

Oliveira, L. E. and M. L. C. P. da Silva. 2013. Relationship between cetane number and calorific value of biodiesel from Tilapia visceral oil blends with mineral diesel. *Renewable Energy and Power Quality Journal* 1: 687–690.

Ong, H. C., H. H. Masjuki, and T. M. I. Mahlia, et al. 2014. Optimization of biodiesel production and engine performance from high free fatty acid *Calophyllum inophyllum* oil in CI diesel engine. *Energy Conversion and Management* 81: 30–40.

Palash, S. M., M. A. Kalam, and H. H. Masjuki, et al. 2014. Impacts of NO_x reducing antioxidant additive on performance and emissions of a multi-cylinder diesel engine fueled with Jatropha biodiesel blends. *Energy Conversion and Management* 77: 577–585.

Qi, D. H., C. F. Lee, C. C. Jia, P. P. Wang, and S. T. Wu. 2014. Experimental investigations of combustion and emission characteristics of rapeseed oil-diesel blends in a two cylinder agricultural diesel engine. *Energy Conversion and Management* 77: 227–232.

Rahman, S. M. A., H. H. Masjuki, and M. A. Kalam, et al. 2013. Production of palm and *Calophyllum inophyllum* based biodiesel and investigation of blend performance and exhaust emission in an unmodified diesel engine at high idling conditions. *Energy Conversion and Management* 76: 362–367.

Rakopoulos, D. C., C. D. Rakopoulos, E. G. Giakoumis, and A. M. Dimaratos. 2013. Studying combustion and cyclic irregularity of diethyl ether as supplement fuel in diesel engine. *Fuel* 109: 325–335.

Rashed, M. M., M. A. Kalam, and H. H. Masjuki, et al. 2016. Improving oxidation stability and NO_x reduction of biodiesel blends using aromatic and synthetic antioxidant in a light duty diesel engine. *Industrial Crops and Products* 89: 273–284.

Shaafi, T. and R. Velraj. 2015. Influence of alumina nanoparticles, ethanol and isopropanol blend as additive with diesel-soybean biodiesel blend fuel: Combustion, engine performance and emissions. *Renewable Energy* 80: 655–663.

Silitonga, A. S., H. H. Masjuki, and T. M. I. Mahlia, et al. 2013. Overview properties of biodiesel diesel blends from edible and non-edible feedstock. *Renewable and Sustainable Energy Reviews* 22: 346–360.

Sivakumar, V., J. Sarangan, and R. B. Anand. 2010. Performance, combustion and emission characteristics of a CI engine using MTBE blended diesel fuel. In *Frontiers in Automobile and Mechanical Engineering-2010*, Chennai, 170–174.

Tamilvanan, A., K. Balamurugan, and M. Vijayakumar. 2019. Effects of nano-copper additive on performance, combustion and emission characteristics of *Calophyllum inophyllum* biodiesel in CI engine. *Journal of Thermal Analysis and Calorimetry* 136: 317–330.

Usta, N. 2005. An experimental study on performance and exhaust emissions of a diesel engine fuelled with tobacco seed oil methyl ester. *Energy Conversion and Management* 46: 2373–2386.

Vairamuthu, G., S. Sundarapandian, C. Kailasanathan, and B. Thangagiri. 2016. Experimental investigation on the effects of cerium oxide nanoparticle on *Calophyllum inophyllum* (Punnai) biodiesel blended with diesel fuel in DI diesel engine modified by nozzle geometry. *Journal of the Energy Institute* 89: 668–682.

Vallinayagam, R., S. Vedharaj, and W. M. Yang, et al. 2013. Emission reduction from a diesel engine fueled by pine oil biofuel using SCR and catalytic converter. *Atmospheric Environment* 80: 190–197.

Varatharajan, K. and M. Cheralathan. 2013. Effect of aromatic amine antioxidants on NO_x emissions from a soybean biodiesel powered DI diesel engine. *Fuel Processing Technology* 106: 526–532.

Varatharajan, K., M. Cheralathan, and R. Velraj. 2011. Mitigation of NO_x emissions from a Jatropha biodiesel fuelled DI diesel engine using antioxidant additives. *Fuel* 90: 2721–2725.

Vedharaj, S., R. Vallinayagam, and W. M. Yang, et al. 2013. Experimental investigation of kapok (*Ceiba pentandra*) oil biodiesel as an alternate fuel for diesel engine. *Energy Conversion and Management* 75: 773–779.

Vedharaj, S., R. Vallinayagam, W. M. Yang, C. G. Saravanan, and P. S. Lee. 2015. Optimization of combustion bowl geometry for the operation of kapok biodiesel-diesel blends in a stationary diesel engine. *Fuel* 139: 561–567.

Wang, X., Y. Ge, L. Yu, and X. Feng. 2013. Comparison of combustion characteristics and brake thermal efficiency of a heavy-duty diesel engine fueled with diesel and biodiesel at high altitude. *Fuel* 107: 852–858.

28 *Moringa oleifera* Oil as a Potential Feedstock for Sustainable Biodiesel Production

S. Niju and *G. Janani*

CONTENTS

28.1   INTRODUCTION

The perpetual upward trend in global primary energy consumption, demand, and environmental pollution challenges society to move towards renewable low-carbon technology such as wind energy, solar energy, electric vehicles, and biofuels. The major part of global oil demand comes from the transport sector of non-OECD countries such as India, China, and the Middle East (OECD, 2011). As per the 2018 energy statistics of British Petroleum, the global share of crude oil production and consumption in India is reported as 0.9 and 5.2% respectively, with 83% of total consumption being met by imports. In addition, the carbon emissions grew by 2% in 2018 which is the fastest growth since 2010 (BP, 2019). Hence, ensuring everyone has access to affordable energy resources in a healthy environment in the future has become of great concern at present. Out of all the renewable energy resources, biofuels like bioethanol and biodiesel are expected to decrease the demand on gasoline and diesel respectively. Consequently, biofuel production has seen the highest growth since 2010 which averaged 9.7% in 2018, with 60.4 Mtoe of bioethanol and 34.9 Mtoe of biodiesel production globally. However, in India, the global share of biofuel production is 1.1%. This highlights the significance of the research on biofuels (BP, 2019).

Biodiesel is a renewable, non-toxic, biodegradable, ecofriendly, carbon neutral fuel as it releases carbon from biomass that is a part of the carbon cycle and reduces particulate emissions with low sulfur and aromatic contents. The high CN and inherent lubricity makes it superior to mineral diesel (Demirbas, 2009a; Sajjadi et al., 2016). It is a mixture of alkyl esters of fatty acids that are derived from vegetable oils, animal fats, or algal oil by transesterification with methanol or ethanol in the presence of a catalyst (Demirbas and Karslioglu, 2007; Pathak et al., 2018). However, methanol is preferred as it is less costly (Dominguez et al., 2019; Musa, 2016). So far, oil seed crops are the most exploited group of feedstock due to their 90% heat content and similarity of 'methyl ester' (ME) properties to that of No. 2 mineral diesel (Kafuku et al., 2010; Kafuku and Mbarawa, 2010).

In addition, the fuel can be used in diesel engines without any modification (Pathak et al., 2018). As per the national biofuel policy of India, it has been targeted to blend 5% of biodiesel with diesel by 2030. However, the biodiesel blend is still less than 0.15% due to the lack of feedstock availability and a proper supply chain (Bradley, 2019). Also, the selection of feedstocks for a sustainable biodiesel production to meet global requirements is a key factor as the commercial viability of the biodiesel market depends solely on its cost and sustainability amidst the food vs. fuel dispute (Sharma et al., 2012). Thus, the exploration of feedstocks is still a great area of current research. This chapter discusses the significance of this feedstock for biodiesel production and a case study is performed on the ultrasound mediated esterification of MOO; it also highlights the significance of ultrasound in biodiesel production.

28.2 FEEDSTOCKS EXPLORED FOR BIODIESEL PRODUCTION

Biodiesel can be categorized into three generations based on the feedstock used for its production.

First generation biodiesel. This includes biodiesel from edible oils like rapeseed, corn, coconut, soybean, sunflower, palm, mustard, and groundnut. Since the demand for food grade oils are increasing, first generation feedstock may lead to food vs. fuel disputes, deforestation, exploitation of arable lands, and be highly expensive. Moreover, it is produced only in developed countries like the USA and China where they exploit corn and sunflower oil to produce first generation biodiesel (Demirbas, 2009a; Sajjadi et al., 2016) and it cannot be a long-term choice.

Second generation biodiesel. This includes biodiesel from nonedible oils like Jatropha, karanja, mahua oil, and waste materials like vegetable oil soapstocks, tall oil, pomace oil, dried distiller's grains, and animal fats like beef tallow, pork lard, and yellow grease. Indeed, biodiesel from these kinds of feedstocks can reduce the deforestation rate as they can be grown in arable lands, be cost effective, and is not in competition with food and agricultural lands. Though waste cooking oil can be a more cost effective and environmentally friendly feedstock, the logistics and an integrated supply chain still remain as barriers as the sources are scattered and the physicochemical properties vary with the source. Similarly, animal fats are also cost effective, but the high level of saturation makes transesterification a complex process (Demirbas 2009b; Pinzi et al., 2014; Sajjadi et al., 2016).

Third generation biodiesel. This comprises biodiesel from algal oil from various macroalgal and microalgal species. Unlike plants, algae has a high growth rate and productivity, has oil content, competes less with land and food, and has a reduced greenhouse effect. But, the difficulties in algal oil extraction, scaling up, and the requirement of high capital investment are the major disadvantages of third generation biodiesel (Ambat et al., 2018).

Ambat et al. (2018) reviewed biodiesel from various feedstocks, and reported the significance of nonedible oils as a promising feedstock for biodiesel production. Several studies have shown the immense potential of nonedible oil as a feedstock for biodiesel, and Jatropha is the current feedstock for biodiesel production in India (Kumar and Sharma, 2011; Taufiq-Yap et al., 2020). However, the challenges with nonedible oil as feedstock were found to be high viscosity, excess methanol requirement, and incapability to meet commercial demand. Hence, discovering a range of feedstocks that can overcome the above challenges could make a great impact on the commercialization of biodiesel at a global level (Ambat et al., 2018).

28.3 MORINGA OLEIFERA AS A PROMISING FEEDSTOCK FOR BIODIESEL PRODUCTION

Though we have more than 350 oil-bearing crops with the potency of biodiesel production, the right choice of feedstock plays a vital role in sustainable and economic biodiesel production with the best specifications. In this context, nonedible oils that can be cultivated in semiarid regions are more recommended as a suitable feedstock for large scale biodiesel production (da Silva et al., 2010). *Moringa oleifera* is a perennial tree belonging to the Moringaceae family that originated in north-east India. It is widely distributed in India, Africa, Asia, Arabia, South America, the Caribbean, and the Pacific Islands (da Silva et al., 2010; Franca et al., 2017). The common name of the tree varies with the region but most commonly recognized as 'drumstick tree' or 'horseradish tree' and the seed oil is popularly known as 'ben oil'. The height of the tree varies from 5 to 10 m. It can be grown even in hot drylands which constitutes about 41% of the global land area, and it is drought resistant. It can be grown in a rainfall ranging between 250 mm and 3,000 mm and a pH of 5.0 to 9.0. Anwar et al. (2007) reported the various nutritional and medicinal value of *Moringa oleifera* for which it is known as the natural nutrition of the tropics. In addition, the seed oil press cake contains various polypeptides that can be used as a natural coagulant in water purification, soil conditioning (Orhevba et al., 2013), and the cosmetic industry (Anwar et al., 2007). The seed oil is used for the extraction of various nutrients like vitamin A, after which it can be turned into biodiesel (Kafuku et al., 2010).

In addition, Franca et al. (2017) has evaluated the potential of *Moringa oleifera* leaf extract as an antioxidant additive for biodiesel. These multipurpose uses can bring a more economic insight to using it as a feedstock for biodiesel production. However, its high acid value may neutralize the catalysts during transesterification by forming soaps that reduce mass transfer, increase viscosity, and make separation difficult, while consuming the catalysts (da Silva et al., 2010; Demirbas, 2009b). Hence, it is essential to follow a two-step method of biodiesel production that includes esterification followed by transesterification.

28.4 IMPROVISATION OF A PROCESS ECONOMY BY ULTRASOUND

Esterification involves the reaction between 'free fatty acids' (FFAs) in oil and a short chain alcohol to form alkyl esters of fatty acid and water in the presence of a catalyst. The immiscibility of oil and alcohol requires extensive mechanical agitation and a high temperature to increase the mass transfer and hence the reaction rate. Ultrasonication makes use of ultrasound waves that form small bubbles in low pressure waves that grow in size and burst, producing high velocity and strong hydrodynamic shear forces in the liquid media. This phenomenon is termed 'acoustic cavitation' (Oliveira et al., 2018). The high velocity, temperature, and pressure produced during ultrasonication overcome the high mass transfer resistance encountered in conventional mechanical stirring (Ambat et al., 2018). This cutting edge technology can reduce the reaction time, amount of solvent and catalyst used, and is cost efficient with less power consumption when compared to conventional mechanical stirring (Borugadda and Goud, 2012; Santos et al., 2010). Joshi et al. (2018) performed a study on ultrasound assisted esterification of karanja oil and compared it with conventional mechanical stirring, wherein they reduced the acid value from 10.5 mg KOH/g to 2.88 mg KOH/g in 45 min. Additionally, the conventional mechanical stirring took 150 min to reduce the acid value to 4.2 mg KOH/g under the same conditions. In addition, the cost of treatment was estimated at 13.35 Rs/L for an ultrasonic flow cell and 14.58 Rs/L for conventional mechanical stirring-based esterification. This shows the efficiency of ultrasonication for commercially exploitable biodiesel production.

28.5 PHYSICOCHEMICAL PROPERTIES OF MORINGA OLEIFERA OIL

Da Silva et al. (2010) studied the characteristics of MOO and its biodiesel, wherein they obtained 39% oil (wt%) from seed kernels, with 19% saturated fatty acids (C16:0, C18: 0, C20:0, C22:0) and 81% unsaturated fatty acids (C16:1, C18:1, C18:2, C18:3, C20:1) with oleic acid (C18:1) dominating by 78%. Several authors have reported the fatty acid profile of MOO as shown in Table 28.1. The minimal fraction of 'polyunsaturated fatty acids' (PUFA) (1%) that is prone to oxidation gives it a higher oxidation stability which is significant for fuel storage for a longer time (da Silva et al., 2010; Franca et al., 2017). The differences in the fatty acid profiles

TABLE 28.1
Fatty Acid Profile of *Moringa oleifera* Oil

Fatty acids[1]	R1	R2	R3	R4
Palmitic C16: 0	6.5	7.6	7	7
Palmitoleic C16:1	—	1.4	2	1.7
Stearic C18: 0	6	5.5	4	5.9
Oleic C18: 1 cis	72.2	66.6	78	71.7
Linoleic C18: 2 cis	1.0	8.1	1	—
Linolenic C18:3 n3	—	0.2	—	—
Arachidic C20: 0	4	5.8	4	3.5
Eicosanoic C20: 1	2	1.7	—	3
Behenic C22:0	7.1	—	4	6
Lignoceric C24:0	—	—	—	1.1
Others	1	—	—	—
Saturated fatty acids	—[2]	—[2]	19	23.5
Monounsaturated fatty acids	—[2]	—[2]	80	76.4
PUFAs	—[2]	—[2]	1	0.1
Oil yield	—[2]	—[2]	39	35.43 ± 1.42%

[1] All values are reported in %;
[2] not specified.
Sources:　R1: Rashid et al. (2008); R2: Kafuku and Mbarawa (2010), Kafuku et al. (2010); R3: da Silva et al. (2010); R4: Dominguez et al. (2019).

may be attributed to biotic factors like genetic variation, as well as abiotic factors like environmental and geological conditions (Dominguez et al., 2019).

The FFA content and fatty acid composition determine the steps involved in biodiesel production and have a great impact on the physicochemical properties of biodiesel. Oil with high FFA (>2%) results in soap formation in direct transesterification that makes the separation process tedious and reduces the 'fatty acid methyl esters' (FAME) yield as the soap formation neutralizes the catalyst. Hence, esterification is performed prior to transesterification to convert FFAs into FAME. Fatty acid composition of the feedstock has a great impact on biodiesel properties such as CN, cloud point, oxidation stability, and pour point. Several authors have reported the physicochemical properties of the '*Moringa oleifera* methyl esters' (MOMEs) which are presented in Table 28.2. The dominance of monounsaturated fatty acid and saturated fatty acids confers a higher CN, and *Moringa oleifera* biodiesel is found to have the highest CN of 64–67. The high level of unsaturation decreases the calorific value that reduces the energy content; the dominance of saturated fatty acids also results in a high cloud point (–9 to 10°C), pour point (–8 to 11°C), and 'cold filter plugging point' (CFPP) (9 to 18°C). This is attributed to the precipitation of saturated fatty acids like stearic and palmitic acids at low temperatures which can clog the filters and result in incomplete combustion. However, biodiesel is generally found to have a higher flash point of about 150°C or more and is found to increase with the chain length of fatty acids. The flash point of MOO varies between 263 and 308°C.

TABLE 28.2
Physicochemical Properties of *Moringa oleifera* Oil

Property	R1	R2	R3	R4	R5	R6	R7	R8	R9	R10	R11	R12	R13	R14	R15
Density (kg/m³)	–	914.4	912[4]	–	897.1[6]	897.5	923.4[3] 906.3[6]	–	897.5	–	876.7	897.50	903.7	914.8[3], 911[4]	–
Kinematic viscosity (mm² s⁻¹)	–	65.81[9]	43.4[5]	–	43.46 & 9.0256[7]	43.33 & 8.91[7]	32.004 & 7.6569[7]	–	43.33	44.5	12.27	43.34	103[2,7,8]	–	–
Dynamic viscosity (MPa s) at 40°C	92.6[4,7]	–	–	–	38.99	38.90[2]	29.003	–	–	–	–	38.90	–	–	–
Specific gravity	0.907	–	–	0.97	–	–	0.9242[3]	–	–	–	–	–	–	–	–
Viscosity at 40°C (mm²s⁻¹)	–	–	–	–	–	–	–	–	–	–	–	–	0.91–1.1827	–	–
Viscosity index	–	–	–	–	195.20	193.1	222	–	–	–	–	–	–	–	–
Flash point (°C)	–	–	–	–	263	268.5	263.5	–	268.5	–	308	268.50	–	–	–
Cloud point (°C)	–	–	–	–	–	10	–[7]	–	10	–	–[9]	10	–	–	–
Pour point (°C)	–	–	–	–	–	11	–[7]	–	11	–	–[8]	11	–	–	–
CFPP (°C)	–	–	–	–	18	–	–	–	18	–	–	–	–	–	–
Cetane number	–	–	–	–	–	–	–	–	–	–	64	–	–	–	–
Iodine number at 38°C	–	–	73	70.50[2,10]	–	–	–	–	–	–	–	–	66–85.3[10]	–	62.25±0.93
Saponification value (mg KOH/g)	192.3	192.3	–	–	–	–	–	172.3	–	–	–	–	171.9–191	–	180.46±1.19
Heating value (MJ/kg)	–	–	–	–	39.76 41.75	38.05	39.868 41.75[2]	–	–	–	–	38.05	–	–	–
Oxidation stability (h at 110°C)	–	–	–	–	–	–	–	–	–	>60	42	–	–	–	–
Acid value (mg KOH/g)	1.194	2.2	4.0	0.97	8.62[8]	8.62	0.8670	3.2	8.62	13.2	0.012	8.62	3.8–5.04	9.04 (mg NaOH/g)	4.23±0.02
Peroxide index (meq O₂ kg⁻¹ oil)	–	–	–	1.37	–	–	–	–	–	10.4	–	–	8.1–15	–	1.57±0.21
Free fatty acid (%)	0.6	1.1	–	–	–	–	–	–	–	–	6.678	–	0.5–2.51	–	2.35±0.03

[1] Not reported; [2] temperature not specified; [3] at 15°C; [4] at 20°C; [5] at 38°C; [6] at 40°C; [7] at 100°C; [8] value reported in Atabani et al. (2013a); [9] reported as viscosity at 40°C (mm²/s); [10] in g I₂/100g oil.

Notes: CFPP: cold filter plugging point. R1: Kafuku and Mbarawa (2010); R2: Kafuku et al. (2010); R3: da Silva et al. (2010); R4: Rashid et al. (2011); R5: Atabani et al. (2013a–b); R6: Mofijur et al. (2014b); R7: Wakil et al. (2014); R8: Zubairu and Ibrahim (2014); R9: Mofijur et al. (2015); R10: Fernandes et al. (2015); R11: Eloka-Eboka and Inambao (2016); R12: Rashed et al. (2016b); R13: Mariod et al. (2017); R14: Boulal et al. (2019); R15: Niju et al. (2019b).

MMO is also reported to have a higher oxidation stability, varying between 41.75 and 60 h due to the presence of PUFAs like linoleic and linolenic acids with bis-allylic methylene carbons in trace amounts and a significant amount of behenic acids. However, the relative density of the fuel increases with the PUFAs, and fuels with high density have great potential energy which varies with temperature and pressure. The variation in the physicochemical properties of MMO is attributed to the methods and conditions during analysis, the fatty acid profile variation due to geographical conditions, the seed variety, and the extraction method.

28.6 BIODIESEL PRODUCTION FROM *MORINGA OLEIFERA* OIL

In past decades, many studies have been performed on evaluating the potential of MOO for biodiesel production, as shown in Table 28.3. Rashid et al. (2011) performed direct transesterification of MOO with an acid value of 0.97 and obtained a 94.30% yield of MOME with a 6.4: 1 methanol to oil molar ratio, 0.8% KOH in 71.08 min at 600 rpm and 50°C. Mofijur et al. (2014a) have reported a two-step biodiesel production from MOO with an acid value of 8.62. They obtained a >90% yield of MOME in 5 h with a 12:1 methanol to oil ratio, 1% (v/v) of concentrated H_2SO_4 in esterification and a 6:1 methanol to oil ratio, and 1% KOH in transesterification at 600 rpm and 60°C.

In another study, the authors reported biodiesel conversion from MOO with a higher acid value of 80.5, where the esterification process, with 1:2 (v/v) methanol to oil ratio and 1.5 % (v/v) concentrated H_2SO_4 in 2 h at 450 rpm and 60°C, reduced the acid value to 2.8. They also employed 8.02 wt% conch shells as a heterogeneous catalyst for transesterification with an 8.66:1 methanol to oil ratio in 2 h 10 min at 450 rpm and 65°C and obtained 97.06% of MOME (Niju et al., 2019a). Kafuku and Mbarawa (2010) optimized the direct transesterification of MOO with an acid value of 1.194 mg KOH/g and obtained an 82% MOME yield with a 30 wt% methanol to oil ratio and 1 wt% KOH at 60°C and 400 rpm in 1 h. From the overall observation, it was evident that the conventional mechanical stirring method takes more time for MOME conversion and consumes a high fraction of methanol. However, ultrasound can be used to reduce reaction time and the amount of solvent used; it has not been utilized for biodiesel production from MOO so far.

28.7 *MORINGA OLEIFERA* BIODIESEL: EVALUATION OF PROPERTIES, PERFORMANCE, AND EMISSION CHARACTERISTICS

Selection of a feedstock is not only important in terms of cost and commercial availability, but also to obtain biodiesel with the best specifications. Three parameters that greatly influence the physicochemical properties include the fatty acid profile, degree of unsaturation, and its molecular weight (Sajjadi et al., 2016). Several authors have reported the fatty acid profile and physicochemical properties of MOME, as shown in Tables 28.4 and 28.5. MOME is found to have high resistance to oxidative degradation. The high level of unsaturated fatty acids confers MOME a high oxidative stability, CN, and low iodine value, but also a high cloud point, pour point, and CFPP, which are undesirable. These undesirable factors also increase the viscosity

TABLE 28.3

Biodiesel Production from *Moringa oleifera* Oil

Process Condition	R1	R2	R3	R4	R5	R6	R7	R8	R9	R10	R11	R12	R13	R14	R15
Acid value	2.9	2.2	1.194	—	0.97	—	0.8670	8.62	0.22	8.62	—	13.2	—	80.5	4.23±0.02
Esterification:															
Methanol to oil ratio				6:1 MR		12:1 (50%v/v oil)		12:1 MR		12:1 (50%v/v oil)	12:1 (50%v/v oil)	10:1 MR		1:2 v/v	
Catalyst concentration				0.5% w/w oil		1% (v/v oil)		1% (v/v)		1% (v/v oil) KOH	1% (v/v oil)	1% (w/w)		1.5% (v/v oil)	
Time				20–30min		3h		3h		3h	3h	1h		2h	
Agitation speed (rpm)				—¹		400		600		600	600	—¹		450	
Temperature (°C)				50		60		60		60	60			60	
Transesterification:	6:1 molar	1:19.5	30 wt% methanol	6:1 MR	6.4:1 MR	6:1 MR	25% (v/v oil) methanol	6:1 MR	6:1 MR KOH	25% (v/v oil) methanol	6:1 MR	6:1	6:1 MR	8.66:1	6:1
Methanol to oil ratio															
Catalyst concentration	1 wt% NaOCH₃	3wt% sulfated tin oxide (SO₄²⁻SnO²⁻SiO₂)	1 wt% KOH	1.0 wt% KOH	0.80% KOH	1% (m/m oil) KOH	1% (m/m oil) KOH	1% (m/m oil) KOH	1% (m/m oil) KOH	1% (m/m oil) KOH	1% (m/m oil) KOH	1% (w/w) KOH	5wt% of CaO	8.02wt% (conch shells as heterogenous catalyst)	0.5% (w/w oil) of NaOH
Time	1h	150min	60 min	60 min	71.08 min	2h	2h	2h	2h	2h	2h	1h	1h	2h 10min	1h
Agitation speed (rpm)		350–360	400	400	600	400	400	600	600	600	600	—¹	—¹	450	500
Temperature (°C)	60	150	60	55–60	55	60	60	60	60	60	60	~ 25	50	65	60
Acid value	0.3914	0.012	0.012	0.16	0.38±0.03							0.21		2.8²	0.67±0.03
Yield		84 wt%	82%		94.30%			>90%³				96.8 wt%	94.2%	97.06%	91.5%

¹ Reaction temperature is not specified but mentioned as heated until boiling;
² corresponds to acid value of esterified MOO; ³ corresponds to value reported by Mofijur et al. (2014a) alone.

Notes: MR: molar ratio. R1: Rashid et al. (2008); R2: Kafuku et al. (2010); R3: Kafuku and Mbarawa (2010); R4: Kivevele et al. (2011); R5: Rashid et al. (2011); R6: Atabani et al. (2013b); R7: Wakil et al. (2014); R8: Rahman et al. (2014); R9: Mofijur et al. (2014a–b); R9: Mofijur et al. (2015); R10: Rashed et al. (2016b); R11: Rashed et al. (2016a); R12: Fernandes et al. (2015); R13: Aziz et al. (2016); R14: Niju et al. (2019a); R15: Dominguez et al. (2019).

TABLE 28.4

Fatty Acid Profile of *Moringa oleifera* Biodiesel

Fatty acid[1]	R1	R2	R3	R4	R5	R6	R7
Myristic C14:0	—	0.2	0.1	—	—	—[2]	—
Palmitic C16: 0	7.01	12.7	7.9	5.2 ± 0.1	6.01	—[2]	6.45
Palmitoleic C16:1	—	1.1	1.7	1.4 ± 0.1	1.36	—[2]	—
Margaric C17:1	—	—	—	—	0.11	—[2]	—
Stearic C18: 0	6.48	4.9	5.5	4.0 ± 0.1	5.46	—[2]	5.5
Oleic C18: 1 cis	74.16	73.0	74.1	81.6 ± 0.5	51.6	—[2]	73.22
Linoleic C18: 2 cis	0.63	2.5	4.1	—	—	—[2]	—
LinolenicC18:3 n3	0	0.8	0.2	—	—	—[2]	—
Arachidic C20: 0	4.38	3.8	2.3	2.1 ± 0.1	3.57	—[2]	4.08
Eicosanoic C20: 1	—	0.9	1.3	—	2.14	—[2]	—
Behenic C22:0	6.70	—	2.8	1.9 ± 0.1	6.70	—[2]	6.16
Lignoceric C24:0	—	—	—	3.8 ± 0.1	1.08	—[2]	—
Saturated fatty acids	25.20	21.6	18.6	17.0 ± 0.1	—[2]	18.6	—[2]
Monounsaturated fatty acids	74.16	75.0	77.1	83.0 ± 0.1	—[2]	81.4	—[2]
PUFAs	0.63	3.3	4.3	—	—[2]	—	—[2]

[1] All values are presented in %;

[2] values not reported

Notes: R1: Boulal et al. (2019); R2: Kivevele et al. (2011); R3: Rashed et al. (2016b), Mofijur et al. (2014a); R4: Fernandes et al. (2015); R5: Eloka-Eboka and Inambao (2016); R6: Rashed et al. (2016a); R7: Rashid et al. (2011).

and lead to plugging in diesel engines with poor starting and performance. Boulal et al. (2019) studied the ternary blend of diesel, bioethanol from dates seed, and biodiesel from *Moringa oleifera* seeds, and observed an appreciable decrease in cloud point, pour point, crystallization onset temperature, and CFPP with the increasing ratio of bioethanol.

Rashed et al. (2016a) carried out a comparative study on performance and emission characteristics of Jatropha (JB), palm (PB), and moringa biodiesel (MB). The authors have evaluated the 'brake power', 'brake specific fuel consumption' (BSFC), and CO, HC, and NO_X emissions as shown in Table 28.6. The BSFC of MB20 has been found to be 8.39% higher compared to that of mineral biodiesel, and the brake power has been found to decrease by 8.03% when compared to normal diesel. However, MB20 biodiesel can reduce CO and HC emissions by 22.93 and 11.84% and increase the NO_X emission by 18.56% when compared to normal diesel. This is mainly attributed by the higher CN, oxygen content, and level of saturation of the feedstock (Mofijur et al., 2014a; Rashed et al., 2016a).

PB is estimated to be superior to JB and MB in its performance and emission characteristics, but it may lead to food insecurity. MB is found to have close characteristics with that of JB which is a highly recommended and commercially exploited nonedible feedstock. A similar study showed the decrease in brake power to be 4.22% in MB10 and due to the high viscosity and low heating value of MOME. But there was an appreciable increase in BSFC, and a lesser one in CO and HC emissions (Mofijur et al., 2014a). Similar results were observed in another study on the comparative evaluation of emission and performance characteristics of PB and MB. The

TABLE 28.5
Physicochemical Properties of *Moringa oleifera* Biodiesel

Property	R1	R2	R3	R4	R5	R6	R7	R8	R9	R10	R11	R12	R13	R14	R15	R16	R17	R18	R19
Density at (Kg m⁻³)	—[1]	877.5[2]	883[3]	875±15.7	890[2]	0.8587[5,8]	869.6	859.6	885.8[2], 867.1[5]	—[1]	859.6	869.6	—	859.6	860.61	866.1[5]	881.9[3], 873.7[3], 870[4], 859.1[5]	865±2.4[2]	—[13]
Kinematic viscosity (mm² s⁻¹)	4.83[5]	4.91[5]	5.4[5]	4.8±0.04[5]	4.78[5]	5.0735[5], 1.9108[6]	5.05[5], 1.84[6]	5.05[5], 1.84[6]	5.05[5], 1.84[6]	4.85[5]	5.05[5]	5.05[5]	4.50[5]	5.05[5]	4.95[5]	4.03[5]	—	5.07±0.46[5]	1.9-6.0
Dynamic viscosity (mPas)						4.3618[5]	4.34[5]	4.34[5]	3.5781[5]						2.62				
Specific gravity at 15°C									0.8858										
Viscosity index						206.7			244.9	135.0	184.6			184.6	190				
Flash point		206		162±3.00	166	176	150.5	150.5	176.5	150.5	150.5	150.5		180.5	170.2	189			130 min
Cloud point	18, 10	10		18.00±0.12	10	21	19	19	18	18	19	19		9	15	0			report
Pour point	17	3		17.00±0.14	3	19	19	19	0	17.0	19	19		18	16	0			—[13]
CFPP				17.00±0.12		18	18	18			18	18		18	17				—[11]
Cetane number	67.07	62.12		67.0±1.52	63		56.3			56.30	56.30	56.30		59	59	54.3	58.36	59.42	47 min
Iodine number			74				77.5[9]			77.5[9]	77.5[9]			75.2	75.2		69.27[9]	68.74	—[13]
Heating value (MJ/kg)				45.28±0.98	38.34	40.115[10]	40.05	40.05	39.888	40.05	40.05	40.05		40.05	40.52	39.90	39.52		—[13]
Acid value (mg KOH/g)		0.012		0.38±0.03	0.16			0.22		0.26	0.22	0.22	0.21			0.24		0.67±0.03	0.50 max
Oxidation stability (h)	3.61			3.52±0.12	5.05[7]	12.64[7]	26.2[7]	26.27			26.2	19.37	19.37		3.05	6.5	7.21		3 min
Saponification number							199				199	199		199	199		197.08	190.9	—[13]
Lubricity (HFRR; μm)	135, 138.5																		—[12]
Sulfur content (%)				0.0124±0.001															0.05 max
Ash content				0.010±0.001															0.02 max
Methanol content (%)				0.165±0.002															—[13]
Free glycerin (%)				0.012±0.001	0.018														0.020 max

[1] Not reported; [2] at 15°C; [3] at 20°C; [4] at 25°C; [5] at 40°C; [6] at 100°C; [7] at 110°C; [8] in kg/cm³; [9] in gI/100g; [10] in kJ/kg; [11] not specified (EN 14214 uses time and location dependent values for the CFPP instead); [12] not specified (maximum wear scar value of 460 and 520 μm are prescribed in petrodiesel standards EN 580 and ASTM D975); [13] not specified.

Notes: R1: Rashid et al. (2008); R2: Kafuku and Mbarawa (2010), Kafuku et al. (2010); R3: da Silva (2010); R4: Rashid et al. (2011); R5: Kivevele et al. (2011); R6: Atabani et al. (2013b); R7: Mofijur et al. (2014b), Rahman et al. (2014); R8: Atabani et al. (2014); R9: Wakil et al. (2014); R10: Zubairu and Ibrahim (2014); R11: Mofijur et al. (2014a); R12: Mofijur et al. (2015); R13: Fernandes et al. (2015); R14: Rashed et al. (2016a); R15: Rashed et al. (2016b); R16: Teoh et al. (2019); R17: Boulal et al. (2019); R18: Rheology; R19: ASTM D6751.

TABLE 28.6
Performance and Emission Characteristics of *Moringa oleifera* Biodiesel

Biodiesel–Diesel Blend	CO Emission (%)[1]	HC Emission (%)[1]	NO$_X$ Emission (%)[2]	CO$_2$ Emission (%)[2]	BSFC (%)[2]	Brake Power (%)[1]	References
MB5	5.37	3.94	3.99	2.25	2.56	2.27	Mofijur et al. (2014)
PB5	13.17	14.47	1.96	5.60	0.69	1.38	
MB10	10.6	9.21	8.46	4.96	5.13	4.22	
PB10	17.36	18.42	3.38	11.73	2.02	3.16	
MB20	22.93	11.84	18.56	—	8.39	8.03	Rashed et al. (2016a)
JB20	27.23	19.73	14.22	—	7.15	8.75	
PB20	32.65	30.26	6.91	—	5.42	6.92	
MB20	28.41	27.64	15.47		7.7	8.04	Rashed et al. (2016b)
MB20 DPPD	17.87	19.07	8.7		6.2	6.9	
MB20 NPPD	15.98	10.04	12.82		6.68	6.4	

[1] Reduction with respect to mineral diesel; [2] increase with respect to mineral diesel.

decrease in brake power and increase in BSFC on the blending of biodiesel with diesel were attributed to high viscosity, density, and low calorific values of blends. Overall, it is evident that blending biodiesel with normal diesel decreases the CO and HC emissions but decreases brake power and increases BSFC on the other hand (Rashed et al., 2016a). It has also been found that the addition of amine antioxidants like 'N, N$_0$-diphenyl-1,4-phenylenediamine' (DPPD) and 'N-phenyl-1,4-phenylenediamine' (NPPD) increased oxidation stability with lower exhaust emissions and increased brake power and brake thermal efficiency (Rashed et al., 2016b).

28.8 CASE STUDY: ULTRASOUND ASSISTED ESTERIFICATION OF *MORINGA OLEIFERA* OIL

28.8.1 MATERIALS AND METHODS

MOO was purchased from a retail supplier, Tamil Traders, Coimbatore, Tamil Nadu, India. The oil was filtered in order to remove the impurities which may increase mass transfer resistance in esterification. Then the moisture was removed by keeping it in a hot air oven at 105°C until it reached a constant weight; this step is essential to avoid the hydrolysis of triglycerides into FFAs by water. Analytical grade methanol (CH$_3$OH, 99.8%), potassium hydroxide (KOH), a sulfuric acid (H$_2$SO$_4$) phenolphthalein indicator, petroleum ether, and ethanol (C$_2$H$_5$OH, 99.9%) were purchased from HiMedia Laboratories Pvt. Ltd., Mumbai.

28.8.2 ESTERIFICATION OF MOO

The acid value of MOO was estimated as 81.5 mg KOH/g of oil which accounts for a high FFA content of 40.75%. However, the acid value of oil for direct

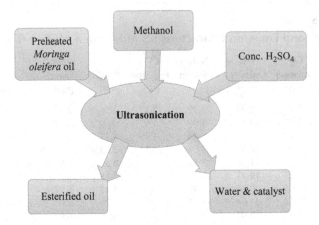

FIGURE 28.1 Schematic representation of esterification.

transesterification has to be less than 5 mg KOH/g as per ASTM and EN standards. Hence, the oil is unfit for direct transesterification and needs to be esterified in a prior step. The schematic representation of esterification is presented in Figure 28.1. In this acid catalyzed esterification, a volumetric ratio of 0.6:1 methanol to MOO and a 1.5 vol% of concentrated H_2SO_4 was added to preheated oil at 60°C for 40 min with ultrasound mediated agitation. After the termination of reaction, excess methanol if any was removed and transferred to a separating funnel for phase separation. Water, the by-product of esterification and catalyst forms the bottom phase which was drained out; the top phase which is an esterified MOO was stored in an airtight container. Further, the acid value was determined using the DGF (C-V-2) method and calculated using Equation 1 (Bockisch, 1998).

$$\text{Acid value} = \frac{V * N * 56.1}{W} \tag{28.1}$$

where N is the normality of KOH, W is the weight of the oil sample, and V is the end point in a burette solution.

The esterification reaction was carried out in a 250 ml batch reactor with a volumetric ratio of 0.6:1 methanol to MOO and a 1.5 vol% of concentrated H_2SO_4 for 40 min at 60°C in the presence of ultrasound at 60Hz, as shown in Figure 28.2. Excess methanol was removed and transferred to a separating funnel for phase separation, as shown in Figure 28.3. The acid value of the esterified MOO was found to be reduced from 81.5 mg KOH/g of oil to 2.78 mg KOH/g of oil. Over the past decades, several tests were done on the esterification of MOO with conventional mechanical stirring. Rashed et al. (2016a–b) have reported the esterification of MOO with a 12:1 methanol to crude MOO (50% v/v oil) ratio and 1% (v/v) of sulfuric acid (H_2SO_4) at 60°C for 3 h with a stirring rate of 600 rpm. Similarly, Niju et al. (2019a) have employed a two-step method of producing biodiesel from MOO wherein they reduced the acid value from 80.5 mg KOH/g of oil to 2.8 mg KOH/g of oil with a volumetric ratio of 1:2 methanol to oil and a 1.5 vol% H_2SO_4 concentration at 60°C in 2 h.

FIGURE 28.2 Experimental setup of ultrasound assisted esterification.

FIGURE 28.3 Phase separation of esterified *Moringa oleifera* oil.

Several authors have performed conventional esterification with a 12:1 molar ratio of methanol to MOO and 1% (v/v) of concentrated H_2SO_4 at 60°C for 3 h (Mofijur et al., 2014a–b; Atabani et al., 2013a–b; Rahman et al., 2014). However, the present study of acid esterification with ultrasonication shows an appreciable reduction in reaction time and methanol concentration, making the overall process economically favorable with less time as well as less energy consumption.

28.9 CONCLUSION

Based on the literature review and the experimental investigation, the following conclusions are drawn.

1. Nonedible oil can be a potential feedstock for economical and sustainable biodiesel production to meet the global demand.
2. *Moringa oleifera*, with a high CN, appreciable oxidation stability, and low iodine value with wide availability, can potentially be utilized for biodiesel production. Though the pour point, cloud point, and CFPP are higher, it is possible to reduce it appreciably with a ternary blend of diesel, bioethanol, and biodiesel.
3. Ultrasonication-mediated esterification has been experimentally proven to be a cutting-edge technology that reduced the acid value from 81.5 mg KOH/g of oil to 2.78 mg KOH/g of oil with a volumetric ratio of 0.6:1 methanol to MOO and 1.5 vol% of concentrated H_2SO_4 for 40 min at 60°C.

ACKNOWLEDGMENT

This work was funded by SERB-DST, New Delhi, Govt. of India (ECR/2015/000036).

REFERENCES

Ambat, I., V. Srivastava, and M. Sillanpaa. 2018. Recent advancement in biodiesel production methodologies using various feedstock: A review. *Renewable and Sustainable Energy Reviews* 90:356–369.

Anwar, F., S. Latif, M. Ashraf, and A. H. Gilani. 2007. *Moringa oleifera*: A food plant with multiple medicinal uses. *Phytotherapy Research* 21: 17–25.

Atabani, A. E., T. M. I. Mahlia, and I. A. Badruddin, et al. 2013a. Investigation of physical and chemical properties of potential edible and non-edible feedstocks for biodiesel production, a comparative analysis. *Renewable and Sustainable Energy Reviews* 21: 749–755.

Atabani, A. E., T. M. I. Mahlia, and H. H. Masjuki, et al. 2013b. A comparative evaluation of physical and chemical properties of biodiesel synthesized from edible and non-edible oils and study on the effect of biodiesel blending. *Energy* 58:296–304.

Aziz, M. A. A., S. Triwahyono, A. A. Jalil, H. A. A. Rapai, and A. E. Atabani. 2016. Transesterification of *Moringa Oleifera* oil to biodiesel over potassium flouride loaded on eggshell as catalyst *Malaysian Journal of Catalysis* 1: 22–26.

Bockisch, M. 1998. Vegetable fats and oils. In *Fats and Oils Handbook,* 803–808. London: Academic Press.

Borugadda, V. B. and V. V. Goud. 2012. Biodiesel production from renewable feedstock: Status and opportunities. *Renewable and Sustainable Energy Reviews* 16: 4763–4784.

Boulal, A., A. E. Atabani, and M. N. Mohammed, et al. 2019. Integrated valorization of *Moringa oleifera* and waste *Phoenix dactylifera* L. dates as potential feedstocks for biofuels production from Algerian Sahara: An experimental perspective. *Biocatalysis and Agricultural Biotechnology* 20: 101234.

BP. 2019. *Statistical Review of World Energy.* London: British Petroleum.

Bradley, A. 2019. *India: Biofuels Annual.* Washington, DC: Department of Agriculture.

da Silva, J. P. V., T. M. Serra, and M. Gossmann, et al. 2010. *Moringa oleifera* oil: Studies of characterization and biodiesel production. *Biomass and Bioenergy* 34: 1527–1530.

Demirbas, A. 2009a. Potential resources of non-edible oils for biodiesel. *Energy Sources, Part B: Economics, Planning, and Policy* 4: 310–314.

Demirbas, A. 2009b. Production of biodiesel fuels from linseed oil using methanol and ethanol in non-catalytic SCF conditions. *Biomass and bioenergy* 33: 113–118.

Demirbas, A., and S. Karslioglu. 2007. Biodiesel production facilities from vegetable oils and animal fats. *Energy Sources, Part A: Recovery, Utilization, and Environmental Effects* 29: 133–141.

Dominguez, Y. D., D. T. Garcia, and L. G. Perez, et al. 2019. Rheological behavior and properties of biodiesel and vegetable oil from *Moringa oleifera* Lam. *Afinidad* 76: 587.

Eloka-Eboka, A. C., and F. L. Inambao. 2016. Hybridization of feedstocks-A new approach in biodiesel development: A case of Moringa and Jatropha seed oils. *Energy Sources, Part A: Recovery, Utilization, and Environmental Effects* 38:1495–1502.

Fernandes, D. M., R. M. F. Sousa, and A. de Oliveira, et al. 2015. Mor*inga oleifera*: A potential source for production of biodiesel and antioxidant additives. *Fuel* 146: 75–80.

Franca, F. R. M., L. dos Santos Freitas, A. L. D. Ramos, G. F. da Silva, and S. T. Brandao. 2017. Storage and oxidation stability of commercial biodiesel using *Moringa oleifera* Lam as an antioxidant additive. *Fuel* 203: 627–632.

Joshi, S. M., P. R. Gogate, and S. S. Kumar. 2018. Intensification of esterification of karanja oil for production of biodiesel using ultrasound assisted approach with optimization using response surface methodology. *Chemical Engineering and Processing-Process Intensification* 124: 186–198.

Kafuku, G. and M. Mbarawa. 2010. Alkaline catalyzed biodiesel production from *Moringa oleifera* oil with optimized production parameters. *Applied Energy* 87: 2561–2565.

Kafuku, G., M. K. Lam, J. Kansedo, K. T. Lee, and M. Mbarawa. 2010. Heterogeneous catalyzed biodiesel production from *Moringa oleifera* oil. *Fuel Processing Technology* 91: 1525–1529.

Kivevele, T. T., M. M. Mbarawa, A. Bereczky, and M. Zoldy. 2011. Evaluation of the oxidation stability of biodiesel produced from *Moringa oleifera* oil. *Energy & Fuels* 25: 5416–5421.

Kumar, A. and S. Sharma. 2011. Potential non-edible oil resources as biodiesel feedstock: An Indian perspective. *Renewable and Sustainable Energy Reviews* 15: 1791–1800.

Mariod, A. A., M. E. S. Mirghani, and I. H. Hussein. (eds.) 2017. *Unconventional Oilseeds and Oil Sources*. London: Academic Press.

Mofijur, M., H. H. Masjuki, and M. A. Kalam, et al. 2014a. Properties and use of *Moringa oleifera* biodiesel and diesel fuel blends in a multi-cylinder diesel engine. *Energy Conversion and Management* 82:169–176.

Mofijur, M., H. H. Masjuki, and M. A. Kalam, et al. 2014b. Comparative evaluation of performance and emission characteristics of *Moringa oleifera* and palm oil based biodiesel in a diesel engine. *Industrial crops and products* 53: 78–84.

Mofijur, M., H. H. Masjuki, and M. A. Kalam, et al. 2015. Effect of biodiesel-diesel blending on physico-chemical properties of biodiesel produced from *Moringa oleifera*. *Procedia Engineering* 105: 665–669.

Musa, I. A. 2016. The effects of alcohol to oil molar ratios and the type of alcohol on biodiesel production using transesterification process. *Egyptian Journal of Petroleum* 25: 21–31.

Niju, S., C. Anushya, and M. Balajii. 2019a. Process optimization for biodiesel production from *Moringa oleifera* oil using conch shells as heterogeneous catalyst. *Environmental Progress & Sustainable Energy* 38: e13015.

Niju, S., M. Balajii, and C. Anushya. 2019b. A comprehensive review on biodiesel production using Moringa oleifera oil. *International Journal of Green Energy* 16: 702–715.

OECD. 2011. *OECD Green Growth Studies: Energy*. Paris: Organization for Economic Co-Operation & Development.

Oliveira, P. A., R. M. Baesso, G. C. Moraes, A. V. Alvarenga, and R. P. B. Costa-Felix. 2018. Ultrasound methods for biodiesel production and analysis. In *Biofuels: State of Development*, ed. K. Biernat, 121. Rijeka: Intech.

Orhevba, B. A., M. O. Sunmonu, and H. I. Iwunze. 2013. Extraction and characterization of *Moringa oleifera* seed oil. *Journal of Food and Dairy Technology* 1:22–27.

Pathak, G., D. Das, K. Rajkumari, and L. Rokhum. 2018. Exploiting waste: Towards a sustainable production of biodiesel using *Musa acuminata* peel ash as a heterogeneous catalyst. *Green Chemistry* 20: 2365–2373.

Pinzi, S., D. Leiva, I. Lopez-Garcia, M. D. Redel-Macias, and M. Pilar Dorado. 2014. Latest trends in feedstocks for biodiesel production. *Biofuels, Bioproducts and Biorefining* 8:126–143.

Rahman, M. M., M. H. Hassan, and M. A. Kalam, et al. 2014. Performance and emission analysis of *Jatropha curcas* and *Moringa oleifera* methyl ester fuel blends in a multi-cylinder diesel engine. *Journal of Cleaner Production* 65: 304–310.

Rashed, M. M., M. A. Kalam, and H. H. Masjuki, et al. 2016a. Performance and emission characteristics of a diesel engine fueled with palm, jatropha, and moringa oil methyl ester. *Industrial Crops and Products* 79: 70–76.

Rashed, M. M., H. H. Masjuki, M. A. Kalam, et al. 2016b. Study of the oxidation stability and exhaust emission analysis of *Moringa olifera* biodiesel in a multi-cylinder diesel engine with aromatic amine antioxidants. *Renewable Energy* 94: 294–303.

Rashid, U., F. Anwar, B. R. Moser, and G. Knothe. 2008. *Moringa oleifera* oil: A possible source of biodiesel. *Bioresource Technology* 99: 8175–8179.

Rashid, U., F. Anwar, M. Ashraf, M. Saleem, and S. Yusup. 2011. Application of response surface methodology for optimizing transesterification of *Moringa oleifera* oil: Biodiesel production. *Energy Conversion and Management* 52: 3034–3042.

Sajjadi, B., A. A. A. Raman, and H. Arandiyan. 2016. A comprehensive review on properties of edible and non-edible vegetable oil-based biodiesel: Composition, specifications and prediction models. *Renewable and Sustainable Energy Reviews* 63: 62–92.

Santos, F. F. P., J. Q. Malveira, M. G. A. Cruz, and F. A. N. Fernandes. 2010. Production of biodiesel by ultrasound assisted esterification of *Oreochromis niloticus* oil. *Fuel* 89: 275–279.

Sharma, M., A. A. Khan, S. K. Puri, and D. K. Tuli. 2012. Wood ash as a potential heterogeneous catalyst for biodiesel synthesis. *Biomass and Bioenergy* 41: 94–106.

Taufiq-Yap, Y. H., M. S. A. Farabi, O. N. Syazwani, M. L. Ibrahim, and T. S. Marliza. 2020. Sustainable production of bioenergy. In *Innovations in Sustainable Energy and Cleaner Environment*, ed. A. K. Gupta, A. De, S. K. Agarwal, A. Kushari, and A. K. Kuchari, 541–561. Singapore: Springer.

Teoh, Y. H., H. G. How, and H. H. Masjuki, et al. 2019. Investigation on particulate emissions and combustion characteristics of a common-rail diesel engine fueled with *Moringa oleifera* biodiesel-diesel blends. *Renewable Energy* 136: 521–534.

Wakil, M. A., M. A. Kalam, H. H. Masjuki, I. M. R. Fattah, and B. M. Masum. 2014. Evaluation of rice bran, sesame and moringa oils as feasible sources of biodiesel and the effect of blending on their physicochemical properties. *RSC Advances* 4: 56984–56991.

Zubairu, A., and F. S. Ibrahim. 2014. *Moringa Oleifera* oilseed as viable feedstock for biodiesel production in Northern Nigeria. *International Journal of Energy Engineering* 4: 21–25.

Part VII

Waste Oil-based Biodiesel Fuels

29 Waste Oil-based Biodiesel Fuels

A Scientometric Review of the Research

Ozcan Konur

CONTENTS

29.1 INTRODUCTION

Crude oils have been primary sources of energy and fuels, such as petrodiesel (Busca et al., 1998; Khalili et al., 1995; Rogge et al., 1993 Schauer et al., 1999). However, significant public concerns about the sustainability, price fluctuations, and adverse environmental impact of crude oils have emerged since the 1970s (Ahmadun et al., 2009; Atlas, 1981; Babich and Moulijn, 2003; Kilian, 2009; Perron, 1989). Thus, biooils have emerged as an alternative to crude oils in recent decades (Bridgwater and Peacocke, 2000; Czernik and Bridgwater, 2004; Gallezot, 2012; Mohan et al., 2006). In this context, WOBD fuels (Canakci and van Gerpen, 2001; Dorado et al., 2003; Ghobadian et al., 2009; Graboski and McCormick, 1998; Kulkarni and Dalai, 2006; Lam et al., 2010; Ozsezen et al., 2009; Usta et al., 2005; Zhang et al., 2003a–b) have also emerged as a viable alternative to crude oil-based petrodiesel fuels (Busca et al., 1998; Khalili et al., 1995; Rogge et al., 1993; Schauer et al., 1999). Nowadays, both petrodiesel and biodiesel fuels are being used at a global scale (Konur, 2021a–ag).

However, for the efficient progression of the research in this field, it is necessary to develop efficient incentive structures for the primary stakeholders and to inform these stakeholders about the research (Konur, 2000, 2002a–c, 2006a–b, 2007a–b; North, 1991a–b).

Scientometric analysis of the research offers ways to evaluate the research in a respective field (Garfield, 1955, 1972; (Konur, 2011, 2012a–n, 2015, 2016a–f, 2017a–f, 2018a–b, 2019a–b). However, there has been no current scientometric study of this field.

This chapter presents a study on the scientometric evaluation of the research in this field using two datasets. The first dataset includes the 100-most-cited papers (n = 100 sample papers) whilst the second set includes population papers (n = over 2,150 population papers) published between 1980 and 2019. This complements the chapters on crude oil-based petrodiesel fuels and other biooil-based biodiesel fuels.

The data on the indices, document types, authors, institutions, funding bodies, source titles, 'Web of Science' subject categories, keywords, research fronts, and citation impact are presented and discussed.

29.2   MATERIALS AND METHODOLOGY

The search for the literature was carried out in the 'Web of Science' (WOS) database in January 2020. It contains the 'Science Citation Index-Expanded' (SCI-E), the Social Sciences Citation Index' (SSCI), the 'Book Citation Index-Science' (BCI-S), the 'Conference Proceedings Citation Index-Science' (CPCI-S), the 'Emerging Sources Citation Index' (ESCI), the 'Book Citation Index-Social Sciences and Humanities' (BCI-SSH), the 'Conference Proceedings Citation Index-Social Sciences and Humanities' (CPCI-SSH), and the 'Arts and Humanities Citation Index' (A&HCI).

The keywords for the search of the literature are collated from the screening of abstract pages for the first 500 highly cited papers. This keyword set is provided in the Appendix.

Two datasets are used for this study. The highly cited 100 papers comprise the first dataset (n = 100 papers) whilst all the papers form the second dataset (population data set, n = over 2,150 papers).

The data on the indices, document types, publication years, institutions, funding bodies, source titles, countries, 'Web of Science' subject categories, citation impact, keywords, and research fronts are collated from these datasets. The key findings are provided in the relevant tables and figure, supplemented with explanatory notes in

the text. The findings are discussed and a number of conclusions are drawn and a number of recommendations for further study are made.

29.3 RESULTS

29.3.1 INDICES AND DOCUMENTS

There are over 2,600 papers in this field in the 'Web of Science' as of January 2020. This original population dataset is refined for the document type (article, review, book chapter, book, editorial material, note, and letter) and language (English), resulting in over 2,150 papers comprising over 80.7% of the original population dataset.

The primary index is SCI-E for both the sample and population papers. About 92.8% of the population papers are indexed by the SCI-E database. Additionally 3.3, 6.1, and 0.9% of these papers are indexed by CPCI-S, ESCI, and BCI-S databases, respectively. The papers on the social and humanitarian aspects of this field are relatively negligible with 0.8 and 0.0% of the population papers indexed by the SSCI and A&HCI, respectively.

Brief information on the document types for both datasets is provided in Table 29.1. The key finding is that article types of documents are the primary documents for both sample and population papers, whilst reviews form 12% of the sample papers.

29.3.2 AUTHORS

Brief information about the most-prolific 13 authors with at least three sample papers each is provided in Table 29.2. Around 290 and 5,700 authors contribute to the sample and population papers, respectively.

The most-prolific author is 'Mustafa Canakci' with six sample papers working primarily on 'biodiesel fuel production' and 'biodiesel fuel properties' using fats and waste oils. All the other top authors have three sample papers each.

On the other hand, a number of authors have a significant presence in the population papers: 'Mohand Tazerout', 'Meisam Tabatabaei', Eilhann E. Kwon',

TABLE 29.1
Document Types

	Document Type	Sample Dataset (%)	Population Dataset (%)	Difference (%)
1	Article	87	95.8	−8.8
2	Review	12	3.5	8.5
3	Book chapter	0	0.9	−0.9
4	Proceeding paper	4	3.3	0.7
5	Editorial matter	0	0.2	−0.2
6	Letter	0	0.2	−0.2
7	Book	0	0.0	0
8	Note	1	0.3	0.7

TABLE 29.2
Authors

	Authors	Sample Papers (%)	Population Papers (%)	Surplus (%)	Institution	Country	Waste Oils	Research Front
1	Canakci, Mustafa	6	0.7	5.3	Kocaeli Univ.	Turkey	Fats, waste oils	Production, properties
2	Al-Widyan, Mohamad I.	3	0.1	2.9	Jordan Univ.	Jordan	Fats	Production
3	Can, Ozer	3	0.1	2.9	Pamukkale Univ.	Turkey	Fats, waste oils	Production, properties
4	Clements, L. Davis*	3	0.1	2.9	Univ. Nebraska	USA	Fats	Production
5	Dalai, Ajay K.	3	0.3	2.7	Univ. Saskatchewan	Canada	Waste oils	Production
6	Dube, Marc A.	3	0.2	2.8	Univ. Ottawa	Canada	Waste oils	Production
7	Haas, Michael J.	3	0.2	2.8	US Dept. Agric.	USA	Waste oils	Production, properties
8	Hanna, Milford A.	3	0.3	2.7	Univ. Nebraska	USA	Fats	Production
9	Kates, Morris	3	0.1	2.9	Univ. Ottawa	Canada	Waste oils	Production
10	McLean, David D.	3	0.1	2.9	Univ. Ottawa	Canada	Waste oils	Production
11	Mittelbach, Martin	3	0.4	2.6	Karl Franzes Univ. Graz	Austria	Waste oils, sludge	Production, properties
12	Orsezen, Ahmet Necati	3	0.3	2.7	Kocaeli Univ.	Turkey	Waste oils	Properties
13	Wang, De-Zheng	3	0.2	2.8	Tsinghua Univ.	China	Waste oils	Production

*'Highly Cited Researchers' in 2019 (Clarivate Analytics, 2019).

'Sivalingam Murugan', 'Stella Bezergianni', 'Yun-Hin Taufiq-Yap', 'Jose Maria Arandes', 'Ayhan Demirbas', 'G. Nagarajan', 'Umer Rashid', 'Javier Bilbao', 'Pedro Castano', 'Shui-Jen Chen', 'Barat Ghobadian', 'Kuo-Lin Huang', 'Jisoo Lee', 'Chih-Chung Lin', and 'Jen-Hsiung Tsai' have at least 0.5% of the population papers each.

The most-prolific institution for these top authors is the 'University of Ottawa' of Canada and the 'Indian Institute of Technology' of India with three authors. The other prolific institutions with two authors are 'Kocaeli University' of Turkey and the 'University of Nebraska' of the USA. Thus, in total, nine institutions house these top authors.

It is notable that only one of these top researchers are listed in the 'Highly Cited Researchers' (HCR) in 2019 (Clarivate Analytics, 2019; Docampo and Cram, 2019).

The most-prolific country for these top authors is Canada with four. The other prolific countries with three authors each are Turkey and the USA. Thus, in total, six countries contribute to these top papers.

There are two key topical research fronts for these top researchers: 'biodiesel fuel production' and 'biodiesel fuel properties' with 12 and 5 authors, respectively. On the other hand, these top authors focus on 'waste oils' and 'animal fats' with 10 and 5 authors, respectively.

It is further notable that there is a significant gender deficit among these top authors as all of them are male (Lariviere et al., 2013; Xie and Shauman, 1998).

The author with the most impact is 'Mustafa Canakci' with a 5.3% publication surplus. On the other hand, the author with the least impact is 'Martin Mittelbach' with a 2.6% publication surplus.

29.3.3 PUBLICATION YEARS

Information about the publication years for both datasets is provided in Figure 29.1. This figure shows that 2, 6, 71, and 21% of the sample papers and 1.0, 2.6, 13.5, and 81% of the population papers were published in the 1980s, 1990s, 2000s, and 2010s, respectively.

Similarly, the most-prolific publication years for the sample dataset are 2008, 2009, 2007, and 2010 with 19, 14, 10, and 8 papers, respectively. On the other hand, the most-prolific publication years for the population dataset are 2019, 2018, 2016, and 2017 with 13.4, 11.3, 11.2, and 9.9% of the population papers, respectively. It is notable that there is a sharply rising trend for the population papers in the 2000s and more strikingly in the 2010s.

29.3.4 INSTITUTIONS

Brief information on the top ten institutions with at least 3% of the sample papers each is provided in Table 29.3. In total, around 120 and 1,700 institutions contribute to the sample and population papers, respectively.

These top institutions publish 34 and 6% of the sample and population papers, respectively. The top institution is 'Kocaeli University' of Turkey with six sample papers and a 5.2% publication surplus. This top institution is followed by the 'University of Nebraska' of the USA with four sample papers and a 3.6% publication surplus.

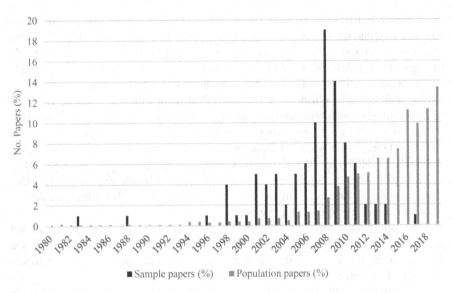

FIGURE 29.1   Research output between 1980 and 2019.

TABLE 29.3
Institutions

	Institution	Country	No. of Sample Papers (%)	No. of Population Papers (%)	Difference (%)
1	Kocaeli Univ.	Turkey	6	0.8	5.2
2	Univ. Nebraska	USA	4	0.4	3.6
3	Iowa State Univ.	USA	3	0.2	2.8
4	Jordan Univ. Sci. Technol.	Jordan	3	0.2	2.8
5	Tsinghua Univ.	China	3	0.4	2.6
6	Dept. Agric.	USA	3	1.0	2
7	Univ. Sains Malaysia	Malaysia	3	1.7	1.3
8	Univ. Graz	Austria	3	0.4	2.6
9	Univ. Ottawa	Canada	3	0.5	2.5
10	Univ. Saskatchewan	Canada	3	0.4	2.6

The most-prolific countries for these top institutions are the USA and Canada with three and two sample papers, respectively. The other countries are Austria, China, Jordan, Malaysia, and Turkey. In total, seven countries house these institutions.

The institutions with the most impact are 'Kocaeli University' and the 'University of Nebraska' with 5.2 and 3.6% publication surpluses, respectively. On the other hand, the institutions with the least impact are the 'University of Sains Malaysia' and the US 'Department of Agriculture' with 1.3 and 2.0% publication surpluses, respectively.

It is notable that some institutions have a heavy presence in the population papers: the 'Indian Institute of Technology', the 'University of Malaya', the 'University of

Putra Malaysia', 'Anna University', the 'Council of Scientific Industrial Research', the ' Chinese Academy of Sciences', the 'National University of Singapore', the 'Technology University of Malaysia', 'Castilla-La Mancha University', the 'Federal University of Rio de Janeiro', the 'Petronas University of Technology', the 'University of Basque Country', and the 'University of Tehran' have at least a 0.9% presence in the population papers each.

29.3.5 FUNDING BODIES

Brief information about the top 11 funding bodies with at least 2% of the sample papers each is provided in Table 29.4. It is significant that only 44.0 and 46.7% of the sample and population papers declare any funding, respectively. Around 40 and 1,300 funding bodies fund the research for the sample and population papers, respectively.

The top funding bodies are the 'National Natural Science Foundation of China' and the 'Scientific and Technological research Council' of Turkey, funding 4% of the sample papers each. These top funding bodies are followed by 'Kocaeli University' of Turkey, the 'Ministry of Energy Science Technology Environment and Climate Change' of Malaysia, and the US 'Department of Energy' with three sample papers each.

TABLE 29.4
Funding Bodies

	Institution	Country	No. of Sample Papers (%)	No. of Population Papers (%)	Difference (%)
1	National Natural Science Foundation of China	China	4	5.2	−1.2
2	Scientific and Technological research Council of Turkey	Turkey	4	0.6	3.4
3	Kocaeli University	Turkey	3	0.5	2.5
4	Ministry of Energy Science Technology Environment and Climate Change	Malaysia	3	0.9	2.1
5	Department of Energy	USA	3	1.0	2.0
6	Ministry of Higher Education	Malaysia	2	0.2	1.8
7	Ministry of Science and Innovation	Spain	2	0.8	1.2
8	National Basic Research Program of China	China	2	0.5	1.5
9	Specialized Research Fund for the Doctoral Program of Higher Education	China	2	0.2	1.8
10	University of Sains Malaysia	Malaysia	2	0.7	1.3
11	Ministry of Science and Environmental Protection	Serbia	2	0.2	1.8

The most-prolific countries for these funding bodies are Malaysia and China with three funding bodies respectively. These countries are followed by Turkey with two funding bodies. In total, these funding bodies are from six countries only.

It is notable that some top funding agencies have a heavy presence in the population studies. Some of them are the 'National Council for Scientific and Technological Development', 'CAPES', and 'Science Technology and Innovation' of Brazil, the 'European Union', the 'Natural Sciences and Engineering Research Council of Canada', the 'National Science Council of Taiwan', the 'Ministry of Science and Technology', the 'Department of Science Technology', the 'University Grants Commission', and the 'Council of Scientific Industrial Research' of India, and the 'Ministry of Education and Science' of Spain with at least 0.8% of the population papers each. These funding bodies are from Brazil, Europe, India, and Spain.

The funding bodies with the most impact are the 'Scientific and Technological Research Council' and 'Kocaeli University' of Turkey, and the 'Ministry of Energy Science Technology Environment and Climate Change' of Malaysia with 3.4, 3.1, and 2.5% publication surpluses, respectively. On the other hand, the funding bodies with the least impact are the 'National Natural Science Foundation' of China, the 'Ministry of Science and Innovation' of Spain, and the 'University of Sains Malaysia' with −1.2, 1.2, and 1.3% publication surpluses, respectively.

29.3.6 Source Titles

Brief information about the top 15 source titles with at least two sample papers each is provided in Table 29.5. In total, 32 and over 500 source titles publish the sample and population papers, respectively. On the other hand, these top 11 journals publish 83 and 35.9% of the sample and population papers, respectively.

The top journal is 'Bioresource Technology', publishing 14 sample papers with a 9.6% publication surplus. This top journal is followed by 'Fuel', 'Fuel Processing Technology', and 'Energy Conversion and Management' with 12, 9, and 8 sample papers, respectively.

Although these journals are indexed by 11 subject categories, the top category is 'Energy Fuels' with ten journals. The other prolific subject categories are 'Engineering Chemical', 'Agricultural Engineering', 'Biotechnology Applied Microbiology', 'Thermodynamics', 'Green Sustainable Science Technology', 'Chemistry Applied', and 'Environmental Sciences' with at least two journals each.

The journals with the most impact are 'Bioresource Technology', 'Fuel Processing Technology', and 'Energy Conversion and Management' with 9.6, 6.3, and 4.5% publication surpluses, respectively. On the other hand, the journals with the least impact are 'Energy Fuels', 'Renewable Energy', and 'Energy' with −1.4, 1.0, and 1.0% publication surpluses, respectively.

It is notable that some journals have a heavy presence in the population papers. Some of them are 'Energy Sources Part A Recovery Utilization and Environmental Effects', 'Journal of Cleaner Production', 'Waste Management', 'Waste and Biomass Valorization', 'Energies', 'International Journal of Green Energy', 'International Journal of Ambient Energy', 'RSC Advances', 'Korean Journal of Chemical Engineering', 'Applied Catalysis B Environmental', 'Environmental Progress

TABLE 29.5
Source Titles

	Source Title	WOS Subject Category	No. of Sample Papers (%)	No. of Population Papers (%)	Difference (%)
1	Bioresource Technology	Agr. Eng., Biot. Appl. Microb., Ener. Fuels	14	4.4	9.6
2	Fuel	Ener. Fuels, Eng. Chem.	12	7.7	4.3
3	Fuel Processing Technology	Chem. Appl., Ener. Fuels, Eng. Chem.	9	2.7	6.3
4	Energy Conversion and Management	Therm., Ener. Fuels, Mechs.	8	3.5	4.5
5	Renewable Energy	Green Sust. Sci. Technol., Ener. Fuels	5	4.0	1.0
6	Renewable Sustainable Energy Reviews	Green Sust. Sci. Technol., Ener. Fuels	5	1.4	3.6
7	Journal of the American Oil Chemists Society	Chem. Appl., Food Sci. Technol.	5	1.3	3.7
8	Applied Energy	Ener. Fuels, Eng. Chem	4	1.6	2.4
9	Biomass Bioenergy	Agr. Eng., Biot. Appl. Microb., Ener. Fuels	4	1.3	2.7
10	Transactions of the ASAE	Agr. Eng.	4	0.3	3.7
11	Energy	Therm., Ener. Fuels	3	2.0	1.0
12	Industrial Engineering Chemistry Research	Eng. Chem.	3	1.4	1.6
13	Resources Conservation and Recycling	Eng. Env., Env. Sci.	3	0.3	2.7
14	Energy Fuels	Ener. Fuels, Eng. Chem.	2	3.4	−1.4
15	Applied Catalysis A General	Chem. Phys., Env. Sci.	2	0.6	1.4

Sustainable Energy', 'Journal of the Energy Institute', 'Aerosol and Air Quality Research', 'Industrial Crops and Products', and 'Journal of Analytical and Applied' with at least a 0.6% presence in the population papers each.

29.3.7 COUNTRIES

Brief information about the top 11 countries with at least three sample papers each is provided in Table 29.6. In total, around 30 and 110 countries contribute to the sample and population papers, respectively.

TABLE 29.6

Countries

	Country	No. of Sample Papers (%)	No. of Population Papers (%)	Difference (%)
1	USA	20	9.3	10.7
2	Turkey	15	5.8	9.2
3	China	13	12.2	0.8
4	Spain	9	6.3	2.7
5	Canada	7	3.3	3.7
6	Malaysia	7	7.8	−0.8
7	India	4	15.3	−11.3
8	Austria	3	0.6	2.4
9	Iran	3	4.1	−1.1
10	Jordan	3	0.6	2.4
11	Serbia	3	0.5	2.5
	Europe-3	15	7.4	7.6
	Asia-5	30	40.0	−10.0

The top country is the USA, publishing 20.0 and 9.3% of the sample and population papers, respectively. Turkey and China follow the USA with 16 and 13% of the sample papers, respectively. The other prolific countries are Spain, Canada, and Malaysia, publishing at least seven sample papers each.

On the other hand, the European and Asian countries represented in this table publish altogether 15 and 30% of the sample papers whilst they publish 7.4 and 40.0% of the population papers, respectively.

It is notable that the publication surplus for the USA and these European and Asian countries is 10.7, 7.6, and −10.0%, respectively. On the other hand, the countries with the most impact are the USA and Turkey with 10.7 and 9.2% publication surplus, respectively. Furthermore, the countries with the least impact are India, Iran, and Malaysia with −11.3, −1.1, and −0.8% publication deficits, respectively.

It is also notable that some countries have a heavy presence in the population papers. The major producers of the population papers are Brazil, South Korea, Taiwan, the UK, Japan, Italy, Thailand, Pakistan, Egypt, France, Indonesia, Saudi Arabia, Australia, Portugal, Greece, Singapore, and Vietnam with at least 1.0% of the population papers each.

29.3.8 'WEB OF SCIENCE' SUBJECT CATEGORIES

Brief information about the top 13 'Web of Science' subject categories with at least three sample papers each is provided in Table 29.7. The sample and population papers are indexed by 20 and 86 subject categories, respectively.

For the sample papers, the top subject is 'Energy Fuels' with 68.0 and 47.9% of the sample and population papers, respectively. This top subject category is followed by 'Engineering Chemical', 'Agricultural Engineering', and 'Biotechnology Applied Microbiology' with 36, 22, and 20% of the sample papers, respectively.

It is notable that the publication surplus is most significant for 'Energy Fuels', 'Agricultural Engineering', 'Biotechnology Applied Microbiology', and 'Chemistry

TABLE 29.7
Web of Science Subject Categories

	Subject	No. of Sample Papers (%)	No. of Population Papers (%)	Difference (%)
1	Energy Fuels	68	47.9	20.1
2	Engineering Chemical	36	32.7	3.3
3	Agricultural Engineering	22	7.0	15.0
4	Biotechnology Applied Microbiology	20	9.9	10.1
5	Chemistry Applied	15	6.3	8.7
6	Thermodynamics	12	8.3	3.7
7	Green Sustainable Science Technology	11	12.2	−1.2
8	Environmental Sciences	8	18.6	−10.6
9	Mechanics	8	4.2	3.8
10	Engineering Environmental	7	10.4	−3.4
11	Chemistry Physical	5	5.4	−0.4
12	Food Science Technology	5	3.7	1.3
13	Chemistry Multidisciplinary	3	6.4	−3.4

Applied' with 20.1, 15.0, 10.1, and 8.7% publication surpluses, respectively. On the other hand, the subjects with least impact are 'Environmental Sciences', 'Engineering Environmental', 'Chemistry Multidisciplinary', and 'Green Sustainable Science Technology' with −10.6, −3.4, −3.4, and −1.2% publication deficits, respectively. This latter group of subject categories are under-represented in the sample papers.

Additionally, some subject categories also have a heavy presence in the population papers: 'Engineering Mechanical', 'Engineering Multidisciplinary', 'Chemistry Analytical', 'Multidisciplinary Sciences', 'Biochemistry Molecular Biology', 'Materials Science Multidisciplinary', and 'Engineering Petroleum' with at least a 1.0% presence in the population papers each.

29.3.9 CITATION IMPACT

These sample and population papers received about 23,500 and 60,000 citations, respectively as of January 2020. Thus, the average numbers of citations per paper are 235 and 28 respectively.

29.3.10 KEYWORDS

Although a number of keywords are listed in the Appendix for the datasets related to this field, some of them are more significant for the sample papers.

The most-prolific keyword for the set related to waste oils is 'waste*' with 70 papers. This top keyword is followed by 'used frying oil*', 'fat', 'fats', and 'tallow' with nine, eight, and seven sample papers, respectively. The other prolific keywords are 'grease', 'used cooking oil*', 'soapstock*', and 'sludge*' with at least two sample papers each.

TABLE 29.8
Research Fronts

Research Front	No. of Sample Papers (%)
Topical research fronts	
Biodiesel production	72
Biodiesel characterization and properties	34
Biomass-based research fronts	
Waste oil-based biodiesel (WOBD) fuels	82
Animal fat-based biodiesel fuels	26
Other WOBD fuels in general	6

On the other hand, the most-prolific keyword related to diesel fuels is '*diesel' with 105 occurrences. This top keyword is followed by 'transester*' and 'methyl ester*' with 16 and 5 papers, respectively.

29.3.11 RESEARCH FRONTS

Brief information about the key research fronts is provided in Table 29.8. There are two major topical research fronts for these sample papers: 'WOBD production' (Canakci and van Gerpen, 2001; Kulkarni and Dalai, 2006; Lam et al., 2010; Zhang et al., 2003a–b) and 'properties and characterization of WOBD fuels' (Dorado et al., 2003; Ghobadian et al., 2009; Graboski and McCormick, 1998; Ozsezen et al., 2009; Usta et al., 2005) with 72 and 34 sample papers, respectively.

On the other hand, there are three primary research fronts for the waste oils used for biodiesel production. The most-prolific research front is 'WOBD fuels' with 82 sample papers (Dorado et al., 2003; Ghobadian et al., 2009; Kulkarni and Dalai, 2006; Lam et al., 2010; Ozsezen et al., 2009; Usta et al., 2005; Zhang et al., 2003a–b). These papers are primarily related to the use of waste cooking oils for biodiesel fuels. This top research front is followed by 'animal fat-based biodiesel fuels' (Canakci and van Gerpen, 2001; Demirbas, 2008; Graboski and McCormick, 1998; Liu et al., 2007; Reddy et al., 2006) and 'other WOBD fuels' (Haas et al., 2001; Mondala et al., 2009) with 26 and 6 sample papers, respectively.

29.4 DISCUSSION

The size of the research in this field has increased to over 2,150 papers as of January 2020. It is expected that the number of the population papers in this field will exceed 6,000 papers by the end of the 2020s.

The research has developed more in the technological aspects of this field, rather than the social and humanitarian pathways, as evidenced by the negligible number of population papers in the indices of the 'Web of Science', SSCI, and A&HCI.

The article types of documents are the primary documents for both datasets and reviews are over-represented by 8.5% in the sample papers, whilst articles are under-represented by 8.8% (Table 29.1). Thus, the contribution of reviews by 12% of the sample papers in this field is highly exceptional (cf. Konur, 2011, 2012a–n, 2015, 2016a–f, 2017a–f, 2018a–b, 2019a–b).

Thirteen authors from nine institutions have at least three sample papers each (Table 29.2). Four, three, and three of these authors are from Canada, Turkey, and the USA, respectively.

These authors focus on 'biodiesel fuel production' and 'biodiesel fuel properties' using animal fats and waste oils. It is significant that there is ample 'gender deficit' among these top authors as all of them are male (Lariviere et al., 2013; Xie and Shauman, 1998).

The population papers have built on the sample papers, primarily published in the 2000s and to a lesser extent in the 2010s (Figure 29.1). Following this rising trend, particularly in the 2000s and 2010s, it is expected that the number of papers will reach 6,000 by the end of the 2020s, nearly tripling their current size.

The engagement of the institutions in this field at the global scale is significant as around 120 and 1,700 institutions contribute to the sample and population papers, respectively.

The top ten institutions publish 34 and 6% of the sample and population papers, respectively (Table 29.3). The top institutions are 'Kocaeli University' of Turkey and the 'University of Nebraska' of the USA with six and four sample papers, respectively.

The most-prolific countries for these top institutions are the USA and Canada. It is notable that some institutions with a heavy presence in the population papers are under-represented in the sample papers.

It is significant that only 44.0 and about 46.7% of the sample and population papers declare any funding, respectively. It is notable that the prolific countries for these funding bodies are Malaysia and China (Table 29.4). It is further notable that some top funding agencies for the population studies do not enter this top funding body list.

However, the presence of substantial Chinese funding bodies in this top funding body table is notable. This finding is in line with the studies showing heavy research funding in China, where the NSFC is the primary funding agency (Wang et al., 2012).

The sample and population papers are published by 32 and over 500 journals, respectively. It is significant that the top 15 journals publish 83.0 and 35.9% of the sample and population papers, respectively (Table 29.5).

The top journal is 'Bioresource Technology' publishing 14 sample papers with a 9.6% publication surplus. This top journal is followed by 'Fuel', 'Fuel Processing Technology', and 'Energy Conversion and Management'.

The top categories for these journals are 'Energy Fuels', 'Engineering Chemical', 'Agricultural Engineering', 'Biotechnology Applied Microbiology', 'Thermodynamics', and 'Green Sustainable Science Technology'. It is notable that some journals with a heavy presence in the population papers are relatively under-represented in the sample papers.

In total, around 30 and 110 countries contribute to the sample and population papers, respectively. The top country is the USA publishing 20.0 and 9.3% of the sample and population papers, respectively (Table 29.6). This finding is in line with the studies arguing that the USA is not losing ground in science and technology (Leydesdorff and Wagner, 2009).

The other prolific countries are Turkey, China, Spain, Canada, and Malaysia. These findings are in line with the studies showing heavy research activity in these countries in recent decades (Bordons et al., 2015; Fu and Ho, 2015; Kumar and Jan, 2014; Leydesdorff and Zhou, 2005).

On the other hand, the European and Asian countries represented in this table publish together 15 and 30% of the sample papers whilst they publish 7.4 and 40.0% of the population papers, respectively. These findings are in line with the studies showing that both European and Asian countries have superior publication performance in science and technology (Bordons et al., 2015; Okubo et al., 1998; Youtie et al., 2008).

It is notable that the publication surplus for the USA and these European and Asian countries is 10.7, 7.6, and −10.0%, respectively. On the other hand, the countries with the most impact are the USA and Turkey. Furthermore, the countries with the least impact are India, Iran, and Malaysia.

China's presence in this top table is notable. This finding is in line with China's efforts to be a leading nation in science and technology (Guan and Ma, 2007; Youtie et al., 2008; Zhou and Leydesdorff, 2006).

It is also notable that some countries have a heavy presence in the population papers. The major producers of the population papers are Brazil, South Korea, Taiwan, the UK, Japan, Italy, Thailand, Pakistan, Egypt, France, Indonesia, Saudi Arabia, Australia, Portugal, Greece, Singapore, and Vietnam (Glanzel et al., 2006; Hassan et al., 2012; Huang et al., 2006; Leydesdorff and Zhou, 2005).

The sample and population papers are indexed by 20 and 86 subject categories, respectively. For the sample papers, the top subject is 'Energy Fuels' with 68.0 and 47.9% of the sample and population papers, respectively (Table 29.7). This top subject category is followed by 'Engineering Chemical', 'Agricultural Engineering', and 'Biotechnology Applied Microbiology'.

It is notable that the publication surplus is most significant for 'Energy Fuels', 'Agricultural Engineering', 'Biotechnology Applied Microbiology', and 'Chemistry Applied'. On the other hand, the subjects with least impact are 'Environmental Sciences', 'Engineering Environmental', 'Chemistry Multidisciplinary', and 'Green Sustainable Science Technology'. This latter group of subject categories are underrepresented in the sample papers.

These sample and population papers receive about 23,500 and 60,000 citations, respectively as of January 2020. Thus, the average number of citations per paper are 235 and 28, respectively. Hence, the citation impact of these top 100 papers in this field has been significant.

Although a number of keywords are listed in the Appendix for the datasets related to this field, some of them are more significant for the sample papers.

The most-prolific keyword for the keyword set related to waste oils is 'waste' with 70 papers. The other prolific keywords are 'used frying oil*' (9), 'fat', 'fats' (8), 'tallow' (7), 'grease' (3), 'used cooking oil*' (3), 'soapstock*' (2), and 'sludge*' (2).

On the other hand, the most-prolific keyword related to diesel fuels is '*diesel' with 105 occurrences. This top keyword is followed by 'transester*' and 'methyl ester*' with 16 and 5 papers, respectively. As expected, these keywords provide valuable information about the pathways of the research in this field.

There are two major topical research fronts for these sample papers: 'WOBD production' and 'properties and characterization of WOBD fuels' with 72 and 34 sample papers, respectively (Table 29.8).

On the other hand, there are three primary research fronts for the waste oils used for biodiesel production. The most-prolific research front is 'WOBD fuels' with 82 sample papers. This top research front is followed by 'animal fat-based biodiesel fuels' and 'other WOBD fuels' in general with 26 and 6 sample papers, respectively.

The key emphasis in these research fronts is the exploration of the structure–processing–property relationships of biomass, waste biooils, and biodiesel (Cheng and Ma, 2011; Konur and Matthews, 1989; Rogers and Hopfinger, 1994; Scherf and List, 2002).

29.5 CONCLUSION

This chapter has mapped the research on WOBD fuels using a scientometric method.

The size of over 2,150 population papers shows the public importance of this interdisciplinary research field. However, it is significant that the research has developed more in the technological aspects in this field, rather than the social and humanitarian pathways.

Articles dominate both the sample and population papers. The population papers, primarily published in the 2010s, build on these sample papers, primarily published in the 2000s.

The data presented in the tables and figure show that a small number of authors, institutions, funding bodies, journals, keywords, research fronts, subject categories, and countries have shaped the research in this field.

It is notable that the authors, institutions, and funding bodies from the USA, Turkey, and China dominate the research in this field. Furthermore, it is also notable that some countries have a heavy presence in the population papers. The major producers of the population papers are Brazil, South Korea, Taiwan, the UK, Japan, Italy, Thailand, Pakistan, Egypt, France, Indonesia, Saudi Arabia, Australia, Portugal, Greece, Singapore, and Vietnam with at least 1.0% of the population papers. Additionally, India and to a lesser extent Malaysia and Iran are under-represented significantly in the sample papers.

These findings show the importance of the development of efficient incentive structures for the development of the research in this field as in other fields. It seems that the Asian countries (such as India, China, and Malaysia) have efficient incentive structures for the development of the research in this field, contrary to Brazil, South Korea, Taiwan, the UK, Japan, Italy, Thailand, Pakistan, Egypt, France, Indonesia, Saudi Arabia, Australia, Portugal, Greece, Singapore, and Vietnam.

It further seems that although the research funding is a significant element of these incentive structures, it might not be a sole solution for increasing the incentives for the research in this field as in the case of Brazil, South Korea, Taiwan, the UK, Japan, Italy, Thailand, Pakistan, Egypt, France, Indonesia, Saudi Arabia, Australia, Portugal, Greece, Singapore, and Vietnam.

On the other hand, it seems there is more to do to reduce the significant gender deficit in this field as in other fields of the science and technology (Lariviere et al., 2013; Xie and Shauman, 1998).

The data on the research fronts, keywords, source titles, and subject categories provide valuable evidence for the interdisciplinary (Lariviere and Gingras, 2010; Morillo et al., 2001) nature of the research in this field.

There is ample justification for the broad search strategy employed in this study due to the interdisciplinary nature of this research field as evidenced by the top subject categories. The search strategy employed in this study is in line with the search strategies employed for related and other research fields (Konur, 2011, 2012a–n, 2015, 2016a–f, 2017a–f, 2018a–b, 2019a–b). It is particularly noted that only 68.0 and 47.9% of the sample and population papers are indexed by the 'Energy Fuels' subject category, respectively.

There are two major topical research fronts for these sample papers: 'WOBD fuel production' and 'properties and characterization of WOBD fuels' (Table 29.8). On the other hand, there are three primary research fronts for the waste oils used for biodiesel production. The most-prolific research front is 'WOBD fuels'. This top research front is followed by 'animal fat-based biodiesel fuels' and 'other WOBD fuels'.

It is recommended that further scientometric studies are carried out for each of these research fronts, building on pioneering studies.

ACKNOWLEDGMENTS

The contribution of the highly cited researchers in the fields of WOBD fuels is greatly acknowledged.

29.A   APPENDIX

The keyword set for WOBD fuels
    Syntax: (1 AND 2) NOT 3

29.A.1   KEYWORD SET FOR WASTE OILS

TI=(*waste or *wastes or fat or fats or tallow or insect* or lard* or grease* or "used oil*" or "used cooking oil*" or "used frying oil*" or soapstock* or wco or sewage or sludge* or tyre* or tires or manure or scrap*).

29.A.2   KEYWORD SET FOR DIESEL FUELS

TI=(*diesel or transester* or "trans-ester*" or "methyl-ester*" or "compression ignition engine*" or methanolysis or "ci engine*" or fame* or hydrotreat* or hydroprocessing or alcoholysis or ethanolysis or hydrodeoxygenation or hydrocracking or "thermal cracking" or *cracking).

29.A.3 EXCLUDING TERMS

NOT TI=(glycerol* or egg* or *shell* or solar or "tallow seed*" or *alga* or "waste heat*").

REFERENCES

Ahmadun, F. R., A. Pendashteh, and L. C. Abdullah, et al. 2009. Review of technologies for oil and gas produced water treatment. *Journal of Hazardous Materials* 170: 530– 551.

Atlas, R. M. 1981. Microbial degradation of petroleum hydrocarbons: An environmental perspective. *Microbiological Reviews* 45: 180–209.

Babich, I. V. and J. A. Moulijn. 2003. Science and technology of novel processes for deep desulfurization of oil refinery streams: A review. *Fuel* 82: 607–631.

Bordons, M., B. Gonzalez-Albo, J. Aparicio, and L. Moreno. 2015. The influence of R & D intensity of countries on the impact of international collaborative research: Evidence from Spain. *Scientometrics* 102: 1385–1400.

Bridgwater, A. V. and G. V. C. Peacocke. 2000. Fast pyrolysis processes for biomass. *Renewable & Sustainable Energy Reviews* 4: 1–73.

Busca, G., L. Lietti, G. Ramis, and F. Berti. 1998. Chemical and mechanistic aspects of the selective catalytic reduction of NO_x by ammonia over oxide catalysts: A review. *Applied Catalysis B-Environmental* 18: 1–36.

Canakci, M. and J. van Gerpen. 2001. Biodiesel production from oils and fats with high free fatty acids. *Transactions of the ASAE* 44: 1429–1436.

Cheng, Y. Q. and E. Ma. 2011. Atomic-level structure and structure–property relationship in metallic glasses. *Progress in Materials Science* 56: 379–473.

Clarivate Analytics. 2019. Highly cited researchers: 2019 Recipients. Philadelphia, PA: Clarivate Analytics. https://recognition.webofsciencegroup.com/awards/highly-cited/2019/ (accessed January, 3, 2020).

Czernik, S. and A. V. Bridgwater. 2004. Overview of applications of biomass fast pyrolysis oil. *Energy & Fuels* 18: 590–598.

Demirbas, A. 2008. Comparison of transesterification methods for production of biodiesel from vegetable oils and fats. *Energy Conversion and Management* 49: 125–130.

Docampo, D. and L. Cram. 2019. Highly cited researchers: A moving target. *Scientometrics* 118: 1011–1025.

Dorado, M. P., E. Ballesteros, J. M. Arnal, J. Gomez, and F. J. Lopez. 2003. Exhaust emissions from a diesel engine fueled with transesterified waste olive oil. *Fuel* 82: 1311–1315.

Fu, H. Z. and Y. S. Ho. 2015. Highly cited Canada articles in Science Citation Index Expanded: a bibliometric analysis. *Canadian Social Science* 11: 50.

Gallezot, P. 2012. Conversion of biomass to selected chemical products. *Chemical Society Reviews* 41: 1538–1558.

Garfield, E. 1955. Citation indexes for science. *Science* 122: 108–111.

Garfield, E. 1972. Citation analysis as a tool in journal evaluation. *Science* 178: 471–479.

Ghobadian, B., H. Rahimi, A. M. Nikbakht, G. Najafi, and T. F. Yusaf. 2009. Diesel engine performance and exhaust emission analysis using waste cooking biodiesel fuel with an artificial neural network. *Renewable Energy* 34: 976–982.

Glanzel, W., J. Leta, and B. Thijs. 2006. Science in Brazil. Part 1: A macro-level comparative study. *Scientometrics* 67: 67–86.

Graboski, M. S. and R. L. McCormick. 1998. Combustion of fat and vegetable oil derived fuels in diesel engines. *Progress in Energy and Combustion Science* 24: 125–164.

Guan, J. C. and N. Ma. 2007. China's emerging presence in nanoscience and nanotechnology: A comparative bibliometric study of several nanoscience giants. *Research Policy* 36: 880–886.

Haas, M. J., K. M. Scott, T. L. Alleman, and R. L. McCormick. 2001. Engine performance of biodiesel fuel prepared from soybean soapstock: A high quality renewable fuel produced from a waste feedstock. *Energy & Fuels* 15: 1207–1212.

Hassan, S. U., P. Haddawy, P. Kuinkel, A. Degelsegger, and C. Blasy. 2012. A bibliometric study of research activity in ASEAN related to the EU in FP7 priority areas. *Scientometrics* 91: 1035–1051.

Huang, M. H., H. W. Chang, and D. Z. Chen. 2006. Research evaluation of research-oriented universities in Taiwan from 1993 to 2003. *Scientometrics* 67: 419–435.

Khalili, N. R., P. A. Scheff, and T. M. Holsen. 1995. PAH source fingerprints for coke ovens, diesel and gasoline-engines, highway tunnels, and wood combustion emissions. *Atmospheric Environment* 29: 533–542.

Kilian, L. 2009. Not all oil price shocks are alike: Disentangling demand and supply shocks in the crude oil market. *American Economic Review* 99: 1053–1069.

Konur, O. 2000. Creating enforceable civil rights for disabled students in higher education: An institutional theory perspective. *Disability & Society* 15: 1041–1063.

Konur, O. 2002a. Access to Nursing Education by disabled students: Rights and duties of nursing programs. *Nurse Education Today* 22: 364–374.

Konur, O. 2002b. Assessment of disabled students in higher education: Current public policy issues. *Assessment and Evaluation in Higher Education* 27: 131–152.

Konur, O. 2002c. Access to employment by disabled people in the UK: Is the Disability Discrimination Act working? *International Journal of Discrimination and the Law* 5: 247–279.

Konur, O. 2006a. Participation of children with dyslexia in compulsory education: Current public policy issues. *Dyslexia* 12: 51–67.

Konur, O. 2006b. Teaching disabled students in Higher Education. *Teaching in Higher Education* 11: 351–363.

Konur, O. 2007a. A judicial outcome analysis of the Disability Discrimination Act: A windfall for the employers? *Disability & Society* 22: 187–204.

Konur, O. 2007b. Computer-assisted teaching and assessment of disabled students in higher education: The interface between academic standards and disability rights. *Journal of Computer Assisted Learning* 23: 207–219.

Konur, O. 2011. The scientometric evaluation of the research on the algae and bio-energy. *Applied Energy* 88: 3532–3540.

Konur, O. 2012a. Evaluation of the research on the social sciences in Turkey: A scientometric approach. *Energy Education Science and Technology Part B: Social and Educational Studies* 4: 1893–1908.

Konur, O. 2012b. Prof. Dr. Ayhan Demirbas' scientometric biography. *Energy Education Science and Technology Part A: Energy Science and Research* 28: 727–738.

Konur, O. 2012c. The evaluation of the biogas research: A scientometric approach. *Energy Education Science and Technology Part A: Energy Science and Research* 29: 1277–1292.

Konur, O. 2012d. The evaluation of the educational research: A scientometric approach. *Energy Education Science and Technology Part B: Social and Educational Studies* 4: 1935–1948.

Konur, O. 2012e. The evaluation of the global energy and fuels research: A scientometric approach. *Energy Education Science and Technology Part A: Energy Science and Research* 30: 613–628.

Konur, O. 2012f. The evaluation of the research on the Arts and Humanities in Turkey: A scientometric approach. *Energy Education Science and Technology Part B: Social and Educational Studies* 4: 1603–1618.

Konur, O. 2012g. The evaluation of the research on the biodiesel: A scientometric approach. *Energy Education Science and Technology Part A: Energy Science and Research* 28: 1003–1014.

Konur, O. 2012h. The evaluation of the research on the bioethanol: A scientometric approach. *Energy Education Science and Technology Part A: Energy Science and Research* 28: 1051–1064.

Konur, O. 2012i. The evaluation of the research on the biofuels: A scientometric approach. *Energy Education Science and Technology Part A: Energy Science and Research* 28: 903–916.

Konur, O. 2012j. The evaluation of the research on the biohydrogen: A scientometric approach. *Energy Education Science and Technology Part A: Energy Science and Research* 29: 323–338.

Konur, O. 2012k. The evaluation of the research on the microbial fuel cells: A scientometric approach. *Energy Education Science and Technology Part A: Energy Science and Research* 29: 309–322.

Konur, O. 2012l. The scientometric evaluation of the research on the production of bioenergy from biomass. *Biomass and Bioenergy* 47: 504–515.

Konur, O. 2012m. The scientometric evaluation of the research on the deaf students in higher education. *Energy Education Science and Technology Part B: Social and Educational Studies* 4: 1573–1588.

Konur, O. 2012n. The scientometric evaluation of the research on the students with ADHD in higher education. *Energy Education Science and Technology Part B: Social and Educational Studies* 4: 1547–1562.

Konur, O. 2015. Current state of research on algal biodiesel. In *Marine Bioenergy: Trends and Developments*, S. K. Kim, and C. G. Lee, ed., 487–512. Boca Raton, FL: CRC Press.

Konur, O. 2016a. Scientometric overview in nanobiodrugs. In *Nanoarchitectonics for Smart Delivery and Drug Targeting*, A. M. Holban and A.M. Grumezescu, ed., 405–428. Amsterdam: Elsevier.

Konur, O. 2016b. Scientometric overview regarding nanoemulsions used in the food industry. In *Emulsions: Nanotechnology in the Agri-Food Industry*, A. M. Grumezescu, ed., 689–711. Amsterdam: Elsevier.

Konur, O. 2016c. Scientometric overview regarding the nanobiomaterials in antimicrobial therapy. In *Nanobiomaterials in Antimicrobial Therapy*, A. M. Grumezescu, ed., 511–535. Amsterdam: Elsevier.

Konur, O. 2016d. Scientometric overview regarding the nanobiomaterials in dentistry. In *Nanobiomaterials in Dentistry*, A. M. Grumezescu, ed., 425–453. Amsterdam: Elsevier.

Konur, O. 2016e. Scientometric overview regarding the surface chemistry of nanobiomaterials. In *Surface Chemistry of Nanobiomaterials*, A. M. Grumezescu, ed., 463–486. Amsterdam: Elsevier.

Konur, O. 2016f. The scientometric overview in cancer targeting. In *Nanoarchitectonics for Smart Delivery and Drug Targeting*, A. M. Holban and A. Grumezescu, ed., 871–895. Amsterdam; Elsevier.

Konur, O. 2017a. Recent citation classics in antimicrobial nanobiomaterials. In *Nanostructures for Antimicrobial Therapy*, A. Ficai and A. M. Grumezescu, ed., 669–685. Amsterdam: Elsevier.

Konur, O. 2017b. Scientometric overview in nanopesticides. In *New Pesticides and Soil Sensors*, A. M. Grumezescu, ed. 719–744. Amsterdam: Elsevier.

Konur, O. 2017c. Scientometric overview regarding oral cancer nanomedicine. In *Nanostructures for Oral Medicine*, E. Andronescu, A. M. Grumezescu, ed., 939–962. Amsterdam: Elsevier.

Konur, O. 2017d. Scientometric overview regarding water nanopurification. In *Water Purification*, A. M. Grumezescu, ed., 693–716. Amsterdam: Elsevier.

Konur, O. 2017e. Scientometric overview in food nanopreservation. In *Food Preservation*, A. M. Grumezescu, ed., 703–729. Amsterdam: Elsevier.

Konur, O. 2017f. The top citation classics in alginates for biomedicine. In *Seaweed Polysaccharides: Isolation, Biological and Biomedical Applications*, J. Venkatesan, S. Anil, S. K. Kim, ed., 223–249. Amsterdam: Elsevier.

Konur, O. 2018a. Scientometric evaluation of the global research in spine: An update on the pioneering study by Wei et al. *European Spine Journal* 27: 525–529.

Konur, O. 2018b. Bioenergy and biofuels science and technology: Scientometric overview and citation classics. In *Bioenergy and Biofuels*, O. Konur, ed., 3–63. Boca Raton: CRC Press.

Konur, O. 2019a. Cyanobacterial bioenergy and biofuels science and technology: A scientometric overview. In *Cyanobacteria: From Basic Science to Applications*, ed. A. K. Mishra, D. N. Tiwari and A. N. Rai, 419–442. Amsterdam: Elsevier.

Konur, O. 2019b. Nanotechnology applications in food: A scientometric overview. In *Nanoscience for Sustainable Agriculture*, R. N., Pudake, N. Chauhan, and C. Kole, ed., 683–711. Cham: Springer.

Konur, O., ed. 2021a. *Handbook of Biodiesel and Petrodiesel Fuels: Science, Technology, Health, and Environment.* Boca Raton, FL: CRC Press.

Konur, O., ed. 2021b. *Handbook of Biodiesel and Petrodiesel Fuels: Science, Technology, Health, and Environment. Volume 1. Biodiesel Fuels: Science, Technology, Health, and Environment.* Boca Raton, FL: CRC Press.

Konur, O., ed. 2021c. *Handbook of Biodiesel and Petrodiesel Fuels: Science, Technology, Health, and Environment. Volume 2. Biodiesel Fuels based on the Edible and Nonedible Feedstocks, Wastes, and Algae: Science, Technology, Health, and Environment.* Boca Raton, FL: CRC Press.

Konur, O., ed. 2021d. *Handbook of Biodiesel and Petrodiesel Fuels: Science, Technology, Health, and Environment. Volume 3. Petrodiesel Fuels: Science, Technology, Health, and Environment.* Boca Raton, FL: CRC Press.

Konur, O. 2021e. Biodiesel and petrodiesel fuels: Science, technology, health, and environment. In *Handbook of Biodiesel and Petrodiesel Fuels: Science, Technology, Health, and Environment. Volume 1. Biodiesel Fuels: Science, Technology, Health, and Environment*, ed. O. Konur. Boca Raton, FL: CRC Press.

Konur, O. 2021f. Biodiesel and petrodiesel fuels: A scientometric review of the research. In *Handbook of Biodiesel and Petrodiesel Fuels: Science, Technology, Health, and Environment. Volume 1. Biodiesel Fuels: Science, Technology, Health, and Environment*, ed. O. Konur. Boca Raton, FL: CRC Press.

Konur, O. 2021g. Biodiesel and petrodiesel fuels: A review of the research. In *Handbook of Biodiesel and Petrodiesel Fuels: Science, Technology, Health, and Environment. Volume 1. Biodiesel Fuels: Science, Technology, Health, and Environment*, ed. O. Konur. Boca Raton, FL: CRC Press.

Konur, O. 2021h Nanotechnology applications in the diesel fuels and the related research fields: A review of the research. In *Handbook of Biodiesel and Petrodiesel Fuels: Science, Technology, Health, and Environment. Volume 1. Biodiesel Fuels: Science, Technology, Health, and Environment*, ed. O. Konur. Boca Raton, FL: CRC Press.

Konur, O. 2021i. Biooils: A scientometric review of the research. In *Handbook of Biodiesel and Petrodiesel Fuels: Science, Technology, Health, and Environment. Volume 1. Biodiesel Fuels: Science, Technology, Health, and Environment*, ed. O. Konur. Boca Raton, FL: CRC Press.

Konur, O. 2021j. Characterization and properties of biooils: A review of the research. In *Handbook of Biodiesel and Petrodiesel Fuels: Science, Technology, Health, and Environment. Volume 1. Biodiesel Fuels: Science, Technology, Health, and Environment*, ed. O. Konur. Boca Raton, FL: CRC Press.

Konur, O. 2021k. Biomass pyrolysis and pyrolysis oils: A review of the research. In *Handbook of Biodiesel and Petrodiesel Fuels: Science, Technology, Health, and Environment. Volume 1. Biodiesel Fuels: Science, Technology, Health, and Environment*, ed. O. Konur. Boca Raton, FL: CRC Press.

Konur, O. 2021l. Biodiesel fuels: A scientometric review of the research. In *Handbook of Biodiesel and Petrodiesel Fuels: Science, Technology, Health, and Environment. Volume 1. Biodiesel Fuels: Science, Technology, Health, and Environment*, ed. O. Konur. Boca Raton, FL: CRC Press.

Konur, O. 2021m. Glycerol: A scientometric review of the research. In *Handbook of Biodiesel and Petrodiesel Fuels: Science, Technology, Health, and Environment. Volume 1. Biodiesel Fuels: Science, Technology, Health, and Environment*, ed. O. Konur. Boca Raton, FL: CRC Press.

Konur, O. 2021n. Propanediol production from glycerol: A review of the research. In *Handbook of Biodiesel and Petrodiesel Fuels: Science, Technology, Health, and Environment. Volume 1. Biodiesel Fuels: Science, Technology, Health, and Environment*, ed. O. Konur. Boca Raton, FL: CRC Press.

Konur, O. 2021o. Edible oil-based biodiesel fuels: A scientometric review of the research. In *Handbook of Biodiesel and Petrodiesel Fuels: Science, Technology, Health, and Environment. Volume 2. Biodiesel Fuels based on the Edible and Nonedible Feedstocks, Wastes, and Algae: Science, Technology, Health, and Environment*, ed. O. Konur. Boca Raton, FL: CRC Press.

Konur, O. 2021p. Palm oil-based biodiesel fuels: A review of the research. In *Handbook of Biodiesel and Petrodiesel Fuels: Science, Technology, Health, and Environment. Volume 2. Biodiesel Fuels based on the Edible and Nonedible Feedstocks, Wastes, and Algae*, ed. O. Konur. Boca Raton, FL: CRC Press.

Konur, O. 2021q. Rapeseed oil-based biodiesel fuels: A review of the research. In *Handbook of Biodiesel and Petrodiesel Fuels: Science, Technology, Health, and Environment. Volume 2. Biodiesel Fuels based on the Edible and Nonedible Feedstocks, Wastes, and Algae*, ed. O. Konur. Boca Raton, FL: CRC Press.

Konur, O. 2021r. Nonedible oil-based biodiesel fuels: A scientometric review of the research. In *Handbook of Biodiesel and Petrodiesel Fuels: Science, Technology, Health, and Environment. Volume 2. Biodiesel Fuels based on the Edible and Nonedible Feedstocks, Wastes, and Algae: Science, Technology, Health, and Environment*, ed. O. Konur. Boca Raton, FL: CRC Press.

Konur, O. 2021s. Waste oil-based biodiesel fuels: A scientometric review of the research. In *Handbook of Biodiesel and Petrodiesel Fuels: Science, Technology, Health, and Environment. Volume 2. Biodiesel Fuels based on the Edible and Nonedible Feedstocks, Wastes, and Algae: Science, Technology, Health, and Environment*, ed. O. Konur. Boca Raton, FL: CRC Press.

Konur, O. 2021t. Algal biodiesel fuels: A scientometric review of the research. In *Handbook of Biodiesel and Petrodiesel Fuels: Science, Technology, Health, and Environment. Volume 2. Biodiesel Fuels based on the Edible and Nonedible Feedstocks, Wastes, and Algae: Science, Technology, Health, and Environment*, ed. O. Konur. Boca Raton, FL: CRC Press.

Konur, O. 2021u. Algal biomass production for biodiesel production: A review of the research. In *Handbook of Biodiesel and Petrodiesel Fuels: Science, Technology, Health, and Environment. Volume 2. Biodiesel Fuels based on the Edible and Nonedible Feedstocks, Wastes, and Algae*, ed. O. Konur. Boca Raton, FL: CRC Press.

Konur, O. 2021v. Algal biomass production in wastewaters for biodiesel production: A review of the research. In *Handbook of Biodiesel and Petrodiesel Fuels: Science, Technology, Health, and Environment. Volume 2. Biodiesel Fuels based on the Edible and Nonedible Feedstocks, Wastes, and Algae*, ed. O. Konur. Boca Raton, FL: CRC Press.

Konur, O. 2021x. Algal lipid production for biodiesel production: A review of the research. In *Handbook of Biodiesel and Petrodiesel Fuels: Science, Technology, Health, and Environment. Volume 2. Biodiesel Fuels based on the Edible and Nonedible Feedstocks, Wastes, and Algae*, ed. O. Konur. Boca Raton, FL: CRC Press.

Konur, O. 2021y. Crude oils: A scientometric review of the research. In *Handbook of Biodiesel and Petrodiesel Fuels: Science, Technology, Health, and Environment. Volume 3. Petrodiesel Fuels: Science, Technology, Health, and Environment*, ed. O. Konur. Boca Raton, FL: CRC Press.

Konur, O. 2021z. Petrodiesel fuels: A scientometric review of the research. In *Handbook of Biodiesel and Petrodiesel Fuels: Science, Technology, Health, and Environment. Volume 3. Petrodiesel Fuels: Science, Technology, Health, and Environment*, ed. O. Konur. Boca Raton, FL: CRC Press.

Konur, O. 2021aa. Bioremediation of petroleum hydrocarbons in the contaminated soils: A review of the research. In *Handbook of Biodiesel and Petrodiesel Fuels: Science, Technology, Health, and Environment. Volume 3. Petrodiesel Fuels: Science, Technology, Health, and Environment*, ed. O. Konur. Boca Raton, FL: CRC Press.

Konur, O. 2021ab. Desulfurization of diesel fuels: A review of the research. In *Handbook of Biodiesel and Petrodiesel Fuels: Science, Technology, Health, and Environment. Volume 3. Petrodiesel Fuels: Science, Technology, Health, and Environment*, ed. O. Konur. Boca Raton, FL: CRC Press.

Konur, O. 2021ac. Diesel fuel exhaust emissions: A scientometric review of the research. In *Handbook of Biodiesel and Petrodiesel Fuels: Science, Technology, Health, and Environment. Volume 3. Petrodiesel Fuels: Science, Technology, Health, and Environment*, ed. O. Konur. Boca Raton, FL: CRC Press.

Konur, O. 2021ad. The adverse health and safety impact of diesel fuels: A scientometric review of the research. In *Handbook of Biodiesel and Petrodiesel Fuels: Science, Technology, Health, and Environment. Volume 3. Petrodiesel Fuels: Science, Technology, Health, and Environment*, ed. O. Konur. Boca Raton, FL: CRC Press.

Konur, O. 2021ae. Respiratory illnesses caused by the diesel fuel exhaust emissions: A review of the research. In *Handbook of Biodiesel and Petrodiesel Fuels: Science, Technology, Health, and Environment. Volume 3. Petrodiesel Fuels: Science, Technology, Health, and Environment*, ed. O. Konur. Boca Raton, FL: CRC Press.

Konur, O. 2021af. Cancer caused by the diesel fuel exhaust emissions: A review of the research. In *Handbook of Biodiesel and Petrodiesel Fuels: Science, Technology, Health, and Environment. Volume 3. Petrodiesel Fuels: Science, Technology, Health, and Environment*, ed. O. Konur. Boca Raton, FL: CRC Press.

Konur, O. 2021ag. Cardiovascular and other illnesses caused by the diesel fuel exhaust emissions: A review of the research. In *Handbook of Biodiesel and Petrodiesel Fuels: Science, Technology, Health, and Environment. Volume 3. Petrodiesel Fuels: Science, Technology, Health, and Environment*, ed. O. Konur. Boca Raton, FL: CRC Press.

Konur, O. and F. L. Matthews. 1989. Effect of the properties of the constituents on the fatigue performance of composites: A review. *Composites* 20: 317–328.

Kulkarni, M. G. and A. K. Dalai. 2006. Waste cooking oil-an economical source for biodiesel: A review. *Industrial & Engineering Chemistry Research* 45: 2901–2913.

Kumar, S. and J. M. Jan. 2014. Research collaboration networks of two OIC nations: Comparative study between Turkey and Malaysia in the field of 'Energy Fuels', 2009–2011. *Scientometrics* 98: 387–414.

Lam, M. K., K. T. Lee, and A. R. Mohamed, AR. 2010. Homogeneous, heterogeneous and enzymatic catalysis for transesterification of high free fatty acid oil (waste cooking oil) to biodiesel: A review. *Biotechnology Advances* 28: 500–518.

Lariviere, V. and Y. Gingras. 2010. On the relationship between interdisciplinarity and scientific impact. *Journal of the American Society for Information Science and Technology* 61: 126–131.

Lariviere, V., C. Ni, Y. Gingras, B. Cronin, and C. R. Sugimoto. 2013. Bibliometrics: Global gender disparities in science. *Nature News* 504: 211–213.

Leydesdorff, L. and C. Wagner. 2009. Is the United States losing ground in science? A global perspective on the world science system. *Scientometrics* 78: 23–36.

Leydesdorff, L. and P. Zhou. 2005. Are the contributions of China and Korea upsetting the world system of science? *Scientometrics* 63: 617–630.

Liu, Y., E. Lotero, J. G. Goodwin, and X. Mo. 2007. Transesterification of poultry fat with methanol using Mg-Al hydrotalcite derived catalysts. *Applied Catalysis A-General* 331: 138–148.

Mohan, D., C. U. Pittman, and P. H. Steele. 2006. Pyrolysis of wood/biomass for bio-oil: A critical review. *Energy & Fuels* 20: 848–889.

Mondala, A., K. W. Liang, H. Toghiani, R. Hernandez, and T. French, 2009. Biodiesel production by in situ transesterification of municipal primary and secondary sludges. *Bioresource Technology* 100: 1203–1210.

Morillo, F., M. Bordons, and I. Gomez. 2001. An approach to interdisciplinarity through bibliometric indicators. *Scientometrics* 51: 203–222.

North, D. C. 1991a. *Institutions, Institutional Change and Economic Performance*. Cambridge, Mass: Cambridge University Press.

North, D.C. 1991b. Institutions. *Journal of Economic Perspectives* 5: 97–112.

Okubo, Y., J. C. Dore, T. Ojasoo, and J. F. Miquel. 1998. A multivariate analysis of publication trends in the 1980s with special reference to South-East Asia. *Scientometrics* 41: 273.

Ozsezen, A.N., M. Canakci, A. Turkcan, and C. Sayin. 2009. Performance and combustion characteristics of a DI diesel engine fueled with waste palm oil and canola oil methyl esters. *Fuel* 88: 629–636.

Perron, P. 1989. The great crash, the oil price shock, and the unit root hypothesis. *Econometrica: Journal of the Econometric Society* 57: 1361–1401.

Reddy, C., V. Reddy, R. Oshel, and J. G. Verkade. 2006. Room-temperature conversion of soybean oil and poultry fat to biodiesel catalyzed by nanocrystalline calcium oxides. *Energy & Fuels* 20: 1310–1314.

Rogers, D. and A. J. Hopfinger. 1994. Application of genetic function approximation to quantitative structure-activity relationships and quantitative structure-property relationships. *Journal of Chemical Information and Computer Sciences* 34: 854–866.

Rogge, W. F., L. M. Hildemann, M. A. Mazurek, G. R. Cass, and B. R. T. Simoneit. 1993. Sources of fine organic aerosol. 2. Noncatalyst and catalyst-equipped automobiles and heavy-duty diesel trucks. *Environmental Science & Technology* 27: 636–651.

Schauer, J. J., M. J. Kleeman, G. R. Cass, and B. R. T. Simoneit. 1999. Measurement of emissions from air pollution sources. 2. C_1 through C_{30} organic compounds from medium duty diesel trucks. *Environmental Science & Technology* 33: 1578–1587.

Scherf, U. and E. J. List. 2002. Semiconducting polyfluorenes-towards reliable structure–property relationships. *Advanced Materials* 14: 477–487.

Usta, N., E. Ozturk, and O. Can, et al. 2005. Combustion of biodiesel fuel produced from hazelnut soapstock/waste sunflower oil mixture in a diesel engine. *Energy Conversion and Management* 46: 741–755.

Wang, X., D. Liu, K. Ding, and X. Wang. 2012. Science funding and research output: A study on 10 countries. *Scientometrics* 91: 591–599.

Xie, Y. and K. A. Shauman. 1998. Sex differences in research productivity: New evidence about an old puzzle. *American Sociological Review* 63: 847–870.

Youtie, J, P. Shapira, and A. L. Porter. 2008. Nanotechnology publications and citations by leading countries and blocs. *Journal of Nanoparticle Research* 10: 981–986.

Zhang, Y., M. A. Dube, D.D. McLean, and M. Kates. 2003a. Biodiesel production from waste cooking oil: 1. Process design and technological assessment. *Bioresource Technology* 89: 1–16.

Zhang, Y., M. A. Dube, D. D. McLean, and M. Kates. 2003b. Biodiesel production from waste cooking oil: 2. *Economic* assessment and sensitivity analysis. *Bioresource Technology* 90: 229–240.

Zhou, P. and L. Leydesdorff. 2006. The emergence of China as a leading nation in science. *Research Policy* 35: 83–104.

30 Biodiesel Production from Municipal Wastewater Sludge
Recent Trends

Muhammad Nurunnabi Siddiquee
Sohrab Rohani

CONTENTS

30.1   INTRODUCTION

The world's energy demand is on an upward trend with the increasing population and improving quality of life. At present, most of the energy comes from the nonrenewable energy sources such as fossil fuel. But the challenge is the limited reserves of the nonrenewable sources to meet the increasing demand for energy. In addition, some nonconventional nonrenewable sources, such as oil sands bitumen or oil shale, involve challenges to extract, upgrade, and refine the oils. Moreover, nonrenewable sources are not evenly distributed worldwide, which greatly involves geopolitics, for example, 63% of the global fossil fuel reserves are in the Middle East (Demirbas and Demirbas, 2011). Another problem with nonrenewable sources is the production of greenhouse gases which lead to environmental pollution.

Research is going on to find alternative energy sources that are renewable, environmentally viable, readily available, and can be used without any engine modification. Alternative energy sources include biodiesel, biooil, biogas, and ethanol. Biooil is very viscous, and its production requires very high temperature and pressure. Ethanol possesses low energy density and requires engine modification. Biodiesel is a potential option for renewable sources that can be used directly or as a blend with petrodiesel without any engine modification. Chemically, biodiesel is known as fatty acid alkyl (mainly methyl) ester which is produced from higher fatty-acid-containing lipid sources. Common sources include vegetable oils such as soybean, palm, canola, castor oil, animal fats, waste cooking oil, lipid from wastewater sludge, and lipid from algae. Although biodiesel is being considered as a potential alternative to petrodiesel, it faces challenges. For example, a higher raw material cost (approximately 70–80% of the cost involves raw material) and the fact that edible vegetable oils and animal fats compete with food materials (Siddiquee and Rohani, 2011a). Hence, it is highly desirable to find an alternative raw material suitable for biodiesel production.

Biodiesel production from lipid sources of 'municipal wastewater sludge' (MWWS) is being considered as a potential option. It would mitigate the raw material cost and could reduce the challenges involved with sludge management. Municipal wastewater treatment plants generate huge amounts of sludge. The European Union generates about 50 million metric tons and the United States generates about 40 million metric tons of sludge every year (Bora et al., 2020). Sludge generation is expected to increase day by day with population increase, industrialization, and urbanization. Based on the chemical composition of the sludge, it could be potentially used as a fertilizer. But its use as a fertilizer is hampered due to the presence of heavy metals and pathogenic microorganisms (Dufreche et al., 2007). MWWS is readily available in some cases with a negative cost as an incentive. Municipal wastewater treatment facilities usually produce primary and secondary sludge. Floating oil, grease, and solids are the key components of 'primary sludge' which is collected after screening and grit removal from the bottom of the primary settling tank of a wastewater treatment plant. Raw primary sludge contains about 4–5% (w/v) solid sludge and about 20 wt% lipid (the dry basis of the sludge). 'Secondary sludge', also known as 'activated sludge', mainly consists of the microbial cells and suspended solids produced during the aerobic biological treatment of wastewater and is collected from the secondary clarifier/settling tank. It contains about 1–2% (w/v) solids, and about 6 wt% lipid (Table 30.1) (Siddiquee and Rohani, 2011a; van Haandel and van der Lubbe, 2007). Both lipids from primary and secondary sludges can be converted to biodiesel.

Few researchers outline biodiesel production from MWWS (Arazo et al., 2017; Capodaglio and Callegari, 2018; Choi et al., 2014; Kargbo, 2010; Siddiquee and Rohani, 2011a). The overall schematic of biodiesel production is illustrated in Figure 30.1. Raw sludge is pretreated first to remove large percentages of water present. Pretreatment techniques include settling, filtration, centrifugation, and drying. Biodiesel is then produced from the sludge either by a two-step process (lipid extraction followed by biodiesel production) or by a single step *in situ* process. Although raw material is readily available free of cost, the overall process is associated with challenges such as expensive pretreatment and a lipid extraction process.

TABLE 30.1

Composition of Primary and Secondary Sludge (as Percentage Weight of Dry Mass)

Component	Primary Sludge			Secondary Sludge	
	(1)	**(2)**	**(3)**	**(4)**	**(5)**
Volatile fraction	79.7	73.5	75.0	59–75	79.0
Lipids	18.6	21.0	10.3	5–12	5.8
Cellulose	18.2	19.9	32.2	7	9.7
Hemicellulose			2.5		
Proteins	17.2	28.7	19.0	32–41	53.7

Source: van Haandel and van der Lubbe (2007)

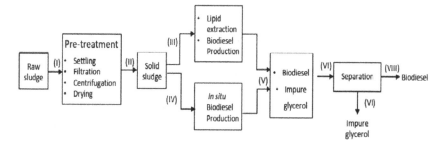

FIGURE 30.1 Overall biodiesel production scheme from municipal wastewater sludge. (*Source:* Adapted from Siddique and Rohani (2011a)).

This chapter provides an overview of the recent advances, potential and challenges, and economic analysis of biodiesel production from municipal sewage sludge.

30.2   OVERVIEW OF BIODIESEL PRODUCTION

Four primary techniques including the direct use and blending of raw oils, microemulsions, thermal cracking, and transesterification are generally used to produce biodiesel (Siddiquee and Rohani, 2011a). The most commonly used method for biodiesel production is transesterification (also known as alcoholysis) reaction in the presence of a catalyst. In the transesterification process, the alkoxy group (RO-) of an ester compound exchanges with another alcohol (Figure 30.2a), where R_1, R_2, and R_3 are higher hydrocarbon chains, also known as fatty acid chains, to produce biodiesel. Primary alcohol, methanol, is widely used for producing biodiesel (fatty acid methyl ester-FAME) because it is the least expensive alcohol. The transesterification reaction follows a series of consecutive, reversible reactions; the triglyceride reacts with primary alcohol to produce diglyceride (Reaction I), monoglyceride (Reaction II), and finally glycerine (Reaction III) (Meher et al., 2006). Raw materials containing

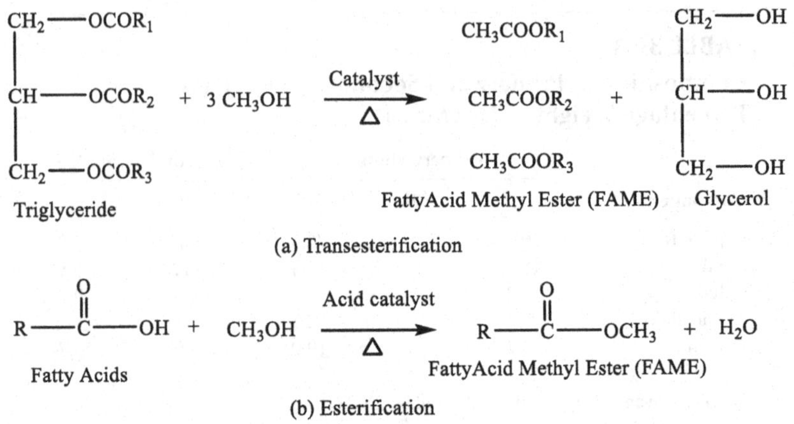

$$CH_2\!-\!OCOR_1$$

$$CH\!-\!OCOR_2 \; + \; 3\,CH_3OH \;\xrightarrow[\Delta]{Catalyst}\; $$

$$CH_2\!-\!OCOR_3$$

Triglyceride

$$CH_3COOR_1$$

$$CH_3COOR_2 \quad + $$

$$CH_3COOR_3$$

FattyAcid Methyl Ester (FAME)

$$CH_2\!-\!OH$$

$$CH\!-\!OH$$

$$CH_2\!-\!OH$$

Glycerol

(a) Transesterification

$$R\!-\!\overset{O}{\overset{\|}{C}}\!-\!OH \;+\; CH_3OH \;\xrightarrow[\Delta]{Acid\ catalyst}\; R\!-\!\overset{O}{\overset{\|}{C}}\!-\!OCH_3 \;+\; H_2O$$

Fatty Acids   FattyAcid Methyl Ester (FAME)

(b) Esterification

FIGURE 30.2 The general form of biodiesel production by (a) transesterification and (b) esterification. (*Source:* Siddiquee and Rohani (2011a)).

'free fatty acids' (FFA) follow the esterification process in the presence of an acid catalyst (Figure 30.2b) and produce FAME (biodiesel).

$$Triglyceride+ROH \rightleftharpoons Diglyceride+RCOOR_1 \qquad \text{Reaction (I)}$$

$$Diglyceride + ROH \rightleftharpoons Monoglyceride + RCOOR_2 \qquad \text{Reaction (II)}$$

$$Monoglyceride + ROH \rightleftharpoons Glyceride + RCOOR_3 \qquad \text{Reaction (III)}$$

Biodiesel production depends on several factors including catalyst type (base, acid, enzyme, or heterogeneous catalyst), alcohol/lipid molar ratio, temperature and reaction time, and quality of lipid (for example, moisture content and FFA content). Excess methanol (60–100% more than required) is usually added to ensure total conversion in the transesterification reaction (Siddiquee and Rohani, 2011a).

Catalyst selection is an important consideration in biodiesel production. Different catalysts are used, for example, homogeneous and heterogeneous base and acid catalysts, zeolite-based catalysts, and enzymatic catalysts. Basic catalytic transesterification is very fast compared to other catalyst types and widely used commercially. But FFAs in the raw materials produce soap in the presence of a base catalyst (Reaction IV). It is not only consumed in the reaction but also inhibits the glycerine separation from biodiesel and contributes to emulsion formation during the water wash. Homogeneous acid (H_2SO_4, for instance) catalyzed transesterification is much slower (by approximately 4,000 times) than base catalyzed transesterification and requires excess alcohol (Demirbas, 2005). But the main advantage of an acid catalyst is that it can catalyze both the esterification and transesterification and produce more biodiesel.

However, the process requires a strict water management technique as the accumulation of water due to esterification (Figure 30.2b) can stop the reaction. Hence, it is important to examine the quality of raw materials to select the appropriate catalyst for the process. In some cases, a two-step process is followed for the raw materials containing higher fatty acids. In the first step, an acid catalyst is used to produce biodiesel via the esterification process, and in the second step, a base catalyst is used to produce biodiesel via the transesterification process at higher reaction rates.

$$\text{Fatty acid}(R_1COOH) + NaOH \rightleftharpoons Soap(r_1COONa) + Water(H_2O) \qquad \text{Reaction (IV)}$$

Heterogeneous catalysts have shown greater promise in biodiesel production as the catalysts can be recovered conveniently from reaction products (Melero et al., 2015; Siddiquee et al., 2011). The undesired saponification reactions can be avoided by using heterogeneous acid catalysts which enable both the transesterification and esterification of raw materials having high content of FFA (Siddiquee and Rohani, 2011a). However, the cost of a solid catalyst is high. The commonly used heterogeneous catalysts used in biodiesel productions are Mg/La mixed oxide, S-ZrO_2 sulfated zirconia, KOH/Nax zeolite, Li/CaO, CaO, KI/Al_2O_3, (ZS/Si) zinc stearate immobilized on silica gel, KNO_3/Al_2O_3, SO_4^-/TiO_2-SiO_2, SBA-15 impregnated with different percentages of heteropolyacid $H_3PO_4.12WO_3.xH_2O$, and a natural zeolite-based catalyst.

Enzymatic transesterification is also an attractive option because of easier product separation and glycerin recovery, reduction in wastewater treatment requirements, and the absence of undesired reactions (Jegannathan et al., 2008). However, the enzymatic process has several disadvantages such as slow reaction rate, product contamination with residual enzymatic activity, and high cost. A noncatalytic 'supercritical methanol' technique is also promising for producing biodiesel by simultaneous transesterification and esterification reactions from the higher FFAs containing oils and fats. This technique provides a higher yield with no saponified products and easier product separation. But it requires extreme operating conditions (350°C and 43 MPa) (Siddiquee and Rohani, 2011a).

MWWS contains lipids from different sources having higher fatty acids. An acid catalytic process or a two-step catalytic process (acid catalytic followed by base catalytic processes) could be a good option for biodiesel production from MWWS. Detailed discussion on the progress of lipid extraction and biodiesel production from MWWS is incorporated in the following sections.

30.3 POTENTIAL AND CHALLENGES OF MUNICIPAL WASTEWATER SLUDGE FOR BIODIESEL PRODUCTION

MWWS is a potential source of biodiesel production. Municipal primary sludge from wastewater treatment plants contains a considerable amount (Table 30.1) of lipids consisting of fats, oil, scums, and greases; and secondary sludge contains a microorganism whose cell wall membrane mostly consists of phospholipids. Lipids from both the primary and secondary sludges can be converted to biodiesel either by following a two-step method (extraction followed by esterification/transesterification) or an *in situ* method as illustrated in Figure 30.1.

MWWS is being considered as a potential feedstock for biodiesel production for the following reasons:

1. Biodiesel can be produced from MWWS which is readily available at practically no cost or even with an incentive. The production cost would be reduced, along with reducing the sludge treatment and disposal cost.
2. Production of biodiesel from edible oils and fats compete with food materials, but MWWS does not.
3. Lipid extraction and biodiesel production from MWWS would facilitate waste material disposal that is environmentally beneficial.
4. Chemical analysis confirms a comparable composition of biodiesel obtained from animal fat/vegetable oil with MWWS.

However, biodiesel production from MWWS presents the following challenges:

1. Sludge collection and pretreatment, especially drying, is an energy intensive process associated with cost.
2. Inconsistency in sludge composition may affect the quality of the lipid and hence the biodiesel. Moreover, it is challenging to separate pharmaceuticals, chemicals, or metallic compounds from the sludge to facilitate the microbial processing.
3. There is a requirement for a huge amount of organic solvents for the lipid extraction. Although it is possible to recycle the solvent, it is energy intensive and would entail large cost for the extraction.
4. There is a requirement for separate esterification steps to overcome the problem associated with higher FFA content. Water management would be important for an efficient esterification process. However, esterification also produces biodiesel and so enhances the overall yield.
5. The reactor design would be challenging due to the variation of lipid and FFA content, product, and by-product separation.
6. Lipid extraction and biodiesel production from wastewater sludge with possible pathogens would involve health-related risks. It would be important to consider proper safety measures.
7. The biodiesel production cost should be lower than conventional petrodiesel. It is challenging to keep the biodiesel production cost lower than the current petrol diesel price.
8. Efficient catalyst selection would be challenging. Because of the higher FFA content, traditional and efficient catalysts such as NaOH cannot be used because of the saponification problem.

Different suggestions/recommendations to overcome the above challenges involved in biodiesel production from wastewater sludge are:

1. Developing an alternative drying or water removal process, and lipid extraction process, to reduce the sludge processing cost.
2. Algae can be cultivated to enhance the lipid content in the sludge, which can also be increased via cultivating the secondary sludge with oleaginous species.

3. Sludge processing can be completed in the wastewater plants whereas lipid extraction/biodiesel production can be performed in a separate plant with access to a few nearby wastewater plants.

4. Sludge may also contain various chemicals, like wax, esters, steroids, terpenoids, polyhydroxyalkanoates, hydrocarbons, linear alkyl benzene, polycyclic aromatic hydrocarbons, and pharmaceutical chemicals, in addition to lipids and phospholipids (Kargbo, 2010; Revellame et al., 2010; Siddiquee and Rohani, 2011a). A method can be developed to extract and convert them to biodiesel via a cracking process that would increase the yield significantly (Carlson et al., 2009; Corma et al., 2007; Revellame et al., 2010).

5. Biooil can be produced from the sludge after lipid extraction. Other value-added chemicals can also be produced from the sludge and/or contaminants.

30.4 ADVANCES IN LIPID EXTRACTION PROCESSES FROM SEWAGE SLUDGE

Lipid extraction is the first of two steps in the biodiesel production process (Step III, Figure 30.1) from MWWS, which is a significant advancement. Lipid extraction efficiency depends on the type and quality of sludge, solvents, sludge-to-solvent ratio, extraction temperature, and extraction time.

The type and quality of sludge are important considerations in the lipid extraction process. The composition of primary and secondary sludges is different and illustrated in Table 30.1. The main lipid sources of the sludge are also different. For example, floating oil, grease, and solids present in the primary sludge are the main sources of lipids which usually are nonpolar in nature. On the other hand, the cell wall membrane of the microorganism present in the activated sludge is the main lipid source, which is polar in nature. The quality of the sludge would also influence the process. For example, the moisture content in the sludge or the processing steps used in the sludge would influence lipid extraction efficiency. The requirements of the other lipid extraction factors such as solvents, sludge-to-solvent ratio, extraction temperature, and extraction time would be different depending on the type and the quality of the sludge.

Different drying techniques are used to facilitate the lipid extraction process. This is considered the greatest challenge and energy intensive stage towards lipid extraction and/or biodiesel production. Common sludge processing techniques include use of the hydromatrix, fluidized bed drying, oven drying, freeze (lyophilization) drying, and vacuum drying. Approximately 42% of the overall cost (shown in Section 30.5) is associated with the drying operation and maintenance. This step is crucial to reduce the extractor and reactor volume as well as solvent requirement for the lipid extraction and biodiesel production.

Organic solvents are usually used to extract the lipid from the MWWS. Common organic solvents used in the lipid extraction processes are methanol, hexane, chloroform, toluene, and ethanol. Polar solvents, such as methanol, extract more polar lipids than nonpolar solvents, such as hexane. A mixture of polar and nonpolar solvents is used to enhance extraction efficiency. Table 30.2 lists different types of lipid extraction from typical MWWS by using different techniques.

TABLE 30.2

Lipid Extraction from MWWS by Organic Solvents

Sludge Type	Sludge Processing	Extraction Conditions/ Methods	Solvents	Lipid Yield (Wt%)	Reference
Dry sewage	—	Soxhlet extraction.	Chloroform, toluene	6 6	Boocock et al. (1992)
Dry sewage	—	Boiling solvent extraction	Chloroform, toluene	18 17	
Primary	Disintegrated with ultrasound at 25 kHz and 50 W power, acidified and dried	Soxhlet extraction. 80 cycles of extractions, 5.5 hours	Hexane	25.3	Olkiewicz et al. (2012)
Secondary			Hexane	9.3	
Secondary	Hydromatrix dried	10.3 MPa, 100°C, 1 h, solvent-to-solid ration 40:1 g/g	Hexane. Methanol HMA[a]	1.94 19.39 21.20	Dufreche et al. (2007)
Secondary	Fluidized bed dryer and oven dried	45°C, 32.5 MPa	$SC\text{-}CO_2$. $SC\text{-}CO_2$ w/ 1.96 wt% methanol. $SC\text{-}CO_2$ w/ 13.04 wt% methanol	3.55 4.19 13.56	Siddiquee and Rohani (2011b)
Primary	Dry under vacuum at ambient temperature for 4 days, 94% solids.	Solvent to sludge: 25 mL/g 75°C, 4 h	Hexane methanol	11.16 14.66	
Secondary		solvent to sludge: 30 mL/g 75°C, 4 h	Hexane methanol	2.95 10.04	
Primary	Dried 98% solid; raw 96% water	1 g total solidequivalent to 10 mL ionic liquid, 500 rpm 100°C, 24 h	$[C_4mim][MeSO_4]$ ionic liquid	18.5 26.9	Olkiewicz et al. (2015)
Secondary	Dried 98 % solid; raw 96% water		$[P(CH_2OH)_4]Cl$ ionic liquid	23.4 27.6	
Secondary	*Trichosporon oleaginosus* cultivated and lyophilization dried	15 min, 25°C, and ultrasonication at 50 Hz and 8,000 W power	Hexane. Methanol. Water chloroform: methanol (1:1 v:v)	40.2[b] 70.8[b] 10.2[b] 99.71[b]	Zhang et al. (2014)
secondary	*Trichosporon* sp. cultivated and oven dried at 60°C	20 min; 30°C, ultrasonication at 520 kHz and 40 W power input	Chloroform: methanol (2:1 v:v)	95–97[c]	Kumar and Banerjee (2019)

[a] HMA: 60% hexane, 20% methanol, and 20% acetone; [c] extraction efficiency of total lipid content; [b] extraction efficiency having total lipid content of 43 ± 0.33%, w/w.

As shown in Table 30.2, the amounts of the lipid content in primary and secondary sludges vary depending on the sludge processing steps, extraction conditions, and the solvents used in the extraction process. Soxhlet extraction and boiling extraction are the two conventional methods used in lipid extraction from sewage sludge. By using chloroform and toluene as solvents, Boocock et al. (1992) obtained 6 wt% lipid via the Soxhlet extraction method and 17–18 wt% lipids via the boiling extraction method from dried sewage sludge by using chloroform and toluene as solvents. The boiling solvent extraction method is more effective than the Soxhlet extraction method. The lipid contains very high amounts (approximately 65 wt%) of FFAs, 7 wt% glyceride fatty acids, and 28 wt% unsaponifiable material (C9 to C16 alkane) (Jarde et al., 2005). Toluene is usually not considered as a suitable solvent as it is not environmentally friendly, has a high boiling point that is not economically feasible for recovery, and above all, the extracted lipid is not suitable for biodiesel production (Mondala et al., 2009).

Dufreche et al. (2007) extracted lipid from the secondary sludge by using hexane, methanol, HMA (60% hexane, 20% methanol, and 20% acetone), supercritical carbon dioxide, and supercritical carbon dioxide with methanol. Among these, HMA provided better lipid yield and solvent extraction was more effective compared to the supercritical CO_2 technique that requires extreme extraction conditions. However, HMA-based extraction was associated with solvent recovery challenges. It is very important to select a judicial solvent amount and solvents that can be used in subsequent esterification/transesterification steps as well (for example, methanol).

Siddiquee and Rohani (2011b) performed a two-level factorial design for lipid extraction from both the primary and secondary sludge using methanol and hexane as solvents to investigate the effects of three key parameters (temperature, solvent to sludge ratio, and extraction time) and their interactions. Temperature was reported as the most significant, at a 5% level of significance, among the three investigated factors. Methanol extraction resulted in a maximum 14.46 (wt/wt) % lipid and 10.04 (wt/wt) % lipid (dry sludge basis) for the primary and secondary sludges, respectively. Hexane, at the identical operating conditions, provided 11.16 (wt/wt) % and 3.04 (wt/wt) % lipid (dry sludge basis), from the primary and secondary sludges, respectively (Table 30.2). Due to the type of lipid present in primary and secondary sludges, hexane provided lipid extraction for the primary sludge whereas methanol worked well for both the primary and secondary sludges. Hexane extracted lipid contain less FFAs compared to the methanol extracted lipid and showed better biodiesel yield (Section 30.5). But methanol is more beneficial as it is also a raw material for the subsequent biodiesel production stage and also for the requirement of a comparatively shorter extraction time. The judicial use of hexane and methanol would be a better option to improve lipid extraction efficiency.

The key advance in the lipid extraction process is the use of an ultrasonication technique that works very well to enhance lipid extraction efficiency (Table 30.2), especially for the secondary sludge. Microorganisms present in the secondary sludge contain phospholipid and lipid. Phospholipid, which is polar in nature, surrounds the nonpolar lipid (cytoplasmic droplets, mainly triglycerides). Ultrasonication facilitates the breakdown of the chain and releases both the phospholipid and the lipid present in the secondary sludge. A combination of the polar and the nonpolar solvents are advantageous in such cases.

The use of methanol and chloroform along with ultrasonication can improve lipid extraction efficiency to greater than 95% within a very short period (15–20 min) at near ambient temperature (Table 30.2). Two levels of ultrasonication are usually used. One is low frequency (i.e. 50 Hz) and high power (i.e. 8,000 W), and the other is high frequency (i.e. 520 kHz) and low power (i.e. 40 W). Both systems work well (Table 30.2), but the low frequency high power (50 Hz and 2,800 W) ultrasonication provided slightly better extraction efficiency (Zhang et al., 2014). Low power and low frequency can also be used to improve efficiency from both the primary and the secondary sludge.

Olkiewicz et al. (2015) used ionic liquid to enhance lipid extraction from the raw and dried primary sludge. [C_4mim][MeSO$_4$] ionic liquid resulted in approximately 18.5% and 26.9% lipids from the dried and raw sludges, respectively. The use of [P(CH$_2$OH)$_4$]Cl ionic liquid enhanced the lipid amount to 23.4% and 27.6%, respectively for dried and raw sludges (Olkiewicz et al., 2015). This is a significant progress in the lipid extraction process from the raw sludges without any organic solvents. It would be beneficial to reduce the sludge processing cost. Choi et al. (2014) used xylene as a solvent as an alternative to hexane and claimed better extraction efficiency.

Another important progression in enhancing the lipid content of secondary sludge is cultivating the sludge with oleaginous substances. The common oleaginous substances that are used in enhancing lipid content include *Trichosporon oleaginosus*, SKF-5, *Mucor circinelloides*, *Chlorella* sp., *Chlorella vulgaris*, and *Nostoc* sp. The lipid content of *Trichosporon oleaginosus* cultivated oven-dried activated (secondary) sludge containing about 43 wt/wt% which is much higher than the other reported lipid content of secondary sludge (Kumar and Banerjee, 2019).

Overall, the selection of a judicial amount of solvent mixture is crucial to improve lipid extraction efficiency. Solvents (for example, methanol) that can used in subsequent esterification/transesterification steps are advantageous. The use of oleaginous substances to cultivate the secondary sludge along with the use of an ultrasonic technique would be promising for biodiesel production from MWWS.

30.5 ADVANCES IN BIODIESEL PRODUCTION PROCESSES FROM SEWAGE SLUDGE

Biodiesel can be produced from MWWS either *ex situ* using extracted lipid from the sludge or *in situ* directly from the solid sludge (Step V, Figure 30.1). There are significant advances in the biodiesel production methods from MWWS. This section summarizes the advances in two-step biodiesel production and *in situ* biodiesel production, separately.

30.5.1 ADVANCES IN TWO-STEP BIODIESEL PRODUCTION

Esterification and transesterification are the most common techniques for producing biodiesel from MWWS and other raw material sources. It is important to use the esterification step before the transesterification step to reduce the impact of the FFA content. It is reported that the esterification step should be used when the FFA content is more than 1% (Mondala et al., 2009). As MWWS contains about 55% lipid, it is important to use the esterification step followed by the transesterification step. Acid catalysts such H_2SO_4 and heterogeneous acids are being used to produce biodiesel from the higher FFA containing lipid extracted from MWWS. The key advantage of such acid catalysts is that they are effective in both the esterification and transesterification processes. However, biodiesel yield and the composition of FAME) are different as the lipid composition and the quality vary, depending on the source (primary or secondary), sludge processing steps, and the solvent used. Table 30.3 summarizes the typical biodiesel yield and maximum FAME composition of biodiesel produced from the lipid extracted from different origins.

As shown in Table 30.3, the maximum biodiesel yield was 85%, significantly higher compared to the listed biodiesel yield (Kumar and Banerjee, 2019). This was for the case of lipid extracted from secondary sludge. It was enzymatically catalyzed (10 U immobilized lipase) at 30°C, 12 h, 200 rpm. The FAME composition was comparable to that of the biodiesel produced from soybean oil. The lipid content of secondary sludge was also enhanced by cultivation with *Trichosporon sp*. Overall, this enzymatic technique of lipid extraction and biodiesel production is promising. The other oleaginous species are also reported to be promising for enhancing lipid content and hence biodiesel production. This biodiesel yield was significantly higher than the conventional biodiesel yields as reported in Table 30.3. However, this strategy is applicable for the secondary sludges.

The biodiesel yield from the hexane and methanol extracted lipids via H_2SO_4 catalyzed esterification–transesterification reactions were 41.25 (wt/wt)% and 38.94 (wt/wt)% (based on lipid) for the primary sludge, and 26.89 (wt/wt)% and 30.28 (wt/wt)% (based on lipid) for the secondary sludge (Table 30.3) (Siddiquee and Rohani, 2011b). The reaction conditions were 24 h, 50°C, and 1% H_2SO_4. However, biodiesel production from the extracted lipid depends on several factors such as the alcohol to lipid ratio, reaction time, temperature, and catalyst loading. Siddiquee and Rohani (2011b) investigated a 'half fraction 2^5 Plackett-Burman factorial design' to investigate the most influential parameters and their interactions during biodiesel production from the hexane and methanol extracted lipid. Five considered factors were methanol-to-lipid ratio (mL/g), reaction time (h), temperature (°C), % acid catalyst (mL/mL), and amount of natural zeolite (mg). The yields would increase by optimizing the acid catalyzed esterification–transesterification reaction. The maximum biodiesel yield at the optimized conditions of a half-fraction 2^5 factorial design was 57.12 (wt/wt)% (based on lipid). The interaction between the methanol-to-lipid ratio and reaction time was the most significant for the developed model.

Olkiewicz et al. (2015) produced biodiesel from the ionic liquid extracted lipids via H_2SO_4 catalyzed esterification–transesterification reactions (Table 30.3). The [C_4mim] [$MeSO_4$] ionic liquid extracted lipid of the dried primary sludge and raw primary sludge resulted in 14.1 and 18.4% biodiesel yields, respectively. The [$P(CH_2OH)_4$]Cl ionic liquid extracted lipid of the dried primary sludge and raw primary sludge resulted in 17.0 and 19.8% biodiesel yields, respectively. The [$P(CH_2OH)_4$]Cl ionic liquid extracted lipid showed better biodiesel production efficiency over the [C_4mim] [$MeSO_4$] ionic liquid extracted lipid.

TABLE 30.3

Biodiesel Production from the Lipid of MWWS

Origin of Lipid	Solvent Used to Extract Lipid	Experimental Conditions	Biodiesel Yield (Wt%)	Maximum Fame Composition	Reference
Primary	Hexane	Lipid in hexane, methanol, 1% H$_2$SO$_4$ overnight heating at 50°C	13.9	C16:0 (47%)	Olkiewicz et al. (2012)
Secondary	Hexane		2.9	C18:1 (28%)	Olkiewicz et al. (2015)
Dried Primary Raw Primary	[C$_4$mim] [MeSO$_4$] ionic liquid	Lipid in hexane, methanol, 1% H$_2$SO$_4$ overnight heating at 50°C	14.1 18.4	–	
Dried Primary Raw Primary	[P(CH$_2$OH)$_4$]Cl ionic liquid		17.0 19.8	–	
Secondary	Hexane methanol HMA[a]	Lipid in hexane, methanol, overnight, 50°C, 1% H$_2$SO$_4$	0.38 2.76 3.44	C16:0 (37%) C16:0 (35%) C18:1 (32%)	Dufreche et al. (2007)
Secondary	SC-CO$_2$ SC-CO$_2$ w/1.96 wt% methanol SC-CO$_2$ w/ 13.04 wt% methanol	Lipid in hexane, 0.5 N sodium –methoxide 10 min	0.28 1.12 2.31	C16:0 (35%) C16:0 (44%) C16:0 (42%)	Siddiquee and Rohani (2011b)
Primary Secondary	Hexane methanol Hexane methanol	Lipid in hexane. Methanol 24 h, 50°C 1% H$_2$SO$_4$	41.25 38.94 30.28 26.89	C16:0 (45%) C16:0 (55%) C16:0 (30%) C16:0 (38%)	
Primary	Methanol	Lipid in hexane. Methanol to lipid = 125 ml/g, 14 h, 60°C 3% H$_2$SO$_4$, natural zeolite 50 mg	57.1	–	
Primary	Methanol	Methanol to lipid = 200 ml/g, 3 h, 130°C, 930 kPa gauge, Acidic mesoporous heterogeneous catalyst SBA-15	30.14	C16:0 (43%)	Siddiquee et al. (2011)
Secondary sludge	Trichosporon sp. cultivated and oven dried at 60°C	Lipid to methanol, (v/v) = 1 : 15 10 U immobilized lipase 30°C, 12 h, 200 rpm	85	C18:0 (37%) C16:0 (26%)	Kumar and Banerjee (2019)

[a] HMA: 60% hexane, 20% methanol, and 20% acetone.

The application of the heterogeneous catalyst also improved the biodiesel yield from the lipid extracted from the primary sludge. For example, mesoporous ordered silica, SBA-15, impregnated with heteropolyacid $H_3PO_4.12WO_3.$ xH_2O (PW_{12}) was used to produce biodiesel from the lipid extracted from MWWS (Siddiquee et al., 2011). The maximum 30.14 (wt/wt)% (based on lipid) biodiesel was produced in 3 h following a methanol-to-lipid ratio of 200 mL/g, 130°C, and 930 kPa gauge pressure (Table 30.3). In all cases, the maximum FAME composition contained palmitic acid (C16:0), followed by stearic acid (C18:0) and oleic acid (C18:1). The FAME composition was comparable to that produced from vegetable oils.

30.5.2 ADVANCES IN *IN SITU* BIODIESEL PRODUCTION

Biodiesel is also successfully produced following an *in situ* esterification and transesterification process. In this process, methanol used as alcohol for esterification and transesterification is also used as a solvent for the lipid extraction and at the same time reacts with it. Hexane is also used to enhance the extraction efficiency and to dissolve nonpolar lipids. Similar to biodiesel production from the extracted lipid, an acid catalyst is also used for the esterification and transesterification processes. Table 30.4. summarizes typical biodiesel yields by using *in situ* esterification and transesterification.

TABLE 30.4
In Situ Biodiesel Production from MWWS

Sludge Type	Sludge Processing	Experimental Conditions	Biodiesel Yield (Wt%)	Maximum Fame Composition	References
Primary	Dry under vacuum at ambient temperature for four days, 94% solids.	Solvent to sludge: 25 mL/g 4% H_2SO_4/ mL 75°C 24 h	12.81	C16:0 (40%)	Siddiquee and Rohani (2011b)
Primary	Dry under vacuum at ambient temperature for four days, 94% solids.	Solvent to sludge: 25 mL/g 4% H_2SO_4/mL 75°C 24 h	12.81	C16:0 (40%)	
Secondary	Fluidized bed dryer and oven dried to 95% solid.	Methanol with 1% H_2SO_4 overnight heating at 50°C	6.23	—	Dufreche et al. (2007)
Primary Secondary	Freeze dried sludge, 94% solids.	Solvent to sludge: 12 g/g 5 % H_2SO_4/mL 75 °C, 24 h	14.3 2.4	C16:0 (43%) C16:0 (36%)	Mondala et al. (2009)
Primary Secondary	Oven dried sludge.	0.25 g sludge/mL methanol, Zr-SBA-15 catalyst, 209°C, 3 h, 2000 rpm	15.5 10.0	— —	Melero et al. (2015)

As shown in Table 30.4 , the maximum reported biodiesel yield was 14.3 wt% for the freeze-dried primary sludge with a solvent-to-sludge ratio of 12 g/g, 5% H_2SO_4/ mL, 75°C, and 24 h. The maximum biodiesel yield from the freeze-dried secondary sludge was 6.3 wt% at a solvent-to-sludge ratio of 12 g/g, 1% H_2SO_4/mL, 50°C, and overnight. *In situ* biodiesel production depended on a few key factors such as solvent-to-sludge ratio, catalyst loading, and temperature.

Siddiquee and Rohani (2011b) investigated a '2^3 Plackett-Burman factorial design' considering the mentioned factors and found a maximum biodiesel yield of 12.81 (wt/wt)% (based on dry sludge). The single most significant factor was the catalyst concentration, and the most significant interaction was that between the methanol-to-sludge ratio and the catalyst concentration for the model developed for the *in situ* biodiesel production. Catalyst concentration was the most single variable for the *in situ* biodiesel production.

Revellame et al. (2010) also optimized the *in situ* biodiesel production from the secondary sludge following a full factorial design of four temperature levels (45, 55, 65, and 75°C), six methanol-to-sludge ratio (5, 10, 15, 20, 25, and 30 v/wt), and five levels of catalyst concentration (0.5, 1, 2, 4, and 6 v%). The maximum biodiesel yield by experimental optimization at 55 °C, a 25 methanol-to-sludge ratio, and 4 v% sulfuric acid was 4.79 wt%. A significant decrease in biodiesel production was observed at temperatures above 60°C that might be due to the acid catalyzed polymerization of unsaturated fatty acids or their ester. It is important to find the optimum conditions for efficient biodiesel production.

In situ biodiesel production from both the primary and secondary sludges showed the maximum FAME composition of C16:0, which is similar to the biodiesel produced following the two-step process of lipid extraction and biodiesel production. It might be feasible to combine the primary and secondary sludges into a single feedstock, also known as blended sludge, for biodiesel production (Siddiquee and Rohani, 2011b). However, it would be important to change the process parameters based on the lipid percentage of the blended sludge. *In situ* biodiesel production is promising as lipid extraction and biodiesel production can be achieved simultaneously. But the requirement of the dried sludge is challenging. If the sludge contains water, it would be difficult to effect the esterification processes as explained in Section 30.2.

Overall, biodiesel production from MWWS is promising using acid catalyzed esterification and transesterification. The FAME composition is comparable to the biodiesel produced from vegetable oils regardless of the *ex situ* or *in situ* production processes.

30.6 ECONOMIC ANALYSIS

Biodiesel is being considered as a most promising alternative to petrodiesel. It possesses all the characteristics and benefits of an ideal fuel, such as being environmentally friendly and renewable. But biodiesel production is not economically viable due to the higher production cost mostly associated with raw materials cost, approximately 70–80%. MWWS offers an alternative to lower this cost. Although MWWS is readily available and free of cost, the process involves challenges to dewater the sludge, extract lipid, and to produce biodiesel.

The biodiesel production cost from MWWS depends on the type of sludge, sludge processing techniques, and biodiesel production methods. Different researchers have performed economic analyses based on different assumptions. Dufreche et al. (2007) provided an economic analysis for *in situ* biodiesel production from secondary sludge. The estimated breakdown of the production cost is shown in Table 30.5.

Overall biodiesel production cost, as shown in Table 30.5, is USD3.11 per gallon which includes USD2.06 per gallon for the centrifuge, drying, and extraction processes, and USD1.05 per gallon for other expenses. One of the key assumptions was the 7.0 wt% overall biodiesel yield. However, as shown in Table 30.4 , *in situ* biodiesel production yield could be higher. But additional costs such as ultrasonication need to be considered. Inclusion of glycerol recovery, biooil production, and other valuable chemical recovery may reduce the production cost as well.

Pokoo-Aikins et al. (2010) also analyzed the overall biodiesel production cost by varying the solvent used in the initial extraction step and found that it was USD3.39 per gallon for ethanol, USD3.37 per gallon for methanol, USD2.89 for hexane, and USD2.79 per gallon for toluene. Toluene is not a potential candidate because of its higher boiling point which would be more energy intensive. However, overall estimated biodiesel production cost, ranging from USD2.79 to USD3.39 per gallon, was lower compared to the vegetable oil-based biodiesel production cost of USD4.00 to USD4.50 per gallon (Siddiquee and Rohani, 2011a). Although the use of ionic liquid is promising to extract lipid from the raw primary sludge, it would be challenging to process a huge volume of sludge using ionic liquid, as raw sludge contains only 1–5% solids.

Recently, Chen et al. (2018) performed an economic assessment of biodiesel production from wastewater sludge by using 'SuperPro Designer' regarding biodiesel production from uncultivated sludge and from the lipid extracted from microorganism-cultivated sludge. The summaries of the estimated costs are listed in Table 30.6. Considering the biodiesel production rate of 8154.3 tons/yr from the uncultivated

TABLE 30.5
Production Cost Estimate for Sludge Biodiesel

	Cost Per Gallon (USD)	Fraction (%)
Centrifuge O&M	0.43	13.83
Drying O&M	1.29	41.48
Extraction O&M	0.34	10.93
Biodiesel processing O&M	0.60	19.93
Labor	0.10	3.22
Insurance	0.03	0.96
Tax	0.02	0.64
Depreciation	0.12	3.86
Capital P&I service	0.18	5.79
Total cost	3.11	100

Notes: Assumes 7.0 wt% overall transesterification yield. O&M: operation and maintenance; P&I: protection and indemnity.
Source: Dufreche et al. (2007).

TABLE 30.6
Production Cost Estimate for Sludge Biodiesel

Item	Name	Biodiesel Production from Uncultivated Sludge		Biodiesel Production From Lipid Extracted from Microorganism Cultivated Sludge	
		Cost Per Gallon (USD)	Fraction (%)	Cost Per Gallon (USD)	Fraction (%)
Raw materials	Reactant (methanol); Lost solvent (hexane, acetone, methanol); catalyst (sulfuric acid); neutralizer (sodium hydroxide); lipid source (sludge: zero cost).	1494,000	27.31	—	—
	Reactant (methanol); lost solvent (chloroform, methanol); catalyst (sodium hydroxide); neutralizer (HCl); nutrient medium (sludge: zero cost).	—	—	6026,00	17.07
Equipment	Dryer; conveyor; grinder; extractor; evaporator; storage tank; transesterification reactor; mixer; centrifuge; distillationcolumns.	1296,00	23.70	10494,000	30.73
Labor	70,819 h per year	779,000	14.24	—	—
	205,091 h per year	—	—	2256,000	6.39
Lab/QC/QA	Laboratory/quality control/quality assurance.	117,000	2.14	338,400	0.96
Utilities	Electricity; steam; cooling water; chilled water	1738,840	32.61	15693,00	44.47
Waste treatment	To treat the wastewater generated after fermentation	—	—	486,200	1.38
Total		5469,84	100.00	35293,600	100.00
Unit biodiesel cost	Biodiesel production rate 8154.3 tonnes/yr			0.67 $/kg (0.59 $/L)	
	Biodiesel production rate 32617.73 tonnes/yr			1.08 US $/kg (0.94 US $/L)	

Source: Chen et al., 2018

sludge, the overall production cost was USD0.57/L (USD2.16/gallon). The overall production cost was USD0.94/L (USD3.56/gallon) regarding production from the lipid extracted from microorganism cultivated sludge. The maximum cost was associated with utilities, which was 32.61% (USD0.70/gallon) of the overall production cost for biodiesel from the uncultivated sludge, and was 44.47% (USD1.58/gallon) of the overall cost from the lipid extracted from microorganism cultivated sludge.

Overall, the cost of biodiesel production from MWWS is lower compared to vegetable oil-based biodiesel production. But the biodiesel production cost is much higher compared to the current low petrodiesel price of USD1.789 as of May 8, 2020 (US Energy Administration, 2020). Considering the environmental benefits and energy recovery from waste sources, governments could provide incentives and reduce/exempt tax associated with biodiesel production from MWWS. In addition, incentives from wastewater treatment plants could also reduce the overall biodiesel production cost. Moreover, biooil production, glycerol recovery, and production of value-added chemicals could reduce the overall cost too.

30.7 CONCLUSIONS AND FUTURE DIRECTIONS

Biodiesel, a renewable and environmentally friendly fuel, is being considered as one of the most promising alternative fuels to petrodiesel. The high production cost is a key obstacle to biodiesel production from vegetable oil sources which has made it uncompetitive compared to petrodiesel. Biodiesel production from MWWS is a promising alternative to reduce the cost, to meet growing future energy demand, and to facilitate the waste management of increasing sludge generation.

Although MWWS is readily available free of cost, biodiesel production from it involves challenges in processing sludge, extracting lipid, and producing biodiesel. However, significant progress has been made in sludge processing, lipid extraction, and biodiesel production processes. The use of oleaginous substances to cultivate the secondary sludge along with the use of an ultrasonic technique would be a promising improvement towards biodiesel production from MWWS. A mixture of polar and nonpolar solvents especially hexane or chloroform and methanol can greatly enhance the lipid extraction efficiency and biodiesel yield.

Most of the reported biodiesel production is currently at the lab scale and some of it on the mL scale (1 mL, for example). It is important to scale up production and characterize the biodiesel produced by following ASTM standards. Biooil and value-added chemical production from the residual sludge after lipid extraction can be a potential strategy to reduce the overall biodiesel production cost.

A comprehensive investigation is needed to study the effect of residual metals, pharmaceuticals in lipid, or the biodiesel produced, and also the P-content in the biodiesel made from the phospholipid containing secondary sludge. Moreover, continuous flow reactor technology and solid catalysts can also be considered to improve further biodiesel production efficiency. However, although biodiesel production cost from MWWS is lower compared to vegetable oil-based biodiesel production, the cost is much higher compared to the current low petrodiesel price. Governments can provide incentives and can reduce/exempt tax associated with biodiesel production from MWWS considering the environmental benefits and energy recovery from the waste sources.

REFERENCES

Arazo, R. O., M. D. G. de Luna, and S. C. Capareda. 2017. Assessing biodiesel production from sewage sludge-derived bio-oil. *Biocatalysis and Agricultural Biotechnology* 10: 189–196.

Boocock, D. G. B., S. K. Konar, A. Leung, and L. D. Ly. 1992. Fuels and chemicals from sewage sludge: 1. The solvent extraction and composition of a lipid from raw sewage sludge. *Fuel* 71: 1283–1289.

Bora, A. P., D. P. Gupta, and K. S. Durbha. 2020. Sewage sludge to bio-fuel: A review on the sustainable approach of transforming sewage waste to alternative fuel. *Fuel* 259: 116262.

Capodaglio, A. G. and A. Callegari. 2018. Feedstock and process influence on biodiesel produced from waste sewage sludge. *Journal of Environmental Management* 216: 176–182.

Carlson, T. R., G. A. Tompsett, W. C. Conner, and G. W. Huber. 2009. Aromatic production from catalytic fast pyrolysis of biomass-derived feedstocks. *Topics in Catalysis* 52: 241.

Chen, J., R. D. Tyagi, and J. Li, et al. 2018. Economic assessment of biodiesel production from wastewater sludge. *Bioresource Technology* 253: 41–48.

Choi, O. K., J. S. Song, D. K. Cha, and J. W. Lee. 2014. Biodiesel production from wet municipal sludge: Evaluation of *in-situ* transesterification using xylene as a cosolvent. *Bioresource Technology* 166: 51–56.

Corma, A., G. W. Huber, L. Sauvanaud, and P. O'Connor. 2007. Processing biomass-derived oxygenates in the oil refinery: Catalytic cracking (FCC) reaction pathways and role of catalyst. *Journal Catalysis* 247: 307–327.

Demirbas, A. 2005. Biodiesel production from vegetable oils via catalytic and non-catalytic supercritical methanol transesterification methods. *Progress in Energy and Combustion Science* 31: 466–487.

Demirbas, A. and M. F. Demirbas. 2011. Importance of algae oil as a source of biodiesel. *Energy Conversion and Management* 52: 163–170.

Dufreche, S., R. Hernandez, and T. French, et al. 2007. Extraction of lipids from municipal wastewater plant micro-organisms for production of biodiesel. *Journal of the American Oil Chemists' Society* 84: 181–187.

Jarde, E., L. Mansuy, and P. Faure. 2005. Organic markers in the lipidic fraction of sewage sludges. *Water Research* 39: 1215–1232.

Jegannathan, K. R., S. Abang, D. Poncelet, E. S. Chan, and P. Ravindra. 2008. Production of biodiesel using immobilized lipase – a critical review. *Critical Reviews in Biotechnology* 28: 253–264.

Kargbo, D. M. 2010. Biodiesel production from municipal sewage sludges. *Energy & Fuels* 24: 2791–2794.

Kumar, S. P. J. and R. Banerjee. 2019. Enhanced lipid extraction from oleaginous yeast biomass using ultrasound assisted extraction: A greener and scalable process. *Ultrasonics Sonochemistry* 52: 25–32.

Meher, L. C., D. V. Sagar, and S. N. Naik. 2006. Technical aspects of biodiesel production by transesterification—a review. *Renewable and Sustainable Energy Reviews* 10: 248–268.

Melero, J. A., R. Sanchez-Vazquez, and I. A. Vasiliadou, et al. 2015. Municipal sewage sludge to biodiesel by simultaneous extraction and conversion of lipids. *Energy Conversion and Management* 103: 111–118.

Mondala, A., K. Liang, H. Toghiani, R. Hernandez, and T. French. 2009. Biodiesel production by *in situ* transesterification of municipal primary and secondary sludges. *Bioresource Technology* 100: 1203–1210.

Olkiewicz, M., A. Fortuny, and F. Stuber, et al. 2012. Evaluation of different sludges from WWTP as a potential source for biodiesel production. *Procedia Engineering* 42: 634–643.

Olkiewicz, M., N. V. Plechkova, and A. Fabregat, et al. 2015. Efficient extraction of lipids from primary sewage sludge using ionic liquids for biodiesel production. *Separation and Purification Technology* 153: 118–125.

Pokoo-Aikins, G., A. Heath, and R. A. Mentzer, et al. 2010. A multi-criteria approach to screening alternatives for converting sewage sludge to biodiesel. *Journal of Loss Prevention in the Process Industries* 23: 412–420.

Revellame, E., R. Hernandez, W. French, W. Holmes, and E. Alley. 2010. Biodiesel from activated sludge through *in situ* transesterification. *Journal of Chemical Technology and Biotechnology* 85: 614–620.

Siddiquee, M. N. and S. Rohani. 2011a. Lipid extraction and biodiesel production from municipal sewage sludges: A review. *Renewable and Sustainable Energy Reviews* 15: 1067–1072.

Siddiquee, M. N. and S. Rohani. 2011b. Experimental analysis of lipid extraction and biodiesel production from wastewater sludge. *Fuel Processing Technology* 92: 2241–2251.

Siddiquee, M. N., H. Kazemian, and S. Rohani. 2011 Biodiesel production from the lipid of wastewater sludge using acidic heterogeneous catalyst. *Chemical Engineering & Technology* 34: 1983–1988.

US Energy Administration. 2020 https://www.eia.gov/petroleum/gasdiesel/ (Access on May 08, 2020).

Van Haandel, A. V. and van der Lubbe, J. 2007. *Handbook of Biological Wastewater Treatment, Design and Optimization of Activated Sludge System*, Second Edition. London: IWA Publishing.

Zhang, X., S. Yan, and R. D. Tyagi, P. Drogui, and R. Y. Surampalli. 2014. Ultrasonication assisted lipid extraction from oleaginous microorganisms. *Bioresource Technology* 158: 253–261.

31 Heterogeneous Acid Catalysts for Biodiesel Production from Waste Cooking Oil

J. E. Castanheiro

CONTENTS

31.1 INTRODUCTION: BACKGROUND AND DRIVING FORCES

Biodiesel is a mixture of 'fatty acid methyl esters' (FAME) or 'fatty acid ethyl esters' (FAEE). It is an environmentally friendly fuel and is biodegradable and not toxic. Biodiesel results from the transesterification reaction between a triglyceride molecule and three alcohol molecules. This reaction yields three FAME molecules and a glycerol molecule. Usually, the transesterification reaction is done using a base (NaOH or KOH) as catalyst. Figure 31.1 exemplifies the transesterification of triglycerides with an alcohol to biodiesel and glycerol (Avhad and Marchetti, 2015).

The raw material for biodiesel production is edible vegetable oils (e.g. sunflower and soybean), nonedible vegetable oils (e.g. Jatropha oil), WCO, and animal fats (Fonseca et al., 2019; Lima and Castanheiro 2018). However, when the raw material presents a high amount of 'free fatty acids' (FFA), these carboxylic acids react with the catalyst (NaOH or KOH), yielding soap. Then, in order to overcome this problem, an esterification step before the transesterification reaction will be necessary. Equation 31.1 shows the scheme of esterification of fatty acid with an alcohol.

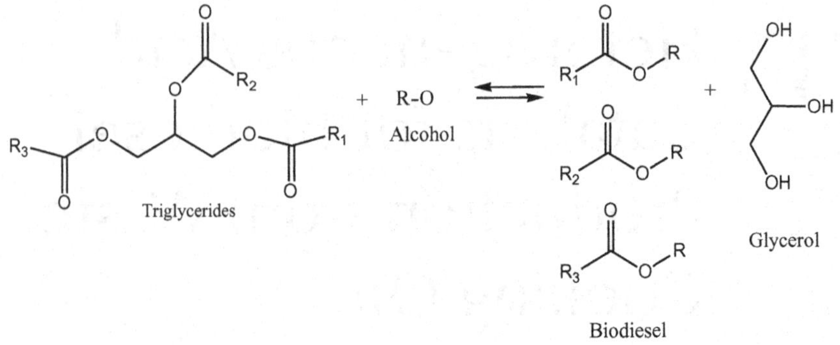

FIGURE 31.1   Scheme of transesterification reaction.

$$R1 - COOH + R2 - OH \Leftrightarrow R1 - COO - R2 + H_2O \tag{31.1}$$

where R1 symbolizes an alkyl group of the carboxylic acid and R2 represents an alkyl group of the alcohol.

The acid catalysts, like sulfuric acid or phosphoric acid, can be used for the esterification and transesterification reactions. However, the homogeneous catalysts have some disadvantages such as the separation of catalysts from the reaction and their corrosiveness. In order to make the biodiesel process an eco-friendly process, the homogeneous catalysts are replaced by heterogeneous catalysts. The heterogeneous solid acids have some advantages: the simultaneous esterification and transesterification reactions and easy separation from the product, and the opportunity to reuse the material (Kiss et al., 2010; Lee and Wilson, 2015).

Biodiesel production from WCO has been performed using zeolites, like ZSM-5 (MFI), mordenite (MOR), faujasite (FAU), beta (BEA) zeolites, and silicalite, Y zeolite, poly(vinyl alcohol) with sulfonic acid groups, acidic ion-exchange resins, $SO_4^{2-}/TiO_2–SiO_2$, 12-tungstosilicic acid immobilized on SBA-15, MCM-41 and Nb_2O_5, sulfonated carbon, and composite catalysts. This chapter reviews biodiesel production from WCO over heterogenous solid-acid materials.

31.2   ZEOLITES

Zeolites are microporous materials, crystalline, and constituted with SiO_4 and AlO_4^- units. These materials have an H^+, NH_4^+, or Na^+ in their networks to counterbalance the negative charge of AlO_4^-. There are many zeolites, like mordenite, ZSM-5, beta, and faujasite. Zeolites have been applied as catalysts in different areas, such as fine chemistry, petrochemistry, and biomass transformation (Serrano et al., 2018).

Chung et al. (2008) studied the esterification of fatty acids with methanol in WCO over zeolites (ZSM-5, mordenite, faujasite, beta). These authors observed that FFA elimination was reduced by the decreasing acidity of the zeolites. Under optimized conditions (oil with an acid number (mg KOH/g) = 1.25, T = 333 K, $m_{catalyst}$ = 1.0 g, t = 180 min), an 81% methyl ester content was achieved over an HMOR(10) zeolite.

Brito et al. (2007) researched the transesterification of WCO with methanol over Y zeolites. Materials with different Si/Al ratios were used. The transesterification was done using a continuous tubular reactor. The viscosities of the oils decreased with the increase of temperature. Under optimal conditions (Y530 zeolite, T = 466°C, space time = 12.35 min, methanol/oil molar ratio = 6) a high conversion of WCO was obtained. The zeolite can be reused.

31.3 ACTIVATED CARBONS

Li et al. (2014) studied the esterification and transesterification of WCO with methanol over activated carbons. These catalysts were prepared from rice husks. The materials were treated with concentrated sulfuric acid. The achieved FFA conversion was 98.17% after 3 h of reaction. After 15 h, the achieved biodiesel yield was 87.57%. The catalyst can be reused.

Tran et al. (2016) prepared activated carbon from xylose by hydrothermal carbonization and sulfonation. These activated carbons, consisting of a sulfonated carbon microsphere, were used as the catalyst in biodiesel production from WCO. Under optimized conditions (T = 100°C, t = 4 h, catalyst amount = 10 wt%, P = 2.3–1.4 bar) a high biodiesel yield (89.6%) was achieved. The catalysts can be reused.

Lou et al. (2008) studied biodiesel production from WCO with 27.8 wt% of FFA over catalysts prepared from several carbohydrates (d-glucose, sucrose, cellulose, and starch). The materials showed higher conversion for both esterification and transesterification, compared to sulfated zirconia and niobic acid. The most active material for these reactions was a starch-derived catalyst. Under optimized conditions (methanol:oil = 20:1; catalyst loading = 10 wt% based on the weight of waste oils; stirring rate = 500 rpm; T = 80°C) a biodiesel yield of 92% was obtained, after 8 h. The starch-derived catalyst is recyclable.

Ahmad et al. (2018) investigated the transesterification of WCO (6 wt% of FFA) over a char-based acidic. Under optimized conditions (T = 65°C, t = 130 min, methanol:oil = 9:1 (molar ratio), 6 wt% catalyst), a yield to biodiesel of 96% was obtained. The catalyst was reused. After five cycles, a yield of 81% was obtained.

31.4 HETEROPOLYACIDS

Heteropolyacids are superacid catalysts used in different reactions. These materials have been immobilized on different supports such as activated carbons, zeolites, silica, and polymers (Patel et al., 2016; Sanchez et al., 2016)

Narkhede et al. (2014) studied biodiesel production by transesterification of WCO using 12-tungstosilicic acid anchored to SBA-15, as catalyst. Under optimized conditions (WCO:alcohol = 1:8 w/w, $m_{catalyst}$ = 0.3 g, T = 65°C, t = 8 h), a maximum conversion (86%) was reached. The catalyst can be reused.

Singh and Patel (2015) researched biodiesel production from WCO with methanol over phosphotungstate anchored on MCM-41, as catalyst. Before the transesterification of WCO, the authors studied the esterification of oleic acid with methanol. The catalyst showed high conversion for biodiesel production by esterification of oleic acid (about 89%). The transesterification of WCO with methanol was also studied over this catalyst. Under optimized conditions (oil:methanol = 1:8 w/w, $m_{catalyst}$ = 0.350 g, T = 65°C, t = 16 h) a 90% conversion was obtained. The catalyst was reused.

Srilatha et al. (2012) studied biodiesel production from WCO (with a high amount of FFA) using a heterogeneous process. This process has two steps: an acid step and a basic step. During the first step, the FFA reacted with methanol over a 25 wt% 12-tungstophosphoric acid (TPA), TPA/Nb_2O_5, as catalyst. In the second step, transesterification of oil with methanol over a ZnO/Na-Y zeolite catalyst was carried out. The catalyst with 20 wt% ZnO loading on Na-Y exhibited high activity. The acid catalyst and the base catalyst can be reused several times.

Cao et al. (2008) studied the transesterification reaction of WCO over a heteropolyacid ($H_3PW_{12}O_{40}.6H_2O$). Under optimized condition (methanol:oil = 70:1, T = 65°C, t = 16 h), there was 87% FAME. The catalyst can be reused.

Talebian-Kiakalaieh et al. (2013) researched biodiesel production from WCO using a heteropolyacid, as catalyst. A conversion of 88.6% was obtained under optimized conditions (t = 14 h, T = 65°C, methanol:oil = 70:1, catalyst loading = 10 wt%).

Jacobson et al. (2008) applied different solid acid catalysts to produce biodiesel from WCO (FFA content of 15%). Zinc stearate supported on silica gel showed high catalytic activity. Under optimized conditions (T = 200°C, oil:alcohol = 1:18, catalyst loading = 3 wt%, t = 10 h) a yield of 98% was obtained.

Li et al. (2009) studied the transesterification of WCO (acidity of 53.8 mg KOH. g^{-1}) over $Zn_{1.2}H_{0.6}PW_{12}O_{40}$, immobilized on nanotubes. This material had Lewis and Bronsted acid sites. The yield to biodiesel was about 95%. The catalyst exhibited good catalysis and can be reused.

31.5   METAL OXIDES

Lam et al. (2009) studied the transesterification of WCO (acidity of 5 mg KOH g^{-1}) over sulfate tin oxide (SO_4^{2-}/SnO_2) as catalyst. The reaction conditions were optimized. Under these conditions (T = 150°C, t = 3 h, methanol:oil = 15:1, SO_4^{2-}/SnO_2-SiO_2 = 3%), a yield of 92.3% in biodiesel was obtained. In another study, Lam and Lee (2011) used SO_4^{2-}/SnO_2-SiO_2 as a catalyst for biodiesel production by transesterification of WCO (content of FFA = 2.54%). These authors studied the effect of the methanol/ethanol mixture on the transesterification. A yield of 81.4% to biodiesel was obtained under optimized conditions (methanol:ethanol:oil = 9:6:1 (molar ratio), T = 150°C, t = 1 h, catalyst amount = 6 wt%).

Peng et al. 2008 studied biodiesel production from WCO over SO_4^{2-}/TiO_2-SiO_2 as catalyst. The optimized reaction conditions were T = 200°C, methanol:oil = 9:1 (molar ratio), and catalyst concentration = 3 wt%. Under this reaction condition, about 90 % of conversion can be obtained. The catalyst can be recycled. This catalyst can be used for esterification and transesterification reactions.

Fu et al. (2009) researched the transesterification of WCO over an SO_4^{2-}/ZrO_2 catalyst. Under optimized reaction conditions (T = 120°C, t = 4 h, methanol:oil = 9:1 (molar ratio), 3 wt% catalyst), a biodiesel yield of 93.6% was obtained.

Komintarachat and Chuepeng (2009) studied the transesterification of WCO (15% FFA) with methanol over WOx supported on Al_2O_3, SiO_2, SnO_2, and ZnO. The WOx/

Al_2O_3 catalyst exhibited the highest activity. Under optimized conditions (T = 383 K, t = 2 h, methanol:WCO ratio = 0.3, catalyst:WCO ratio = 1.0%), a yield of 97.5% was obtained. The WO_x/Al_2O_3 catalyst can be reused.

Elhassan et al. (2015) studied the transesterification of WCO (FFA content of 17.5%) over nanoparticles of sulfated zirconia doped with manganese sulfate as a catalyst. A biodiesel yield of 96.5% was reached under optimized conditions (T = 180°C, t = 6 h, 3 wt% catalyst, methanol:oil = 20:1 (molar ratio)). The material was recycled.

31.6 POLYMERS WITH ACTIVE SITES

Ozbay et al. (2008) researched biodiesel production from WCO with a high amount of FFA over acidic ion-exchange resins: Amberlyst-15 (A-15), Amberlyst-35 (A-35), Amberlyst-16 (A-16), and Dowex (HCR-W2). All material tests were active for esterification. When the reaction temperature and the catalyst amount increased, the FFA conversion increased as well. Resin A-15 showed the highest FFA conversion (about 45.7%) at 60°C with 2 wt% catalyst amounts.

Tropecelo et al. (2016) studied biodiesel conversion from WCO with methanol over poly(vinyl alcohol) with sulfonic acid groups ($PVA-SO_3H$) and polystyrene with sulfonic acid groups ($PS-SO_3H$), at 60°C. The $PVA-SO_3H$ material exhibited greater activity than the $PS-SO_3H$ material. At T = 60°C, when the amount of $PVA-SO_3H$ increased, the WCO conversion into FAME increased. A fall in biodiesel production was obtained when the reaction of WCO with ethanol was performed over $PVA-SO_3H$ at 60°C. The WCO conversion into FAEE increased when the temperature was increased from 60 to 80°C. The $PVA-SO_3H$ catalyst can be reused and recycled.

Caiado and Castanheiro (2018) studied biodiesel production from WCO with methanol over $SBA-15-SO_3H$ dispersed in poly (vinyl alcohol) as an acid catalyst. The activity of composite increased with the amount of $SBA-15-SO_3H$ immobilized on PVA, due to the amount of the sulfonic acid group on the polymeric matrix. The composite catalyst showed good stability.

31.7 CONCLUSIONS

Biodiesel can be produced from edible oils with methanol or ethanol using homogeneous catalysts. This biofuel is a renewable energy, with low toxicity. However, in order to reduce the costs of biodiesel production, other raw materials have been used: WCO, animal fats, and nonedible oils (second-generation biofuel). Biodiesel production is performed using homogeneous catalysts, which can cause some problems such as difficulty to remove the catalyst from the reaction mixture and the possibility of soap production and corrosion. The use of heterogeneous catalysts can overcome these problems. Different heterogeneous acid catalysts (zeolites, heteropolyacids, metal oxides, materials with sulfonic acid groups) have been employed in biodiesel production from WCO.

REFERENCES

Ahmad, J., U. Rashid, F. Patuzzi, M. Baratieri, and Y. H. Taufiq-Yap. 2018. Synthesis of char-based acidic catalyst for methanolysis of waste cooking oil: An insight into a possible valorization pathway for the solid by-product of gasification. *Energy Conversion and Management* 158: 186–192.

Avhad, M. R. and J. M. Marchetti. 2015. A review on recent advancement in catalytic materials for biodiesel production. *Renewable and Sustainable Energy Reviews* 50: 696–718.

Brito, A., M. E. Borges, and N. Otero. 2007. Zeolite Y as a heterogeneous catalyst in biodiesel fuel production from used vegetable oil. *Energy & Fuels* 21: 3280–3283.

Caiado, M. and J. E. Castanheiro 2018. Biodiesel production from waste cooking oil in the presence of composite catalysts. *Advances in Chemistry Research* 43: 245–258.

Cao, F., Y. Chen, and F. Zhai, et al. 2008. Biodiesel production from high acid value waste frying oil catalyzed by superacid heteropolyacid. *Biotechnology and Bioengineering* 101: 93–100.

Chung, K. H., D. R. Chang, and B. G. Park. 2008. Removal of free fatty acid in waste frying oil with methanol on zeolite catalysts. *Bioresource Technology* 99: 7438–7443.

Elhassan, F. H., U. Rashid, and Y. H. Taufiq-Yap. 2015. Synthesis of waste cooking oil-based biodiesel via effectual recyclable bi-functional Fe_2O_3-MnO-SO_4^{2-}/ZrO_2 nanoparticle solid catalyst. *Fuel* 142: 38–45.

Fonseca, J. M., J. G. Teleken, V. de Cinque Almeida, and C. da Silva. 2019. Biodiesel from waste frying oils: Methods of production and purification. *Energy Conversion and Management* 184: 205–218.

Fu, B., L. Gao, L. Niu, R. Wei, and G. Xiao. 2009. Biodiesel from waste cooking oil via heterogeneous superacid catalyst SO_4^{2-}/ZrO_2. *Energy Fuels* 23: 569–572.

Jacobson, K., R. Gopinath, L. C. Meher, and A. K. Dalai. 2008. Solid acid catalyzed biodiesel production from waste cooking oil. *Applied Catalysis B: Environmental* 85: 86–91.

Kiss, F. E., M. Jovanovic, and G. C. Boskovic. 2010. Economic and ecological aspects of biodiesel production over homogeneous and heterogeneous catalysts. *Fuel Processing Technology* 91:1316–1320.

Komintarachat, C. and S. Chuepeng. 2009. Solid acid catalyst for biodiesel production from waste used cooking oils. *Industrial & Engineering Chemistry Research* 48: 9350–9353.

Lam, M. K. and K. T. Lee. 2011. Mixed methanol-ethanol technology to produce greener biodiesel from waste cooking oil: A breakthrough for SO_4^{2-}/SnO_2-SiO_2 catalyst. *Fuel Processing Technology* 92: 1639–1645.

Lam, M. K., K. T. Lee, and A. R. Mohamed. 2009. Sulfated tin oxide as solid superacid catalyst for transesterification of waste cooking oil: An optimization study. *Applied Catalysis B: Environmental* 93: 134–139.

Lee, A. F. and K. Wilson. 2015. Recent developments in heterogeneous catalysis for the sustainable production of biodiesel. *Catalysis Today* 242: 3–18.

Li, J., X. Wang, W. Zhu, and F. Cao. 2009. $Zn_{1.2}H_{0.6}PW_{12}O_{40}$ nanotubes with double acid sites as heterogeneous catalysts for the production of biodiesel from waste cooking oil. *Chem Sus Chem* 2: 177–183.

Li, M., Y. Zheng, Y. Chen, and X. Zhu. 2014. Biodiesel production from waste cooking oil using a heterogeneous catalyst from pyrolyzed rice husk. *Bioresource Technology* 154: 345–348.

Lima, H. and J. E. Castanheiro. 2018. Valorization of waste cooking oil into biodiesel over heterogeneous catalysts. In *Recycled Cooking Oil: Processing and Uses*, ed. K. Garner, pp. 69–88. New York, NY: Nova Science Publishers.

Lou, W. Y., M. H. Zong, and Z. Q. Duan. 2008. Efficient production of biodiesel from free fatty acid-containing waste oils using various carbohydrates-derived solid acid catalysts. *Bioresource Technology* 99: 8752–8758.

Narkhede, N., V. Brahmkhatri, and A. Patel. 2014. Efficient synthesis of biodiesel from waste cooking oil using solid acid catalyst comprising 12-tungstosilicic acid and SBA-15. *Fuel* 135: 253–261.

Ozbay, N., N. Oktar, and N. A. Tapan. 2008. Esterification of free fatty acids in waste cooking oils (WCO): Role of ion-exchange resins. *Fuel* 87: 1789–1798.

Patel, A., N. Narkhede, S. Singh, and S. Pathan. 2016. Keggin-type lacunary and transition metal substituted polyoxometalates as heterogeneous catalysts: A recent progress. *Catalysis Reviews* 58: 337–370.

Peng, B. X., Q. Shu, and J. F. Wang, et al. 2008. Biodiesel production from waste oil feedstocks by solid acid catalysis. *Process Safety and Environment Protection* 86: 441–447.

Sanchez, L. M., H. J. Thomas, M. J. Climent, G. P. Romanelli, and S. Iborra. 2016. Heteropolycompounds as catalysts for biomass product transformations. *Catalysis Reviews* 58: 497–586.

Serrano, D. P., J. A. Melero, G. Morales, J. Iglesias, and P. Pizarro. 2018. Progress in the design of zeolite catalysts for biomass conversion into biofuels and bio-based chemicals. *Catalysis Reviews* 60:1–70.

Singh, S. and A. Patel. 2015. Mono lacunary phosphotungstate anchored to MCM-41 as recyclable catalyst for biodiesel production via transesterification of waste cooking oil. *Fuel* 159: 720–727.

Srilatha, K., B. L. A. P. Devi, N. Lingaiah, R. B. N. Prasad, and P. S. S. Prasad. 2012. Biodiesel production from used cooking oil by two-step heterogeneous catalyzed process. *Bioresource Technology* 119: 306–311.

Talebian-Kiakalaieh, A., N. A. S. Amin, A. Zarei, and I. Noshadi. 2013. Transesterification of waste cooking oil by heteropoly acid (HPA) catalyst: Optimization and kinetic model. *Applied Energy* 102: 283–292.

Tran, T. T. V., S. Kaiprommarat, and S. Kongparakul, et al. 2016. Green biodiesel production from waste cooking oil using an environmentally benign acid catalyst. *Waste Management* 52: 367–374.

Tropecelo, A. I., C. S. Caetano, M. Caiado, and J. E. Castanheiro. 2016. Biodiesel production from waste cooking oil over sulfonated catalysts. *Energy Sources, Part A: Recovery, Utilization and Environmental Effects* 38: 174–182.

32 Microbial Biodiesel Production using Microbes in General

K. V. V. Satyannarayana

Randeep Singh

I. Ganesh Moorthy

R. Vinoth Kumar

CONTENTS

32.1   INTRODUCTION

Biodiesel is well accepted in the world due to its properties, such as renewability, non-toxicity, high flash point, biodegradability, and eco-friendly nature, compared to conventional fossil fuels. Furthermore, biodiesel can be blended with petrodiesels for use in the transport sector. Feedstock selection is a crucial step in biodiesel production, as it affects various factors, such as cost, yield, composition, and purity. Generally, feedstocks are selected based on local availability and type of source (waste or edible or nonedible). Importantly, the selection of feedstocks is done on the basis of parameters like oil content availability, chemical composition, suitability, and physical properties.

Recently, oil-accumulating microbes have been reported as a potential feedstock for biodiesel production. Microalgae, yeasts, bacteria, and fungi are typically called third-generation feedstocks for biodiesel production. These microbial lipids are more beneficial than plant-derived oils due to their availability and cultivation across the seasons on non-fertile lands. Also, the rate of oil production is normally two to three times higher than conventional energy crops. Furthermore, lipid oil composition derived from microbes is almost similar to vegetable oils.

Microbial sources for lipid production have these advantages:

1. Quell food versus fuel controversies;
2. Microorganisms grow faster in comparison to plants and animals;
3. Lipid production is not restricted to the season, climate, and geographical location;
4. Lipid composition and heat content is similar to oil extracted from plant and animal resources.

These three types of microbes, namely algae, yeasts, and bacteria, are discussed below.

32.1.1 ALGAE

Third-generation biofuels have been explored since the middle of the 19th century. Algae are promising feedstocks for the third-generation of biodiesel. They belong to the category of eukaryotic photosynthetic microorganisms (Musolino, 2016). These microorganisms grow rapidly (Wen and Johnson, 2009) and are viable in harsh conditions. The use of algae is advantageous as they require only minimum nutrients (CO_2, H_2O, and sunlight) to grow. Further, some of the algae are capable of growing in wastewaters that again reduce the cost of algal growth and wastewater treatment, along with reduced consumption of freshwater. It is estimated that up till now only 30,000 algal species out of 50,000 have been studied. Furthermore, literature reports suggest that algae can yield energy per acre 30 times more than other feedstocks, such as soybeans. However, the commercial potential has not been explored much.

The lipid content of different species of microalgae is given in Table 32.1. Along with lipids, microalgae also include carbohydrates, proteins, and unique products like carotenoids, fatty acids, polymers, enzymes, and peptides.

For the production of biodiesel microbial sources must have a lipid content, so that the process can be economical. Some of the important microbial sources which are used for biodiesel production are listed in Table 32.2.

32.1.2 BIODIESEL FROM YEASTS

Microalgae are principally phototrophic in nature and able to accumulate large amounts of lipids swiftly. Microalgae are capable of producing 'single cell oils' (SCO) effectively. However, they require a large area for cultivation and are subject to seasonal variations. Therefore, oleaginous yeasts are a better choice for SCO

TABLE 32.1

Lipid Content of Different Microalgae Species

Microalgae	Lipid Content (%)
Botryococcus braunii	25–75
Chlorella protothecoides	14–57
Crypthecodinium cohnii	20–51
Dunaliella tertiolecta	16–71
Nannochloris sp.	20–56
Neochloris oleoabundans	29–6
Phaeodactylum tricornutum	18–57
Schizochytrium sp.	50–77
Skeletonema coastatum	13–51

Source: Dahman et al. (2019)

production compared to algae. Further, the growth rate of yeasts is higher than algae which results in their low duplication time, sometimes as low as less than an hour. Furthermore, unlike phototrophic algae, yeast growth is not affected by seasonal variations. Also, yeasts are less prone to microbial contamination. Lignocellulosic biomass for saccharification is obtained from different sources, namely agricultural residues, nonedible or energy crops, and forest and industrial residues. The crop residues are the remains following crop collection, such as corn stover, bagasse, rice husk, and wheat straws. Important forest residues are deadwood and broken tree branches. Further, wood sources for lipids are sawdust, recycled woods (obtained from the demolition of buildings), cardboard boxes, and packing crates. Industrial residues like palm oil, paper and pulp, and rubber are also suitable for biodiesel production. Furthermore, wastewaters containing organic content without toxic compounds can also be used. The general process of production of biodiesel from yeasts or bacteria is shown in Figure 32.1.

To boost the digestion of lignocellulosic materials, different pretreatments are available. These are divided into mechanical, thermal, acid and alkali, oxidative, steam explosion, liquid hot water, CO_2, ozonolysis, green solvents, microwave, and biological methods (Passos et al., 2014).

Yeasts are the storehouse of different types of lipids (Blagovic et al., 2001; Losel, 1988; Rattray, 1988) that include monoacylglycerol, triacylglycerols, fatty acids, free sterols, steryl esters, glycerophospholipids, cardiolipins, glycolipids, sphingolipids and hydrocarbons, long-chain alcohols, polyprenols, waxes, and isoprenoid quinines. Oleaginous yeasts accumulate lipids in triacylglycerols (TAG) form in intracellular lipid bodies. These yeast species are the most effective and efficient for lipid production. For example, *Cryptococcus curvatus* reportedly accumulates approximately 60% lipids (Meesters et al., 1996; Moon et al., 1978). Further, these yeast species can grow to a cell density of 118 gL^{-1} under well-maintained culture conditions.

TABLE 32.2
Microbial Sources of Lipids for Biodiesel Production

Sl. No.	Genus	Species	Lipid (%w/w)
1	Myxozyma	mellibiosi, udenii	23.00, 20.30
2	Lipomyces	lipofer, tetrasorus, starkeyi, doorenjongii, kockii	43.00, 66.50, 62.00,72.30, 77.80
3	Kodamaea	ohmeri	53.28
4	Metschnikowia	pulcheriima, gruessii	30.00, 34.00
5	Cyberlindnera	jadinii, saturnus	20.00–30.00, 25.00
6	Candida	freyschussii	32
7	Wickerhamomyces	ciferrii	22
8	Schwanniomyces	occidentalis	23
9	Kurtzmaniella	cleridarum	33
10	Candida	diddensiae, tenuis, tropicalis	37.00, 20.4–56.58, 23.00
11	Geotrichum	fermentans	19.5
12	Magnusiomyces	magnusii	27
13	Galactomyces	pseudocandidus	28.04
14	Torulaspora	delbrueckii	40
15	Trigonopsis	variabilis	20.00–43.70
16	Yarrowia	lipolytica	37.60–54.80
17	Rhodotorula	minuta, mucilaginosa, colostri, graminis, glutinis, babjevae, diobovatum, toruloides, fluvial, glacialis, terpenoidalis	24.62, 28.00, 2.89, 41.10, 35.40, 65.00, 20.00–41.00, 19.00–51.10, 25.00, 68.00, 27.00
18	Leucosporidiella	creatinivora	61.00–62.00
19	Rhodosporidium	paludigenum, sphaerocarpum	31.00, 43.00
20	Cryptococcus	terreus, terricola, aerius, adeliensis, albidus, wieringae, oerirensis	51.00, 69.30, 63.30, 32.10, 40.10, 23.00–53.00, 25.00
21	Trichosporon	coremiiforme, domesticum, montevideense,brassicae	37.80, 35.16, 25.60, 20.34
22	Aspergillus	niger, terreus	9.60, 37.40
23	Cunninghamella	elegans	33.6
24	Mucor	circinelloides, plumbeus	23.80, 20.60
25	Neosartorya	fischeri	8.8
26	Rhizopus	oryzae	34.8
27	Thermomyces	lanuginosus	21
28	Mortierella	isabellina, vinacea	67.00, 51.90
29	Chaetomium	globosum	5.1
30	Botryococcus	braunii	25.00–75.00
31	Chaetoceros	muelleri, calcitrans	33.60, 14.60–16.40
32	Chlorella	emersonii, protothecoides, sorokiniana, vulgaris, pyrenoidosa	25.00–63.00, 14.60–57.80, 19.00–22.00, 5.00–58.00, 2.00
33	Crypthecodinium	cohnii	20.00–51.10
34	Dunaliella	salina, primolecta, tertiolecta	6.00–25.00, 23.10, 16.70–71.00
35	Euglena	gracilis	14.00–20.00
36	Haematococcus	pluvialis	25

Source: Gujjala et al. (2017)

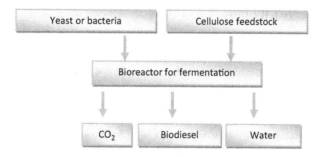

FIGURE 32.1   Production of biodiesel from yeasts or bacteria.

There are some yeast species well known for their capability to accumulate SCO, for example, *Lipomyces starkeyi*, *Cryptococcus curvatus*, *Rhodosporidium toruloides*, and *Rhodotorula glutinis*. *R. glutinis* is ideal for SCO production as it can be cultured in low-cost carbon sources. Further, it is very effective and efficient in the accumulation of SCO with a total oil percentage up to 72%. It is also capable of producing carotenoids, such as carotene, torularhodin, and torulene. The carotenoid composition and amount depend on the culture conditions.

The cultivation of *R. glutinis* in paper and pulp wastewater produces microbial lipids. Results show that *R. glutinis* yeast used in this study was successfully able to convert glycerol and a mixture of xylose and glucose to biomass very efficiently. Also, the fatty acid produced from these yeasts is similar in composition to the fatty acids produced from vegetable oils and are suitable for biodiesel production. The 'cetane number' (CN), higher heating value (FAMEs), and saponification number are quite similar to biodiesel produced from vegetable oils (Amirsadeghi et al.,2015).

Ricotta cheese whey is also a profitable use of a dairy by-product for microbial diesel production (Carota et al., 2017). RCW contains an appreciable amount of sugar that makes it a good growth media candidate for microbial diesel production.

In a recent study, yeast cells were successfully immobilized on sugarcane bagasse and a maximum biodiesel yield of 85.29% was reported. Further, parameters like overall biocatalyst loading, oil to methyl acetate molar ratio, water content, temperature, and reaction time were also studied (Surendhiram et al., 2014). This study proposes that the use of sugarcane bagasse immobilized yeast is cost-effective, eco-friendly, and an alternative method for carrying out enzymatic production of biodiesel on a large scale. The ideal characteristics or requirements of lipid producing yeast strains are high cell density, high oil production, rapid growth, osmotolerance, complete and simultaneous consumption of carbohydrates available, growth independent of added vitamins, and high resistance to inhibitors present in the hydrolysate (Sitepu et al., 2014).

The factor important for the type of oleaginous yeasts to be used depends on the following characteristics: (1) the potential for high lipid productivity; (2) the maximum TAG amount in the produced lipids; (3) easy downstream and upstream processing.

32.1.3   Bacterial Biodiesel

Hydrocarbons (alkenes/alkanes) are produced as metabolic by-products by various bacterial species (Bharti et al., 2014a). The hydrocarbons are produced from fatty acid TAG. TAGs are trimester fatty acids, with different properties that depend on the composition of the fatty acids. They are energy reserves for various eukaryotes, for example, fungi, yeasts, plants, and animals. The bacterial species that belong to the *Actinomycetes* family shows TAG production, such as bacteria belonging to the *Nocardia, Streptomyces, Rhodococcus,* and *Mycobacterium* category (Alvarez et al., 2002). The biosynthesis of lipids occurs during the exponential phase of growth of the bacteria and this process utilizes fatty acids as precursors. Further, the bacterial TAG can also be used as a raw material for biodiesel production (Bharti et al., 2014b).

The following equation can be used to estimate the percentage of 'polycyclic aromatic hydrocarbon' (PAH) degradation:

$$\% \text{PAH degradation} = \frac{(C_0 - C_f) - C_n}{C_0} \times 100 \qquad (32.1)$$

where C_0 is the initial PAHs concentration; C_f is the final concentration in test flasks; and C_n is the final concentration in control flasks.

$$\text{Specific growth rate} (\mu) = \frac{1}{X} \frac{dX}{dt} \qquad (32.2)$$

The specific PAH uptake rate and specific lipid accumulation rates by bacteria can be measured as:

$$\text{Specific PAH uptake rate} (q) = \frac{1}{C_p} \frac{dC_p}{dt} \qquad (32.3)$$

$$\text{Specific lipid accumulation rate} (L) = \frac{1}{C_L} \frac{dC_L}{dt} \qquad (32.4)$$

where μ represents the specific biomass growth rate (min^{-1}), q the specific PAH uptake rate (min^{-1}), L the specific accumulation rate (min^{-1}), X, C_p, and C_L the concentrations (mg L^{-1}) of biomass, and t the time (min).

Biokinetic models used for the measurement of bacterial parameters, such as the specific growth rate (min^{-1}), are given by:

$$\text{Monod model} (\mu) = \frac{\mu_{max}[S]}{K_s + [S]} \qquad (32.5)$$

$$\text{Haldane model} (\mu) = \frac{\mu_{max}[S]}{K_s + [S] + \dfrac{[S^2]}{K_i}} \qquad (32.6)$$

where μ represents the specific growth rate (min^{-1}), μ$_{max}$ the maximum specific growth rate (min^{-1}), [S] the PAH concentration (mg L^{-1}), Ks the half-saturation constant (mg L^{-1}), K$_i$ the inhibition constant (mg L^{-1}), and I the PAH inhibitory concentration (mg L^{-1}).

Chemolithotrophic bacteria perform a significant role in CO_2 sequestration and produce several products like lipids, exopolysaccharides, bioplastics, and fatty acid (Kumar and Thakur, 2018; Thakur et al., 2018). As compared to photosynthetic organisms (cyanobacteria), bacterial systems are advantageous as they can grow in large-scale systems with high growth rate, good metabolism, and genetic manipulation (Bharti et al., 2014a). Goswami et al. (2017a–b) reported that *R. opacus* was capable of PAH degradation. Further, the increase in *R. opacus* inoculum size shows a positive effect on biomass and lipid accumulation. In a simultaneous work, *R. opacus* shows anthracene degradation with high cell density and accumulation of lipid-rich biomass.

Fatty acid content and composition of biodiesel product mainly depends on the type of bacteria, carbon and nitrogen sources, and cultivation time (Sharma and Singh, 2017; Tasic et al., 2016; Zhao et al., 2016). The *Rhodococcus* type of bacteria has well been reported as a bacterial microbe with the capability of holding greater than 20% of TAG with respect to its total weight. Further, it can grow in nitrogen-deficient environments using available carbon sources (such as sugars, hydrocarbons, and organic acids). Also, it carries the capability of using different aromatic compounds (Kurosawa et al., 2015). The *Rhodococcus* species grown in oil palm biomass hydrolysate accumulates a high amount of fatty acid content and the produced biodiesel shows improved performance compared to algal biodiesel and petrodiesel (Bhatia et al., 2017).

With increasing industrialization and urbanization, the volume of sludge produced from wastewaters has increased drastically (Kumar et al., 2016). The inherent lipid content of sludge makes it ideal feedstock for oleaginous bacteria (Olkiewicz et al., 2014). The biodiesel obtained from oleaginous bacteria shows sludge is a promising option for biodiesel production. Analysis of 5% blends of this biodiesel shows better fuel properties and that it can be used in existing engines without any modifications (Kumar et al., 2018). The use of wastewater for lipid accumulation using *Rhodococcus opacus* bacteria is a prominent idea for biodiesel production. The bacterial high cell density and lipid-rich biomass result in high biomass content and 'chemical oxygen demand' (COD) removal (Goswami et al., 2017a–b).

32.2 TRANSESTERIFICATION REACTION

The transesterification reaction between triglycerides and alcohol in the presence of a catalyst produces biodiesel (Kumar et al., 2017). The reaction was first proposed by Rochieder before 1846 while analyzing the production of glycerol from castor oil, known as ethanolysis (Demirbas, 2009). In this process, three successive reversible reactions take place. In the first reaction, triglycerides are converted into diglycerides, then these diglycerides are converted into monoglycerides, that are finally converted into glycerol. In each step of the process, one ester molecule and three 'fatty acid methyl esters' (FAME) molecules are produced from a single triglyceride molecule (Sharma and Singh, 2008). The conversion of monoglycerides into alkyl esters is the rate-limiting step that defines the rate of reaction. Therefore, in simple terms, esterification is the process that takes place in the presence of a catalyst (alkali or acid) that splits fatty acid and alcohol molecules to make them react with the separated ester (Kumar et al., 2018). This is a well-known process for biodiesel production as the viscosity of the obtained end product is low (Demirbas, 2009). Further, glycerol, the end product of this process, carries high commercial value and is used in many applications. The general process of transesterification process is shown in Figure 32.2.

Lipase is an example of a biological catalyst that is commonly used in the esterification process. It is advantageous compared to acid and alkali catalysts; however, commercially it is not viable and cost-effective. Further, transesterification can also take place with or without catalysts by using one or two-degree monohydric aliphatic alcohols that contain one to eight carbon atoms. Therefore, ethanol and methanol are employed extensively. Furthermore, ethanol is preferred over methanol since it can be produced from agricultural wastes (Demirbas, 2005). However, sometimes methanol is preferred due to its better physical and chemical properties, such as having the shortest chain length and polarity. The variables affecting the esterification reaction are temperature, lipid to alcohol molar ratio, water, and 'free fatty acids' (FFA) content (Bharti et al., 2014a–b; Kumar et al., 2017). Further, it is reported that increase in the reaction temperatures and alcohol to oil molar ratio increases the production of FAMEs (Bharti et al., 2014a; Kumar et al., 2016; Ma and Hanna, 1999).

32.2.1 Catalytic Transesterification Methods

The conversion of oils and lipids into biodiesel occurs at a certain temperature using a transesterification reaction in the presence of excess methanol and a catalyst (acid or alkali). The catalyst could be alkali (Bharti et al., 2014a), acid (Kumar et al., 2016; Mondala et al., 2009), or an enzyme (Shieh et al., 2003).

32.2.1.1 Acid-Catalyzed Transesterification

The acid-catalyzed transesterification reaction uses acid catalysts, such as sulfuric acid (Demirbas, 2005; Kumar et al., 2015, 2016; Ma and Hanna, 1999), ferric sulfate

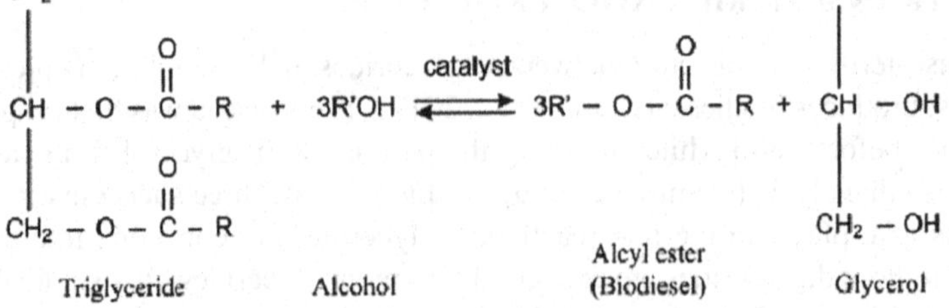

FIGURE 32.2 Transesterification of triglyceride with alcohol. (*Source:* Tabernero et al. (2012)).

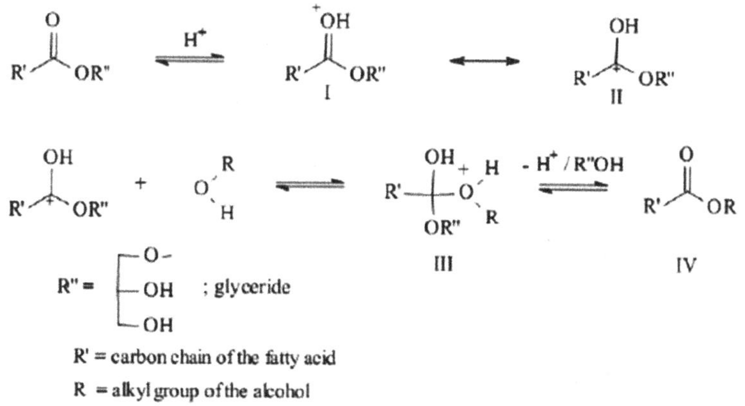

FIGURE 32.3 Mechanism of acid-based transesterification reaction. (*Source:* Ambat et al. (2018)).

(Wang et al., 2007), sulfonic acid (Guerreiro et al., 2006), boron trifluoride, and hydrogen chloride (Darnoko and Cheryan, 2000). Firstly, the acid catalyst is mixed appropriately with methanol. This acidified methanol is then taken with the raw materials (oils or lipids) for a transesterification reaction. This acid-catalyzed reaction results in high yields; however, it has a slow reaction rate. Further, the oil to alcohol molar ratio defines the yield of the transesterification reaction (Kumar et al., 2020a). The general process of acid-catalyzed transesterification reaction is represented in Figure 32.3.

32.2.1.2 Alkali-Catalyzed Transesterification

Alkali-catalyzed transesterification preferably uses potassium or sodium hydroxide as a catalyst. Firstly, as in the case of acid-catalyzed transesterification, the catalyst is mixed with methanol to form the alkaline methanol. This alkaline methanol is added to the reaction vessel along with oil, as for the transesterification reaction. This reaction mixture is then vigorously stirred at 340 K for 2 h at atmospheric pressure (Demirbas, 2009). On completion of the reaction, phase separation takes place and the mixture constituents settle, based on their density differences. The final separated phases are ester and crude glycerol. Further, the alkali catalysts, for example, alkaline metal alkoxides, provide higher yields (>98%) due to their higher activity (Thliveros et al. 2014). The general process of an alkali-catalyzed transesterification reaction is shown in Figure 32.4.

32.2.1.3 Enzyme-Catalyzed Transesterification

The use of biological catalysts or enzymes is the latest introduction to biodiesel production using a transesterification reaction (Khosla et al., 2017; Shieh et al., 2003). Recently, a lipase enzyme from *Chromobacterium viscosum*, *Candida rugosa*, and *Porcine pancreas* was chosen for transesterification reaction. The process of an enzyme-catalyzed transesterification reaction is represented in Figure 32.5. Lipase

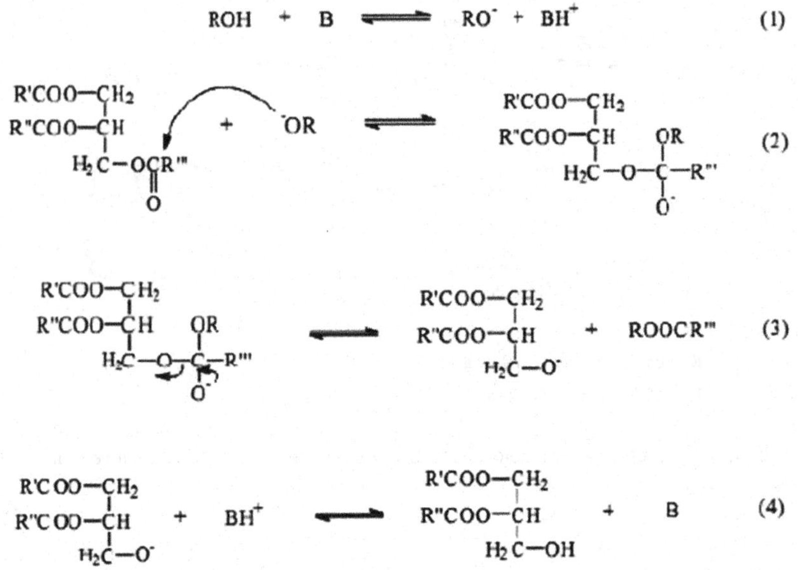
ROH + B ⇌ RO⁻ + BH⁺ (1)
(2)
ROOCR''' (3)
BH⁺ B (4)

FIGURE 32.4 Mechanism of an alkali-based transesterification reaction. (*Source:* Ambat et al. (2018)).

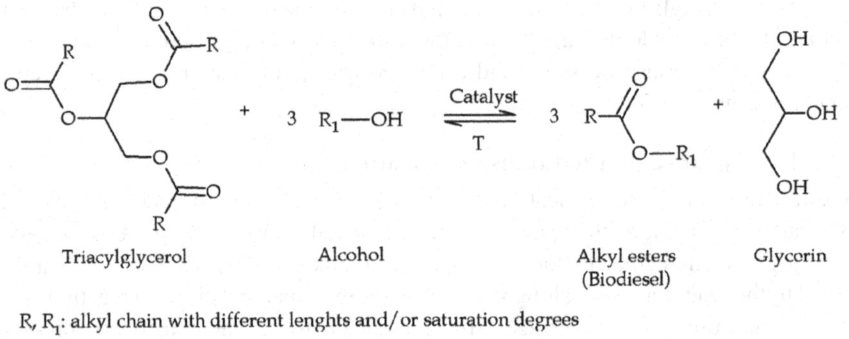
Catalyst
T
3
Triacylglycerol
Alcohol
Alkyl esters
(Biodiesel)
Glycerin
R, R1: alkyl chain with different lenghts and/or saturation degrees

FIGURE 32.5 Mechanism of an enzyme-based transesterification reaction. (*Source:* Ambat et al. (2018)).

from *C. viscosum* was the only one that produced higher yields (Shah et al., 2004). Transesterification by enzymes has been developed and reported in the literature; however, its commercial use is still not viable. The yield and sustainability of enzyme-catalyzed reactions are still low in comparison to alkali-catalyzed transesterification (Kumar et al., 2020a).

There are some advantages and disadvantages in the production of biodiesel using various technologies. Each technique has its own advantages and disadvantages.

TABLE 32.3
Pros and Cons of Various Biodiesel Production Techniques

S. no	Production Technologies	Advantageous	Limitations
1	Pyrolysis	Simple, pollution-free process	Requires high temperature, needs an expensive apparatus, low purity due to intolerable amount of carbon residues, clinker
2	Micro-emulsion	Simple process	High viscosity, poor volatility and stability.
3	Dilution	Simple process	Incomplete combustion, the formation of carbon in the engine
4	Microwave technology	Short reaction time, limited heat loss	Removal of the catalyst after the process is required, process conversion depends on catalyst activity. Not suitable for the feed stocks with solids
5	Reactive distillation	Can be used for feedstock with high FFA content, simple process, less usage of methanol, easy separation of products	High energy requirements, process conversion depends on catalyst efficiency
6	Catalytic distillation	Easy product separation	Conversion rate as well as solvent usage for post-treatment depends on catalyst recovery
7	Supercritical fluid extraction method	Less reaction time, high conversion, no catalyst required	High energy consumption and apparatus cost.
8	Transesterification	Suitable for industrialized production, fuel properties are comparable to diesel	Conversion efficiency, cost, catalyst reusability, applicability for feedstock with water and high FFA content depends on the type of catalyst used.

Source: Ambat et al. (2018).

Some of the pros and cons of the various production technologies are shown in Table 32.3.

32.3   BIODIESEL CHARACTERISTICS AND ANALYTICAL METHODS

Biodiesels have superior lubricating characteristics compared to diesel fuels because of high viscosity (Knothe, 2005). Characteristics include biodegradability, non-toxicity, renewability, and water immiscibility. Also, color varies depending upon the feedstock used in between dark and gold. Further, important sought after biodiesel characteristics are high CN and low sulfur content. The flashpoint of biodiesel (>130°C, >266°F) is higher compared to diesel fuel (−45°C, −52°F) which makes it

TABLE 32.4
Chemical Characteristics of Biodiesel and its Significance

Characteristics	Definition	Significance
Melt point	Temperature at which the oil in solid form starts to melt	Important physical property and used for the indication of purity
Cloud point	Temperature at which waxes starts to crystallize	Indicates the lowest temperature of the fuel for operability
Flash point	Minimum temperature at which the fuel will ignite on application of ignition source	Used as a safe index for biofuels and indicates the level of purification
Iodine value	Amount of iodine, measured in grams, absorbed by 100 g of given oil	Used as a measure of the chemical stability properties of different biodiesel fuels
Viscosity	Measuring the amount of time taken for a given measure of oil to pass through an orifice of a specified size	The higher the viscosity, the higher is the tendency of the fuel to form engine deposits
Aniline point/CN	Relative measure of the interval between the beginning of injection and auto-ignition of the fuel	The higher the CN of the fuel, the shorter the delay interval and the greater its combustibility
Density	Weight per unit volume	Oils that are denser contain more energy
Ash content	Measure of the amount of metals contained in the fuel	High concentrations of these materials can cause injector tip plugging, combustion deposits, and injection system wear

Source: Kumar et al. (2020b).

safe to handle and store. Biodiesel shows low vapor pressure and a high boiling point. The calorific value and density vary around 33.27 mg L^{-1} and 0.88 g cm^{-3}, respectively. Furthermore, there are some quality parameters that define biodiesel characteristics such as viscosity, specific gravity, flash point, ash content, refractive index, calorific value, iodine number, saponification value, fatty acid content, and acid value (Kumar et al., 2020b). The chemical characteristics of biodiesel and their significance is shown in Table 32.4.

The properties required to estimate the quality of biodiesel are viscosity (η), pour point (PP), cloud point (CP), CN, saponification value (SV), iodine value (IV), and degree of unsaturation (DU). These parameters are analyzed by using the following empirical equations:

$$\eta = 0.235\,M_{CN} - 0.468\,M_{ds} \tag{32.7}$$

$$CN = 3.930\,M_{CN} - 15.93\,M_{ds} \tag{32.8}$$

$$CP = 0.526\left(P_{FAME}\right) - 4.992 \tag{32.9}$$

$$PP = 18.80_{WC} - 1.0\,M_{UF} \tag{32.10}$$

$$SV = \frac{\sum^{560} X P_{CF}}{M_r} \qquad (32.11)$$

$$IV = \frac{\sum^{256 X} PCF \ X \ N_{db}^0}{Mr} \qquad (32.12)$$

$$DU = M_{MUFA} + \left(2 \times M_{PUFA}\right) \qquad (32.13)$$

where M_{CN} represents the weighted average number of carbon units in the FAMEs, M_{ds} the weighted average number of double bonds, M_{UF} the total unsaturated FAME content (% weight), P_{FAME} the percentage content of palmitic acid methyl ester, P_{CF} the percentage of each fatty acid, M_r the molecular mass of the individual fatty acids, N_{db}^0 the number of double bonds, M_{MUFA} the monosaturated fatty acids (% weight), and M_{PUFA} the polyunsaturated fatty acids (% weight).

In general, two types of standard characterization techniques are used for the analysis of biodiesel, namely chromatographic and spectroscopic analysis. Further, the chromatographic technique is of different types, namely liquid chromatography, gas chromatography, gel permeation chromatography, size exclusion chromatography, supercritical fluid chromatography, and thin-layer chromatography. On the other hand, different types of spectroscopic techniques are infra-red, fluorescence, ultraviolet, proton nuclear magnetic resonance, and inductively coupled plasma mass spectrometry. Other methods and techniques include viscometry, refractive index, titration, and enzymatic and wet chemical methods. Out of all the techniques, gas chromatography due to its accuracy is the most preferred method for biodiesel analysis.

32.4   FUTURE PROSPECTS AND CHALLENGES

Biodiesel is a potential alternative source of renewable energy; however, there are many hurdles for its commercial use. Presently, biodiesel cost is double that of fossil fuels, therefore there is a need to improve the available biodiesel production methods and technology. For example, in the algal biodiesel production process, the dewatering step is one of the costliest steps. Therefore, technology should be developed to lower the cost of this step as well as of the overall process. Furthermore, research should be focused on the mass cultivation of microbes for biodiesel production using low-cost raw materials; for example, mass production of algae using industrial and municipal wastewaters. The survival ability of microalgae or other microbes could also be enhanced by a genetic engineering approach. The main focus of the research should be to develop efficient production techniques that can help to reduce the overall cost of biodiesel production. In addition to biodiesel, future research should also consider the effective and efficient extraction of many by-products from microbial biomass. Altogether, there is a need to develop efficient engines for the effective use of biodiesels.

32.5 CONCLUSION

The present chapter has discussed the potential sources for the production of biodiesel using microbes like algae, fungi, and yeasts. Algae, yeasts, and fungi all have their own advantages and disadvantages that have been discussed in detail in this chapter. The selection of microbial species depends on the type of feedstock used and their culture conditions. These microbial sources promise effective biodiesel production in less time, and the use of biodiesel leads to a safe environment. The chapter also focused on methods to accumulate high lipids and lipid extraction techniques. Further, selection and analysis of various essential biodiesel parameters like pH, salinity, temperature, and feedstock have been discussed. However, there is much scope for future research on the optimization of these parameters for getting high-quality biodiesel.

REFERENCES

Alvarez, H. M., H. Luftmann, and R. A. Silva, et al. 2002. Identification of phenyldecanoic acid as a constituent of triacylglycerols and wax ester produced by *Rhodococcus opacus* PD630. *Microbiology* 148: 1407–1412.

Ambat, I., V. Srivastava, and M. Sillanpaa. 2018. Recent advancement in biodiesel production methodologies using various feedstock: A review. *Renewable and Sustainable Energy Reviews* 90: 356–369.

Amirsadeghi, M., S. Shields-Menard, W. T. French, and R. Hernandez. 2015. Lipid production by *Rhodotorula glutinis* from pulp and paper wastewater for biodiesel production. *Journal of Sustainable Bioenergy Systems* 5: 114–125.

Bharti, R. K., S. Srivastava, and I. S. Thakur. 2014a. Production and characterization of biodiesel from carbon dioxide concentrating chemolithotrophic bacteria, *Serratia* sp. ISTD04. *Bioresource Technology* 153: 189–197.

Bharti, R. K., S. Srivastava, and I. S. Thakur. 2014b. Extraction of extracellular lipids from chemoautotrophic bacteria *Serratia* sp. ISTD04 for production of biodiesel. *Bioresource Technology* 165: 201–204.

Bhatia, S. K., H. Kim, and H.-S. Song, et al. 2017. Microbial biodiesel production from oil palm biomass hydrolysate using marine *Rhodococcus* sp. YHY01. *Bioresource Technology* 233: 99–109.

Blagovic, V., J. Rupcic, M. Mesaric, K. Georgiu, and V. Maric. 2001. Lipid composition of brewer's yeast. *Food Technology and Biotechnology* 39: 175–182.

Carota, E., S. Crognale, and A. D'Annibale, et al. 2017. A sustainable use of Ricotta cheese whey for microbial biodiesel production. *Science of the Total Environment* 584: 554–560.

Dahman, Y., K. Syed, S. Begum, P. Roy, and B. Mohtasebi. 2019. Biofuels: Their characteristics and analysis. In *Biomass, Biopolymer-Based Materials, and Bioenergy*, eds. D. Verma, E. Fortunati, S. Jain, X. Zhang, 277–325. Cambridge: Woodhead Publishing.

Darnoko, D. and M. Cheryan. 2000. Kinetics of palm oil transesterification in a batch reactor. *Journal of the American Oil Chemists' Society* 77: 1263–1267.

Demirbas, A. 2005. Biodiesel production from vegetable oils via catalytic and non-catalytic supercritical methanol transesterification methods. *Progress in Energy and Combustion Science* 31: 466–487.

Demirbas, A. 2009. Progress and recent trends in biodiesel fuels. *Energy Conversion and Management* 50: 14–34.

Goswami, L., R. V. Kumar, N. A. Manikandan, K. Pakshirajan, and G. Pugazhenthi. 2017a. Anthracene biodegradation by oleaginous *Rhodococcus opacus* for biodiesel production and its characterization. *Polycyclic Aromatic Compounds* 39: 207–219.

Goswami, L., R. V. Kumar, N. A. Manikandan, K. Pakshirajan, and G. Pugazhenthi. 2017b. Simultaneous polycyclic aromatic hydrocarbon degradation and lipid accumulation by *Rhodococcus opacus* for potential biodiesel production. *Journal of Water Process Engineering* 17: 1–10.

Guerreiro, L., J. E. Castanheiro, and I. M. Fonseca, et al. 2006. Transesterification of soybean oil over sulfonic acid functionalised polymeric membranes. *Catalysis Today* 118: 166–171.

Gujjala, L. K. S., S. P. J. Kumar, and B. Talukdar, et al. 2017. Biodiesel from oleaginous microbes: Opportunities and challenges. *Biofuels* 10: 45–59.

Khosla, K., R. Rathour, R. Maurya, et al. 2017. Biodiesel production from lipid of carbon dioxide sequestrating bacterium and lipase of psychrotolerant *Pseudomonas* sp. ISTPL3 immobilized on biochar. *Bioresource Technology* 245743–245750.

Knothe, G. 2005. Dependence of biodiesel fuel properties on the structure of fatty acid alkyl esters. *Fuel Processing Technology* 86: 1059–1070.

Kumar, M., and I. S. Thakur. 2018. Municipal secondary sludge as carbon source for production and characterization of biodiesel from oleaginous bacteria. *Bioresource Technology Reports* 4: 106–113.

Kumar, S., N. Gupta, and K. Pakshirajan. 2015. Simultaneous lipid production and dairy wastewater treatment using *Rhodococcus opacus* in a batch bioreactor for potential biodiesel application. *Journal of Environmental Chemical Engineering* 3: 1630–1636.

Kumar, M., P. Ghosh, K. Khosla, and I. S. Thakur. 2016. Biodiesel production from municipal secondary sludge. *Bioresource Technology* 216: 165–171.

Kumar, M., R. Morya, E. Gnansounou, C. Larroche, and I. S. Thakur. 2017. Characterization of carbon dioxide concentrating chemolithotrophic bacterium *Serratia* sp. ISTD04 for production of biodiesel. *Bioresource Technology* 243: 893–897.

Kumar, M., S. Sundaram, E. Gnansounou, C. Larroche, and I. S. Thakur. 2018. Carbon dioxide capture, storage and production of biofuel and biomaterials by bacteria: A review. *Bioresource Technology* 247: 1059–1068.

Kumar, M., R. Rathour, and J. Gupta, et al. 2020a. Bacterial production of fatty acid and biodiesel: Opportunity and challenges. In *Refining Biomass Residues for Sustainable Energy and Bioproducts*, eds. R. P. Kumar, E. Gnansounou, J. K. Raman, and G. Baskar, 21–49. London: Academic Press.

Kumar, R. V., I. G. Moorthy, L. Goswami, et al. 2020b. Analytical methods in biodiesel production. In *Biomass Valorization to Bioenergy*, eds. R. P. Kumar, B. Bharathiraja, R. Kataki, and V. S. Moholkar, 197–219. Singapore: Springer.

Kurosawa, K., A. Radek, J. K. Plassmeier, and A. J. Sinskey. 2015. Improved glycerol utilization by a triacylglycerol-producing *Rhodococcus opacus* strain for renewable fuels. *Biotechnology for Biofuels* 8: 31.

Losel, D. M. 1988. Fungal lipids. In *Microbial Lipids*, vol. 1, eds. C. Ratledge, S. Wilkinson, 699–806. London: Academic Press.

Ma, F. and M. A. Hanna. 1999. Biodiesel production: A review. *Bioresource Technology* 70: 1–15.

Meesters, P. A. E. P., G. N. M. Huijberts, and G. Eggink. 1996. High-cell-density cultivation of the lipid accumulating yeast *Cryptococcus curvatus* using glycerol as a carbon source. *Applied Microbiology and Biotechnology* 45: 575–579.

Mondala, A., K. Liang, H. Toghiani, R. Hernandez, and T. French. 2009. Biodiesel production by *in situ* transesterification of municipal primary and secondary sludges. *Bioresource Technology* 100: 1203–1210.

Moon, N. J., E. G. Hammond, and B. A. Glatz. 1978. Conversion of cheese whey and whey permeate to oil and single-cell protein. *Journal of Dairy Science* 61: 1537–1547.

Musolino, V. M. 2016. *Anaerobic digestion for nutrient recycling in industrial microalgae cultivation: Experiments and process simulation.* Padua: University of Padua.

Olkiewicz, M., M. P. Caporgno, and A. Fortuny, et al. 2014. Direct liquid-liquid extraction of lipid from municipal sewage sludge for biodiesel production. *Fuel Processing Technology* 128.331–338.

Passos, F., E. Uggetti, H. Carrere, and I. Ferrer. 2014. Pretreatment of microalgae to improve biogas production: A review. *Bioresource Technology* 172: 403–412.

Rattray, J. 1988. Yeast. In *Microbial Lipids,* eds. C. Ratledge, S. G. Wilkinson, 555–597. London: Academic Press.

Shah, S., S. Sharma, and M. N. Gupta. 2004. Biodiesel preparation by lipase-catalyzed trans-esterification of Jatropha oil. *Energy & Fuels* 18: 154–159.

Sharma, Y. C. and B. Singh. 2008. Development of biodiesel from karanja, a tree found in rural India. *Fuel* 87: 1740–1742.

Sharma, Y. C., and V. Singh. 2017. Microalgal biodiesel: A possible solution for India's energy security. *Renewable and Sustainable Energy Reviews* 67: 72–88.

Shieh, C-J., H-F. Liao, and C-C. Lee. 2003. Optimization of lipase-catalyzed biodiesel by response surface methodology. *Bioresource Technology* 88: 103–106.

Sitepu, I. R., L. A. Garay, and R. Sestric, et al. 2014. Oleaginous yeasts for biodiesel: Current and future trends in biology and production. *Biotechnology Advances* 32: 1336–1360.

Surendhiran, D., M. Vijay, and A. R. Sirajunnisa. 2014. Biodiesel production from marine microalga *Chlorella salina* using whole cell yeast immobilized on sugarcane bagasse. *Journal of Environmental Chemical Engineering* 2: 1294–1300.

Tabernero, A., E. M. M. del Valle, and M. A. Galan. 2012. Evaluating the industrial potential of biodiesel from a microalgae heterotrophic culture: Scale-up and economics. *Biochemical Engineering Journal* 63: 104–115.

Tasic, M. B., L. F. R. Pinto, and B. C. Klein, et al. 2016. *Botryococcus braunii* for biodiesel production. *Renewable and Sustainable Energy Reviews* 64: 260–270.

Thakur, I. S., M. Kumar, and S. J. Varjani, et al. 2018. Sequestration and utilization of carbon dioxide by chemical and biological methods for biofuels and biomaterials by chemoautotrophs: Opportunities and challenges. *Bioresource Technology* 256: 478–490.

Thliveros, P., E. U. Kiran, and C. Webb. 2014. Microbial biodiesel production by direct methanolysis of oleaginous biomass. *Bioresource Technology* 157: 181–187.

Wang, Y., S. Ou, P. Liu, and Z. Zhang. 2007. Preparation of biodiesel from waste cooking oil via two-step catalyzed process. *Energy Conversion and Management* 48: 184–188.

Wen, Z., and M. B. Johnson. 2009. *Microalgae as a feedstock for biofuel production.* Virginia, VA: Virginia State University.

Zhao, W., Y. Xue, and P. Ma, et al. 2016. Improving the cold flow properties of high-proportional waste cooking oil biodiesel blends with mixed cold flow improvers. *RSC Advances* 6: 13365–13370.

Part VIII

Algal Biodiesel Fuels

33 Algal Biodiesel Fuels
A Scientometric Review of the Research

Ozcan Konur

CONTENTS

33.1 INTRODUCTION

Crude oils have been primary sources of energy and fuels, such as petrodiesel (Busca et al., 1998; Chisti, 2007, 2008; Khalili et al., 1995; Lapuerta et al., 2008; Konur, 2012g, 2015; Rogge et al., 1993; Schauer et al., 1999; van Gerpen, 2005). However, significant public concerns about the sustainability, price fluctuations, and adverse environmental impact of crude oils have emerged since the 1970s (Ahmadun et al., 2009; Atlas, 1981; Babich and Moulijn, 2003; Kilian, 2009; Moldowan et al., 1985;

Perron, 1989). Thus, biooils have emerged as an alternative to crude oils in recent decades (Bridgwater and Peacocke, 2000; Czernik and Bridgwater, 2004; Evans and Milne, 1987; Gallezot, 2012; Mohan et al., 2006; Yaman, 2004; Zhang et al., 2007). In this context, algal oil-based biodiesel fuels (Brennan and Owende, 2010; Chisti, 2007, 2008; Hu et al., 2008; Mata et al., 2010; Rodolfi et al., 2009; Schenk et al., 2008) have emerged as a viable alternative to crude oil-based petrodiesel fuels (Busca et al., 1998; Khalili et al., 1995; Rogge et al., 1993; Schauer et al., 1999). Nowadays, both petrodiesel fuels and algal oil-based biodiesel fuels are being used extensively at a global scale (Konur, 2021a–ag).

However, for the efficient progression of the research in this field, it is necessary to develop efficient incentive structures for the primary stakeholders and to inform these stakeholders about the research (Konur, 2000, 2002a–c, 2006a–b, 2007a–b; North, 1991a–b).

Scientometric analysis offers ways to evaluate the research in a respective field (Garfield, 1955, 1972; Konur, 2011, 2012a–n, 2015, 2016a–f, 2017a–f, 2018a–b, 2019a–b). However, there has been no current scientometric study of this field.

This chapter presents a study of the scientometric evaluation of the research in this field using two datasets. The first dataset includes the 100-most-cited papers (n = 100 sample papers) whilst the second set includes population papers (n = over 15,000 population papers) published between 1980 and 2019. It complements the other chapters on crude oil-based petrodiesel fuels and biodiesel fuels (Konur, 2021e–ag).

The data on the indices, document types, authors, institutions, funding bodies, source titles, 'Web of Science' subject categories, keywords, research fronts, and citation impact are presented and discussed.

33.2   MATERIALS AND METHODOLOGY

The search for the literature was carried out in the 'Web of Science' (WOS) database in January 2020. It contains the 'Science Citation Index-Expanded' (SCI-E), the Social Sciences Citation Index' (SSCI), the 'Book Citation Index-Science' (BCI-S), the 'Conference Proceedings Citation Index-Science' (CPCI-S), the 'Emerging Sources Citation Index' (ESCI), the 'Book Citation Index-Social Sciences and Humanities' (BCI-SSH), the 'Conference Proceedings Citation Index-Social Sciences and Humanities' (CPCI-SSH), and the 'Arts and Humanities Citation Index' (A&HCI).

The keywords for the search of the literature are collated from the screening of abstract pages for the first 1,000 highly cited papers. This keyword set is provided in the Appendix.

Two datasets are used for this study. The 100 highly cited papers comprise the first dataset (sample dataset, n = 100 papers) whilst all the papers form the second dataset (population dataset, over 15,000 papers).

The data on the indices, document types, publication years, institutions, funding bodies, source titles, countries, 'Web of Science' subject categories, citation impact, keywords, and research fronts are collated from these datasets. The key findings are provided in the relevant tables and figure, supplemented with explanatory notes in the text. The findings are discussed and a number of conclusions are drawn and a number of recommendations for further study are made.

33.3 RESULTS

33.3.1 INDICES AND DOCUMENTS

There are over 17,300 papers in this field in the 'Web of Science' as of January 2020. This original population dataset is refined for the document type (article, review, book chapter, book, editorial material, note, and letter) and language (English), resulting in over 15,100 papers comprising over 87% of the original population dataset.

The primary index is the SCI-E for both the sample and population papers. About 96% of the population papers are indexed by the SCI-E database. Additionally 3.8, 2.3, and 1.6% of these papers are indexed by the CPCI-S, ESCI, and BCI-S databases, respectively. The papers on the social and humanitarian aspects of this field are relatively negligible with 0.4 and 0.0.% of the population papers indexed by the SSCI and A&HCI, respectively.

Brief information on the document types for both datasets is provided in Table 33.1. The key finding is that article types are the primary documents for the population papers whilst reviews form 35% of the sample papers. Articles are under-represented by −28.9% whilst reviews are over-represented by 30.4% in the sample papers.

33.3.2 AUTHORS

Brief information about the most-prolific 11 authors with at least three sample papers each is provided in Table 33.2. Around 330 and 31,500 authors contribute to the sample and population papers, respectively.

The most-prolific author is 'Qingyu Wu' with six sample papers, working primarily on 'algal biodiesel production', 'algal biomass pyrolysis', 'algal biomass production', and 'algal lipid production'. This top author is followed by 'Yusuf Chisti' and

TABLE 33.1
Document Types

	Document Type	Sample Dataset (%)	Population Dataset (%)	Difference (%)
1	Article	65	93.9	−28.9
2	Review	35	4.6	30.4
3	Book chapter	0	1.6	−1.6
4	Proceeding paper	5	3.8	1.2
5	Editorial material	0	0.5	−0.5
6	Letter	0	0.2	−0.2
7	Book	0	0.1	−0.1
8	Note	0	0.9	−0.9

TABLE 33.2
Authors

	Author	Sample Papers (%)	Population Papers (%)	Surplus (%)	Institution	Country	Research Front
1	Wu, Qingyu	6	0.2	5.8	Tsinghua Univ.	China	Biodiesel production, biomass pyrolysis, biomass production, lipid production
2	Chisti, Yusuf	5	0.3	4.7	Massey Univ.	New Zealand	Biodiesel production, biomass pyrolysis
3	Miao, Xiaoling*	5	0.1	4.9	Shanghai Jiao Tong Univ.	China	Biodiesel production, biomass pyrolysis, CO2 biofixation, lipid production
4	Danquah, Michael K.	4	0.2	3.8	Monash Univ.	Australia	Biomass production, lipid extraction
5	Chen, Paul	3	0.3	2.7	Univ. Minnesota	USA	Biomass production in wws, biodiesel production in wws
6	Chen, Yifeng	3	0.1	2.9	Univ. Minnesota	USA	Biomass production in wws, biodiesel production in ww
7	Li, Yecong	3	0.1	2.9	Univ. Minnesota	USA	Biomass production in wws, biodiesel production in ww
8	Malcata, F. Xavier	3	0.1	2.9	Univ. New Lisbon	Portugal	Biomass production CO2 biofixation, biodiesel production
9	Min, Min*	3	0.1	2.9	Univ. Minnesota	USA	Biomass production in wws, biodiesel production in ww
10	Ross, Andrew B.	3	0.1	2.9	Univ. Leeds	UK	Biomass liquefaction, biomass pyrolysis
11	Sommerfeld, Milton	3	0.1	2.9	Arizona State Univ.	US	Lipid production, lipid characterization, biodiesel production

*Female.
Note: wws: wastewaters.

'Xiaoling Miao' with five sample papers each. These top three authors have the most impact with 15.2% publication surpluses altogether.

On the other hand, a number of authors have a significant presence in the population papers: 'Jo-Shu Chang', 'Rene H. Wijffels', 'Emilio Molina-Grima', 'F. Gabriel Acien', 'Faizal Bux', 'Jorge A. V. Costa', 'You-Kwan Oh', 'Roger R. Ruan', 'Navid Reza Moheimani', 'Koenraad Muylaert', 'Hee-Mock Oh', and 'Jeremy Pruvost' with at least 0.3% of the population papers each.

The most-prolific institution for these top authors is the 'University of Minnesota' of the USA with four. In total, eight institutions house these top authors.

It is notable that none of these top researchers are listed in the 'Highly Cited Researchers' (HCR) in 2019 (Clarivate Analytics, 2019; Docampo and Cram, 2019).

The most-prolific country for these top authors is the USA with five. The other prolific country is China with two authors. In total, six countries contribute to these top papers.

There are ten key research fronts for these top researchers: 'algal biomass production', 'algal biomass production in wastewaters', 'algal lipid extraction', 'algal biodiesel production', 'algal biodiesel production in wastewaters', 'algal biomass pyrolysis', 'algal CO_2 biofixation', 'algal lipid production', 'algal biomass liquefaction', and 'algal lipid characterization'.

The top research front is 'algal biodiesel production' with five authors. The other prolific research fronts are 'algal biomass pyrolysis', 'algal biomass production in wastewaters', and 'algal biodiesel production in wastewaters' with four authors each.

It is further notable that there is a significant gender deficit among these top authors as only two of them are female (Lariviere et al., 2013; Xie and Shauman, 1998).

33.3.3 PUBLICATION YEARS

Information about the publication years for both datasets is provided in Figure 33.1.

This figure shows that 3, 4, 37, and 56% of the sample papers and 8.3, 10.1, 11.6, and 69.8% of the population papers were published in the 1980s, 1990s, 2000s, and 2010s, respectively.

Similarly, the most-prolific publication years for the sample dataset are 2010, 2011, 2009, and 2008 with 23, 19, 12, and 11 papers, respectively. On the other hand, the most-prolific publication years for the population dataset are 2019, 2018, 2017, and 2016 with at least 9% of the population papers each. It is notable that there is a sharply rising trend for the population papers, particularly in the 2010s.

33.3.4 INSTITUTIONS

Brief information on the top nine institutions with at least 3% of the sample papers each is provided in Table 33.3. In total, around 125 and 5,600 institutions contribute to the sample and population papers, respectively.

These top institutions publish 37 and 4.7% of the sample and population papers, respectively. The top institution is 'Tsinghua University' of China with seven sample

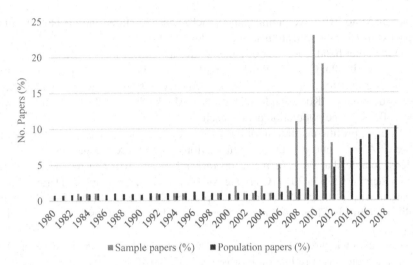

FIGURE 33.1   Research output between 1980 and 2019.

TABLE 33.3
Institutions

	Institution	Country	No. of Sample Papers (%)	No. of Population Papers (%)	Difference (%)
1	Tsinghua Univ.	China	7	0.9	6.1
2	Massey Univ.	China	6	0.5	5.5
3	Arizona State Univ.	USA	4	0.8	3.2
4	Monash Univ.	Australia	4	0.5	3.5
5	Natl. Renew. Ener. Lab.	USA	4	0.4	3.6
6	Chinese Acad. Sci.	USA	3	0.8	2.2
7	Ningde Normal Univ.	China	3	0.1	2.9
8	Univ. Leeds	UK	3	0.1	2.9
9	Univ. Minnesota	USA	3	0.6	2.4

papers and a 6.1% publication surplus. This top institution is closely followed by 'Massey University' of New Zealand with six sample papers and a 5.5% publication surplus. The other top institutions are 'Monash University' of Australia, 'Arizona State University', and the 'National Renewable Energy Laboratory' of the USA with four sample papers each.

The most-prolific country for these top institutions is the USA with four. This top country is closely followed by China with three institutions. In total four countries house these institutions.

The institutions with the most impact are 'Tsinghua University', 'Massey University', the 'National Renewable Energy Laboratory', and 'Monash University' with at least a 3.5% publication surplus each. On the other hand, the institutions with the least impact are the 'Chinese Academy of Sciences' and 'Ningde Normal

University' of China and the 'University of Minnesota' with at least a 2.2% publication surplus each.

It is notable that some institutions have a heavy presence in the population papers: the 'Scientific Research National Center' of France, the 'Indian Institute of Technology' of India, the 'Council of Scientific Industrial Research' of India, the 'Korea Advanced Institute of Science Technology' of South Korea, the 'University of Almeria' of Spain, 'Wageningen University Research' of the Netherlands, the 'National Cheng Kung University' of Taiwan, and the 'Superior Council of Scientific Research' of Spain have at least a 1% presence in the population papers each.

33.3.5 FUNDING BODIES

Brief information about the top 12 funding bodies with at least 2% of the sample papers each is provided in Table 33.4. It is significant that only 51.0 and 59.2% of the sample and population papers declare any funding, respectively.

The top funding body is the 'Engineering Physical Sciences Research Council' of the UK and the 'National Science Council of Taiwan', funding 4% of the sample papers and with a 3.4% publication surplus each. These top funding bodies are closely followed by the 'Australian Research Council', the 'French National Research Agency', and the 'National Science Foundation' of the USA with three sample papers each.

It is notable that some top funding agencies have a heavy presence in the population studies. Some of them are the 'National Natural Science Foundation of China', the US 'Department of Energy', the 'European Union', the 'National Council for Scientific and Technological Development' of Brazil, the 'Fundamental Research Funds for the Central Universities' and the 'National Basic Research Program of China' of China, the 'Ministry of Education Culture Sports Science and Technology' of Japan, the 'Natural Sciences and Engineering Research Council of Canada', and

TABLE 33.4
Funding Bodies

	Institution	Country	No. of Sample Papers (%)	No. of Population Papers (%)	Difference (%)
1	Eng. Phys. Sci. Res. Counc.	UK	4	0.6	3.4
2	Natl. Sci. Counc. Taiwan	Taiwan	4	0.6	3.4
3	Australian Res. Counc.	Australia	3	0.6	2.4
4	French Natl. Res. Agcy.	France	3	0.8	2.2
5	Natl. Sci. Found.	USA	3	2.9	0.1
6	Air Force Off. Sci. Res.	USA	2	0.3	1.7
7	Natl. Counc. Sci. Technol.	Mexico	2	0.8	1.1
8	Leverhulme Trust	USA	2	0.2	1.8
9	Min. Educ. Sci. Technol.	S. Korea	2	0.6	1.4
10	Univ. Michigan	USA	2	0.2	1.8
11	Univ. Minnesota	USA	2	0.2	1.8
12	Massachusetts Inst. Technol.	USA	2	0.2	1.8

the 'National High Technology Research and Development Program of China' with
at least 1% of the population papers each.

It is notable that the most-prolific country for these top funding bodies is the USA
with six.

The 'Engineering Physical Sciences Research Council' of the UK and the
'National Science Council of Taiwan' are the funding bodies with the most impact
whilst the 'National Science Foundation' of the USA and the 'National Science
Council of Taiwan' are those with the least impact.

33.3.6 SOURCE TITLES

Brief information about the top 16 source titles with at least two sample papers each
is provided in Table 33.5. In total, 44 and over 1,500 source titles publish the sample

TABLE 33.5
Source Titles

Source Title	WOS Subject Category	No. of Sample Papers (%)	No. of Population Papers (%)	Difference (%)
1 Bioresource Technology	Agr. Eng., Biot. Appl. Microb., Ener. Fuels	24	10.9	13.1
2 Biotechnology Advances	Biot. Appl. Microb.	6	0.2	5.8
3 Applied Microbiology and Biotechnology	Biot. Appl. Microb.	5	0.9	4.1
4 Renewable Sustainable Energy Reviews	Green Sust. Sci. Technol., Ener. Fuels	5	0.7	4.3
5 Applied Energy	Ener. Fuels, Eng. Chem.	5	0.7	4.3
6 Journal of Biotechnology	Biot. Appl. Microb.	4	0.5	3.5
7 Journal of Phycology	Plant Sci., Mar. Fresh. Biol.	3	2.5	0.5
8 Energy Fuels	Ener. Fuels, Eng. Chem.	3	0.5	2.5
9 Trends in Biotechnology	Biot. Appl. Microb.	3	0.2	2.8
10 Biomass Bioenergy	Agr. Eng., Biot. Appl. Microb., Ener. Fuels	2	1.1	0.9
11 Biotechnology and Bioengineering	Biot. Appl. Microb.	2	1.1	0.9
12 Energy Conversion and Management	Therm., Ener. Fuels, Mechs.	2	0.8	1.2
13 Fuel	Ener. Fuels, Eng. Chem.	2	0.7	1.3
14 Environmental Science Technology	Eng. Env., Env. Sci.	2	0.4	1.6
15 Current Opinion in Biotechnology	Bioch. Res. Meth., Biot. Appl. Microb.	2	0.2	1.8
16 Eukaryotic Cell	Microbiol., Mycol.	2	0.2	1.8
		72	21.6	50.4

and population papers, respectively. On the other hand, these top 13 journals publish 72.0 and 21.6% of the sample and population papers, respectively.

The top journal is 'Bioresource Technology', publishing 24 sample papers with a 11.9% publication surplus. The other top journals are 'Biotechnology Advances', 'Applied Microbiology and Biotechnology', 'Renewable Sustainable Energy Reviews', and 'Applied Energy' with at least five sample papers and a 4.1% publication surplus each.

Although these journals are indexed by 14 subject categories, the top categories are 'Biotechnology Applied Microbiology' and 'Energy Fuels' with eight and seven journals, respectively. The other subject categories are 'Agricultural Engineering', 'Engineering Chemical', and 'Mechanics' with at least two journals each.

The journals with the most impact are 'Bioresource Technology', 'Biotechnology Advances', 'Renewable Sustainable Energy Reviews', and 'Applied Energy' with at least a 4.3% publication surplus each. On the other hand, the journals with the least impact are the 'Journal of Phycology', 'Biomass Bioenergy', 'Biotechnology and Bioengineering', and 'Energy Conversion and Management' with at least a 0.5% publication surplus each.

It is notable that some journals have a heavy presence in the population papers. Some of them are the 'Journal of Applied Phycology', 'Algal Research Biomass Biofuels and Bioproducts', 'Water Science and Technology', 'Hydrobiologia', 'Water Research', 'Botanica Marina', 'Biotechnology for Biofuels', 'Applied Biochemistry and Biotechnology', 'Plos One', and 'Plant Physiology' with at least a 0.7% presence in the population papers each.

It is also notable that the journals primarily related to algae publish 15.5% and 4.0% of the population and sample papers, respectively.

33.3.7 Countries

Brief information about the top 13 countries with at least three sample papers each is provided in Table 33.6. In total, 30 and over 125 countries contribute to the sample and population papers, respectively.

The top country is the USA, publishing 26.0 and 18.6% of the sample and population papers, respectively. China follows the USA with 14.0 and 16.2% of the sample and population papers, respectively. The other prolific countries are the UK, Australia, New Zealand, Taiwan, and Germany, publishing at least five sample papers each.

On the other hand, the European and Asian countries represented in this table publish altogether 29 and 35% of the sample papers and 19.1 and 29.7% of the population papers, respectively.

It is notable that the publication surplus for the USA and these European and Asian countries is 7.4, 9.9, and 5.5%, respectively.

The countries with the most impact are the USA, New Zealand, the UK, Australia, and Taiwan with 7.4, 4.9, 4.0, 2.9, and 2.8% publication surpluses, respectively. On the other hand, the countries with the least impact are Japan, China, France, and Germany with –3.1, –2.2, –0.3, and 0.1% publication deficits, respectively.

It is also notable that some countries have a heavy presence in the population papers. The major producers of these papers are India, South Korea, Spain, Canada,

TABLE 33.6
Countries

	Country	No. of Sample Papers (%)	No. of Population Papers (%)	Difference (%)
1	USA	26	18.6	7.4
2	China	14	16.2	−2.2
3	UK	9	5.0	4
4	Australia	7	4.1	2.9
5	New Zealand	6	1.1	4.9
6	Taiwan	5	2.2	2.8
7	Germany	5	4.9	0.1
8	France	4	4.3	−0.3
9	Netherlands	4	2.3	1.7
10	Portugal	4	1.3	2.7
11	Belgium	3	1.3	1.7
12	Japan	3	6.1	−3.1
13	South Africa	3	1.2	1.8
	Europe-6	29	19.1	9.9
	Asia-3	35	29.7	5.3

and Brazil with 7.9, 5.5, 5.2, 3.5, and 3.2%, respectively. The other producers of the population papers are Italy, Malaysia, Mexico, Israel, Iran, Thailand, Sweden, and Turkey with at least a 1.1% presence each.

33.3.8 'WEB OF SCIENCE' SUBJECT CATEGORIES

Brief information about the top 12 'Web of Science' subject categories with at least three sample papers each is provided in Table 33.7. The sample and population papers are indexed by 25 and over 100 subject categories, respectively.

TABLE 33.7
Web of Science Subject Categories

	Subject	No. of Sample Papers (%)	No. of Population Papers (%)	Difference (%)
1	Biotechnology Applied Microbiology	56	39.6	16.4
2	Energy Fuels	49	22.1	26.9
3	Agricultural Engineering	26	12.5	13.5
4	Engineering Chemical	15	11.0	4.0
5	Plant Sciences	7	13.1	−6.1
6	Environmental Sciences	6	11.2	−5.2
7	Green Sustainable Science Technology	6	3.7	2.3
8	Marine Freshwater Biology	5	15.3	−10.3
9	Engineering Environmental	4	6.9	−2.9
10	Biochemical Research Methods	3	1.3	1.7
11	Microbiology	3	5.9	−2.9
12	Multidisciplinary Sciences	3	3.3	−0.3

For the sample papers, the top subject is 'Biotechnology Applied Microbiology' with 56.0 and 39.6% of the sample and population papers, respectively. This top subject category is followed by 'Energy Fuels' with 49.0 and 22.1% of the sample and population papers, respectively. The other prolific subjects are 'Agricultural Engineering', 'Engineering Chemical', and 'Plant Sciences' with 26, 15, and 7 sample papers, respectively.

It is notable that the publication surplus is most significant for 'Energy Fuels', 'Biotechnology Applied Microbiology', and 'Agricultural Engineering' with 26.9, 19.4, and 13.5% publication surpluses, respectively. On the other hand, the subjects with least impact are 'Marine Freshwater Biology', 'Plant Sciences', 'Environmental Sciences', 'Engineering Environmental', and 'Microbiology' with at least a −10.3% publication deficit each. This latter group of subject categories are under-represented in the sample papers.

Additionally, some subject categories with a heavy presence in the population papers do not have a place in this top subject table: 'Biochemistry Molecular Biology', 'Water Resources', 'Chemistry Multidisciplinary', 'Ecology', 'Food Science Technology', 'Cell Biology', 'Thermodynamics', 'Biophysics', 'Chemistry Applied', 'Biology', and 'Chemistry Analytical' have at least a 1.1% presence each.

33.3.9 CITATION IMPACT

These sample papers received about 57,000 citations as of January 2010. Thus, the average number of citations per paper is about 570.

33.3.10 KEYWORDS

Although a number of keywords are listed in the Appendix for the datasets related to this field, some of them are more significant for the sample papers.

The most-prolific keyword for the set related to algae is '*alga*' with 90 papers. Furthermore, 64 and 31 papers have 'microalga*' and plain 'alga*'.The other prolific algal keyword is 'chlorella' with 13 papers. The other keywords are 'nannochloropsis', 'neochloris', chlamydomonas', 'scenedesmus', 'dunaliella', and 'enteromorpha'.

The most-prolific keywords related to biodiesel fuels are '*diesel*', 'oil*', 'lipid*', '*fuel*', 'cultivat', 'biomass*', 'wastewater*', 'nitrogen*', 'CO_2', 'growth', 'liquefaction', 'nutrient*', '*reactor*', and 'fatty-acid*' with 26, 13, 21, 28, 10, 11, 11, 8, 8, 8, 6, 6, 11, and 5 papers, respectively.

33.3.11 RESEARCH FRONTS

Brief information about the key research fronts is provided in Table 33.8. There are three major research fronts for these sample papers: 'algal biomass production', 'algal biooil production', and 'algal biodiesel production'.

The most-prolific research front is 'algal biomass production' with 69 sample papers. This top research front is followed by 'algal biooil production' with 50

TABLE 33.8
Research Fronts

	Research Front	No. of Sample Papers (%)
1	Algal biodiesel production	27
2	Algal biomass production	69
2.1	Algal biomass production	48
2.2	Algal biomass production in wastewaters	14
2.3	CO_2 biofixation	7
3	Algal biooil production	50
3.1	Algal lipid production	34
3.2.	Algal lipid production in wastewaters	1
3.3.	Algal biomass pyrolysis	3
3.4.	Algal biomass liquefaction	8
3.5	Algal lipid extraction	3
3.6	Algal lipid characterization	1

sample papers. Additionally, 27 sample papers are related to 'algal biodiesel production'.

There are three research subfronts for the top research front: 'algal biomass production in general' (Grima et al., 2003; Williams and Lauren, 2010), 'algal biomass production and nutrient removal in wastewaters' (Markou and Georgakakis, 2011; Wang et al., 2010), and 'CO_2 bioremediation by algal biomass' (Chiu et al., 2009; Wang et al., 2008) with 48, 14, and 7 sample papers, respectively.

There are six research subfronts for 'algal biooil production': 'algal lipid production in general' (Converti et al., 2009; Griffiths and Harrison, 2009), 'algal lipid production in wastewaters' (Li et al., 2011; Zhou et al., 2011), 'algal biomass pyrolysis' (Miao and Wu, 2004; Miao et al., 2004), 'algal biomass liquefaction' (Biller and Ross, 2011; Brown et al., 2010), 'algal lipid extraction' (Halim et al., 2012; Lee et al., 2010), and 'algal lipid characterization' (Siaut et al., 2011; Wang et al., 2009).

33.4 DISCUSSION

The size of the research in this field has increased to over 15,000 papers as of January 2020. It is expected that the number of the population papers in this field will exceed 30,000 papers by the end of the 2020s.

The research has developed more in the technological aspects of this field, rather than the social and humanitarian pathways, as evidenced by the negligible number of population papers in the indices of the 'Web of Science', SSCI, and A&HCI.

The article types of documents are the primary documents for both datasets and reviews are over-represented by 30.4% in the sample papers (Table 33.1). Thus, the contribution of reviews by 35% of the sample papers in this field is highly exceptional (cf. Konur, 2011, 2012a–n, 2015, 2016a–f, 2017a–f, 2018a–b, 2019a–b).

Eleven authors from six institutions have at least three sample papers each (Table 33.2). Five of these authors are from the USA. The other prolific country is China with two authors.

These authors focus on 'algal biodiesel production', 'algal biomass pyrolysis', 'algal biomass production in wastewaters', and 'algal biodiesel production in wastewaters'.

There is a significant gender deficit among these top authors as only two of them are female (Lariviere et al., 2013; Xie and Shauman, 1998).

The population papers have built on the sample papers, primarily published in the 1990s 2000s (Figure 33.1). Following this rising trend, particularly in the 2010s, it is expected that the number of papers will reach 30,000 by the end of the 2020s, more than doubling the current size.

The engagement of the institutions in this field at the global scale is significant, as over 125 and over 5,600 institutions contribute to the sample and population papers, respectively.

Nine top institutions publish 37.0 and 4.7% of the sample and population papers, respectively (Table 33.3). The top institution is 'Tsinghua University' of China with seven sample papers and a 6.1% publication surplus. The other top institutions are 'Massey University' of New Zealand, 'Arizona State University' and the 'National Renewable Energy Laboratory' of the USA, and 'Monash University' of Australia.

As in the case of the top authors, the most-prolific countries for these top institutions are the USA and China. It is notable that some institutions with a heavy presence in the population papers are under-represented in the sample papers.

It is significant that only 51.0 and about 59.2% of the sample and population papers declare any funding, respectively. It is notable that the most-prolific country for these top funding bodies is the USA with six (Table 33.4). It is further notable that some top funding agencies for the population studies do not enter this top funding body list.

However, Chinese funding bodies dominate the top funding body table for the population papers. This finding is in line with studies showing the heavy research funding in China, where the NSFC is the primary funding agency (Wang et al., 2012).

The sample and population papers are published by 44 and over 1,500 journals, respectively. It is significant that the top 13 journals publish 72.0 and 21.6% of the sample and population papers, respectively (Table 33.5).

The top journal is 'Bioresource Technology', publishing 24 sample papers with an 11.9% publication surplus. The other top journals are 'Biotechnology Advances', 'Applied Microbiology and Biotechnology', 'Renewable Sustainable Energy Reviews', and 'Applied Energy'.

The top categories for these journals are 'Biotechnology Applied Microbiology', 'Energy Fuels', 'Agricultural Engineering', 'Engineering Chemical', and 'Mechanics'. It is notable that some journals with a heavy presence in the population papers are relatively under-represented in the sample papers.

In total, 30 and over 120 countries contribute to the sample and population papers, respectively. The top country is the USA publishing 26.0 and 18.6% of the sample and population papers, respectively. China follows the USA with 14.0 and 16.2% of the sample and population papers, respectively (Table 33.6). This finding is in line with the studies arguing that the USA is not losing ground in science and technology (Leydesdorff and Wagner, 2009).

The other prolific countries are the UK, Australia, New Zealand, Taiwan, and Germany, publishing at least five sample papers each. These findings are in line with the studies showing that European countries have superior publication performance in science and technology (Bordons et al., 2015; Youtie et al., 2008).

On the other hand, the European and Asian countries represented in this table publish altogether 29 and 35% of the sample papers and 19.1 and 29.7% of the population papers, respectively.

It is notable that the publication surplus for the USA and these European and Asian countries is 7.4, 9.9, and 5.5%, respectively.

It is further notable that China has a significant publication deficit (–2.2%). This finding is in contrast with China's efforts to be a leading nation in science and technology (Zhou and Leydesdorff, 2006) but it is in line with the findings of Guan and Ma (2007) and Youtie et al. (2008) relating to China's performance in nanotechnology.

It is also notable that some countries have a presence in the population papers. Some of them are India, South Korea, Spain, Canada, and Brazil. The other producers of population papers are Italy, Malaysia, Mexico, Israel, Iran, Thailand, Sweden, and Turkey (Bhattacharya et al., 2012; Bordons et al., 2015; Glanzel et al., 2006; Leydesdorff and Zhou, 2005; Oleinik, 2012).

The sample and population papers are indexed by 25 and over 100 subject categories, respectively.

For the sample papers, the top subject is 'Biotechnology Applied Microbiology' with 56.0 and 39.6% of the sample and population papers, respectively (Table 33.7). This top subject category is followed by 'Energy Fuels'. The other prolific subjects are 'Agricultural Engineering', 'Engineering Chemical', and 'Plant Sciences'.

It is notable that the publication surplus is most significant for 'Energy Fuels', 'Biotechnology Applied Microbiology', and 'Agricultural Engineering'. On the other hand, the subjects with least impact are 'Marine Freshwater Biology', 'Plant Sciences', 'Environmental Sciences', 'Engineering Environmental', and 'Microbiology'. This latter group of subject categories are under-represented in the sample papers.

These sample papers receive about 57,000 citations as of January 2010. Thus, the average number of citations per paper is about 570. Hence, the citation impact of these 100 top papers in this field has been significant.

Although a number of keywords are listed in the Appendix for the datasets related to this field, some of them are more significant for the sample papers.

The most-prolific keywords for algae are '*alga*', 'microalga*', and plain 'alga*'. The other prolific algal keywords are 'chlorella', 'nannochloropsis', 'neochloris', 'chlamydomonas', 'scenedesmus', 'dunaliella', and 'enteromorpha'.

The most-prolific keywords related to biodiesel fuels are '*diesel*', 'oil*', 'lipid*', '*fuel*', 'cultivat', 'biomass*', 'wastewater*', 'nitrogen*', 'CO_2', 'growth', 'liquefaction', 'nutrient*', '*reactor*', and 'fatty-acid*'. As expected, these keywords provide valuable information about the pathways of the research in this field.

There are three major research fronts for these sample papers: 'algal biomass production', 'algal biooil production', and 'algal biodiesel production' (Table 33.8).

The most-prolific research front is 'algal biomass production'. This top research front is followed by 'algal biooil production' and 'algal biodiesel production'.

There are three research subfronts for the top research front: 'algal biomass production in general', 'algal biomass production and nutrient removal in wastewaters', and 'CO_2 bioremediation by algal biomass'.

There are six research subfronts for 'algal biooil production': 'algal lipid production in general', 'algal lipid production in wastewaters', 'algal biomass pyrolysis', 'algal biomass liquefaction', 'algal lipid extraction', and 'algal lipid characterization'.

The key emphasis in these research fronts is the exploration of the structure–processing–property relationships of algal biomass, algal biooils, and algal biodiesel (Cheng and Ma, 2011; Konur and Matthews, 1989; Rogers and Hopfinger, 1994; Scherf and List, 2002).

33.5 CONCLUSION

This chapter has mapped the research on algal biomass, algal biooils, and algal biodiesel fuels using a scientometric method.

The size of over 15,000 population papers shows the public importance of this interdisciplinary research field. However, it is significant that the research has developed more in the technological aspects in this field, rather than the social and humanitarian pathways.

Articles and reviews dominate the sample papers, primarily published in the 1990s and 2000s. The population papers, primarily published in the 2010s, build on these sample papers.

The data presented in the tables and figure show that a small number of authors, institutions, funding bodies, journals, keywords, research fronts, subject categories, and countries have shaped the research in this field.

It is notable that the authors, institutions, and funding bodies from the USA, Europe, China, Australia, New Zealand, and Taiwan dominate the research. Furthermore, China, Japan, India, South Korea, Spain, Canada, Brazil, Italy, Malaysia, Mexico, Israel, Iran, Thailand, Sweden, and Turkey are under-represented significantly in the sample papers.

These findings show the importance of the progression of efficient incentive structures for the development of the research in this field as in other fields. It seems that the USA and European countries (such as the UK and Germany) have efficient incentive structures for the development of the research in this field, contrary to China, Japan, India, South Korea, Spain, Canada, Brazil, Italy, Malaysia, Mexico, Israel, Iran, Thailand, Sweden, and Turkey.

It further seems that although the research funding is a significant element of these incentive structures, it might not be a sole solution for increasing the incentives for the research in this field, as is the case of China, Japan, India, South Korea, Spain, Canada, Brazil, Italy, Malaysia, Mexico, Israel, Iran, Thailand, Sweden, and Turkey.

On the other hand, it seems there is more to do to reduce the significant gender deficit in this field as in other fields of science and technology (Lariviere et al., 2013; Xie and Shauman, 1998).

The data on the research fronts, keywords, source titles, and subject categories provide valuable evidence for the interdisciplinary (Lariviere and Gingras, 2010; Morillo et al., 20001) nature of the research in this field.

There is ample justification for the broad search strategy employed in this study due to the interdisciplinary nature of this research field as evidenced by the top subject categories. The search strategy employed in this study is in line with the search strategies employed for related and other research fields (Konur, 2011, 2012a–n, 2015, 2016a–f, 2017a–f, 2018a–b, 2019a–b). It is particularly noted that only 49.0 and 22.1% of the sample and population papers are indexed by the 'Energy Fuels' subject category, respectively.

There are three major research fronts for these sample papers: 'algal biomass production', 'algal biooil production', and 'algal biodiesel production' (Table 33.8). There are three research subfronts for the top research front: 'algal biomass production in general', 'algal biomass production and nutrient removal in wastewaters', and 'CO$_2$ bioremediation by algal biomass' with 48, 14, and 7 sample papers, respectively.

There are six research subfronts for the 'algal biooil production' research front: 'algal lipid production in general', 'algal lipid production in wastewaters', 'algal biomass pyrolysis', 'algal biomass liquefaction', 'algal lipid extraction', and 'algal lipid characterization'.

It is recommended that further scientometric studies are carried out for each of these research fronts, building on the pioneering studies in these fields.

ACKNOWLEDGMENTS

The contribution of the highly cited researchers in the fields of algal biomass, algal biooils, and algal biodiesel fuels is greatly acknowledged.

33.A   APPENDIX

The keyword set for the algal biodiesel fuels
   Syntax: ((1 AND 2) OR 3) NOT 4

33.A.1   BIODIESEL FUEL-RELATED KEYWORDS

TI=(*pyroly* or "bio-oil*" or *liquefaction or "thermal-decomposition" or hydro-thermal or thermochemical or torrefaction or *deoxygenation or *cracking or Thermogravimet* or "thermal-conversion" or biocrude* or "bio-crude*" or Biooil* or lipid* or triacylglycerol* or "fatty-acid*" or oil or triglyceride* or triacylglycer-ide* or *diesel* or biodiesel or "methyl-ester*" or transester* or "trans-ester*" or "biomass-product*" or "biomass-recovery" or "biomass-accumulation" or *cultivat* or harvest* or *reactor* or dewatering or *floccul* or *filtrat* or ultrasound or coag-ulat* or "cell-density" or *flotat* or "light-emitting diode*" or "cell-disruption" or ultrasonic or microwave* or sonication or "forward-osmosis" or "membrane-foul-ing"   or   "flue-gas*"   or   "metabolic-eng*"   or   "metabolic-pathw*"   or "systems-biology" or "synthetic-biology" or wastewater* or (nutrient* and (*removal* or uptake or recovery or recycling or availability or wastewater*)) or "waste-water*" or effluent* or waste* or manure or sewage or fame* or residue* or ((CO2 or "carbon-dioxide" or "greenhouse-gas*") and (capture or *fixation or *mitigation or *remediation or *sequestration or utilization or removal or recycling or concentration or reduction)) or alkane* or "compression-ignition-engine*" or "ci-engine*" or growth or "downstream-processing" or "thermal-degradation" or drying or "Pulsed-electric-field" or "catalytic-upgrading" or hydrotreatment or "Heterotrophic-culture*" or *refining or *refiner* or "py-gc/ms" or "life-cycle*" or "techno-econ*" or starchless or processing or extraction* or ponds or engineer* or ((nitrogen or phosphorus or *phosphate or ammoni* or nitrate or sulfur) and (depriv* or starv* or limitation or removal or replete or availability or stress* or uptake or deficiency or supply)) or raceway* or "genetically modified" or conversion or "mass culture" or separation or immobiliz* or "carbon capture" or decomposition or oleaginous or alveolar or recovery or density or milking or "mass production" or Transcriptom* or "Heterotrophic culture*" or batch*) OR AU=("chisti y" or "ross ab" or "grima em" or "molina-grima e" or "molina e").

33.A.2 ALGAE-RELATED KEYWORDS

TI=(algae or algal or alga or chlorarachniophy* or phycol*) OR TI=(dinoflagellat* or *coccolith* or dinophy* or alexandrium or emiliania or gambier* or *gonyau* or *gymnodini* or haptophyt* or prorocentr* or prymnesi* or zooxanthella* or amphidin* or akashiwo or isochrysis or karenia* or phaeocystis or symbiodinium or chrysophyt* or chrysophyc* or raphidophy* or ochromonas or peridin* or pfiesteria) OR TI=(chlamydomon* or *chlorella or microalga* or chlorophyt* or "green-alga*" or chlorophyc* or euglen* or "micro-alga*" or chrysophy* or dunaliella or haematococcus or nannochloropsis or scenedesmus or cryptophy* or porphyridium or volvoc* or acetabularia or botryococcus or chlorococc* or phormidium or prototheca or tetraselmis or volvox or prasinophy*) OR TI=(macroalga* or rhodophy* or seaweed* or "red-alga*" or "brown-alga*" or gracilar* or kelp* or phaeophy* or porphyra or ulva* or caulerpa* or corallina* or fucus or gigartina* or laminaria* or saccharina or sargassum or nitell* or characea* or charophyt* or dictyota* or enteromorpha or fucale* or fucoxanthin* or halocynthia* or zygnema* or ascophyllum or bangia* or chondrus or cladophor* or codium or cystoseira or ecklonia or gelidium or kappaphycus or laurencia* or macrocystis or ectocarp* or ceramiale* or pyropia* or rhodomela* or spirogyra or undaria) OR TI=(diatoms or diatom or bacillarioph* or thalassiosira* or *nitzschia or phaeodactylum or chaetoceros or navicula or skeletonema or cyclotell*) OR TI=(*cyanobact* or *synechoc* or *cylindrospermops* or "blue-green-alga*" or *anabaen* or cyanophy* or nostoc* or *oscillatoria* or spirul* or arthrospira or *lyngbya* or aphanizomenon or planktothrix or trichodesmium) OR SO=("Algal Research*" or "European Journal of Phycology" or "Journal of Applied Phycology" or "Journal of Phycology" or Phycologia or "British Phycological Journal" or "Diatom Research" or "Phycological Research" or Algae or "Cryptogamie Algologie" or Fottea*)

33.A.3 Algal biodiesel fuel-related keywords

TI=(photobioreactor* or "photo-bioreactor*" or "*alga*growth" or "*alga*product*" or "open-pond*" or "raceway*pond*" or "*alga*oil*" or oilgae or "*alga*pond*" or "*alga-cultur*" or "*alga*biomass")

33.A.4 NOT

WC=(nutr* or pharm* or meteor* or geo* or materials* or polymer* or "engineering geol*" or metall* or "engineering biomed*" or physiol* or physics* or "chemistry organic" or econ* or business* or ocean* or fish* or limnol* or toxic* or endoc* or oncol* or optic* or veter* or integr* or med* or parasit*) OR TI=("light-harvest*" or photosy* or *sorp* or *sorb* or herbivor* or aqua* or biostimulant* or "heavy-metal*" or ion* or diet* or coral* or biogas or microcysti* or alkenone* or chrom* or megasc* or cadmium or lead or symbio* or ethanol or bioethanol or ocean* or food* or triclosan or *toxi* or caroten* or dye or bloom* or *butan* or lake* or sea* or acclim* or *plankton* or uranium or coast* or chilling or mariculture or "mem-brane-lipid*" or neuron* or disturbance or chlorosis or virus or macrophyte* or desa-tur* or biodegrad* or "h-2" or *hydrogen or antioxid* or astaxanth* or crust* or apic* or omega or arachidonic or "nitrogen-fix*" or *plastic or nutraceut* or dha or lectin* or edible or "plant-growth" or limestone or morphogenesis or therap* or bar-ley or gelat* or bioactive* or docosahexaenoic or eicosapentaenoic or polyglucan* or carbohydrat* or wave* or epa or isoprenoid or polyunsaturated or ultraviolet or pufa or "n-2 fixation" or methane or bivalve* or forest or estuar* or sterol or crop* or shale or fucoid* or *arctic or atmosph* or nit2 or "stress-tolerance" or *inhibitory or respir* or mat or carbonization or periphyton or binding or escherichia or pigment* or zeaxanthin or phycocyanin or benthic or anhydrase or ecol* or antibacter* or "valuable-product*" or ecosystem* or contaminant* or "concentrating-mechanism*" or eutrophication or phycoerythrin or lutein or inhibitory or alken* or circadian or habitat* or antifouling or silicon) OR SO=("marine biology" or "marine poll*" or "harmful algae") OR AU=("raven ja").

REFERENCES

Ahmadun, F. R., A. Pendashteh, and L. C. Abdullah, et al. 2009. Review of technologies for oil and gas produced water treatment. *Journal of Hazardous Materials* 170: 530–551.

Atlas, R. M. 1981. Microbial degradation of petroleum hydrocarbons: An environmental per-spective. *Microbiological Reviews* 45: 180–209.

Babich, I. V. and J. A. Moulijn. 2003. Science and technology of novel processes for deep desulfurization of oil refinery streams: A review. *Fuel* 82: 607–631.

Bhattacharya, S., Shilpa and M. Bhati. 2012. China and India: The two new players in the nanotechnology race. *Scientometrics* 93: 59–87.

Biller, P. and A. B. Ross. 2011. Potential yields and properties of oil from the hydrothermal liquefaction of microalgae with different biochemical content. *Bioresource Technology* 102: 215–225.

Bordons, M., B. Gonzalez-Albo, J. Aparicio and L. Moreno. 2015. The influence of R & D intensity of countries on the impact of international collaborative research: Evidence from Spain. *Scientometrics* 102: 1385–1400.

Brennan, L. and P. Owende. 2010. Biofuels from microalgae: A review of technologies for production, processing, and extractions of biofuels and co-products. *Renewable & Sustainable Energy Reviews* 14: 557–577.

Bridgwater, A. V. and G. V. C. Peacocke. 2000. Fast pyrolysis processes for biomass. *Renewable & Sustainable Energy Reviews* 4: 1–73.

Brown, T. M., P. G. Duan and P. E. Savage. 2010. Hydrothermal liquefaction and gasification of *Nannochloropsis* sp. *Energy & Fuels* 24: 3639–3646.

Busca, G., L. Lietti, G. Ramis, and F. Berti. 1998. Chemical and mechanistic aspects of the selective catalytic reduction of NO_x by ammonia over oxide catalysts: A review. *Applied Catalysis B-Environmental* 18: 1–36.

Cheng, Y. Q. and E. Ma. 2011. Atomic-level structure and structure–property relationship in metallic glasses. *Progress in Materials Science* 56: 379–473.

Chisti, Y. 2007. Biodiesel from microalgae. *Biotechnology Advances* 25: 294–306.

Chisti, Y. 2008. Biodiesel from microalgae beats bioethanol. *Trends in Biotechnology* 26: 126–131.

Chiu, SY, C. Y. Kao and M. T. Tsai et al. 2009. Lipid accumulation and CO_2 utilization of *Nannochloropsis oculata* in response to CO_2 aeration. *Bioresource Technology* 100: 833–838.

Clarivate Analytics. 2019. *Highly cited researchers: 2019 Recipients.* Philadelphia, PA: Clarivate Analytics. https://recognition.webofsciencegroup.com/awards/highly-cited/2019/ (accessed January, 3, 2020).

Converti, A., A.A. Casazza, E. Y. Ortiz, P. Perego and M. Del Borghi. 2009. Effect of temperature and nitrogen concentration on the growth and lipid content of *Nannochloropsis* oculata and *Chlorella vulgaris* for biodiesel production. *Chemical Engineering and Processing-Process Intensification* 48: 1146–1151.

Czernik, S. and A. V. Bridgwater. 2004. Overview of applications of biomass fast pyrolysis oil. *Energy & Fuels* 18: 590–598.

Docampo, D. and L. Cram. 2019. Highly cited researchers: A moving target. *Scientometrics* 118: 1011–1025.

Evans, R. J. and T. A. Milne. 1987. Molecular characterization of the pyrolysis of biomass. 1. Fundamentals. *Energy & Fuels* 1: 123–137.

Gallezot, P. 2012. Conversion of biomass to selected chemical products. *Chemical Society Reviews* 41: 1538–1558.

Garfield, E. 1955. Citation indexes for science. *Science* 122: 108–111.

Garfield, E. 1972. Citation analysis as a tool in journal evaluation. *Science* 178: 471–479.

Glanzel, W., J. Leta and B. Thijs. 2006. Science in Brazil. Part 1: A macro-level comparative study. *Scientometrics* 67: 67–86.

Griffiths, M.J. and S. T. L. Harrison. 2009. Lipid productivity as a key characteristic for choosing algal species for biodiesel production. *Journal of Applied Phycology* 21: 493–507.

Grima, E. M., E. H. Belarbi, F. G. A. Fernandez, A. R. Medina and Y. Chisti. 2003. Recovery of microalgal biomass and metabolites: Process options and economics. *Biotechnology Advances* 20: 491–515.

Guan, J. C. and N. Ma. 2007. China's emerging presence in nanoscience and nanotechnology: A comparative bibliometric study of several nanoscience 'giants'. *Research Policy* 36: 880–886.

Halim, R., M. K. Danquah and P. A. Webley. 2012. Extraction of oil from microalgae for biodiesel production: A review. *Biotechnology Advances* 30: 709–732.

Hu, Q., M. Sommerfeld, and E. Jarvis, et al. 2008. Microalgal triacylglycerols as feedstocks for biofuel production: perspectives and advances. *Plant Journal* 54: 621–639.

Khalili, N. R., P. A. Scheff, and T. M. Holsen. 1995. PAH source fingerprints for coke ovens, diesel and gasoline-engines, highway tunnels, and wood combustion emissions. *Atmospheric Environment* 29: 533–542.

Kilian, L. 2009. Not all oil price shocks are alike: Disentangling demand and supply shocks in the crude oil market. *American Economic Review* 99: 1053–1069.

Konur, O. 2000. Creating enforceable civil rights for disabled students in higher education: An institutional theory perspective. *Disability & Society* 15: 1041–1063.

Konur, O. 2002a. Access to nursing education by disabled students: Rights and duties of nursing programs. *Nurse Education Today* 22: 364–374.

Konur, O. 2002b. Assessment of disabled students in higher education: Current public policy issues. *Assessment and Evaluation in Higher Education* 27: 131–152.

Konur, O. 2002c. Access to employment by disabled people in the UK: Is the Disability Discrimination Act working? *International Journal of Discrimination and the Law* 5: 247–279.

Konur, O. 2006a. Participation of children with dyslexia in compulsory education: Current public policy issues. *Dyslexia* 12: 51–67.

Konur, O. 2006b. Teaching disabled students in higher education. *Teaching in Higher Education* 11: 351–363.

Konur, O. 2007a. A judicial outcome analysis of the Disability Discrimination Act: A windfall for the employers? *Disability & Society* 22: 187–204.

Konur, O. 2007b. Computer-assisted teaching and assessment of disabled students in higher education: The interface between academic standards and disability rights. *Journal of Computer Assisted Learning* 23: 207–219.

Konur, O. 2011. The scientometric evaluation of the research on the algae and bio-energy. *Applied Energy* 88: 3532–3540.

Konur, O. 2012a. Evaluation of the research on the social sciences in Turkey: A scientometric approach. *Energy Education Science and Technology Part B: Social and Educational Studies* 4: 1893–1908.

Konur, O. 2012b. Prof. Dr. Ayhan Demirbas' scientometric biography. *Energy Education Science and Technology Part A: Energy Science and Research* 28: 727–738.

Konur, O. 2012c. The evaluation of the biogas research: A scientometric approach. *Energy Education Science and Technology Part A: Energy Science and Research* 29: 1277–1292.

Konur, O. 2012d. The evaluation of the educational research: A scientometric approach. *Energy Education Science and Technology Part B: Social and Educational Studies* 4: 1935–1948.

Konur, O. 2012e. The evaluation of the global energy and fuels research: A scientometric approach. *Energy Education Science and Technology Part A: Energy Science and Research* 30: 613–628.

Konur, O. 2012f. The evaluation of the research on the Arts and Humanities in Turkey: A scientometric approach. *Energy Education Science and Technology Part B: Social and Educational Studies* 4: 1603–1618.

Konur, O. 2012g. The evaluation of the research on the biodiesel: A scientometric approach. *Energy Education Science and Technology Part A: Energy Science and Research* 28: 1003–1014.

Konur, O. 2012h. The evaluation of the research on the bioethanol: A scientometric approach. *Energy Education Science and Technology Part A: Energy Science and Research* 28: 1051–1064.

Konur, O. 2012i. The evaluation of the research on the biofuels: A scientometric approach. *Energy Education Science and Technology Part A: Energy Science and Research* 28: 903–916.

Konur, O. 2012j. The evaluation of the research on the biohydrogen: A scientometric approach. *Energy Education Science and Technology Part A: Energy Science and Research* 29: 323–338.

Konur, O. 2012k. The evaluation of the research on the microbial fuel cells: A scientometric approach. *Energy Education Science and Technology Part A: Energy Science and Research* 29: 309–322.

Konur, O. 2012l. The scientometric evaluation of the research on the production of bioenergy from biomass. *Biomass and Bioenergy* 47: 504–515.

Konur, O. 2012m. The scientometric evaluation of the research on the deaf students in higher education. *Energy Education Science and Technology Part B: Social and Educational Studies* 4: 1573–1588.

Konur, O. 2012n. The scientometric evaluation of the research on the students with ADHD in higher education. *Energy Education Science and Technology Part B: Social and Educational Studies* 4: 1547–1562.

Konur, O. 2015. Current state of research on algal biodiesel. In *Marine Bioenergy: Trends and Developments*, S. K. Kim, and C. G. Lee, ed., 487–512. Boca Raton, FL: CRC Press.

Konur, O. 2016a. Scientometric overview in nanobiodrugs. In *Nanoarchitectonics for Smart Delivery and Drug Targeting*, A. M. Holban and A.M. Grumezescu, ed., 405–428. Amsterdam: Elsevier.

Konur, O. 2016b. Scientometric overview regarding nanoemulsions used in the food industry. In *Emulsions: Nanotechnology in the Agri-Food Industry*, A. M. Grumezescu, ed., 689–711. Amsterdam: Elsevier.

Konur, O. 2016c. Scientometric overview regarding the nanobiomaterials in antimicrobial therapy. In *Nanobiomaterials in Antimicrobial Therapy*, A. M. Grumezescu, ed., 511–535. Amsterdam: Elsevier.

Konur, O. 2016d. Scientometric overview regarding the nanobiomaterials in dentistry. In *Nanobiomaterials in Dentistry*, A. M. Grumezescu, ed., 425–453. Amsterdam: Elsevier.

Konur, O. 2016e. Scientometric overview regarding the surface chemistry of nanobiomaterials. In *Surface Chemistry of Nanobiomaterials*, A. M. Grumezescu, ed., 463–486. Amsterdam: Elsevier.

Konur, O. 2016f. The scientometric overview in cancer targeting. In *Nanoarchitectonics for Smart Delivery and Drug Targeting*, A. M. Holban and A. Grumezescu, ed., 871–895. Amsterdam; Elsevier.

Konur, O. 2017a. Recent citation classics in antimicrobial nanobiomaterials. In *Nanostructures for Antimicrobial Therapy*, A. Ficai and A. M. Grumezescu, ed., 669–685. Amsterdam: Elsevier.

Konur, O. 2017b. Scientometric overview in nanopesticides. In *New Pesticides and Soil Sensors*, A. M. Grumezescu, ed. 719–744. Amsterdam: Elsevier.

Konur, O. 2017c. Scientometric overview regarding oral cancer nanomedicine. In *Nanostructures for Oral Medicine*, E. Andronescu, A. M. Grumezescu, ed., 939–962. Amsterdam: Elsevier.

Konur, O. 2017d. Scientometric overview regarding water nanopurification. In *Water Purification*, A. M. Grumezescu, ed., 693–716. Amsterdam: Elsevier.

Konur, O. 2017e. Scientometric overview in food nanopreservation. In *Food Preservation*, A. M. Grumezescu, ed., 703–729. Amsterdam: Elsevier.

Konur, O. 2017f. The top citation classics in alginates for biomedicine. In *Seaweed Polysaccharides: Isolation, Biological and Biomedical Applications*, J. Venkatesan, S. Anil, S. K. Kim, ed., 223–249. Amsterdam: Elsevier.

Konur, O. 2018a. Scientometric evaluation of the global research in spine: An update on the pioneering study by Wei et al. *European Spine Journal* 27: 525–529.

Konur, O. 2018b. Bioenergy and biofuels science and technology: Scientometric overview and citation classics. In *Bioenergy and Biofuels*, O. Konur, ed., 3–63. Boca Raton: CRC Press.

Konur, O. 2019a. Cyanobacterial bioenergy and biofuels science and technology: A sciento-metric overview. In *Cyanobacteria: From Basic Science to Applications*, ed. A. K. Mishra, D. N. Tiwari and A. N. Rai, 419–442. Amsterdam: Elsevier.

Konur, O. 2019b. Nanotechnology applications in food: A scientometric overview. In *Nanoscience for Sustainable Agriculture*, R. N. Pudake, N. Chauhan, and C. Kole, ed., 683–711. Cham: Springer.

Konur, O., ed. 2021a. *Handbook of Biodiesel and Petrodiesel Fuels: Science, Technology, Health, and Environment*. Boca Raton, FL: CRC Press.

Konur, O., ed. 2021b. *Handbook of Biodiesel and Petrodiesel Fuels: Science, Technology, Health, and Environment. Volume 1. Biodiesel Fuels: Science, Technology, Health, and Environment*. Boca Raton, FL: CRC Press.

Konur, O., ed. 2021c. *Handbook of Biodiesel and Petrodiesel Fuels: Science, Technology, Health, and Environment. Volume 2. Biodiesel Fuels based on the Edible and Nonedible Feedstocks, Wastes, and Algae: Science, Technology, Health, and Environment*. Boca Raton, FL: CRC Press.

Konur, O., ed. 2021d. *Handbook of Biodiesel and Petrodiesel Fuels: Science, Technology, Health, and Environment. Volume 3. Petrodiesel Fuels: Science, Technology, Health, and Environment*. Boca Raton, FL: CRC Press.

Konur, O. 2021e. Biodiesel and petrodiesel fuels: Science, technology, health, and environment. In *Handbook of Biodiesel and Petrodiesel Fuels: Science, Technology, Health, and Environment. Volume 1. Biodiesel Fuels: Science, Technology, Health, and Environment*, ed. O. Konur. Boca Raton, FL: CRC Press.

Konur, O. 2021f. Biodiesel and petrodiesel fuels: A scientometric review of the research. In *Handbook of Biodiesel and Petrodiesel Fuels: Science, Technology, Health, and Environment. Volume 1. Biodiesel Fuels: Science, Technology, Health, and Environment*, ed. O. Konur. Boca Raton, FL: CRC Press.

Konur, O. 2021g. Biodiesel and petrodiesel fuels: A review of the research. In *Handbook of Biodiesel and Petrodiesel Fuels: Science, Technology, Health, and Environment. Volume 1. Biodiesel Fuels: Science, Technology, Health, and Environment*, ed. O. Konur. Boca Raton, FL: CRC Press.

Konur, O. 2021h Nanotechnology applications in the diesel fuels and the related research fields: A review of the research. In *Handbook of Biodiesel and Petrodiesel Fuels: Science, Technology, Health, and Environment. Volume 1. Biodiesel Fuels: Science, Technology, Health, and Environment*, ed. O. Konur. Boca Raton, FL: CRC Press.

Konur, O. 2021i. Biooils: A scientometric review of the research. In *Handbook of Biodiesel and Petrodiesel Fuels: Science, Technology, Health, and Environment. Volume 1. Biodiesel Fuels: Science, Technology, Health, and Environment*, ed. O. Konur. Boca Raton, FL: CRC Press.

Konur, O. 2021j. Characterization and properties of biooils: A review of the research. In *Handbook of Biodiesel and Petrodiesel Fuels: Science, Technology, Health, and Environment. Volume 1. Biodiesel Fuels: Science, Technology, Health, and Environment*, ed. O. Konur. Boca Raton, FL: CRC Press.

Konur, O. 2021k. Biomass pyrolysis and pyrolysis oils: A review of the research. In *Handbook of Biodiesel and Petrodiesel Fuels: Science, Technology, Health, and Environment. Volume 1. Biodiesel Fuels: Science, Technology, Health, and Environment*, ed. O. Konur. Boca Raton, FL: CRC Press.

Konur, O. 2021l. Biodiesel fuels: A scientometric review of the research. In *Handbook of Biodiesel and Petrodiesel Fuels: Science, Technology, Health, and Environment. Volume 1. Biodiesel Fuels: Science, Technology, Health, and Environment*, ed. O. Konur. Boca Raton, FL: CRC Press.

Konur, O. 2021m. Glycerol: A scientometric review of the research. In *Handbook of Biodiesel and Petrodiesel Fuels: Science, Technology, Health, and Environment. Volume 1.*

Biodiesel Fuels: Science, Technology, Health, and Environment, ed. O. Konur. Boca Raton, FL: CRC Press.

Konur, O. 2021n. Propanediol production from glycerol: A review of the research. In *Handbook of Biodiesel and Petrodiesel Fuels: Science, Technology, Health, and Environment. Volume 1. Biodiesel Fuels: Science, Technology, Health, and Environment*, ed. O. Konur. Boca Raton, FL: CRC Press.

Konur, O. 2021o. Edible oil-based biodiesel fuels: A scientometric review of the research. *In Handbook of Biodiesel and Petrodiesel Fuels: Science, Technology, Health, and Environment. Volume 2. Biodiesel Fuels based on the Edible and Nonedible Feedstocks, Wastes, and Algae: Science, Technology, Health, and Environment*, ed. O. Konur. Boca Raton, FL: CRC Press.

Konur, O. 2021p. Palm oil-based biodiesel fuels: A review of the research. In *Handbook of Biodiesel and Petrodiesel Fuels: Science, Technology, Health, and Environment. Volume 2. Biodiesel Fuels based on the Edible and Nonedible Feedstocks, Wastes, and Algae*, ed. O. Konur. Boca Raton, FL: CRC Press.

Konur, O. 2021q. Rapeseed oil-based biodiesel fuels: A review of the research. In *Handbook of Biodiesel and Petrodiesel Fuels: Science, Technology, Health, and Environment. Volume 2. Biodiesel Fuels based on the Edible and Nonedible Feedstocks, Wastes, and Algae*, ed. O. Konur. Boca Raton, FL: CRC Press.

Konur, O. 2021r. Nonedible oil-based biodiesel fuels: A scientometric review of the research. In *Handbook of Biodiesel and Petrodiesel Fuels: Science, Technology, Health, and Environment. Volume 2. Biodiesel Fuels based on the Edible and Nonedible Feedstocks, Wastes, and Algae: Science, Technology, Health, and Environment*, ed. O. Konur. Boca Raton, FL: CRC Press.

Konur, O. 2021s. Waste oil-based biodiesel fuels: A scientometric review of the research. In *Handbook of Biodiesel and Petrodiesel Fuels: Science, Technology, Health, and Environment. Volume 2. Biodiesel Fuels based on the Edible and Nonedible Feedstocks, Wastes, and Algae: Science, Technology, Health, and Environment*, ed. O. Konur. Boca Raton, FL: CRC Press.

Konur, O. 2021t. Algal biodiesel fuels: A scientometric review of the research. In *Handbook of Biodiesel and Petrodiesel Fuels: Science, Technology, Health, and Environment. Volume 2. Biodiesel Fuels based on the Edible and Nonedible Feedstocks, Wastes, and Algae: Science, Technology, Health, and Environment*, ed. O. Konur. Boca Raton, FL: CRC Press.

Konur, O. 2021u. Algal biomass production for biodiesel production: A review of the research. In *Handbook of Biodiesel and Petrodiesel Fuels: Science, Technology, Health, and Environment. Volume 2. Biodiesel Fuels based on the Edible and Nonedible Feedstocks, Wastes, and Algae*, ed. O. Konur. Boca Raton, FL: CRC Press.

Konur, O. 2021v. Algal biomass production in wastewaters for biodiesel production: A review of the research. In *Handbook of Biodiesel and Petrodiesel Fuels: Science, Technology, Health, and Environment. Volume 2. Biodiesel Fuels based on the Edible and Nonedible Feedstocks, Wastes, and Algae*, ed. O. Konur. Boca Raton, FL: CRC Press.

Konur, O. 2021x. Algal lipid production for biodiesel production: A review of the research. In *Handbook of Biodiesel and Petrodiesel Fuels: Science, Technology, Health, and Environment. Volume 2. Biodiesel Fuels based on the Edible and Nonedible Feedstocks, Wastes, and Algae*, ed. O. Konur. Boca Raton, FL: CRC Press.

Konur, O. 2021y. Crude oils: A scientometric review of the research. In *Handbook of Biodiesel and Petrodiesel Fuels: Science, Technology, Health, and Environment. Volume 3. Petrodiesel Fuels: Science, Technology, Health, and Environment*, ed. O. Konur. Boca Raton, FL: CRC Press.

Konur, O. 2021z. Petrodiesel fuels: A scientometric review of the research. In *Handbook of Biodiesel and Petrodiesel Fuels: Science, Technology, Health, and Environment. Volume*

3. Petrodiesel Fuels: Science, Technology, Health, and Environment, ed. O. Konur. Boca Raton, FL: CRC Press.

Konur, O. 2021aa. Bioremediation of petroleum hydrocarbons in the contaminated soils: A review of the research. In *Handbook of Biodiesel and Petrodiesel Fuels: Science, Technology, Health, and Environment. Volume 3. Petrodiesel Fuels: Science, Technology, Health, and Environment*, ed. O. Konur. Boca Raton, FL: CRC Press.

Konur, O. 2021ab. Desulfurization of diesel fuels: A review of the research. In *Handbook of Biodiesel and Petrodiesel Fuels: Science, Technology, Health, and Environment. Volume 3. Petrodiesel Fuels: Science, Technology, Health, and Environment*, ed. O. Konur. Boca Raton, FL: CRC Press.

Konur, O. 2021ac. Diesel fuel exhaust emissions: A scientometric review of the research. In *Handbook of Biodiesel and Petrodiesel Fuels: Science, Technology, Health, and Environment. Volume 3. Petrodiesel Fuels: Science, Technology, Health, and Environment*, ed. O. Konur. Boca Raton, FL: CRC Press.

Konur, O. 2021ad. The adverse health and safety impact of diesel fuels: A scientometric review of the research. In *Handbook of Biodiesel and Petrodiesel Fuels: Science, Technology, Health, and Environment. Volume 3. Petrodiesel Fuels: Science, Technology, Health, and Environment*, ed. O. Konur. Boca Raton, FL: CRC Press.

Konur, O. 2021ae. Respiratory illnesses caused by the diesel fuel exhaust emissions: A review of the research. In *Handbook of Biodiesel and Petrodiesel Fuels: Science, Technology, Health, and Environment. Volume 3. Petrodiesel Fuels: Science, Technology, Health, and Environment*, ed. O. Konur. Boca Raton, FL: CRC Press.

Konur, O. 2021af. Cancer caused by the diesel fuel exhaust emissions: A review of the research. In *Handbook of Biodiesel and Petrodiesel Fuels: Science, Technology, Health, and Environment. Volume 3. Petrodiesel Fuels: Science, Technology, Health, and Environment*, ed. O. Konur. Boca Raton, FL: CRC Press.

Konur, O. 2021ag. Cardiovascular and other illnesses caused by the diesel fuel exhaust emissions: A review of the research. In *Handbook of Biodiesel and Petrodiesel Fuels: Science, Technology, Health, and Environment. Volume 3. Petrodiesel Fuels: Science, Technology, Health, and Environment*, ed. O. Konur. Boca Raton, FL: CRC Press.

Konur, O. and F. L. Matthews. 1989. Effect of the properties of the constituents on the fatigue performance of composites: A review. *Composites* 20: 317–328.

Lapuerta, M., O. Armas, and J. Rodriguez-Fernandez. 2008. Effect of biodiesel fuels on diesel engine emissions. *Progress in Energy and Combustion Science* 34: 198–223.

Lariviere, V. and Y. Gingras. 2010. On the relationship between interdisciplinarity and scientific impact. *Journal of the American Society for Information Science and Technology* 61: 126–131.

Lariviere, V., C. Ni, Y. Gingras, B. Cronin, and C.R. Sugimoto. 2013. Bibliometrics: Global gender disparities in science. *Nature News* 504: 211–213.

Lee, J. Y., C. Yoo, S. Y. Jun, C. Y. Ahn, and H. M. Oh. 2010. Comparison of several methods for effective lipid extraction from microalgae. *Bioresource Technology* 101: S75–S77.

Leydesdorff, L. and P. Zhou. 2005. Are the contributions of China and Korea upsetting the world system of science? *Scientometrics* 63: 617–630.

Leydesdorff, L. and Wagner, C. 2009. Is the United States losing ground in science? A global perspective on the world science system. *Scientometrics* 78: 23–36.

Li, Y. C., Y. F. Chen, P. Chen, et al. 2011. Characterization of a microalga *Chlorella* sp. well adapted to highly concentrated municipal wastewater for nutrient removal and biodiesel production. *Bioresource Technology* 102: 5138–5144.

Markou, G. and D. Georgakakis. 2011. Cultivation of filamentous cyanobacteria (blue-green algae) in agro-industrial wastes and wastewaters: A review. *Applied Energy* 88: 3389–3401.

Mata, T. M., A. A. Martins, and N. S. Caetano. 2010. Microalgae for biodiesel production and other applications: A review. *Renewable & Sustainable Energy Reviews* 14: 217–232.

Miao, X. L. and Q. Y. Wu. 2004. High yield bio-oil production from fast pyrolysis by metabolic controlling of *Chlorella protothecoides*. *Journal of Biotechnology* 110: 85–93.

Miao, X. L., Q. Y. Wu and C. Y. Yang. 2004. Fast pyrolysis of microalgae to produce renewable fuels. *Journal of Analytical and Applied Pyrolysis* 71: 855–863.

Mohan, D., C. U. Pittman, and P. H. Steele. 2006. Pyrolysis of wood/biomass for bio-oil: A critical review. *Energy & Fuels* 20: 848–889.

Moldowan, J. M., W. K. Seifert, and E. J. Gallegos. 1985. Relationship between petroleum composition and depositional environment of petroleum source rocks. *AAPG Bulletin-American Association of Petroleum Geologists* 69: 1255–1268.

Morillo, F., M. Bordons and I. Gomez. 2001. An approach to interdisciplinarity through bibliometric indicators. *Scientometrics* 51: 203–222.

North, D. C. 1991a. *Institutions, Institutional Change and Economic Performance.* Cambridge, Mass.: Cambridge University Press.

North, D.C. 1991b. Institutions. *Journal of Economic Perspectives* 5: 97–112.

Oleinik, A. 2012. Publication patterns in Russia and the West compared. *Scientometrics* 93: 533–551.

Perron, P. 1989. The great crash, the oil price shock, and the unit root hypothesis. *Econometrica: Journal of the Econometric Society* 57: 1361–1401.

Rodolfi, L., G. C. Zittelli, and N. Bassi et al. 2009. Microalgae for oil: Strain selection, induction of lipid synthesis and outdoor mass cultivation in a low-cost photobioreactor. *Biotechnology and Bioengineering* 102: 100–112.

Rogers, D., and A. J. Hopfinger. 1994. Application of genetic function approximation to quantitative structure-activity relationships and quantitative structure-property relationships. *Journal of Chemical Information and Computer Sciences* 34: 854–866.

Rogge, W. F., L. M. Hildemann, M. A. Mazurek, G. R. Cass, and B. R. T. Simoneit. 1993. Sources of fine organic aerosol. 2. Noncatalyst and catalyst-equipped automobiles and heavy-duty diesel trucks. *Environmental Science & Technology* 27: 636–651.

Schauer, J. J., M. J. Kleeman, G. R. Cass, and B. R. T. Simoneit. 1999. Measurement of emissions from air pollution sources. 2. C_1 through C_{30} organic compounds from medium duty diesel trucks. *Environmental Science & Technology* 33: 1578–1587.

Schenk, P. M., S. R. Thomas-Hall, and E. Stephens et al. 2008. Second generation biofuels: high-efficiency microalgae for biodiesel production. *Bioenergy Research* 1: 20–43.

Scherf, U. and E. J. List. 2002. Semiconducting polyfluorenes-towards reliable structure–property relationships. *Advanced Materials* 14: 477–487.

Siaut, M., S. Cuine, and C. Cagnon et al. 2011. Oil accumulation in the model green alga *Chlamydomonas reinhardtii*: characterization, variability between common laboratory strains and relationship with starch reserves. *BMC Biotechnology* 11: 7.

Van Gerpen, J. 2005. Biodiesel processing and production. *Fuel Processing Technology* 86: 1097–1107.

Wang, B., Y. Q. Li, N. Wu and C. Q. Lan. 2008. CO_2 bio-mitigation using microalgae. *Applied Microbiology and Biotechnology* 79: 707–718.

Wang, Z. T., N. Ullrich, S. Joo, S. Waffenschmidt, and U. Goodenough. 2009. Algal lipid bodies: stress induction, purification, and biochemical characterization in wild-type and starchless *Chlamydomonas reinhardtii*. *Eukaryotic Cell* 8: 1856–1868.

Wang, L. A., M. Min and Y. C. Li et al. 2010. Cultivation of green algae *Chlorella* sp. in different wastewaters from municipal wastewater treatment plant. *Applied Biochemistry And Biotechnology* 162: 1174–1186.

Wang, X., D. Liu, K. Ding. and X. Wang. 2012. Science funding and research output: A study on 10 countries. *Scientometrics* 91: 591–599.

Williams, P. J. L. and L. M. L. Laurens. 2010. Microalgae as biodiesel & biomass feedstocks: Review & analysis of the biochemistry, energetics & economics. *Energy & Environmental Science* 3: 554–590.

Xie, Y. and K. A. Shauman. 1998. Sex differences in research productivity: New evidence about an old puzzle. *American Sociological Review* 63: 847–870.

Yaman, S. 2004. Pyrolysis of biomass to produce fuels and chemical feedstocks. *Energy Conversion and Management* 45: 651–671.

Youtie, J, P. Shapira, and A. L. Porter. 2008. Nanotechnology publications and citations by leading countries and blocs. *Journal of Nanoparticle Research* 10: 981–986.

Zhang, Q., J. Chang, T. J. Wang, and Y. Xu. 2007. Review of biomass pyrolysis oil properties and upgrading research. *Energy Conversion and Management* 48: 87–92.

Zhou, P. and L. Leydesdorff. 2006. The emergence of China as a leading nation in science. *Research Policy* 35: 83–104.

Zhou, W. G., Y. C. Li, M. Min et al. 2011. Local bioprospecting for high-lipid producing microalgal strains to be grown on concentrated municipal wastewater for biofuel production. *Bioresource Technology* 102: 6909–6919.

34 Algal Biomass Production for Biodiesel Production

A Review of the Research

Ozcan Konur

CONTENTS

34.1   INTRODUCTION

Crude oils have been primary sources of energy and fuels, such as petrodiesel. However, significant public concerns about the sustainability, price fluctuations, and adverse environmental impact of crude oils have emerged since the 1970s (Ahmadun et al., 2009; Atlas, 1981; Babich and Moulijn, 2003; Kilian, 2009; Perron, 1989). Thus, bioo-ils (Bridgwater et al., 1999; Bridgwater and Peacocke, 2000; Czernik and Bridgwater, 2004; Mohan et al., 2006; Zhang et al., 2007) and biooil-based biodiesel fuels (Chisti, 2007; Hill et al., 2006; Hu et al., 2008; Mata et al., 2010; Rodolfi et al., 2009; Schenk et al., 2008) have emerged as alternatives to crude oils and crude oil-based petrodiesel fuels, respectively, in recent decades. Nowadays, although petrodiesel fuels are still used extensively, biodiesel fuels are being used increasingly in the transportation and power sectors (Konur, 2021a–ag). Therefore, there has been great public interest in the development of algal biodiesel fuels as the fourth generation of biodiesel fuels (Chisti, 2007; Hu et al., 2008; Mata et al., 2010; Rodolfi et al., 2009; Schenk et al., 2008). However, it is necessary to reduce the total cost of biodiesel production by reducing the feedstock cost through the improvement of the biomass productivity of algal biomass (Borowitzka, 1999; Chen et al., 2011; Christenson and Sims, 2011; Perez-Garcia et al., 2011; Radakovits et al., 2012; Ugwu et al., 2008; Wang et al., 2008).

Furthermore, for the efficient progression of the research in this field, it is necessary to develop efficient incentive structures for the primary stakeholders and to inform these stakeholders about the research (Konur, 2000, 2002a–c, 2006a–b, 2007a–b; North, 1991a–b).

Although there have been a number of reviews and book chapters in this field (Borowitzka, 1999; Chen et al., 2011; Christenson and Sims, 2011; Perez-Garcia et al., 2011; Radakovits et al., 2012; Ugwu et al., 2008; Wang et al., 2008), there has been no review of the 25-most-cited articles. Thus, this chapter reviews these articles by highlighting the key findings of these most-prolific studies on algal biomass production for algal biodiesel production. Then, it discusses these key findings.

34.2   MATERIALS AND METHODOLOGY

The search for the literature was carried out in the 'Web of Science' (WOS) database in May 2020. It contains the 'Science Citation Index-Expanded' (SCI-E), the Social Sciences Citation Index' (SSCI), the 'Book Citation Index-Science' (BCI-S), the 'Conference Proceedings Citation Index-Science' (CPCI-S), the 'Emerging Sources Citation Index' (ESCI), the 'Book Citation Index-Social Sciences and Humanities' (BCI-SSH), the 'Conference Proceedings Citation Index-Social Sciences and Humanities' (CPCI-SSH), and the 'Arts and Humanities Citation Index' (A&HCI).

The keywords for the search of the literature are collated from the screening of abstract pages for the first 500 highly cited papers on algal biomass production. This keyword set is provided in the Appendix.

The 25-most-cited articles are selected for this review and the key findings are presented and discussed briefly.

34.3   RESULTS

34.3.1   ALGAL CULTIVATION ENGINEERING IN GENERAL

There are five research papers on the cultivation engineering of algae in general. Irradiance, N deprivation, P deprivation, mixothropic cultivation, iron repletion, and 'two-phase cultivation' are the key parameters studied in these papers. All are confined to microalgae (keyword set II in the Appendix).

Rodolfi et al. (2009) study the effect of N and P deprivation and irradiances on microalgal biomass and lipid productivity in a low-cost photobioreactor (PBR) in a seminal paper with 1,613 citations. They screened 30 microalgal strains for their biomass productivity and lipid content. They selected four strains with a relatively high lipid content and cultivated them under N deprivation in 0.6 L bubbled tubes. They find that only two of them accumulated lipid under such conditions. They then grew *Nannochloropsis* sp., which attained 60% lipid content after N starvation, in a 20 L flat alveolar panel PBR. They find that fatty acid content increased with high irradiances (up to 32.5% of dry biomass) and following both N and P deprivation (up to about 50%). They then grew this strain outdoors in 110 L green wall panel PBRs under nutrient sufficient and deficient conditions. They observed that lipid productivity increased from 117 mg/L/day in nutrient sufficient media (with an average biomass productivity of 0.36 g/L/day and 32% lipid content) to 204 mg/L/day (with an average biomass productivity of 0.30 g/L/day and more than 60% final lipid content) in N-deprived media. In a two-phase cultivation process (a nutrient-sufficient phase to produce the inoculum followed by an N-deprived phase to boost lipid synthesis) they project the oil production potential as more than 90 kg per hectare per day. They estimate that this microalgal strain has the potential for an annual production of 20 tons of lipid per hectare in the Mediterranean climate and of more than 30 tons of lipid per hectare in sunny tropical areas.

Li et al. (2008) study the effect of N sources and their concentrations on cell growth and lipid accumulation of *Neochloris oleoabundans* in a paper with 657 citations. They observed that whilst the highest lipid cell content of 0.40 g/g was obtained at the lowest sodium nitrate ($NaNO_3$) concentration (3 mM), a remarkable lipid productivity of 0.133 g l^{-1} day^{-1} was achieved at 5 mM with a lipid cell content of 0.34 g/g and a biomass productivity of 0.40 g l^{-1} day^{-1}. On the other hand, the highest biomass productivity was obtained at 10 mM sodium nitrate, with a biomass concentration of 3.2 g l^{-1} and a biomass productivity of 0.63 g l^{-1} day^{-1}. They observed that cell growth continued after the exhaustion of the external N pool, hypothetically supported by the consumption of intracellular N pools such as chlorophyll molecules.

Illman et al. (2000) study biomass production in algae under N deficiency in a paper with 587 citations. They grew algae in small (2L) stirred tank bioreactors and obtained the best growth with *Chlorella vulgaris* with a growth rate of 0.99/d^{-1} and obtained the highest calorific value (29 KJ/g) with *C. emersonii*. They propose that the calorific value of algae is related to the lipid content rather than any other component.

Liang et al. (2009) study the biomass and lipid productivities of *Chlorella vulgaris* under autotrophic, heterotrophic, and mixotrophic growth conditions in a paper with 530 citations. They observed that whilst autotrophic growth provided higher cellular lipid content (38%), the lipid productivity was much lower compared with those from heterotrophic growth with acetate, glucose, or glycerol. On the other hand, optimal cell growth (2 g l^{-1}) and lipid productivity (54 mg L^{-1} day^{-1}) were attained using glucose at 1% (w/v) whereas higher concentrations were inhibitory. Additionally, the growth of *C. vulgaris* on glycerol had similar dose effects as those from glucose. They conclude that *C. vulgaris* is mixotrophic.

Liu et al. (2008) study the effect of iron on growth and lipid accumulation in *Chlorella vulgaris* in a paper with 468 citations. They found that supplementing the growth media with chelated iron chloride ($FeCl_3$) in the late growth phase increased the final cell density but did not induce lipid accumulation in cells. They collected cells in the late-exponential growth phase by centrifugation and reinoculation into new media supplemented with five levels of ferric ion (Fe^{3+}) concentration. They observed that total lipid content in cultures supplemented with 1.2×10^{-5} mol L^{-1} $FeCl_3$ was up to 56.6% biomass by dry weight and was three to seven-fold that in other media supplemented with a lower iron concentration.

34.3.2 ALGAL CULTIVATION ENGINEERING: WASTEWATER AND CARBON DIOXIDE BIOREMEDIATION

34.3.2.1 Algal Cultivation and Wastewater Bioremediation

There are six research papers on the cultivation engineering of microalgae in wastewaters and nutrient removal.

Li et al. (2010) study the effects of different N and P concentrations on the growth, nutrient uptake, and lipid accumulation of *Scenedesmus* sp. in wastewaters in a paper with 498 citations. They observed that microalgal growth was in accordance with the 'Monod model', the N- and P-saturated maximum growth rate was 2.21×10^6 cells mL^{-1} d^{-1}, and the half-saturation constants of N and P uptake were 12.1 mg L^{-1} and 0.27 mg L^{-1}, respectively. They found that in the N/P ratio of 5: 1–12:1, 83–99% N and 99% P could be removed. In conditions of N (2.5 mg L^{-1}) or P (0.1 mg L^{-1}) limitation, microalgae could accumulate lipids to as much as 30 and 53%, respectively, of its algal biomass. However, the lipid productivity/unit volume of culture was not enhanced.

Wang et al. (2010a) study the cultivation of *Chlorella* sp. in wastewaters and nutrient removal in a paper with 497 citations. They sampled wastewaters from four different points of the treatment process flow of a local municipal wastewater treatment plant (MWTP). These wastewaters were before primary settling (#1 wastewater), after primary settling (#2 wastewater), after activated sludge tank (#3 wastewater), and centrate (#4 wastewater), which is the wastewater generated in a sludge centrifuge. They observed that the average specific growth rates in the exponential period were 0.412, 0.429, 0.343, and 0.948 day^{-1} for wastewaters #1, #2, #3, and #4, respectively. On the other hand, the removal rates of NH_4-N were 82.4, 74.7, and 78.3% for wastewaters #1, #2, and #4, respectively. For #3 wastewater, 62.5% of NO_3-N was removed with 6.3-fold of NO_2-N generated. From wastewaters #1, #2, and #4, 83.2, 90.6, and 85.6% of P and 50.9, 56.5, and 83.0% of 'chemical oxygen demand' (COD) were removed, respectively. Only 4.7% was removed in #3 wastewater and the COD increased slightly after algal growth, probably due to the excretion of small photosynthetic organic molecules by algae. Additionally, metal ions, especially Al, Ca, Fe, Mg, and Mn in centrate, were removed very efficiently.

Chinnasamy et al. (2010) study microalgal biomass production in wastewaters and nutrient removal in a paper with 409 citations. They used wastewater containing 85–90% carpet industry effluents with 10–15% municipal sewage. They observed that both freshwater and marine microalgae showed good growth in wastewaters. A consortium of 15 native algal isolates showed >96% nutrient removal in treated wastewater. The biomass production potential and lipid content of this consortium cultivated in treated wastewater were around 9.2–17.8 tons ha^{-1} year^{-1} and 6.82%, respectively.

Li et al. (2011) study microalgal biomass production and nutrient removal in wastewaters with 372 citations. They grew *Chlorella* sp. in the centrate, a highly concentrated municipal wastewater stream generated from an activated sludge thickening process. They additionally used two culture media, autoclaved centrate and raw centrate for comparison. They observed that by the end of a 14-day batch culture, algae could remove ammonia, total N, total P, and COD by 93.9, 89.1, 80.9, and 90.8%, respectively from the raw centrate. The 'fatty acid methyl ester' (FAME) content was 11.04% of dry biomass, providing a biodiesel yield of 0.12 g-biodiesel/L-algae culture solution. They further found that the system could be successfully scaled up and continuously operated at 50% of the daily harvesting rate, providing a net biomass productivity of 0.92 g-algae/(L day).

Aslan and Kapdan (2006) study microalgal biomass production and nutrient removal using *Chlorella vulgaris* in a paper with 362 citations. They performed the experiments at pH 7.0 and at room temperature (20°C) with artificial illumination (4,100 lux). They observed that effluent water quality decreased with increasing nutrient concentrations; the algae culture could remove N more effectively compared to P. Biokinetic coefficients for N were $k = 1.5$ mg NH$_4$-N mg^{-1} chl a d^{-1}, K$_m$ = 31.5 mg l^{-1}, Y_N = 0.15 mg chl a mg^{-1} NH$_4$-N for nitrogen. On the other hand, the values for P were $k = 0.5$ mg PO$_4$-P mg^{-1} chl a d^{-1}, K$_m$ = 10.5 mg l^{-1}, Y_N = 0.14 mg chl a mg^{-1} PO$_4$-P.

Wang et al. (2010b) study microalgal biomass production and nutrient removal in wastewaters in a paper with 359 citations. They used digested dairy manure as a nutrient supplement for cultivation of *Chlorella* sp. They applied different dilution multiples of 10, 15, 20, and 25 to the digested manure. They observed slower growth rates with less diluted manure samples with higher turbidities in the initial cultivation days. They found a reverse linear relationship between the average specific growth rate of the first seven days and the initial turbidities. Algae removed ammonia, total N, total P, and COD by 100, 75.7–82.5, 62.5–74.7, and 27.4–38.4%, respectively, in differently diluted dairy manure. COD in digested dairy manure, as well as CO$_2$, was another carbon source for mixotrophic *Chlorella*. Octadecadienoic acid (08:2) and hexadecanoic acid (C16:0) were the two most abundant fatty acids in the algae. The total fatty acid content of the dry weight increased from 9.0 to 13.7% along with the increasing dilution multiples.

34.3.2.2 Microalgal Cultivation and Carbon Dioxide Bioremediation

There are five research papers on microalgal biomass production and carbon dioxide bioremediation. The CO$_2$ concentration, growth phases, culture approaches, light intensity, and N deprivation are the key parameters studied in these papers.

Chiu et al. (2009) study the effects of the concentration of CO$_2$ aeration on biomass production and lipid accumulation of *Nannochloropsis oculata* in a semicontinuous culture in a paper with 443 citations. They also explored the lipid content of these microalgal cells at different growth phases. They observed that the lipid accumulation from the logarithmic phase to stationary phase of microalgae was significantly increased from 30.8 to 50.4%. In the microalgal cultures aerated with 2, 5, 10, and 15% CO_2, the maximal biomass and lipid productivity in the semicontinuous system were 0.480 and 0.142 g L^{-1} d^{-1} with 2% CO_2 aeration, respectively. On the other hand, in the microalgae cultured in the semicontinuous system aerated with 15% CO_2, biomass and lipid productivity could reach 0.372 and 0.084 g L^{-1} d^{-1}, respectively. They next observed that the biomass and lipid productivity of microalgae were 0.497 and 0.151 g L^{-1} d^{-1} in one-day replacement (half the broth was replaced each day), and were 0.296 and 0.121 g L^{-1} d^{-1} in three-day replacement (three-fifths of the broth was replaced every three days), respectively. They present the optimized conditions as the growth in the semicontinuous system aerated with 2% CO_2 and operated by one-day replacement.

Yoo et al. (2010) study the selection of microalgae for lipid production under high levels of CO_2 in a paper with 382 citations. They cultivate *Botryococcus braunii*, *Chlorella vulgaris*, and *Scenedesmus* sp. with ambient air containing 10% CO_2 and flue gas. They found that the biomass and lipid productivity for *S.* sp. with 10% CO_2 were 217.50 and 20.65 mg L^{-1} d^{-1} (9% of biomass), while those for *B. braunii* were 26.55 and 5.51 mg L^{-1} d^{-1} (21% of biomass). Furthermore, with flue gas, the lipid productivity for *S.* sp. and *B. braunii* increased 1.9-fold (39.44 mg L^{-1} d^{-1}) and 3.7-fold (20.65 mg L^{-1} d^{-1}), respectively. Oleic acid occupied 55% of the fatty acids in *B. braunii*. They observed that *S.* sp. is appropriate for mitigating CO_2, due to its high biomass productivity and C-fixation ability, whereas *B. braunii* is appropriate for producing biodiesel, due to its high lipid content and oleic acid proportion.

Ho et al. (2012) study the effect of light intensity, N starvation, and CO_2 fixation on the lipid and carbohydrate production of *Scenedesmus obliquus* in a paper with 372 citations. They determined the light intensity that promotes cell growth, carbohydrate/lipid productivity, and CO_2 fixation efficiency. They observed that the highest productivity of biomass and lipid and carbohydrate was 840.57 mg L^{-1} d^{-1} and 140.35 mg L^{-1} d^{-1}, respectively. The highest lipid and carbohydrate content was 22.4% (five-day N-starvation) and 46.65% (one-day N-starvation), respectively, whereas the optimal CO_2 consumption rate was 1420.6 mg L^{-1} d^{-1}. They then found that under N starvation, the microalgal lipid was mainly composed of C_{16}/C_{18} fatty acid (around 90%), which is suitable for biodiesel synthesis. The carbohydrate present in the biomass was mainly glucose, accounting for 77–80% of total carbohydrates.

Tang et al. (2011) study the CO_2 biofixation and fatty acid composition of *Scenedesmus obliquus* and *Chlorella pyrenoidosa* in response to different CO_2 levels in a paper with 372 citations. They cultivated these microalgae with 0.03, 5, 10, 20, 30, and 50% CO_2. They observed that these microalgae could grow at 50% CO_2 (>0.69 g L^{-1}) and grew well (>1.22 g L^{-1}) under CO_2 concentrations ranging from 5 to 20%. However, these microalgae showed best growth potential at 10% CO_2. The maximum biomass concentration and CO_2 biofixation rate were 1.84 g L^{-1} and 0.288 g L^{-1} d^{-1} for *S. obliquus* and 1.55 g L^{-1} and 0.260 g L^{-1} d^{-1} for *C. pyrenoidosa*, respectively. The main fatty acid compositions of these microalgae were fatty acids with C_{16}-C_{18} (>94%) under different CO_2 levels. High CO_2 levels (30–50%) were favorable for the accumulation of total lipids and polyunsaturated fatty acids. They propose that these microalgae are appropriate for mitigating CO_2 in flue gases and biodiesel production.

Chiu et al. (2008) study the biofixation of CO_2 by *Chlorella* sp. in a semicontinuous PBR in a paper with 361 citations. During an eight-day interval, in cultures in semicontinuous cultivation, they observed that the specific growth rate and biomass of algal cultures in the condition of aerated 2–15% CO_2 were 0.58–0.66 d^{-1} and 0.76–0.87g L^{-1}, respectively. At CO_2 concentrations of 2, 5, 10, and 15%, the rate of CO_2 biofixation was 0.261, 0.316, 0.466, and 0.573 g h^{-1}, and the efficiency of CO_2 removal was 58, 27, 20, and 16%, respectively. Furthermore, the efficiency of CO_2 removal was similar in the single PBR and in the six-parallel PBR. However, CO_2 biofixation, the production of biomass, and the production of lipid were six times greater in the six-parallel PBR than those in the single PBR. They propose a system with high CO_2 (10–15%) aeration via a high-density culture of microalgal inoculum that was adapted to 2% CO_2.

34.3.3 LIFE-CYCLE ANALYSIS (LCA) OF ALGAL BIOMASS PRODUCTION

There are five research papers on the 'life-cycle analysis' (LCA) of algal biomass production.

Lardon et al. (2009) carry out the LCA of microalgal biomass production for biodiesel production, providing an analysis of the potential environmental impacts in a seminal paper with 852 citations. They undertook a comparative LCA study of a virtual facility to assess the energetic balance and potential environmental impacts of the whole process chain, from biomass production to biodiesel combustion. They tested two different culture conditions, N replete or N starvation, as well as two different extraction options, dry or wet. They then compared the best scenario to food crop-based biodiesel and petrodiesel. They confirmed the potential of microalgae as an energy source but highlighted the imperative necessity of decreasing the energy and nutrient consumption. Therefore, they propose the control of N stress during the culture and optimization of wet extraction as valuable options. They also emphasize the potential of the anaerobic digestion of oilcakes as a way to reduce external energy demand and to recycle a part of the nutrients.

Clarens et al. (2010) compare the environmental life cycle of algae to other bioenergy feedstocks in a paper with 644 citations. They determined the impacts associated with algal production using a 'stochastic life cycle model' and compared with switchgrass, canola, and corn production. They observed that these conventional crops have lower environmental impacts than algae in energy use, greenhouse gas emissions, and water, regardless of cultivation location. Only in total land use and eutrophication potential algae performed favorably. They reason that the large environmental footprint of algal cultivation is driven predominantly by upstream impacts, such as the demand for CO_2 and nutrients. To reduce these impacts, they propose that flue gas and, to a greater extent, wastewater could be used to offset most of the environmental burdens associated with algae. To demonstrate the benefits of algal production coupled with wastewater treatment, they expanded the life-cycle model to include three different municipal wastewater effluents as sources of N and P. They found that each effluent provided a significant reduction in the burdens of algal cultivation, and the use of source-separated urine made algae more environmentally beneficial than the conventional crops.

Yang et al. (2011) carry out the LCA of microalgal biomass production for biodiesel production, examining the life-cycle of water and nutrient usage in a paper with 435 citations. They analyzed the influence of water types, operation with and without recycling, algal species, and geographic distributions. They confirm the competitiveness of microalgal biodiesel and highlight the necessity of recycling harvested water and using sea/wastewater as the water source. They calculate that to generate 1 kg biodiesel, 3,726 kg water, 0.33 kg N, and 0.71 kg P are required if freshwater is used without recycling. Recycling harvest water reduced the water and nutrients usage by 84 and 55%, respectively. Furthermore, using sea/wastewater decreased water requirement by 90% and eliminated the need of all the nutrients except P.

Jorquera et al. (2010) carry out comparative energy LCAs of microalgal biomass production in 'raceway ponds' and tubular and flat-plate PBRs using *Nannochloropsis* sp. in a paper with 425 citations. They calculated the 'net energy ratio' (NER) for each process. They observed that the use of horizontal tubular PBRs is not economically feasible ([NER] < 1) and that the estimated NERs for flat-plate PBRs and raceway ponds is >1. They propose that the NER for ponds and flat-plate PBRs could be raised to significantly higher values if the lipid content of the biomass was increased to 60% dw/cwd.

Stephenson et al. (2010) carry out the LCA of algal biomass production, comparing raceways and air-lift tubular PBRs in a paper with 341 citations. They focus on the 'global warming potential' (GWP) and fossil-energy requirement of a hypothetical operation in which biodiesel is produced from *Chlorella vulgaris*, grown using flue gas as the carbon source. They considered cultivation using a two-stage method, whereby the cells were initially grown to a high concentration of biomass under N-replete conditions, before the supply of N was discontinued, whereupon the cells accumulated triacylglycerides. They next considered cultivation in typical raceway ponds and air-lift tubular PBRs, as well as different methods of downstream processing. They found that if the future target for the productivity of lipids of around 40 tons ha^{-1} year^{-1} could be achieved, cultivation in typical raceways would be significantly more environmentally sustainable than in closed air-lift tubular bioreactors. Next, while microalgal biodiesel cultivated in raceway ponds would have a GWP similar to 80% lower than petrodiesel (on the basis of the net energy content), if air-lift tubular PBRs were used, the GWP of the biodiesel would be significantly greater than the energetically equivalent amount of petrodiesel. They finally observed that the GWP and fossil-energy requirement in this operation were particularly sensitive to the yield of oil achieved during cultivation, the velocity of circulation of the algae in the cultivation facility, whether the culture media could be recycled or not, and the concentration of CO_2 in the flue gas.

34.3.4 Techno-Economics of Algal Biomass Production

There are four research papers on the techno-economics of algal biomass production.

Williams and Laurens (2010) carry out a techno-economic analysis of microalgal biomass production in a seminal paper with 633 citations. They show that the maximum conversion efficiency of total solar energy into primary photosynthetic organic products falls in the region of 10%, where the algal biomass biochemical composition further conditions this yield as 30 and 50% of the primary product, mass is lost on producing cell protein and lipid, respectively, and obtained yields that are one-third to one-tenth of the theoretical ones. They considered that wasted energy from captured photons is a major loss and a major challenge in maximizing mass algal production. They produced a simple model of algal biomass production and its variation with latitude and lipid content. Their key conclusions were that the biochemical composition of the biomass influences the economics, in particular, increased lipid content reduces other valuable compounds in the biomass; the 'biofuel only' option is unlikely to be economically viable; and among the hardest problems in assessing the economics are the cost of the CO_2 supply and uncertain nature of downstream processing.

Davis et al. (2011) carry out a techno-economic analysis of autotrophic microalgae for biodiesel production in a paper with 500 citations. They establish baseline economics for two microalgae pathways, by performing a comprehensive analysis using a set of assumptions for what can plausibly be achieved within a five-year time frame. These specific pathways include autotrophic production via both open pond and closed tubular PBR systems. They set the production scales at 10 million gallons per year of raw algal oil, subsequently upgraded to a 'green diesel' blend stock via hydrotreating. Upon completing the base case scenarios, they determined the cost of lipid production to achieve a 10% return at $8.52/gal for open ponds and $18.10/gal for PBRs. Hydrotreating to produce a diesel blend stock added onto this marginally, bringing the totals to $9.84/gal and $20.53/gal of diesel, for the respective cases. They reason that these costs have potential for significant improvement in the future if better microalgal strains can be identified that would be capable of sustaining high growth rates at high lipid content. Given that it is difficult to maximize both of these parameters simultaneously, they propose that the near-term research should focus on maximizing lipid content as it offers more substantial cost reduction potential relative to an improved algal growth rate.

Norsker et al. (2011) study the techno-economics of microalgal biomass production in a paper with 401 citations. They calculated the microalgal biomass production costs for three different microalgal production systems of open ponds, horizontal tubular PBRs, and flat panel PBRs. For these systems, they found that the resulting biomass production costs including dewatering were 4.95, 4.15, and 5.96 euros per kg, respectively. The important cost factors are irradiation conditions, mixing, photosynthetic efficiency of systems, nutrient, and CO_2 costs. Optimizing production with respect to these factors resulted in a price of 0.68 euros per kg. They argue that at this cost level microalgae could become a promising feedstock for biodiesel and value-added products.

Slade and Bauen (2013) study the techno-economics of algal biomass production in a paper with 377 citations. They considered the energy and carbon balance, environmental impacts, and production cost. They observed that achieving a positive energy balance would require technological advances and highly optimized production systems. They recommend the focus should be on the energy required for pumping, the embodied energy required for construction, the embodied energy in nutrients, and the energy required for drying and dewatering. They stress that the conceptual and often incomplete nature of algal production systems in the published studies, together with the limited sources of primary data for process and scale-up assumptions, highlight future uncertainties around microalgal biodiesel production. They further reason that environmental impacts from water management, CO_2 handling, and nutrient supply could constrain system design and implementation options. They propose that significant (>50%) cost reductions might be achieved if CO_2, nutrients, and water could be obtained at low cost.

34.4 DISCUSSION

The location of the high-impact studies in algal biomass production requires a carefully designed keyword set and a collection of databases covering a wide range of academic disciplines. It is apparent that the keyword set for biomass production in general (keyword set I in the Appendix) fairly collects the relevant papers in this area. The keyword set for algae (keyword set II in the Appendix), a modified copy of the keyword sets provided in the relevant algal studies (Konur, 2020a–aq), also collects the relevant papers in this area. The third keyword set (keyword set III in the Appendix) for algal biomass also collects the relevant papers in this area. The final keyword set (keyword set IV in the Appendix) helps in locating a core set of papers mainly related to biomass production for algal biodiesel production, omitting the papers on biomass production for food, medicine, fisheries, and animal husbandry (Conquer and Holub, 1996; Dawczynski et al., 2007; Patil et al., 2007; Volkman et al., 1989).

Initially, all the databases contained in the 'Web of Science Core Collection' were searched using these keyword sets. However, it is apparent that the primary database is SCI-E, supported by ESCI and BCI-S.

Apparently, the core set of papers on the algal biomass production are collected using these databases and keyword sets. Hence, the papers presented in this chapter are all relevant to algal biomass production for biodiesel production, rather than for food, biomedicine, fisheries, and animal husbandry (Becker, 2007; Ogbonna et al., 1997; Vigani et al., 2015). It is notable that the review papers (Borowitzka, 1999; Chen et al., 2011; Christenson and Sims, 2011; Perez-Garcia et al., 2011; Radakovits et al., 2012; Ugwu et al., 2008; Wang et al., 2008) are not presented in this chapter since the focus is on the presentation of primary findings from the experimental or analytical studies (such as LCA and 'techno-economic analysis'). It is notable however that these review papers do play a critical role for the key stakeholders to engage with the key research issues on algal biomass production (Konur, 2000, 2002a–c, 2006a–b, 2007a–b; North, 1991a–b).

It appears that there are pragmatically seven distinct primary research fronts for the research on algal biomass production. These are 'algal cultivation engineering in general', 'algal cultivation and wastewater bioremediation', 'algal cultivation and carbon dioxide bioremediation', 'LCA of algal biomass production', 'techno-economics of algal biomass production', 'PBRs', and 'omics and cell biology' studies in algal biomass production with five, six, three, five, four, zero, and zero papers, respectively. Furthermore, some papers classed in one section are also relevant for the other sections, as in the case of life-cycle and techno-economics studies which are related to wastewater-based biomass production.

For the first research front, a number of parameters emerge, as studied in these papers. Nitrogen (N) deficiency is the primary tool to boost the lipid content and productivity of the microalgae in the cultivation engineering of algae (Illman et al., 2000; Li et al., 2008; Rodolfi et al., 2009). In general, nutrient deprivation boosts both biomass and lipid productivity in these studies. These papers have opened the way for a significant stream of research in this area.

The seminal study by Rodolfi et al. (2009) sets out the key strategy to boost both biomass and lipid productivity: a two-phase cultivation process where a nutrient sufficient phase produces the inoculum, followed by a nitrogen-deprived phase to boost lipid synthesis. This paper has opened the way for a significant stream of research in this area.

The studies on wastewater bioremediation and algal biomass production confirm biomass production and nutrient (mainly P and N) removal in wastewaters (Aslan and Kapdan, 2006; Chinnasamy et al., 2010; Li et al., 2010, 2011; Wang et al., 2010a–b). These papers have opened the way for a significant stream of research in this area.

The studies on algal biomass production and CO_2 bioremediation confirm the positive impact of CO_2 repletion on algal biomass production (Chiu et al., 2008, 2009; Ho et al., 2012; Tang et al., 2011; Yoo et al., 2010). Similarly, these papers have opened the way for a significant stream of research in this area.

The studies on the LCA of algal biomass production present valuable data for the key stakeholders on the environmental impact of algal biomass production, compared to edible and nonedible oil-based biofuel production (Clarens et al., 2010; Jorquera et al., 2010; Lardon et al., 2009; Stephenson et al., 2010; Yang et al., 2011). These papers have opened the way for a significant stream of research in this area.

It is significant to note that to generate 1 kg biodiesel, 3,726 kg water, 0.33 kg N, and 0.71 kg P are required if freshwater is used without recycling (Stephenson et al., 2010). Thus, water, N, P, and CO_2 emerge as the key resources with heavy environmental effects due to algal biomass production.

The studies on the techno-economic analysis of algal biomass production present valuable data for the key stakeholders on these aspects of the impact of algal biomass production, compared to edible and nonedible oil-based biofuel production (Davis et al., 2011; Norsker et al., 2011; Slade and Bauen, 2013; Williams and Laurens, 2010). These highly cited studies also shaped the research papers published in this field in the 2010s.

Following the two-phase cultivation strategy of Rodolfi et al. (2009), Davis et al. (2011) confirm that microalgae with high growth rates at high lipid content would reduce production costs, whilst maximizing lipid content, which offers a more substantial cost reduction potential relative to an improved algal growth rate.

Slade and Bauen (2013) build on the key findings of the LCA of algal biomass production and confirm that algal biomass cultivation in wastewaters and nutrient removal could substantially reduce production cost, at the same time reducing the adverse environmental impact of algal biomass production. Additionally, this study confirms the substantial cost reduction obtained through the use of CO_2 from the flue gases from power plants.

It is apparent from these studies on both the LCA and techno-economics of algal biomass production that cultivation of algae using wastewaters from wastewater treatment plants and CO_2 from flue gases could substantially reduce the adverse environmental impact and production cost.

In addition to the five research streams outlined briefly above, two more distinct research fronts emerge from the research studies on algal biomass production: PBRs and open ponds, as well as omics and cell biology studies.

The studies involving open ponds and PBRs have already been key issues engaged with in some papers (Davis et al., 2011; Illman et al., 2000; Jorquera et al., 2010; Norsker et al., 2011; Stephens et al., 2010). There have been a number of review papers in this field (Borowitzka, 1999; Chen et al., 2011; Pulz, 2001; Ugwu et al., 2008). These review papers have opened the way for a distinct research stream for the cultivation of algae in the 2000s and 2010s (Janssen et al., 2003; Molina et al., 2001; Sierra et al., 2008). Apparently, the focus of these studies has been on the development of efficient PBRs with high algal biomass and lipid productivity.

With advances in cell biology and omics technologies in recent years, a distinct research stream on these topics regarding algal biomass production for biodiesel production has emerged (Geider et al., 1993; Miller et al., 2010; Radakovits et al., 2012; Wykoff et al., 1998). These studies help in understanding the impact of nutrient deprivation at the cell and genomic levels and provide ways to engineer these algal cells to increase biomass and lipid productivity for biodiesel production.

It is also notable that although there are a number of algae such as microalgae, cyanobacteria, macroalgae, diatoms, dinoflagellates, and coccolithophores (keyword set II in the Appendix), these studies mostly use microalgal species for enhancing biomass productivity.

In this context, the most-prolific microalgal species are *Chlorella*, *Scenedesmus*, and *Nannochloropsis* with 16, 9, and 8 papers, respectively. Thus, for a relatively core set of papers on algal biomass production, the keyword set for microalgae could be used more conveniently (within keyword set II in the Appendix).

It is also notable that the prolific lead authors are 'Roger R. Ruan' of the University of Minnesota, 'Chih-Sheng Lin' of the National Chia Tung University of Taiwan, and 'Rene H. Wijffels' of Wageningen University of the Netherlands with three, two, and two papers, respectively. Similarly, the most-prolific country for these lead authors is the USA with ten papers. The Netherlands, the UK, China, and Taiwan are the other prolific countries for these lead authors.

As around 85% of the total cost of biodiesel production emanates from the feedstock costs (Davis et al., 2011; Norsker et al., 2011; Williams and Laurens, 2010; Zhang et al., 2003), it is helpful to reduce the total cost of biodiesel production by improving biomass and lipid productivity through engineering the cultivation processes as discussed briefly above. This would make algal biodiesel fuels more competitive with petrodiesel fuels (Davis et al., 2011; Norsker et al., 2011; Williams and Laurens, 2010; Zhang et al., 2003).

This would also help with the treatment of wastewater where it is used for the cultivation and lipid production of algae (Bigda, 1995; Chen, 2004; Rozendal et al.,

2008; Speece, 1983). Another beneficial effect of using wastewater for algal biomass and lipid accumulation as well nutrient removal would be the lessening of the volume and harmful effects of algal blooms, which have been a significant ecological disaster in recent years (Anderson et al., 2002; Hallegraeff, 1993; Michalak et al., 2013; O'Neil et al., 2012; Paerl, 1997).

CO_2 emissions have been a great source of public concerns (Bazzaz, 1990; Hansen et al., 2008; Tans and Takahashi, 1990). The studies on algal biomass production and CO_2 bioremediation confirm the positive impact of CO_2 repletion on algal biomass production (Chiu et al., 2008, 2009; Ho et al., 2012; Tang et al., 2011; Yoo et al., 2010). Thus, the biofixation of CO_2 from flue gases and the atmosphere would be tremendously helpful in meeting these public concerns.

The production of algal biodiesel fuels in competition with petrodiesel fuels would also help in shifting biodiesel production from edible oil-based biodiesel fuels. There has been significant competition between biodiesel production from edible oils and the household consumption of edible oils, resulting in significant public concerns about 'food security' (Ajanovic, 2011; Godfray et al., 2010; Lam et al., 2009; Lobell et al., 2008; Naylor et al., 2007; Rosegrant and Cline, 2003; Schmidhuber and Tubiello, 2007; Tenenbaum, 2008).

This would also help in dealing with the public concerns emanating from the fact that palm plantations expand at the expense of forests as oil palm plantations have replaced large areas of forests in Southeast Asia and other places (Carlson et al., 2012, 2013; Fitzherbert et al., 2008; Koh and Wilcove, 2008; Koh et al., 2011). This has resulted in public concerns about the destruction of forests (Fitzherbert et al., 2008; Koh and Wilcove, 2008; Koh et al., 2011), destruction of ecological biodiversity (Fitzherbert et al., 2008; Koh and Wilcove, 2008; Koh et al., 2011), the significant increase in CO_2 emissions (Carlson et al., 2012, 2013), and finally, exploitation of local communities (McCarthy, 2010; Obidzinski et al., 20012, Rist et al., 2010).

The basis of these studies is related to the exploration and assessment of structure–processing–property relationships for algal biomass production (Borowitzka, 1999; Chen et al., 2011; Christenson and Sims, 2011; Konur and Matthews, 1989; Perez-Garcia et al., 2011; Radakovits et al., 2012; Ugwu et al., 2008; Wang et al., 2008).

34.5 CONCLUSION

This chapter has presented the key findings of the 25-most-cited article papers in five pragmatically distinct research streams in this field.

These prolific studies provide valuable evidence on cultivation engineering in general, cultivation engineering in wastewaters and nutrient removal, cultivation engineering and CO_2 bioremediation, LCA of algal biomass production, and techno-economic studies of algal biomass production. There are also two more distinct research fields: PBRs and cell biology and omics studies of algal biomass production, although there are no papers for these research streams in the top 25 paper table.

All these research streams contribute significantly towards improving algal biomass productivity, mostly in microalgae. It is apparent that it is highly desirable to enhance both biomass productivity and lipid productivity such as through the use of a two-phase cultivation strategy where first biomass growth is ensured in a nutrient (N, P) replete phase and then followed by a nutrient deficient phase.

It is then apparent that the use of wastewaters and flue gases are of critical importance to reduce production costs as well as to reduce the adverse environmental impact of algal biomass production. It is also apparent that the studies on omics and cell biology complement primary cultivation processes by proving to be valuable tools to understand and assess the primary cultivation processes.

The ecological, techno-economic, and environmental benefits using wastewaters for these purposes have been significant. Their use would make biodiesel production more competitive in relation to petrodiesel fuels in reducing the feedstock costs significantly.

This would also help with the treatment of wastewaters. Another beneficial effect of using wastewaters for algal biomass accumulation and nutrient removal would be the lessening of the volume and harmful effects of algal blooms, which have been a significant ecological disaster in recent years.

The production of algal biodiesel fuels in competition with petrodiesel fuels would also help in shifting biodiesel production from edible oil-based biodiesel fuels, dealing with public concerns about food security and concerns emanating from the destruction of forests for the expansion of palm oil plantations, the destruction of ecological biodiversity, the significant increase in CO_2 emissions, and finally, the exploitation of local communities.

It is recommended that similar studies are carried out for each of the seven major research streams. Scientometric studies on each of these major research streams as well as for the whole field of algal biomass production would complement these studies, as well as other published studies on the scientometrics of algae (Konur, 2021a–ag).

34.A APPENDIX

The keyword set for the search on algal biomass production
    Syntax: ((I and II) or III) not IV

34.A.1 BIOMASS-RELATED KEYWORDS

TI=(biomass* or *cultivat* or reactor* or nutrient* or growth* or pond* or nitrogen or phosphorus or nitrate or phosphate or culture or production or feedstock* or ammoni* or wastewater* or effluent* or manure or waste* or "cheese whey" or brewery or sewage or "flue gas*" or co2* or "carbon-dioxide*" or "metabolic-eng*" or "metabolic-pathw*" or "systems-biology" or "synthetic-biology" or "life-cycle*" or "techno-econ*" or engineering or "genetically modified" or "carbon capture" or alveolar or *economic* or "light intensity" or salt or engineered or glucose or mixothropic or photoperiod or diode* or heterotrophic*).

34.A.2 ALGAE-RELATED KEYWORDS

A modified keyword set of Keyword set II in Konur (2021t).

34.A.3 Algal biomass keyword set

TI=(photobioreactor* or "photo-bioreactor*" or "*alga*growth" or "*alga*production" or "open-pond*" or "raceway*pond*" or "*alga*pond*" or "*alga-cultur*" or "*alga*biomass").

34.A.4 Excluding keyword set

TI=(bloom* or eutrophication* or lake* or *toxic* or concentating or phytoplankton or *Sorp* or *sorb* or coast* or hydrogen or dewater* or *ethanol or phospholipid* or utex or aqua* or herbivor* or ocean* or zooplank* or extraction or remote or coral* or plant* or *methane or halogen* or "heavy metal*" or dyes or bacterial or floc* or photosynth* or conversion or caroten* or *gas or *refinery or pyroly* or liquefact* or diet* or shelf* or food* or forest or sea* or microcystin* or toxin* or harvest* or leuconostoc* or sediment*).

ACKNOWLEDGMENTS

The contribution of the highly cited researchers in this field is greatly acknowledged.

REFERENCES

Ahmadun, F. R., A. Pendashteh, and L. C. Abdullah, et al. 2009. Review of technologies for oil and gas produced water treatment. *Journal of Hazardous Materials* 170: 530–551.

Ajanovic, A. 2011. Biofuels versus food production: Does biofuels production increase food prices? *Energy* 36: 2070–2076.

Anderson, D. M. P. M. Glibert, and J. M. Burkholder. 2002. Harmful algal blooms and eutrophication: Nutrient sources, composition, and consequences. *Estuaries* 25: 704–726.

Aslan, S. and I. K. Kapdan. 2006. Batch kinetics of nitrogen and phosphorus removal from synthetic wastewater by algae. *Ecological Engineering* 28: 64–70.

Atlas, R. M. 1981. Microbial degradation of petroleum hydrocarbons: An environmental perspective. *Microbiological Reviews* 45: 180–209.

Babich, I. V. and J. A. Moulijn. 2003. Science and technology of novel processes for deep desulfurization of oil refinery streams: A review. *Fuel* 82: 607–631.

Bazzaz, F. A. 1990. The response of natural ecosystems to the rising global CO_2 levels. *Annual Review of Ecology and Systematics* 21: 167–196.

Becker, E. W. 2007. Micro-algae as a source of protein. *Biotechnology Advances* 25: 207–210.

Bigda, R. J. 1995. Consider Fenton's chemistry for wastewater treatment. *Chemical Engineering Progress* 91: 62–66.

Borowitzka, M. A. 1999. Commercial production of microalgae: Ponds, tanks, tubes and fermenters. *Journal of Biotechnology* 70: 313–321.

Bridgwater, A. V. and G. V. C. Peacocke. 2000. Fast pyrolysis processes for biomass. *Renewable & Sustainable Energy Reviews* 4: 1–73.

Bridgwater, A. V., D. Meier, and D. Radlein. 1999. An overview of fast pyrolysis of biomass. *Organic Geochemistry* 30: 1479–1493.

Carlson, K. M., L. M. Curran, and D. Ratnasari, et al. 2012. Committed carbon emissions, deforestation, and community land conversion from oil palm plantation expansion in

West Kalimantan, Indonesia. *Proceedings of the National Academy of Sciences of the United States of America* 109: 7559–7564.

Carlson, K. M., L. M. Curran, and G. P. Asner, et al. 2013. Carbon emissions from forest conversion by Kalimantan oil palm plantations. *Nature Climate Change* 3: 283–287.

Chen, G. 2004. Electrochemical technologies in wastewater treatment. *Separation and Purification Technology* 38(1), 11–41.

Chen, C. Y., K. L. Yeh, R. Aisyah, D. J. Lee, and J. S. Chang. 2011. Cultivation, photobioreactor design and harvesting of microalgae for biodiesel production: A critical review. *Bioresource Technology* 102: 71–81.

Chinnasamy, S., A. Bhatnagar, R. W. Hunt, and K. C. Das. 2010. Microalgae cultivation in a wastewater dominated by carpet mill effluents for biofuel applications. *Bioresource Technology* 101: 3097–3105.

Christenson, L. and R. Sims. 2011. Production and harvesting of microalgae for wastewater treatment, biofuels, and bioproducts. *Biotechnology Advances* 29: 686–702.

Chisti, Y. 2007. Biodiesel from microalgae. *Biotechnology Advances* 25: 294–306.

Chiu, S. Y., C. Y. Kao, and C. H. Chen, et al. 2008. Reduction of CO_2 by a high-density culture of *Chlorella* sp. in a semicontinuous photobioreactor. *Bioresource Technology* 99: 3389–3396.

Chiu, S. Y., C. Y. Kao, and M. T. Tsai, et al. 2009. Lipid accumulation and CO_2 utilization of *Nannochloropsis oculata* in response to CO_2 aeration. *Bioresource Technology* 100: 833–838.

Clarens, A. F., E. P. Resurreccion, M. A. White, and L. M. Colosi. 2010. Environmental life cycle comparison of algae to other bioenergy feedstocks. *Environmental Science & Technology* 44: 1813–1819.

Conquer, J. A. and B. J. Holub. 1996. Supplementation with an algae source of docosahexaenoic acid increases (3–3) fatty acid status and alters selected risk factors for heart disease in vegetarian subjects. *Journal of Nutrition* 126: 3032–3039.

Czernik, S. and A. V. Bridgwater. 2004. Overview of applications of biomass fast pyrolysis oil. *Energy & Fuels* 18: 590–598.

Davis, R., A. Aden, and P. T. Pienkos. 2011. Techno-economic analysis of autotrophic microalgae for fuel production. *Applied Energy* 88: 3524–3531.

Dawczynski, C., R. Schubert, and G. Jahreis. 2007. Amino acids, fatty acids, and dietary fibre in edible seaweed products. *Food Chemistry* 103: 891–899.

Fitzherbert, E. B., M. J. Struebig, and A. Morel, et al. 2008. How will oil palm expansion affect biodiversity? *Trends n Ecology & Evolution* 23: 538–545.

Geider, R. J., J. Laroche, R. M. Greene, and M. Olaizola. 1993. Response of the photosynthetic apparatus of *Phaeodactylum-tricornutum* (bacillariophyceae) to nitrate, phosphate, or iron starvation. *Journal of Phycology* 29: 755–766.

Godfray, H. C. J., J. R. Beddington, and I. R. Crute, et al., 2010. Food security: The challenge of feeding 9 billion people. *Science* 327: 812–818.

Hallegraeff, G. M. 1993. A review of harmful algal blooms and their apparent global increase. *Phycologia* 32: 79–99.

Hansen, J., M. Sato, and P. Kharecha, et al. 2008. Target Atmospheric CO_2: Where should humanity aim? *Open Atmospheric Science Journal* 2: 21731.

Hill, J., E. Nelson, D. Tilman, S. Polasky, and D. Tiffany. 2006. Environmental, economic, and energetic costs and benefits of biodiesel and ethanol biofuels. *Proceedings of the National Academy of Sciences of the United States of America* 103: 11206–11210.

Ho, S. H., C. Y. Chen, and J. S. Chang. 2012. Effect of light intensity and nitrogen starvation on CO_2 fixation and lipid/carbohydrate production of an indigenous microalga *Scenedesmus obliquus* CNW-N. *Bioresource Technology* 113: 244–252.

Hu, Q., M. Sommerfeld, and E. Jarvis, et al. 2008. Microalgal triacylglycerols as feedstocks for biofuel production: Perspectives and advances. *Plant Journal* 54: 621–639.

Illman, A. M., A. H. Scragg, and S. W. Shales. 2000. Increase in Chlorella strains calorific values when grown in low nitrogen medium. *Enzyme and Microbial Technology* 27: 631–635.

Janssen, M., J. Tramper, L. R. Mur, and R. H. Wijffels. 2003. Enclosed outdoor photobioreactors: Light regime, photosynthetic efficiency, scale-up, and future prospects. *Biotechnology and Bioengineering* 81: 193–210.

Jorquera, O., A. Kiperstok, E. A. Sales, M. Embirucu, and M. L. Ghirardi. 2010. Comparative energy life-cycle analyses of microalgal biomass production in open ponds and photobioreactors. *Bioresource Technology* 101: 1406–1413.

Kilian, L. 2009. Not all oil price shocks are alike: Disentangling demand and supply shocks in the crude oil market. *American Economic Review* 99: 1053–1069.

Koh, L. P. and D. S. Wilcove. 2008. Is oil palm agriculture really destroying tropical biodiversity? *Conservation Letters* 1: 60–64.

Koh, L. P., J. Miettinen, S. C. Liew, and J. Ghazoul. 2011. Remotely sensed evidence of tropical peatland conversion to oil palm. *Proceedings of the National Academy of Sciences of the United States of America* 108: 5127–5132.

Konur, O. 2000. Creating enforceable civil rights for disabled students in higher education: An institutional theory perspective. *Disability & Society* 15: 1041–1063.

Konur, O. 2002a. Access to Nursing Education by disabled students: Rights and duties of nursing programs. *Nurse Education Today* 22: 364–374.

Konur, O. 2002b. Assessment of disabled students in higher education: Current public policy issues. *Assessment and Evaluation in Higher Education* 27: 131–152.

Konur, O. 2002c. Access to employment by disabled people in the UK: Is the Disability Discrimination Act working? *International Journal of Discrimination and the Law* 5: 247–279.

Konur, O. 2006a. Participation of children with dyslexia in compulsory education: Current public policy issues. *Dyslexia* 12: 51–67.

Konur, O. 2006b. Teaching disabled students in Higher Education. *Teaching in Higher Education* 11: 351–363.

Konur, O. 2007a. A judicial outcome analysis of the Disability Discrimination Act: A windfall for the employers? *Disability & Society* 22:187–204.

Konur, O. 2007b. Computer-assisted teaching and assessment of disabled students in higher education: The interface between academic standards and disability rights. *Journal of Computer Assisted Learning* 23: 207–219.

Konur, O., 2020a. The scientometric analysis of the research on the algal science, technology, and medicine. In *Handbook of Algal Science, Technology and Medicine*, ed. O. Konur, 3–18. London: Academic Press.

Konur, O., 2020b. 100 citation classics in the algal science, technology, and medicine: a scientometric analysis. In *Handbook of Algal Science, Technology and Medicine*, ed. O. Konur, 19–38. London: Academic Press.

Konur, O., 2020c. The scientometric analysis of the research on the algal structures. In In *Handbook of Algal Science, Technology and Medicine*, ed. O. Konur, 41–60. London: Academic Press.

Konur, O., 2020d. The scientometric analysis of the research on the algal genomics. In *Handbook of Algal Science, Technology and Medicine*, ed. O. Konur, 105–125. London: Academic Press.

Konur, O., 2020e. The scientometric analysis of the research on the algal photosystems and phtosythesis. In *Handbook of Algal Science, Technology and Medicine*, ed. O. Konur, 195–215. London: Academic Press.

Konur, O., 2020f. The pioneering research on the cyanobacterial photosystems and photosynthesis. In *Handbook of Algal Science, Technology and Medicine*, ed. O. Konur, 231–343. London: Academic Press.

Konur, O., 2020g. The scientometric analysis of the research on the algal ecology. In *Handbook of Algal Science, Technology and Medicine*, ed. O. Konur, 257–276. London: Academic Press.

Konur, O., 2020h. The scientometric analysis of the research on the algal bioenergy and biofuels. In *Handbook of Algal Science, Technology and Medicine*, ed. O. Konur, 319–341. London: Academic Press.

Konur, O., 2020i. The scientometric analysis of the research on the bioethanol production from green macroalgae. In *Handbook of Algal Science, Technology and Medicine*, ed. O. Konur, 385–401. London: Academic Press.

Konur, O., 2020j. The scientometric analysis of the research on the algal biomedicine. In *Handbook of Algal Science, Technology and Medicine*, ed. O. Konur, 405–427. London: Academic Press.

Konur, O., 2020k. The pioneering research on the wound care by alginates. In *Handbook of Algal Science, Technology and Medicine*, ed. O. Konur, 467–481. London: Academic Press.

Konur, O., 2020l. The scientometric analysis of the research on the algal foods. In *Handbook of Algal Science, Technology and Medicine*, ed. O. Konur, 485–506. London: Academic Press.

Konur, O., 2020m. The scientometric analysis of the research on the algal toxicology. In *Handbook of Algal Science, Technology and Medicine*, ed. O. Konur, 521–541. London: Academic Press.

Konur, O., 2020n. The scientometric analysis of the research on the algal bioremediation. In *Handbook of Algal Science, Technology and Medicine*, ed. O. Konur, 607–627. London: Academic Press.

Konur, O., 2020o. Algal drugs: The state of the research. In *Encyclopedia of Marine Biotechnology*. Vol. 1, ed. S.K. Kim. Oxford: Wiley-Blackwell.

Konur, O., 2020p. Algal genomics. In *Encyclopedia of Marine Biotechnology*, Vol. 3, ed. S.K. Kim. Oxford: Wiley-Blackwell.

Konur, O., ed., 2020q. *Handbook of Algal Science, Technology and Medicine*. London: Academic Press.

Konur, O., ed. 2021a. *Handbook of Biodiesel and Petrodiesel Fuels: Science, Technology, Health, and Environment*. Boca Raton, FL: CRC Press.

Konur, O., ed. 2021b. *Handbook of Biodiesel and Petrodiesel Fuels: Science, Technology, Health, and Environment. Volume 1. Biodiesel Fuels: Science, Technology, Health, and Environment*. Boca Raton, FL: CRC Press.

Konur, O., ed. 2021c. *Handbook of Biodiesel and Petrodiesel Fuels: Science, Technology, Health, and Environment. Volume 2. Biodiesel Fuels based on the Edible and Nonedible Feedstocks, Wastes, and Algae: Science, Technology, Health, and Environment*. Boca Raton, FL: CRC Press.

Konur, O., ed. 2021d. *Handbook of Biodiesel and Petrodiesel Fuels: Science, Technology, Health, and Environment. Volume 3. Petrodiesel Fuels: Science, Technology, Health, and Environment*. Boca Raton, FL: CRC Press.

Konur, O. 2021e. Biodiesel and petrodiesel fuels: Science, technology, health, and environment. In *Handbook of Biodiesel and Petrodiesel Fuels: Science, Technology, Health, and Environment. Volume 1. Biodiesel Fuels: Science, Technology, Health, and Environment*, ed. O. Konur. Boca Raton, FL: CRC Press.

Konur, O. 2021f. Biodiesel and petrodiesel fuels: A scientometric review of the research. In *Handbook of Biodiesel and Petrodiesel Fuels: Science, Technology, Health, and Environment. Volume 1. Biodiesel Fuels: Science, Technology, Health, and Environment*, ed. O. Konur. Boca Raton, FL: CRC Press.

Konur, O. 2021g. Biodiesel and petrodiesel fuels: A review of the research. In *Handbook of Biodiesel and Petrodiesel Fuels: Science, Technology, Health, and Environment. Volume 1. Biodiesel Fuels: Science, Technology, Health, and Environment*, ed. O. Konur. Boca Raton, FL: CRC Press.

Konur, O. 2021h Nanotechnology applications in the diesel fuels and the related research fields: A review of the research. In *Handbook of Biodiesel and Petrodiesel Fuels: Science, Technology, Health, and Environment. Volume 1. Biodiesel Fuels: Science, Technology, Health, and Environment*, ed. O. Konur. Boca Raton, FL: CRC Press.

Konur, O. 2021i. Biooils: A scientometric review of the research. In *Handbook of Biodiesel and Petrodiesel Fuels: Science, Technology, Health, and Environment. Volume 1. Biodiesel Fuels: Science, Technology, Health, and Environment*, ed. O. Konur. Boca Raton, FL: CRC Press.

Konur, O. 2021j. Characterization and properties of biooils: A review of the research. In *Handbook of Biodiesel and Petrodiesel Fuels: Science, Technology, Health, and Environment. Volume 1. Biodiesel Fuels: Science, Technology, Health, and Environment*, ed. O. Konur. Boca Raton, FL: CRC Press.

Konur, O. 2021k. Biomass pyrolysis and pyrolysis oils: A review of the research. In *Handbook of Biodiesel and Petrodiesel Fuels: Science, Technology, Health, and Environment. Volume 1. Biodiesel Fuels: Science, Technology, Health, and Environment*, ed. O. Konur. Boca Raton, FL: CRC Press.

Konur, O. 2021l. Biodiesel fuels: A scientometric review of the research. In *Handbook of Biodiesel and Petrodiesel Fuels: Science, Technology, Health, and Environment. Volume 1. Biodiesel Fuels: Science, Technology, Health, and Environment*, ed. O. Konur. Boca Raton, FL: CRC Press.

Konur, O. 2021m. Glycerol: A scientometric review of the research. In *Handbook of Biodiesel and Petrodiesel Fuels: Science, Technology, Health, and Environment. Volume 1. Biodiesel Fuels: Science, Technology, Health, and Environment*, ed. O. Konur. Boca Raton, FL: CRC Press.

Konur, O. 2021n. Propanediol production from glycerol: A review of the research. In *Handbook of Biodiesel and Petrodiesel Fuels: Science, Technology, Health, and Environment. Volume 1. Biodiesel Fuels: Science, Technology, Health, and Environment*, ed. O. Konur. Boca Raton, FL: CRC Press.

Konur, O. 2021o. Edible oil-based biodiesel fuels: A scientometric review of the research. In *Handbook of Biodiesel and Petrodiesel Fuels: Science, Technology, Health, and Environment. Volume 2. Biodiesel Fuels based on the Edible and Nonedible Feedstocks, Wastes, and Algae: Science, Technology, Health, and Environment*, ed. O. Konur. Boca Raton, FL: CRC Press.

Konur, O. 2021p. Palm oil-based biodiesel fuels: A review of the research. In *Handbook of Biodiesel and Petrodiesel Fuels: Science, Technology, Health, and Environment. Volume 2. Biodiesel Fuels based on the Edible and Nonedible Feedstocks, Wastes, and Algae*, ed. O. Konur. Boca Raton, FL: CRC Press.

Konur, O. 2021q. Rapeseed oil-based biodiesel fuels: A review of the research. In *Handbook of Biodiesel and Petrodiesel Fuels: Science, Technology, Health, and Environment. Volume 2. Biodiesel Fuels based on the Edible and Nonedible Feedstocks, Wastes, and Algae*, ed. O. Konur. Boca Raton, FL: CRC Press.

Konur, O. 2021r. Nonedible oil-based biodiesel fuels: A scientometric review of the research. In *Handbook of Biodiesel and Petrodiesel Fuels: Science, Technology, Health, and Environment. Volume 2. Biodiesel Fuels based on the Edible and Nonedible Feedstocks, Wastes, and Algae: Science, Technology, Health, and Environment*, ed. O. Konur. Boca Raton, FL: CRC Press.

Konur, O. 2021s. Waste oil-based biodiesel fuels: A scientometric review of the research. In *Handbook of Biodiesel and Petrodiesel Fuels: Science, Technology, Health, and Environment. Volume 2. Biodiesel Fuels based on the Edible and Nonedible Feedstocks,*

Wastes, and Algae: Science, Technology, Health, and Environment, ed. O. Konur. Boca Raton, FL: CRC Press.

Konur, O. 2021t. Algal biodiesel fuels: A scientometric review of the research. In *Handbook of Biodiesel and Petrodiesel Fuels: Science, Technology, Health, and Environment. Volume 2. Biodiesel Fuels based on the Edible and Nonedible Feedstocks, Wastes, and Algae: Science, Technology, Health, and Environment*, ed. O. Konur. Boca Raton, FL: CRC Press.

Konur, O. 2021u. Algal biomass production for biodiesel production: A review of the research. In *Handbook of Biodiesel and Petrodiesel Fuels: Science, Technology, Health, and Environment. Volume 2. Biodiesel Fuels based on the Edible and Nonedible Feedstocks, Wastes, and Algae*, ed. O. Konur. Boca Raton, FL: CRC Press.

Konur, O. 2021v. Algal biomass production in wastewaters for biodiesel production: A review of the research. In *Handbook of Biodiesel and Petrodiesel Fuels: Science, Technology, Health, and Environment. Volume 2. Biodiesel Fuels based on the Edible and Nonedible Feedstocks, Wastes, and Algae*, ed. O. Konur. Boca Raton, FL: CRC Press.

Konur, O. 2021x. Algal lipid production for biodiesel production: A review of the research. In *Handbook of Biodiesel and Petrodiesel Fuels: Science, Technology, Health, and Environment. Volume 2. Biodiesel Fuels based on the Edible and Nonedible Feedstocks, Wastes, and Algae*, ed. O. Konur. Boca Raton, FL: CRC Press.

Konur, O. 2021y. Crude oils: A scientometric review of the research. In *Handbook of Biodiesel and Petrodiesel Fuels: Science, Technology, Health, and Environment. Volume 3. Petrodiesel Fuels: Science, Technology, Health, and Environment*, ed. O. Konur. Boca Raton, FL: CRC Press.

Konur, O. 2021z. Petrodiesel fuels: A scientometric review of the research. In *Handbook of Biodiesel and Petrodiesel Fuels: Science, Technology, Health, and Environment. Volume 3. Petrodiesel Fuels: Science, Technology, Health, and Environment*, ed. O. Konur. Boca Raton, FL: CRC Press.

Konur, O. 2021aa. Bioremediation of petroleum hydrocarbons in the contaminated soils: A review of the research. In *Handbook of Biodiesel and Petrodiesel Fuels: Science, Technology, Health, and Environment. Volume 3. Petrodiesel Fuels: Science, Technology, Health, and Environment*, ed. O. Konur. Boca Raton, FL: CRC Press.

Konur, O. 2021ab. Desulfurization of diesel fuels: A review of the research. In *Handbook of Biodiesel and Petrodiesel Fuels: Science, Technology, Health, and Environment. Volume 3. Petrodiesel Fuels: Science, Technology, Health, and Environment*, ed. O. Konur. Boca Raton, FL: CRC Press.

Konur, O. 2021ac. Diesel fuel exhaust emissions: A scientometric review of the research. In *Handbook of Biodiesel and Petrodiesel Fuels: Science, Technology, Health, and Environment. Volume 3. Petrodiesel Fuels: Science, Technology, Health, and Environment*, ed. O. Konur. Boca Raton, FL: CRC Press

Konur, O. 2021ad. The adverse health and safety impact of diesel fuels: A scientometric review of the research. In *Handbook of Biodiesel and Petrodiesel Fuels: Science, Technology, Health, and Environment. Volume 3. Petrodiesel Fuels: Science, Technology, Health, and Environment*, ed. O. Konur. Boca Raton, FL: CRC Press.

Konur, O. 2021ae. Respiratory illnesses caused by the diesel fuel exhaust emissions: A review of the research. In *Handbook of Biodiesel and Petrodiesel Fuels: Science, Technology, Health, and Environment. Volume 3. Petrodiesel Fuels: Science, Technology, Health, and Environment*, ed. O. Konur. Boca Raton, FL: CRC Press.

Konur, O. 2021af. Cancer caused by the diesel fuel exhaust emissions: A review of the research. In *Handbook of Biodiesel and Petrodiesel Fuels: Science, Technology, Health, and Environment. Volume 3. Petrodiesel Fuels: Science, Technology, Health, and Environment*, ed. O. Konur. Boca Raton, FL: CRC Press.

Konur, O. 2021ag. Cardiovascular and other illnesses caused by the diesel fuel exhaust emissions: A review of the research. In *Handbook of Biodiesel and Petrodiesel Fuels: Science, Technology, Health, and Environment. Volume 3. Petrodiesel Fuels: Science, Technology, Health, and Environment*, ed. O. Konur. Boca Raton, FL: CRC Press.

Konur, O. and F. L. Matthews. 1989. Effect of the properties of the constituents on the fatigue performance of composites: A review. *Composites* 20: 317–328.

Lam, M. K., K. T. Tan, K. T. Lee, and A. R. Mohamed. 2009. Malaysian palm oil: Surviving the food versus fuel dispute for a sustainable future. *Renewable and Sustainable Energy Reviews* 13: 1456–1464.

Lardon, L., A. Helias, B. Sialve, J. P. Steyer, and O. Bernard. 2009. Life-cycle assessment of biodiesel production from microalgae. *Environmental Science & Technology* 43: 6475–6481

Li, Y. Q., M. Horsman, B. Wang, N. Wu, and C. Q. Lan. 2008. Effects of nitrogen sources on cell growth and lipid accumulation of green alga *Neochloris oleoabundans*. *Applied Microbiology and Biotechnology* 81: 629–636.

Li, X., H. Y. Hu, K. Gan, and Y. X. Sun. 2010. Effects of different nitrogen and phosphorus concentrations on the growth, nutrient uptake, and lipid accumulation of a freshwater microalga *Scenedesmus* sp. *Bioresource Technology* 101: 5494–5500.

Li, Y. C., Y. F. Chen, and P. Chen, et al. 2011. Characterization of a microalga *Chlorella* sp. well adapted to highly concentrated municipal wastewater for nutrient removal and biodiesel production. *Bioresource Technology* 102: 5138–5144.

Liang, Y. N., N. Sarkany, and Y. Cui. 2009. Biomass and lipid productivities of *Chlorella vulgaris* under autotrophic, heterotrophic and mixotrophic growth conditions. *Biotechnology Letters* 31: 1043–1049.

Liu, Z. Y., G. C. Wang, and B. C. Zhou. 2008. Effect of iron on growth and lipid accumulation in *Chlorella vulgaris*. *Bioresource Technology* 99: 4717–4722.

Lobell, D. B., M. B. Burke, and C. Tebaldi, et al. 2008. Prioritizing climate change adaptation needs for food security in 2030. *Science* 319: 607–610.

Mata, T. M., A. A. Martins, and N. S. Caetano. 2010. Microalgae for biodiesel production and other applications: A review. *Renewable & Sustainable Energy Reviews* 14: 217–232.

McCarthy, J. F. 2010. Processes of inclusion and adverse incorporation: oil palm and agrarian change in Sumatra, Indonesia. *Journal of Peasant Studies* 37: 821–850.

Michalak, A. M., E. J. Anderson, and D. Beletsky, et al. 2013. Record-setting algal bloom in Lake Erie caused by agricultural and meteorological trends consistent with expected future conditions. *Proceedings of the National Academy of Sciences* 110: 6448–6452.

Miller, R., G. X. Wu, and R. R. Deshpande, et al. 2010. Changes in Transcript abundance in *Chlamydomonas reinhardtii* following nitrogen deprivation predict diversion of metabolism. *Plant Physiology* 154: 1737-1752.

Mohan, D., C. U. Pittman Jr, and P. H. Steele. 2006. Pyrolysis of wood/biomass for bio-oil: A critical review. *Energy & Fuels* 20: 848-889.

Molina, E., J. Fernandez, F. G. Acien, and Y. Chisti. 2001. Tubular photobioreactor design for algal cultures. *Journal of Biotechnology* 92:113-131.

Naylor, R. L., A. J. Liska, M. B. Burke, et al. 2007. The ripple effect: Biofuels, food security, and the environment. *Environment: Science and Policy for Sustainable Development* 49: 30-43.

Norsker, N. H., M. J. Barbosa, M. H. Vermue, and R. H. Wijffels. 2011. Microalgal production: A close look at the economics. *Biotechnology Advances* 29: 24-27.

North, D. C. 1991a. *Institutions, Institutional Change and Economic Performance*. Cambridge, MA: Cambridge University Press.

North, D.C. 1991b. Institutions. *Journal of Economic Perspectives* 5: 97-112.

O'Neil, J. M., T. W. Davis, M. A. Burford, and C. J. Gobler. 2012. The rise of harmful cyano-bacteria blooms: The potential roles of eutrophication and climate change. *Harmful Algae* 14: 313-334.

Obidzinski, K., R. Andriani, H. Komarudin, and A. Andrianto. 2012. Environmental and social impacts of oil palm plantations and their implications for biofuel production in Indonesia. *Ecology and Society* 17: 25.

Ogbonna, J. C., H. Masui, and H. Tanaka. 1997. Sequential heterotrophic/autotrophic cultiva-tion–an efficient method of producing *Chlorella* biomass for health food and animal feed. *Journal of Applied Phycology* 9: 359–366.

Paerl, H. W. 1997. Coastal eutrophication and harmful algal blooms: Importance of atmo-spheric deposition and groundwater as "new" nitrogen and other nutrient sources. *Limnology and Oceanography* 42:1154–1165.

Patil, V., T. Kallqvist, E. Olsen, G. Vogt, and H. R. Gislerod. 2007. Fatty acid composition of 12 microalgae for possible use in aquaculture feed. *Aquaculture International* 15: 1–9.

Perez-Garcia, O., F. M. E. Escalante, L. E. de Bashan, and Y. Bashan. 2011. Heterotrophic cultures of microalgae: Metabolism and potential products. *Water Research* 45: 11–36.

Perron, P. 1989. The great crash, the oil price shock, and the unit root hypothesis. *Econometrica: Journal of the Econometric Society* 57: 1361–1401.

Pulz, O. 2001. Photobioreactors: Production systems for phototrophic microorganisms. *Applied Microbiology and Biotechnology* 57: 287–293.

Radakovits, R., R. E. Jinkerson, and S. I. Fuerstenberg, et al. 2012. Draft genome sequence and genetic transformation of the oleaginous alga *Nannochloropis gaditana*. *Nature Communications* 4: 686.

Rist, L., L. Feintrenie, and P. Levang. 2010. The livelihood impacts of oil palm: Smallholders in Indonesia. *Biodiversity and Conservation* 19: 1009–1024.

Rodolfi, L., G. C. Zittelli, and N. Bassi, et al. 2009. Microalgae for oil: Strain selection, induc-tion of lipid synthesis and outdoor mass cultivation in a low-cost photobioreactor. *Biotechnology and Bioengineering* 102:100–112.

Rosegrant, M. W. and S. A. Cline. 2003. Global food security: Challenges and policies. *Science* 302:1917–1919.

Rozendal, R. A., H. V. Hamelers, K. Rabaey, J. Keller, C. J. Buisman. 2008. Towards practical implementation of bioelectrochemical wastewater treatment. *Trends in Biotechnology* 26: 450–459.

Schenk, P. M., S. R. Thomas-Hall, and E. Stephens, et al. 2008. Second generation biofuels: High-efficiency microalgae for biodiesel production. *Bioenergy Research* 1: 20–43.

Schmidhuber, J. and F. N. Tubiello. 2007. Global food security under climate change. *Proceedings of the National Academy of Sciences* 104: 19703–19708.

Sierra, E., F. G. Acien, and J. M. Fernandez, et al. 2008. Characterization of a flat plate photo-bioreactor for the production of microalgae. *Chemical Engineering Journal* 138: 136–147.

Slade, R. and A. Bauen. 2013. Micro-algae cultivation for biofuels: Cost, energy balance, environmental impacts and future prospects. *Biomass & Bioenergy* 53: 29–38.

Speece, R. E. 1983. Anaerobic biotechnology for industrial wastewater treatment. *Environmental Science & Technology* 17: 416A–427A.

Stephenson, A. L., E. Kazamia, and J. S. Dennis, et al. 2010. Life-cycle assessment of poten-tial algal biodiesel production in the United Kingdom: A comparison of raceways and air-lift tubular bioreactors. *Energy & Fuels* 24: 4062–4077.

Tang, D. H., W. Han, P.L. Li, X. L. Miao, and J. J. Zhong. 2011. CO_2 biofixation and fatty acid composition of *Scenedesmus obliquus* and *Chlorella pyrenoidosa* in response to differ-ent CO_2 levels. *Bioresource Technology* 102: 3071–3076.

Tans, P. P. and T. Takahashi. 1990. Observational constraints on the global atmospheric CO_2 budget. *Science* 247: 1431–1438.

Tenenbaum, D. J. 2008. Food vs. fuel: Diversion of crops could cause more hunger. *Environmental Health Perspectives* 116: A254–A2A7.

Ugwu, C. U., H. Aoyagi, and H. Uchiyama. 2008. Photobioreactors for mass cultivation of algae. *Bioresource Technology* 99: 4021–4028.

Vigani, M., C. Parisi, and E. Rodriguez-Cerezo, et al. 2015. Food and feed products from micro-algae: Market opportunities and challenges for the EU. *Trends in Food Science & Technology* 42: 81–92.

Volkman, J. K., S. W. Jeffrey, P. D. Nichols, G. I. Rogers, and C. D. Garland. 1989. Fatty-acid and lipid-composition of 10 species of microalgae used in mariculture. *Journal of Experimental Marine Biology and Ecology* 128 219–240.

Wang, B., Y. Q. Li, N. Wu, and C. Q. Lan. 2008. CO_2 bio-mitigation using microalgae. *Applied Microbiology and Biotechnology* 79: 707–718.

Wang, L. A., M. Min, and Y. C. Li, et al. 2010a. Cultivation of green algae *Chlorella* sp. in different wastewaters from municipal wastewater treatment plant. *Applied Biochemistry and Biotechnology* 162:1174–1186.

Wang, L., Y. C. Li, and P. Chen, et al. 2010b. Anaerobic digested dairy manure as a nutrient supplement for cultivation of oil-rich green microalgae *Chlorella* sp. *Bioresource Technology* 101: 2623–2628.

Williams, P. J. L. and L. M. L. Laurens. 2010. Microalgae as biodiesel & biomass feedstocks: Review & analysis of the biochemistry, energetics & economics. *Energy & Environmental Science* 3: 554–590.

Wykoff, D. D., J. P. Davies, A. Melis, and A. R. Grossman. 1998. The regulation of photosynthetic electron transport during nutrient deprivation in Chlamydomonas reinhardtii. *Plant Physiology* 117: 129–139.

Yang, J., M. Xu, and X. Z. Zhang, et al. 2011. Life-cycle analysis on biodiesel production from microalgae: Water footprint and nutrients balance. *Bioresource Technology* 102:159–165.

Yoo, C., S. Y. Jun, J. Y. Lee, C. Y. Ahn, and H. M. Oh. 2010. Selection of microalgae for lipid production under high levels carbon dioxide. *Bioresource Technology* 101: S71–S74.

Zhang, Y., M. A. Dube, D. D. McLean, and M. Kates. 2003. Biodiesel production from waste cooking oil. 2. Economic assessment and sensitivity analysis. *Bioresource Technology* 90: 229–240.

Zhang, Q., J. Chang, T. J. Wang, and Y. Xu. 2007. Review of biomass pyrolysis oil properties and upgrading research. *Energy Conversion and Management* 48: 87–92.

35 Algal Biomass Production in Wastewaters for Biodiesel Production
A Review of the Research

Ozcan Konur

CONTENTS

35.1 INTRODUCTION

Crude oils have been primary sources of energy and fuels, such as petrodiesel. However, significant public concerns about the sustainability, price fluctuations, and adverse environmental impact of crude oils have emerged since the 1970s (Ahmadun et al., 2009; Atlas, 1981; Babich and Moulijn, 2003; Kilian, 2009; Perron, 1989). Thus, biooils (Bridgwater et al., 1999; Bridgwater and Peacocke, 2000; Czernik and Bridgwater, 2004; Mohan et al., 2006; Zhang et al., 2007) and biooil-based biodiesel fuels (Chisti, 2007; Hill et al., 2006; Hu et al., 2008; Mata et al., 2010; Rodolfi et al., 2009; Schenk et al., 2008) have emerged as alternatives to crude oils and crude oil-based petrodiesel fuels, respectively, in recent decades. Nowadays, although petrodiesel fuels are still used extensively, biodiesel fuels are being used increasingly in the transportation and power sectors (Konur, 2021a–ag). Therefore, there has been great public interest in the development of algal biodiesel fuels as the fourth generation of biodiesel fuels (Chisti, 2007; Hu et al., 2008; Mata et al., 2010; Rodolfi et al.,

2009; Schenk et al., 2008). However, it is necessary to reduce the total cost of bio-diesel production by reducing the feedstock cost through the use of wastewater instead of freshwater (Bhatnagar et al., 2011; Chinnasamy et al., 2010; de-Bashan et al., 2004; Feng et al., 2011; Li et al., 2010).

Furthermore, for the efficient progression of the research in this field, it is neces-sary to develop efficient incentive structures for the primary stakeholders and to inform these stakeholders about the research (Konur, 2000, 2002a–c, 2006a–b, 2007a–b; North, 1991a–b).

Although there have been a number of reviews and book chapters in this field (Cai et al., 2013; Christenson and Sims, 2011; Park et al., 2011; Pittman et al., 2011; Rawat et al., 2011), there has been no review of the 25-most-cited articles. Thus, this chapter reviews these articles by highlighting the key findings of these most-prolific studies on algal biomass production and nutrient removal in wastewaters. Then, it discusses these key findings.

35.2   MATERIALS AND METHODOLOGY

The search for the literature was carried out in the 'Web of Science' (WOS) database in February 2020. It contains the 'Science Citation Index-Expanded' (SCI-E), the 'Social Sciences Citation Index' (SSCI), the 'Book Citation Index-Science' (BCI-S), the 'Conference Proceedings Citation Index-Science' (CPCI-S), the 'Emerging Sources Citation Index' (ESCI), the 'Book Citation Index-Social Sciences and Humanities' (BCI-SSH), the 'Conference Proceedings Citation Index-Social Sciences and Humanities' (CPCI-SSH), and the 'Arts and Humanities Citation Index' (A&HCI).

The keywords for the search of the literature are collated from the screening of abstract pages for the first 1,000 highly cited papers on algal biodiesel fuels. These keyword sets are provided in the Appendix of the related chapters (Konur, 2021t–u).

The 25-most-cited articles are selected for this review and the key findings are presented and discussed briefly.

35.3   RESULTS

Li et al. (2010) study the effects of Nitrogen (N) and Phosphorus (P) concentrations on growth, nutrient uptake, and lipid accumulation of *Scenedesmus* sp. LX1 in waste-waters in a paper with 482 citations. They observed that *Scenedesmus* sp. LX1's growth was in accordance with the Monod model. The N and P-saturated maximum growth rate was 2.21×10^6 cells $mL^{-1}d^{-1}$, and the half-saturation constants of N and P uptake were 12.1 and 0.27 mgL^{-1}, respectively. In the N/P ratio of 5: 1–12: 1, 83–99% N and 99% P could be removed. In conditions of N (2.5 mg L^{-1}) or P (0.1 mg L^{-1}) limitation, *Scenedesmus* sp. LX1 could accumulate lipids to as much as 30 and 53%, respectively, of its algal biomass. The lipid productivity/unit volume of the culture, however, was not enhanced. They conclude that using wastewaters was ben-eficial in algal biomass accumulation, nutrient removal, and lipid production.

Wang et al. (2010a) evaluate the growth of *Chlorella* sp. on wastewaters sampled from four different points of the treatment process flow of a local municipal wastewater treatment plant and how well the algal growth removed N, P, 'chemical oxygen demand' (COD), and metal ions from the wastewaters in a paper with 478 citations. The four wastewaters were wastewater before primary settling (WW1), wastewater after primary settling (WW2), wastewater after activated sludge tank (WW3), and centrate (WW4), which was the wastewater generated in sludge centrifuge. They observe that the average specific growth rates in the exponential period were 0.412, 0.429, 0.343, and 0.948 day^{-1} for WW1, WW2, WW3, and WW4, respectively. The removal rates of NH_4-N were 82.4, 74.7, and 78.3% for wastewaters WW1, WW2, and WW4, respectively. For WW3, 62.5% of NO_3-N, the major inorganic N form, was removed with 6.3-fold of NO_2-N generated. From wastewaters WW1, WW2, and WW4, 83.2, 90.6, and 85.6% P and 50.9, 56.5, and 83.0% COD were removed, respectively. Only 4.7% was removed in WW3 and the COD in WW3 increased slightly after algal growth, probably due to the excretion of small photosynthetic organic molecules by algae. Metal ions, especially Al, Ca, Fe, Mg, and Mn in centrate, were removed very efficiently. They conclude that growing algae in nutrient-rich centrate offers a new option of applying algal process in wastewater treatment plants to manage the nutrient load for the aeration tank to which the centrate is returned, serving the dual roles of nutrient reduction and valuable biofuel feedstock production. They further conclude that using wastewaters was beneficial in algal biomass accumulation and nutrient removal.

Yang et al. (2011) examine the life-cycle water and nutrient usage of microalgal biodiesel production in a paper with 425 citations. They analyze the influence of water types, operation with and without recycling, algal species, and geographic distributions. They confirm the competitiveness of microalgal biofuels and highlight the necessity of recycling harvested water and using sea/wastewater as water source. To generate 1 kg biodiesel, 3726 kg water, 0.33 kg N, and 0.71 kg phosphate were required if freshwater is used without recycling. Recycling harvest water reduced the water and nutrients usage by 84 and 55%. Using sea/wastewater decreased 90% water requirement and eliminated the need of all the nutrients except phosphate. They also analyze the variation in microalgae species and geographic distribution to reflect microalgal biofuel development in the US. They then discuss the impacts of current federal and state renewable energy programs in the US to suggest suitable microalgae biofuel implementation pathways and identify potential bottlenecks. They conclude that using sea/wastewater was beneficial in reducing water requirement and eliminating the need of all the nutrients except phosphate. They further conclude that using wastewaters was beneficial in algal biomass accumulation and nutrient removal.

Chinnasamy et al. (2010) perform a study using wastewater containing 85–90% carpet industry effluents with 10—15% municipal sewage to evaluate the feasibility of algal biomass and biodiesel production in a paper with 398 citations. They isolate native algal strains from carpet wastewater. They observe that both freshwater and marine algae showed good growth in wastewaters. A consortium of 15 native algal isolates showed >96% nutrient removal in treated wastewater. Biomass production potential and lipid content of this consortium cultivated in treated wastewater were similar to 9.2–17.8 tons ha^{-1} year^{-1} and 6.82%, respectively. They obtain about 63.9% of algal oil from the consortium which could be converted into biodiesel. However, further studies on anaerobic digestion and thermochemical liquefaction were required to make this consortium approach economically viable for producing algae biofuels. They conclude that using wastewaters was beneficial in algal biomass accumulation, nutrient removal, and lipid production.

Li et al. (2011) test the feasibility of growing *Chlorella* sp. in the centrate, a highly concentrated municipal wastewater stream generated from activated sludge thickening process, for simultaneous wastewater treatment and biodiesel production in a paper with 358 citations. They examine the characteristics of algal growth, biodiesel production, wastewater nutrient removal and the viability of scale-up and the stability of continuous operation. They use two culture media of 'autoclaved centrate' and 'raw centrate' for comparison. They observe that by the end of a 14-day batch culture, algae could remove ammonia, total N (TN), total P (TP), and COD by 93.9, 89.1, 80.9, and 90.8%, respectively from RC. The 'fatty acid methyl ester' (FAME) content was 11.04% of dry biomass providing a biodiesel yield of 0.12 g-biodiesel/L-algae culture solution. The system could be successfully scaled up, and continuously operated at 50% daily harvesting rate, providing a net biomass productivity of 0.92 g-algae/(L day). They conclude that using wastewater concentrate was beneficial in algal biomass accumulation, nutrient removal, and biodiesel production.

Aslan and Kapdan (2006) study the effect of the initial N and P concentrations on nutrient removal performance of *Chlorella Vulgaris* in a paper with 351 citations. They determine biokinetic coefficients such as k; 'reaction rate constant', K_m, 'half saturation constant', and Y, 'yield coefficient' by using 'Michaelis–Menten rate expression'. They observe that the NH_4-N concentration was varied between 13.2–410 mgL^{-1} while PO_4-P concentration was between 7.7–199 mgL^{-1} by keeping N/P ratio around 2/1 in the synthetic wastewater. They perform the experiments at pH 7.0 and at room temperature (20°C) with artificial illumination (4100 lux). They observe that effluent water quality decreased with increasing nutrient concentrations and algal culture could remove N more effectively compared to P. They determine biokinetic coefficients as k=1.5 mg NH_4-N mg^{-1} chl a d^{-1}, K_m=31.5 mgL^{-1}, Y_N=0.15 mg chl a mg^{-1} NH_4-N for N and k=0.5 mg PO_4-P mg^{-1} chl a d^{-1}, K_m=10.5 mgL^{-1}, Y_P=0.14 mg chl a mg^{-1} PO_4-P for P. They conclude that using wastewaters was beneficial in algal biomass accumulation and nutrient removal.

Wang et al. (2010b) study the effectiveness of using digested dairy manure as a nutrient supplement for cultivation of *Chlorella* sp. in a paper with 341 citations. They apply different dilution multiples of 10, 15, 20, and 25 to the digested manure and compare algal growth in regard to growth rate, nutrient removal efficiency, final algal fatty acids content, and composition. They observe slower growth rates with less diluted manure samples with higher turbidities in the initial cultivation days. They find a reverse linear relationship between the average specific growth rate of the beginning 7 days and the initial turbidities. Algae removed ammonia, total N, total P, and COD by 100%, 75.7–82.5%, 62.5–74.7%, and 27.4–38.4%, respectively, in differently diluted dairy manure. COD in digested dairy manure, beside CO_2, was another carbon source for mixotrophic *Chlorella*. Fatty acid profiles derived from triacylglyceride (TAG), phospholipid and free fatty acids showed that 'octadecadienoic acid' (08:2) and 'hexadecanoic acid' (C16:0) were the two most abundant fatty acids in the algae. The total fatty acid content of the dry weight increased from 9.00

to 13.7% along with the increasing dilution multiples. They propose a process combining anaerobic digestion and algae cultivation as an effective way to convert high strength dairy manure into profitable byproducts as well as to reduce contaminations to environment. They conclude that using wastewaters was beneficial in algal biomass accumulation, nutrient removal, and lipid production.

Ruiz-Marin et al. (2010) compare two species of microalgae growing as immobilized and free-cells to test its ability to remove N and P in batch cultures of urban wastewater in a paper with 293 citations. They selected the best microalgal cell growth configuration to test in a bioreactor operated in semicontinuous mode. They observe that *Scenedesmus obliquus* showed a higher N and P uptake rate in urban wastewater than *Chlorella vulgaris*. When tested in semicontinuous mode and with the recalcification of beads, *S. obliquus* was more effective in removing N and P for longer periods (181 h) than batch cultures; fecal coliforms removal was good (95%) although the final concentration was still unsuitable for discharge to natural water bodies. Immobilized systems could facilitate the separation of the biomass from the treated wastewater, although in terms of nutritional value of the biomass, immobilized systems did not represent an advantage over free-cell systems. They conclude that using wastewaters was beneficial in algal biomass accumulation and nutrient removal.

Woertz et al. (2009) study lipid productivity and nutrient removal by green algae grown during treatment of dairy farm and municipal wastewaters supplemented with CO_2 in a paper with 268 citations. They treated dairy wastewater outdoors in bench-scale batch cultures. They observed that the lipid content of the volatile solids peaked at day 6, during exponential growth, and declined thereafter. Peak lipid content ranged from 14 to 29%, depending on wastewater concentration. Maximum lipid productivity also peaked at day 6 of batch growth, with a volumetric productivity of 17 mg/day/L of reactor and an areal productivity of 2.8 g/m^2/day, which would be equivalent to 11,000 L/ha/year (1,200 gal/acre/year) if sustained year round. After 12 days, ammonium and orthophosphate removals were 96 and >99%, respectively. They also treated municipal wastewater in semicontinuous indoor cultures with two to four day hydraulic residence times (HRTs). Maximum lipid productivity for the municipal wastewater was 24 mg/day/L, observed in the three-day HRT cultures. They observed over 99% removal of ammonium and orthophosphate. They conclude that CO_2-supplemented algal cultures could simultaneously remove dissolved N and P to low levels while generating a feedstock potentially useful for liquid biofuel production. They further conclude that using wastewaters was beneficial in algal biomass accumulation, nutrient removal, and lipid production.

Martinez et al. (2000) study the removal of P and N by *Scenedesmus obliquus*, cultured in urban wastewater, previously submitted to secondary sewage treatment, under different conditions of stirring and temperature in a paper with 258 citations. In all cases, they evaluated the amount of NH_3 lost, as well as biomass productivity and its biochemical composition. They observed that the specific growth rate was greatest in the stirred cultures, the highest μ value being 0.0438 h^{-1} at 30°C. The stirring increased biomass productivity (P-B) in the linear growth phase after exponential growth, with the optimum appearing at 25°C. For the temperatures studied stirring was not necessary to provide the highest percentage of P elimination ($\%P_{max}$), but did reduce the time needed to reach that percentage (t_{max}). The highest $\%P_{max}$ value, 98%, within the shortest time period, $t_{max} = 94.33$ h, was present in the culture with stirring at 25°C. They determined ammonium removal by two factors of the consumption of ammonium for growth and elimination by desorption as ammonia. The highest percentage of ammonium removal ($\%N_{max}$), 100%, resulted at the final culture time (t_f) of 188.33 h, in the stirred culture at 25°C. The biochemical composition of the biomass gave the normal values for this microalga reported by other authors. The protein content was notably low, around 11.8% by weight, and the polyunsaturated fatty acid content was high. The N:P ratio of the culture medium was 12.9. Finally, they proposed a dilution factor for the treated wastewater (f) to be dumped in order to regulate operation conditions and time for an optimal removal of N and P. They conclude that using wastewaters was beneficial in algal biomass accumulation and nutrient removal.

Zhou et al. (2011) study 60 algae-like microorganisms collected from different sampling sites in Minnesota using multi-step screening and acclimation procedures to select high-lipid producing facultative heterotrophic microalgae strains capable of growing on concentrated municipal wastewater (CMW) for simultaneous algal biomass accumulation and wastewater treatment in a paper with 224 citations. They determined 27 facultative heterotrophic microalgal strains, among which 17 strains were tolerant to CMW. They identified these 17 top-performing strains through morphological observation and DNA sequencing as *Chlorella* sp., *Heynigia* sp., *Hindakia* sp., *Micractinium* sp., and *Scenedesmus* sp. They select five strains for other studies because of their ability to adapt to CMW, high growth rates (0.455–0.498 d^{-1}), and higher lipid productivities (74.5–77.8 mg L^{-1} d^{-1}). They consider these strains highly promising compared with other strains reported in the literature. They conclude that using wastewaters was beneficial in algal biomass accumulation, nutrient removal, and lipid production.

Zhu et al. (2013) study an integrated approach, which combined *Chlorella zofingiensis* cultivation with piggery wastewater treatment, in a paper with 204 citations. They examined the characteristics of algal growth, lipid and biodiesel production, and nutrient removal by using 'tubular bubble column photobioreactors' to cultivate *C. zofingiensis* in piggery wastewater with six different concentrations. They observed that pollutants in piggery wastewater were efficiently removed among all the treatments. The specific growth rate and biomass productivity were different among all the cultures. As the initial nutrient concentration increased, the lipid content of *C. zofingiensis* decreased. The differences in lipid and biodiesel productivity of *C. zofingiensis* among all the treatments mainly resulted from the differences in biomass productivity. The diluted piggery wastewater with 1,900 mg L^{-1} COD provided an optimal nutrient concentration for *C. zofingiensis* cultivation, where the advantageous nutrient removal and the highest productivities of biomass, lipid, and biodiesel were presented. They conclude that using wastewaters is beneficial in algal biomass accumulation, nutrient removal, as well as lipid and biodiesel production.

Feng et al. (2011) use *Chlorella vulgaris* to study algal lipid production with wastewater treatment in a paper with 204 citations. They used artificial wastewater to cultivate *C. vulgaris* in a column aeration photobioreactor (CAP) under batch and semicontinuous cultivation with various daily culture replacements (0.51–1.51 per

21 reactor). They observed that the cell density was decreased from 0.89 g/L with the daily replacement of 0.51 to 0.28 g/L with 1.51 replacement. However, *C. vulgaris* culture achieved the highest lipid content (42%, average value of the phase) and lipid productivity (147 mg/L d^{-1}) with the daily replacement of 1.01. The nutrient removal efficiency was 86% (COD), 97% (NH^{4+}), and 96% (TP), respectively. The net energy ratio (NER) for lipid production with daily replacement of 1.01 (1.25) was higher than the other volume replacement protocols. The algal biomass could be competitive with crude oil at US\$63.97 per barrel with potential credit for wastewater treatment. They conclude that the present research would lead to an economical technology of algal lipid production: using wastewaters was beneficial in algal biomass accumulation, nutrient removal, and lipid production.

Jiang et al. (2011) study the feasibility of using the mixture of seawater and municipal wastewater as a culture medium and CO$_2$ from flue gas for the cultivation of marine microalgae in a paper with 203 citations They examined the effects of different ratios of municipal wastewater and 15% CO$_2$ aeration on the growth of *Nannochloropsis* sp., and also studied the lipid accumulation of microalgae under N starvation and high lighting. They observed that optimal growth of microalgae occurred in 50% of municipal wastewater, and the growth was further significantly enhanced by aeration with 15% CO$_2$. When *Nannochloropsis* sp. cells were transferred from the first growth phase to the second lipid accumulation phase under the combination of N deprivation and high lighting, both biomass and lipid production of *Nannochloropsis* sp. were significantly increased. After 12 days of the second-phase cultivation, the biomass concentration and total lipid content increased from 0.71 to 2.23 g L^{-1} and 33.8 to 59.9%, respectively. They conclude that it is possible to utilize municipal wastewater to replace nutrients in seawater medium and use flue gas to provide CO$_2$ in the cultivation of microalgae for biodiesel production.

Mulbry et al. (2008) determine values for productivity, nutrient content, and nutrient recovery using filamentous green algae grown in outdoor raceways at different loading rates of raw and anaerobically digested dairy manure effluents in a paper with 201 citations. They operated 'algal turf scrubber raceways' (30 m^2 each) in central Maryland for approximately 270 days each year (roughly April–December). They harvest algal biomass every 4–12 days from the raceways after daily additions of manure effluent corresponding to loading rates of 0.3 to 2.5 g TN and 0.08 to 0.42 g TP m^{-2} d^{-1}. They observed that mean algal productivity values increased from approximately 2.5 gDWm^{-2}d^{-1} at the lowest loading rate (0.3 g TN m^{-2}d^{-1}) to 25 g DWm^{-2}d^{-1} at the highest loading rate (2.5 g TN m^{-2}d^{-1}). Mean N and P contents in the dried biomass increased 1.5 to 2.0-fold with increasing loading rates up to maximums of 7% N and 1% P (dry weight basis). Although variable, algal N and P accounted for roughly 70–90% of input N and P at loading rates below 1 g TN, 0.15 g TP m^{-2}d^{-1}. N and P recovery rates decreased to 50–80% at higher loading rates. There were no significant differences in algal productivity, algal N and P content, or N and P recovery values from raceways with CO$_2$ supplementation compared to values from raceways without added CO$_2$. Projected annual operational costs were very high on a per animal basis (\$780 per cow). However, within the context of reducing nutrient inputs in sensitive watersheds such as the Chesapeake Bay, projected operational costs of \$11 per kg N were well below the costs cited for upgrading existing water treatment plants. They conclude that using wastewaters is beneficial in algal biomass accumulation and nutrient removal.

De-Bashan et al. (2004) develop a combination of microalgae (*Chlorella vulgaris* or *C. sorokiniana*) and a microalgae growth-promoting bacterium (MGPB, *Azospirillum brasilense* strain Cd), coimmobilized in small alginate beads, to remove nutrients (P and N) from municipal wastewater in a paper with 198 citations. They outline the most recent technical details necessary for successful coimmobilization of the two microorganisms, and the usefulness of the approach in cleaning the municipal wastewater of the city of La Paz, Mexico. They observed that *A. brasilense* Cd significantly enhanced the growth of both *Chlorella* species when the coimmobilized microorganisms were grown in wastewater. *A. brasilense* was incapable of significant removal of nutrients from the wastewater, whereas both microalgae could. Coimmobilization of the two microorganisms was superior to removal by the microalgae alone, reaching removal of up to 100% ammonium, 15% nitrate, and 36% P within six days (varied with the source of the wastewater), compared to 75% ammonium, 6% nitrate, and 19% P by the microalgae alone. They show the potential of coimmobilization of microorganisms in small beads to serve as a treatment for wastewater in tropical areas. They conclude that using wastewaters is beneficial in algal biomass accumulation and nutrient removal.

Bhatnagar et al. (2011) evaluate the mixotrophic growth potential of native microalgae in media supplemented with different organic carbon substrates and wastewaters in a paper with 194 citations. They isolated three robust mixotrophic microalgae of *Chlamydomonas globosa*, *Chlorella minutissima*, and *Scenedesmus bijuga* after long term enrichments from industrial wastewaters. They observed that the mixotrophic growth of these microalgae resulted in 3–10 times more biomass production relative to phototrophy. Glucose, sucrose, and acetate supported significant mixotrophic growth. Poultry litter extract (PLE) as a growth medium recorded up to 180% more biomass growth compared to standard growth medium (BG11), while treated and untreated carpet industry wastewaters also supported higher biomass, compared to BG11 growth, with no significant effect of additional N supplementation. Supplementing treated wastewater and PLE with glucose and N resulted in a 2–7 times increase in biomass relative to the unaltered wastewaters or PLE. The consortia of *Chlamydomonas-Chlorella* and *Scenedesmus-Chlorella* were the best for PLE and untreated wastewater respectively, while a combination of all three strains was suitable for both PLE and wastewaters. These algae could be good candidates for biofuel feedstock generation as they would not require freshwater or fertilizers. Such mixotrophic algal consortia offer great promise for the production of renewable biomass for bioenergy applications using wastewaters. They conclude that using wastewaters was beneficial in algal biomass accumulation and nutrient removal.

Gonzalez et al. (1997) evaluate the ammonia and P removal efficiencies of *Chlorella vulgaris* and *Scenedesmus dimorphus* during biotreatment of secondary effluents from the agroindustrial wastewater of a dairy industry and pig farming in a paper with 192 citations. They isolated these microalgae from a wastewater stabilization pond near Santafe de Bogota, Colombia. They made batch cultures using both species in 4-l cylindrical glass bioreactors each containing 2l of culture. They also cultivated *Chlorella vulgaris* on wastewater in a triangular bioreactor. They ran three

216 h experimental cycles for each microalga and in each bioreactor. They observed that in the cylindrical bioreactor, *S. dimorphus* was more efficient in removing ammonia than *C. vulgaris*. However, the final efficiency of both microalgae at the end of each cycle was similar. Both microalgae removed P from the wastewater to the same extent in a cylindrical bioreactor. Using *C. vulgaris*, the triangular bioreactor was superior for removing ammonia and the cylindrical bioreactor was superior for removing P. They conclude that using wastewaters was beneficial in algal biomass accumulation and nutrient removal.

Wilkie and Mulbry (2002) assess the ability of benthic freshwater algae to recover nutrients from dairy manure and evaluate nutrient uptake rates and dry matter/crude protein yields in comparison to a conventional cropping system in a paper with 190 citations. They operated benthic algal growth chambers in semibatch mode by continuously recycling wastewater and adding manure inputs daily. They observed that using TN loading rates of 0.64–1.03 gm^{-2}d^{-1}, the dried algal yields were 5.3–5.5 gm^{-2}d^{-1}. The dried algae contained 1.5–2.1% P and 4.9–7.1% N. At a TN loading rate of 1.03 gm^{-2}d^{-1}, algal biomass contained 7.1% N compared to only 4.9% N at a TN loading rate of 0.64 gm^{-2}d^{-1}. In the best case, algal biomass had a crude protein content of 44%, compared to a typical corn silage protein content of 7%. At a dry matter yield of 5.5 gm^{-2}d^{-1}, this was equivalent to an annual N uptake rate of $1,430$ kgha^{-1}yr^{-1}. Compared to a conventional corn/rye rotation, such benthic algal production rates would require 26% of the land area requirements for equivalent N uptake rates and 23% of the land area requirements on a P uptake basis. They conclude that using wastewaters was beneficial in algal biomass accumulation and nutrient removal.

Shi et al. (2007) study the removal of N and P from wastewater by *Chlorella vulgaris* and *Scenedesmus rubescens* using algal cell immobilization and the twin-layer system, in a paper with 188 citations. In this system, they immobilized microalgae by self-adhesion on a wet, microporous, ultrathin substrate (the substrate layer). Subtending the substrate layer, a second layer, consisting of a macroporous fibrous tissue (the source layer), provided the growth medium. Twin layers effectively separated the microalgae from the bulk of their growth medium, yet allowed diffusion of nutrients. In the twin-layer system, algae remained 100% immobilized, which compared favorably with gel entrapment methods for cell immobilization. They observed that both microalgae removed nitrate efficiently from municipal wastewater. Using secondary, synthetic wastewater, the two algae also removed phosphate, ammonium, and nitrate to less than 10% of their initial concentration within nine days. They conclude that immobilization of *C. vulgaris* and *S. rubescens* on twin layers is an effective means to reduce N and P levels in wastewater. They further conclude that using wastewaters is beneficial in algal biomass accumulation and nutrient removal.

Kong et al. (2010) develop large-scale technologies to produce oil-rich algal biomass from wastewater in a paper with 184 citations. They performed the experiments using Erlenmeyer flasks and a biocoil photobioreactor. They grew *Chlamydomonas reinhardtii* in artificial media and wastewaters taken from three different stages of the treatment process, namely, influent, effluent, and centrate. Each of the wastewaters contained different levels of nutrients. They monitored the specific growth rate of *C. reinhardtii* in different cultures over a period of ten days and evaluated the biomass yield of microalgae and associated N and P removal, whilst they also studied the effects of CO_2 and pH on the growth. They observed that the level of nutrients greatly influenced algal growth. High levels of nutrients inhibited algal growth in the beginning, but provided sustained growth to a high degree. The optimal pH for *C. reinhardtii* was in the range of 7.5. An injection of air and a moderate amount of CO_2 promoted algae growth. However, too much CO_2 inhibited algal growth due to a significant decrease in pH. Algal dry biomass yield reached a maximum of 2.0 $gL^{-1}day^{-1}$ in the biocoil and the oil content of *C. reinhardtii* was 25.25% (w/w) in dry biomass weight. In the biocoil, 55.8 mg N and 17.4 mg P per Liter per day were effectively removed from the centrate wastewater. They conclude that using wastewaters was beneficial in algal biomass accumulation, nutrient removal, and lipid production.

De-Bashan et al. (2002) study the removal of ammonium and P ions from synthetic wastewater by *Chlorella vulgaris* coimmobilized in alginate beads with *Azospirillum brasilense* in a paper with 180 citations. They observed that coimmobilization of *C. vulgaris* in alginate beads with *A. brasilense* under semicontinuous synthetic wastewater culture conditions significantly increased the removal of ammonium and soluble P ions compared to immobilization of the microalgae alone. In continuous or batch cultures, removal of these ions followed a similar trend but was less efficient than in semicontinuous culture. They proposed that coimmobilization of a microalgae with *A. brasilense* could serve as a tool in devising novel wastewater treatments where using wastewaters was beneficial in algal biomass accumulation and nutrient removal.

Yun et al. (1997) study the CO_2 fixation by algal cultivation using wastewater nutrients in a paper with 180 citations. They cultivated *Chlorella vulgaris* in wastewater discharged from a steelmaking plant to develop an economically feasible system to remove ammonia from wastewater and CO_2 from flue gas simultaneously. Since no P compounds existed in the wastewater, they added external phosphate ($15.3–46.0$ gm^{-3}). After adaptation to 5% (v/v) CO_2, they observed that the growth of *C. vulgaris* was significantly improved at a typical concentration of CO_2, in flue gas of 15% (v/v). Growth of *C. vulgaris* in raw wastewater was better than that in wastewater buffered with HEPES at 15% (v/v) CO_2. CO_2 fixation and ammonia removal rates were 26.0 g CO_2 $m^{-3}h^{-1}$ and 0.92 g NH_3 $m^{-3}h^{-1}$, respectively, when the alga was cultivated in wastewater supplemented with 46.0 g PO_4^3 m^{-3} without pH control at 15% (v/v) CO_2. They conclude that using wastewaters was beneficial in algal biomass accumulation and nutrient removal.

Park et al. (2010) study *Scenedesmus* sp. for its ability to remove N from anaerobic digestion effluent possessing livestock waste with high ammonium content and alkalinity in addition to its growth characteristics in a paper with 167 citations. Nitrate and ammonium were indistinguishable as an N source when the ammonium concentration was at normal cultivation levels. They observed that ammonium up to 100 ppm NH_4-N did not inhibit cell growth, but did decrease final cell density by up to 70% at a concentration of 200–500 ppm NH_4-N. The inorganic C of alkalinity in the form of bicarbonate was consumed rapidly, in turn causing the attenuation of cell growth. Therefore, maintaining a certain level of inorganic C was necessary in order to prolong ammonia removal. A moderate degree of aeration was beneficial to ammonia removal, not only due to the stripping of ammonium to ammonia gas but also due to the stripping of oxygen, which was an inhibitor of regular photosynthesis. Maintaining appropriate levels of alkalinity, Mg, aeration, along with an optimal initial NH^{4+}/cell ratio were all necessary for long term semicontinuous ammonium removal and cell growth. They conclude that using wastewaters was beneficial in algal biomass accumulation and nutrient removal.

De Godos et al. (2009) evaluate the performance of two 464-L 'high rate algal' ponds (HRAPs) treating 20- and 10-fold diluted swine manure at ten days of 'hydraulic residence time' under continental climatic conditions in Castilla y Leon (Spain) from January to October in a paper with 160 citations. They observed that under optimum environmental conditions (from July to September), both HRAPs supported a stable and efficient C and N oxidation performance, with average COD and 'total Kjeldahl nitrogen' (TKN) removal efficiencies of 76 and 88%, respectively, and biomass productivities ranging from 21 to 28 g/m^2d. They identified nitrification as the main TKN removal mechanism at dissolved oxygen concentrations higher than 2 mg/L (accounting for 80–86% of the TKN removed from January to May and for 54% from July to September). On the other hand, they found empirical evidence of a simultaneous nitrification-denitrification process at dissolved oxygen concentrations lower than 0.5 mg/L (high organic loading rates). However, despite the achievement of excellent COD and N oxidation performance, P removal efficiencies were lower than 10% in both HRAPs, probably due to the high buffer capacity of the piggery wastewater treated. They conclude that using wastewaters was beneficial in algal biomass accumulation and nutrient removal.

35.4   DISCUSSION

Li et al. (2010) study the effects of N and P concentrations on growth, nutrient uptake, and lipid accumulation of *Scenedesmus* sp. LX1 in wastewaters in a paper with 482 citations. They conclude that using wastewaters was beneficial in algal biomass accumulation, nutrient removal, and lipid production.

Wang et al. (2010a) evaluate the growth of *Chlorella* sp. on wastewaters sampled from four different points of the treatment process flow of a local municipal wastewater treatment plant and how well the algal growth removed N, P, COD, and metal ions from the wastewaters in a paper with 478 citations. They conclude that using wastewaters was beneficial in algal biomass accumulation and nutrient removal.

Yang et al. (2011) examine the life-cycle water and nutrients usage of microalgae-based biodiesel production in a paper with 425 citations. They conclude that using sea/wastewaters was beneficial in reducing water requirement and eliminating the need of all the nutrients except phosphate. They further conclude that using wastewater was beneficial in algal biomass accumulation and nutrient removal.

Chinnasamy et al. (2010) perform a study using wastewaters containing 85–90% carpet industry effluents with 10–15% municipal sewage to evaluate the feasibility of algal biomass and biodiesel production in a paper with 398 citations. They conclude that using wastewater was beneficial in algal biomass accumulation, nutrient removal, and biodiesel production.

Li et al. (2011) test the feasibility of growing *Chlorella* sp. in the centrate, a highly concentrated   municipal   wastewater   stream   generated   from   activated   sludge thickening process, for simultaneous wastewater treatment and biodiesel production in a paper with 358 citations. They conclude that using wastewater concentrate was beneficial in algal biomass accumulation, nutrient removal, and biodiesel production.

Aslan and Kapdan (2006) study the effect of the initial N and P concentrations on nutrient removal performance of *Chlorella Vulgaris* in a paper with 351 citations. They conclude that using wastewaters was beneficial in algal biomass accumulation and nutrient removal.

Wang et al. (2010b) study the effectiveness of using digested dairy manure as a nutrient supplement for cultivation of *Chlorella* sp. in a paper with 341 citations. They conclude that using wastewaters was beneficial in algal biomass accumulation, nutrient removal, and lipid production.

Ruiz-Marin et al. (2010) compare two species of microalgae growing as immobilized and free cells to test its ability to remove N and P in batch cultures of urban wastewater in a paper with 293 citations. They conclude that using wastewaters was beneficial in algal biomass accumulation and nutrient removal.

Woertz et al. (2009) study lipid productivity and nutrient removal by green algae grown during treatment of dairy farm and municipal wastewaters supplemented with CO_2 in a paper with 268 citations. They conclude that using wastewaters was beneficial in algal biomass accumulation and nutrient removal.

Martinez et al. (2000) study the removal of P and N by *Scenedesmus obliquus*, cultured in urban wastewater, previously submitted to secondary sewage treatment, under different conditions of stirring and temperature in a paper with 258 citations. They conclude that using wastewaters was beneficial in algal biomass accumulation and nutrient removal.

Zhou et al. (2011) study 60 algae-like microorganisms collected from different sampling sites in Minnesota using multi-step screening and acclimation procedures to select high-lipid producing facultative heterotrophic microalgae strains capable of growing on 'concentrated municipal wastewater' for simultaneous algal biomass accumulation and wastewater treatment in a paper with 224 citations. They conclude that using wastewaters was beneficial in algal biomass accumulation, nutrient removal, and lipid production.

Zhu et al. (2013) study an integrated approach, which combined *Chlorella zofingiensis* cultivation with piggery wastewater treatment, in a paper with 204 citations. They conclude that using wastewaters was beneficial in algal biomass accumulation, nutrient removal, as well as lipid and biodiesel production.

Feng et al. (2011) use *Chlorella vulgaris* to study algal lipid production with wastewater treatment in a paper with 204 citations. They conclude using wastewaters was beneficial in algal biomass accumulation, nutrient removal, and lipid production.

Jiang et al. (2011) study the feasibility of using the mixture of seawater and municipal wastewater as culture medium and CO_2 from flue gas for the cultivation of marine microalgae in a paper with 203 citations. They conclude that it was possible to utilize municipal wastewater to replace nutrients in seawater medium and use flue gas to provide CO_2 in the cultivation of microalgae for biodiesel production.

Mulbry et al. (2008) determine values for productivity, nutrient content, and nutrient recovery using filamentous green algae grown in outdoor raceways at different loading rates of raw and anaerobically digested dairy manure effluents' in a paper with 201 citations. They conclude that using wastewaters was beneficial in algal biomass accumulation and nutrient removal.

De-Bashan et al. (2004) develop a combination of microalgae (*Chlorella vulgaris* or *C. sorokiniana*) and a microalgae growth-promoting bacterium, *Azospirillum brasilense* strain Cd, coimmobilized in small alginate beads, to remove nutrients (P and N) from municipal wastewater in a paper with 198 citations. They conclude that using wastewaters was beneficial in algal biomass accumulation and nutrient removal.

Bhatnagar et al. (2011) evaluate mixotrophic growth potential of native microalgae in media supplemented with different organic carbon substrates and wastewaters in a paper with 194 citations. They conclude that using wastewaters was beneficial in algal biomass accumulation and nutrient removal.

Gonzalez et al. (1997) evaluate the ammonia and P removal efficiencies of *Chlorella vulgaris* and *Scenedesmus dimorphus* during biotreatment of secondary effluents from an agroindustrial wastewater of a dairy industry and pig farming in a paper with 192 citations. They conclude that using wastewaters was beneficial in algal biomass accumulation and nutrient removal.

Wilkie and Mulbry (2002) assess the ability of benthic freshwater algae to recover nutrients from dairy manure and evaluate nutrient uptake rates and dry matter/crude protein yields in comparison to a conventional cropping system in a paper with 190 citations. They conclude that using wastewaters was beneficial in algal biomass accumulation and nutrient removal.

Shi et al. (2007) study the removal of N and P from wastewater by *Chlorella vulgaris* and *Scenedesmus rubescens* using algal cell immobilization, the twin-layer system, in a paper with 188 citations. They conclude that immobilization of *C. vulgaris* and *S. rubescens* on twin layers was an effective means to reduce N and P levels in wastewater. They further conclude that using wastewaters was beneficial in algal biomass accumulation and nutrient removal.

Kong et al. (2010) develop large-scale technologies to produce oil-rich algal biomass from wastewater in a paper with 184 citations. They conclude that using wastewaters was beneficial in algal biomass accumulation, nutrient removal, and lipid production.

De-Bashan et al. (2002) study the removal of ammonium and P ions from synthetic wastewater by *Chlorella vulgaris* coimmobilized in alginate beads with *Azospirillum brasilense* in a paper with 180 citations. They conclude that using wastewaters was beneficial in algal biomass accumulation and nutrient removal.

Yun et al. (1997) study the CO_2 fixation by algal cultivation using wastewater nutrients in a paper with 180 citations. They conclude that using wastewater was beneficial in algal biomass accumulation and nutrient removal.

Park et al. (2010) study *Scenedesmus* sp. for its ability to remove N from anaerobic digestion effluent possessing livestock waste with high ammonium content and alkalinity in addition to its growth characteristics in a paper with 167 citations. They conclude that using wastewaters was beneficial in algal biomass accumulation and nutrient removal.

De Godos et al. (2009) evaluate the performance of two 464-L high rate algal ponds treating 20- and 10-fold diluted swine manure at ten days of hydraulic residence time under continental climatic conditions in Castilla y Leon (Spain) from January to October in a paper with 160 citations. They conclude that using wastewaters was beneficial in algal biomass accumulation and nutrient removal.

These prolific studies highlight the use of wastewater for algal biomass accumulation, wastewater treatment, and nutrient removal from these wastewaters as well as lipid production and biodiesel production. As Yang et al. (2011) show, 3,726 kg water, 0.33 kg N, and 0.71 kg phosphate were required if freshwater is used without recycling to produce 1 kg biodiesel. They further show that using sea/wastewater decreased by 90% the water requirement and eliminated the need for all the nutrients except phosphate. As around 85% of the total cost of biodiesel production emanates from the feedstock costs (Davis et al., 2011; Norsker et al., 2011; Williams and Laurens, 2010; Zhang et al., 2003), it is helpful to reduce the total cost of biodiesel production by supplying water, N, and P cheaply from wastewaters. This would make the algal biodiesel fuels more competitive with petrodiesel fuels (Davis et al., 2011; Norsker et al., 2011; Williams and Laurens, 2010; Zhang et al., 2003).

This would also help with the treatment of wastewaters (Bigda, 1995; Chen, 2004; Rozendal et al., 2008; Speece, 1983). Another beneficial effect of using wastewaters for algal biomass accumulation and nutrient removal would be the lessening of the volume and harmful effects of the harmful algal blooms, which have been a significant ecological disaster in recent years (Anderson et al., 2002; Hallegraeff, 1993; Michalak et al., 2013; O'Neill et al., 2012; Paerl, 1997).

The production of the algal biodiesel fuels in competition with petrodiesel fuels would also help in shifting biodiesel production from edible oil-based biodiesel fuels. There has been significant competition between biodiesel production from edible oils and the household consumption of edible oils, resulting in significant public concerns about food security (Ajanovic, 2011; Godfray et al., 2010; Lam et al., 2009; Lobell et al., 2008; Naylor et al., 2007; Rosegrant and Cline, 2003; Schmidhuber and Tubiello, 2007; Tenenbaum, 2008).

This would also help in dealing with public concerns emanating from the fact that palm plantations expand at the expense of forests as oil palm plantations have replaced large areas of forests in Southeast Asia and other places (Carlson et al., 2012, 2013; Fitzherbert et al., 2008; Koh and Wilcove, 2008; Koh et al., 2011). This has resulted in public concerns about the destruction of forests (Fitzherbert et al., 2008; Koh and Wilcove, 2008; Koh et al., 2011), destruction of ecological biodiversity (Fitzherbert et al., 2008; Koh and Wilcove, 2008; Koh et al., 2011), significant increase in CO_2 emissions (Carlson et al., 2012, 2013), and finally, exploitation of local communities (McCarthy, 2010; Obidzinski et al., 20012, Rist et al., 2010).

35.5 CONCLUSION

This chapter has presented the key findings of the 25-most-cited article papers in this field.

These prolific studies provide valuable evidence on algal biomass production and nutrient removal from these wastewaters by algae as well as lipid production and biodiesel production.

The ecological, technoeconomic, and environmental benefits from using wastewaters for these purposes is significant. The use of these wastewaters would make biodiesel production more competitive in relation to petrodiesel fuels by reducing the feedstock costs significantly.

This would also help with the treatment of wastewaters. Another beneficial effect of using wastewaters for algal biomass accumulation and nutrient removal would be the lessening of the volume and harmful effects of algal blooms, which have been a significant ecological disaster in recent years.

The production of algal biodiesel fuels in competition with petrodiesel fuels would also help in shifting biodiesel production from edible oil-based biodiesel fuels and in dealing with public concerns about food security and the destruction of forests for the expansion of oil palm plantations, the destruction of ecological biodiversity, the significant increase in CO_2 emissions, and finally, the exploitation of local communities.

ACKNOWLEDGMENTS

The contribution of the highly cited researchers in this field is greatly acknowledged.

REFERENCES

Ahmadun, F. R., A. Pendashteh, and L. C. Abdullah, et al. 2009. Review of technologies for oil and gas produced water treatment. *Journal of Hazardous Materials* 170: 530–551.

Ajanovic, A. 2011. Biofuels versus food production: Does biofuels production increase food prices? *Energy* 36: 2070–2076.

Anderson, D. M. P. M. Glibert, and J. M. Burkholder. 2002. Harmful algal blooms and eutrophication: Nutrient sources, composition, and consequences. *Estuaries* 25: 704–726.

Aslan, S. and I. K. Kapdan. 2006. Batch kinetics of nitrogen and phosphorus removal from synthetic wastewater by algae. *Ecological Engineering* 28: 64–70.

Atlas, R. M. 1981. Microbial degradation of petroleum hydrocarbons: An environmental perspective. *Microbiological Reviews* 45: 180–209.

Babich, I. V. and J. A. Moulijn. 2003. Science and technology of novel processes for deep desulfurization of oil refinery streams: A review. *Fuel* 82: 607–631.

Bhatnagar, A., S. Chinnasamy, M. Singh, and K. C. Das. 2011. Renewable biomass production by mixotrophic algae in the presence of various carbon sources and wastewaters. *Applied Energy* 88: 3425–3431.

Bigda, R. J. 1995. Consider Fentons chemistry for wastewater treatment. *Chemical Engineering Progress* 91: 62–66.

Bridgwater, A. V. and G. V. C. Peacocke. 2000. Fast pyrolysis processes for biomass. *Renewable & Sustainable Energy Reviews* 4: 1–73.

Bridgwater, A. V., D. Meier, and D. Radlein. 1999. An overview of fast pyrolysis of biomass. *Organic Geochemistry* 30: 1479–1493.

Cai, T., S. Y. Park, and Y. B. Li. 2013. Nutrient recovery from wastewater streams by microalgae: Status and prospects. *Renewable & Sustainable Energy Reviews* 19: 360–369.

Carlson, K. M., L. M. Curran, and D. Ratnasari, et al. 2012. Committed carbon emissions, deforestation, and community land conversion from oil palm plantation expansion in West Kalimantan, Indonesia. *Proceedings of the National Academy of Sciences of the United States of America* 109: 7559–7564.

Carlson, K. M., L. M. Curran, and G. P. Asner, et al. 2013. Carbon emissions from forest conversion by Kalimantan oil palm plantations. *Nature Climate Change* 3: 283–287.

Chen, G. 2004. Electrochemical technologies in wastewater treatment. *Separation and Purification Technology* 38(1), 11–41.

Chinnasamy, S., A. Bhatnagar, R. W. Hunt, and K. C. Das. 2010. Microalgae cultivation in a wastewater dominated by carpet mill effluents for biofuel applications. *Bioresource Technology* 101: 3097–3105.

Chisti, Y. 2007. Biodiesel from microalgae. *Biotechnology Advances* 25: 294–306.

Christenson, L. and R. Sims. 2011. Production and harvesting of microalgae for wastewater treatment, biofuels, and bioproducts. *Biotechnology Advances* 29: 686–702.

Czernik, S. and A. V. Bridgwater. 2004. Overview of applications of biomass fast pyrolysis oil. *Energy & Fuels* 18: 590–598.

Davis, R., A. Aden, and P. T. Pienkos. 2011. Techno-economic analysis of autotrophic microalgae for fuel production. *Applied Energy* 88: 3524–3531.

De Godos, I., S. Blanco, P. A. Garcia-Encina, E. Becares, and R. Munoz. 2009. Long-term operation of high rate algal ponds for the bioremediation of piggery wastewaters at high loading rates. *Bioresource Technology* 100: 4332–4339.

De-Bashan, L. E., M. Moreno, J. P. Hernandez, and Y. Bashan. 2002. Removal of ammonium and phosphorus ions from synthetic wastewater by the microalgae *Chlorella vulgaris* coimmobilized in alginate beads with the microalgae growth-promoting bacterium *Azospirillum brasilense*. *Water Research* 36: 2941–2948.

De-Bashan, L. E., J.P. Hernandez, T. Morey, and Y. Bashan. 2004. Microalgae growth-promoting bacteria as "helpers" for microalgae: A novel approach for removing ammonium and phosphorus from municipal wastewater. *Water Research* 38: 466–474.

Feng, Y. J., C. Li, and D. W. Zhang. 2011. Lipid production of *Chlorella vulgaris* cultured in artificial wastewater medium. *Bioresource Technology* 102: 101–105.

Fitzherbert, E. B., M. J. Struebig, and A. Morel, et al. 2008. How will oil palm expansion affect biodiversity? *Trends n Ecology & Evolution* 23: 538–545.

Godfray, H. C. J., J. R. Beddington, and I. R. Crute, et al., 2010. Food security: The challenge of feeding 9 billion people. *Science* 327: 812–818.

Gonzalez, L.E., R. O. Canizares, and S. Baena. 1997. Efficiency of ammonia and phosphorus removal from a Colombian agroindustrial wastewater by the microalgae *Chlorella vulgaris* and *Scenedesmus dimorphus*. *Bioresource Technology* 60: 259–262.

Hallegraeff, G. M. 1993. A review of harmful algal blooms and their apparent global increase. *Phycologia* 32: 79–99.

Hill, J., E. Nelson, D. Tilman, S. Polasky, and D. Tiffany. 2006. Environmental, economic, and energetic costs and benefits of biodiesel and ethanol biofuels. *Proceedings of the National Academy of Sciences of the United States of America* 103: 11206–11210.

Hu, Q., M. Sommerfeld, and E. Jarvis, et al. 2008. Microalgal triacylglycerols as feedstocks for biofuel production: Perspectives and advances. *Plant Journal* 54: 621–639.

Jiang, L. L., S. J. Luo, and X. L. Fan, Z. M. Yang, and R. B. Guo. 2011. Biomass and lipid production of marine microalgae using municipal wastewater and high concentration of CO_2. *Applied Energy* 88: 3336–3341.

Kilian, L. 2009. Not all oil price shocks are alike: Disentangling demand and supply shocks in the crude oil market. *American Economic Review* 99: 1053–1069.

Koh, L. P. and D. S. Wilcove. 2008. Is oil palm agriculture really destroying tropical biodiversity? *Conservation Letters* 1: 60–64.

Koh, L. P., J. Miettinen, S. C. Liew, and J. Ghazoul. 2011. Remotely sensed evidence of tropical peatland conversion to oil palm. *Proceedings of the National Academy of Sciences of the United States of America* 108: 5127–5132.

Kong, Q. X., L. Li, B. Martinez, P. Chen, and R. Ruan. 2010. Culture of microalgae *Chlamydomonas reinhardtii* in wastewater for biomass feedstock production. *Applied Biochemistry and Biotechnology* 160: 9–18.

Konur, O. 2000. Creating enforceable civil rights for disabled students in higher education: An institutional theory perspective. *Disability & Society* 15: 1041–1063.

Konur, O. 2002a. Access to Nursing Education by disabled students: Rights and duties of nursing programs. *Nurse Education Today* 22: 364–374.

Konur, O. 2002b. Assessment of disabled students in higher education: Current public policy issues. *Assessment and Evaluation in Higher Education* 27: 131–152.

Konur, O. 2002c. Access to employment by disabled people in the UK: Is the Disability Discrimination Act working? *International Journal of Discrimination and the Law* 5: 247–279.

Konur, O. 2006a. Participation of children with dyslexia in compulsory education: Current public policy issues. *Dyslexia* 12: 51–67.

Konur, O. 2006b. Teaching disabled students in Higher Education. *Teaching in Higher Education* 11: 351–363.

Konur, O. 2007a. A judicial outcome analysis of the Disability Discrimination Act: A windfall for the employers? *Disability & Society* 22: 187–204.

Konur, O. 2007b. Computer-assisted teaching and assessment of disabled students in higher education: The interface between academic standards and disability rights. *Journal of Computer Assisted Learning* 23: 207–219.

Konur, O., ed. 2021a. *Handbook of Biodiesel and Petrodiesel Fuels: Science, Technology, Health, and Environment.* Boca Raton, FL: CRC Press.

Konur, O., ed. 2021b. *Handbook of Biodiesel and Petrodiesel Fuels: Science, Technology, Health, and Environment. Volume 1. Biodiesel Fuels: Science, Technology, Health, and Environment.* Boca Raton, FL: CRC Press.

Konur, O., ed. 2021c. *Handbook of Biodiesel and Petrodiesel Fuels: Science, Technology, Health, and Environment. Volume 2. Biodiesel Fuels based on the Edible and Nonedible Feedstocks, Wastes, and Algae: Science, Technology, Health, and Environment.* Boca Raton, FL: CRC Press.

Konur, O., ed. 2021d. *Handbook of Biodiesel and Petrodiesel Fuels: Science, Technology, Health, and Environment. Volume 3. Petrodiesel Fuels: Science, Technology, Health, and Environment.* Boca Raton, FL: CRC Press.

Konur, O. 2021e. Biodiesel and petrodiesel fuels: Science, technology, health, and environment. In *Handbook of Biodiesel and Petrodiesel Fuels: Science, Technology, Health, and Environment. Volume 1. Biodiesel Fuels: Science, Technology, Health, and Environment*, ed. O. Konur. Boca Raton, FL: CRC Press.

Konur, O. 2021f. Biodiesel and petrodiesel fuels: A scientometric review of the research. In *Handbook of Biodiesel and Petrodiesel Fuels: Science, Technology, Health, and Environment. Volume 1. Biodiesel Fuels: Science, Technology, Health, and Environment*, ed. O. Konur. Boca Raton, FL: CRC Press.

Konur, O. 2021g. Biodiesel and petrodiesel fuels: A review of the research. In *Handbook of Biodiesel and Petrodiesel Fuels: Science, Technology, Health, and Environment. Volume 1. Biodiesel Fuels: Science, Technology, Health, and Environment*, ed. O. Konur. Boca Raton, FL: CRC Press.

Konur, O. 2021h Nanotechnology applications in the diesel fuels and the related research fields: A review of the research. In *Handbook of Biodiesel and Petrodiesel Fuels: Science, Technology, Health, and Environment. Volume 1. Biodiesel Fuels: Science, Technology, Health, and Environment*, ed. O. Konur. Boca Raton, FL: CRC Press.

Konur, O. 2021i. Biooils: A scientometric review of the research. In *Handbook of Biodiesel and Petrodiesel Fuels: Science, Technology, Health, and Environment. Volume 1. Biodiesel Fuels: Science, Technology, Health, and Environment*, ed. O. Konur. Boca Raton, FL: CRC Press.

Konur, O. 2021j. Characterization and properties of biooils: A review of the research. In *Handbook of Biodiesel and Petrodiesel Fuels: Science, Technology, Health, and Environment. Volume 1. Biodiesel Fuels: Science, Technology, Health, and Environment*, ed. O. Konur. Boca Raton, FL: CRC Press.

Konur, O. 2021k. Biomass pyrolysis and pyrolysis oils: A review of the research. In *Handbook of Biodiesel and Petrodiesel Fuels: Science, Technology, Health, and Environment. Volume 1. Biodiesel Fuels: Science, Technology, Health, and Environment*, ed. O. Konur. Boca Raton, FL: CRC Press.

Konur, O. 2021l. Biodiesel fuels: A scientometric review of the research. In *Handbook of Biodiesel and Petrodiesel Fuels: Science, Technology, Health, and Environment. Volume 1. Biodiesel Fuels: Science, Technology, Health, and Environment*, ed. O. Konur. Boca Raton, FL: CRC Press.

Konur, O. 2021m. Glycerol: A scientometric review of the research. In *Handbook of Biodiesel and Petrodiesel Fuels: Science, Technology, Health, and Environment. Volume 1. Biodiesel Fuels: Science, Technology, Health, and Environment*, ed. O. Konur. Boca Raton, FL: CRC Press.

Konur, O. 2021n. Propanediol production from glycerol: A review of the research. In *Handbook of Biodiesel and Petrodiesel Fuels: Science, Technology, Health, and Environment. Volume 1. Biodiesel Fuels: Science, Technology, Health, and Environment*, ed. O. Konur. Boca Raton, FL: CRC Press.

Konur, O. 2021o. Edible oil-based biodiesel fuels: A scientometric review of the research. In *Handbook of Biodiesel and Petrodiesel Fuels: Science, Technology, Health, and Environment. Volume 2. Biodiesel Fuels based on the Edible and Nonedible Feedstocks, Wastes, and Algae: Science, Technology, Health, and Environment*, ed. O. Konur. Boca Raton, FL: CRC Press.

Konur, O. 2021p. Palm oil-based biodiesel fuels: A review of the research. In *Handbook of Biodiesel and Petrodiesel Fuels: Science, Technology, Health, and Environment. Volume 2. Biodiesel Fuels based on the Edible and Nonedible Feedstocks, Wastes, and Algae*, ed. O. Konur. Boca Raton, FL: CRC Press.

Konur, O. 2021q. Rapeseed oil-based biodiesel fuels: A review of the research. In *Handbook of Biodiesel and Petrodiesel Fuels: Science, Technology, Health, and Environment. Volume 2. Biodiesel Fuels based on the Edible and Nonedible Feedstocks, Wastes, and Algae*, ed. O. Konur. Boca Raton, FL: CRC Press.

Konur, O. 2021r. Nonedible oil-based biodiesel fuels: A scientometric review of the research. In *Handbook of Biodiesel and Petrodiesel Fuels: Science, Technology, Health, and Environment. Volume 2. Biodiesel Fuels based on the Edible and Nonedible Feedstocks, Wastes, and Algae: Science, Technology, Health, and Environment*, ed. O. Konur. Boca Raton, FL: CRC Press.

Konur, O. 2021s. Waste oil-based biodiesel fuels: A scientometric review of the research. In *Handbook of Biodiesel and Petrodiesel Fuels: Science, Technology, Health, and Environment. Volume 2. Biodiesel Fuels based on the Edible and Nonedible Feedstocks, Wastes, and Algae: Science, Technology, Health, and Environment*, ed. O. Konur. Boca Raton, FL: CRC Press.

Konur, O. 2021t. Algal biodiesel fuels: A scientometric review of the research. In *Handbook of Biodiesel and Petrodiesel Fuels: Science, Technology, Health, and Environment. Volume 2. Biodiesel Fuels based on the Edible and Nonedible Feedstocks, Wastes, and Algae: Science, Technology, Health, and Environment*, ed. O. Konur. Boca Raton, FL: CRC Press.

Konur, O. 2021u. Algal biomass production for biodiesel production: A review of the research. In *Handbook of Biodiesel and Petrodiesel Fuels: Science, Technology, Health, and Environment. Volume 2. Biodiesel Fuels based on the Edible and Nonedible Feedstocks, Wastes, and Algae*, ed. O. Konur. Boca Raton, FL: CRC Press.

Konur, O. 2021v. Algal biomass production in wastewaters for biodiesel production: A review of the research. In *Handbook of Biodiesel and Petrodiesel Fuels: Science, Technology, Health, and Environment. Volume 2. Biodiesel Fuels based on the Edible and Nonedible Feedstocks, Wastes, and Algae*, ed. O. Konur. Boca Raton, FL: CRC Press.

Konur, O. 2021x. Algal lipid production for biodiesel production: A review of the research. In *Handbook of Biodiesel and Petrodiesel Fuels: Science, Technology, Health, and Environment. Volume 2. Biodiesel Fuels based on the Edible and Nonedible Feedstocks, Wastes, and Algae*, ed. O. Konur. Boca Raton, FL: CRC Press.

Konur, O. 2021y. Crude oils: A scientometric review of the research. In *Handbook of Biodiesel and Petrodiesel Fuels: Science, Technology, Health, and Environment. Volume 3. Petrodiesel Fuels: Science, Technology, Health, and Environment*, ed. O. Konur. Boca Raton, FL: CRC Press.

Konur, O. 2021z. Petrodiesel fuels: A scientometric review of the research. In *Handbook of Biodiesel and Petrodiesel Fuels: Science, Technology, Health, and Environment. Volume 3. Petrodiesel Fuels: Science, Technology, Health, and Environment*, ed. O. Konur. Boca Raton, FL: CRC Press.

Konur, O. 2021aa. Bioremediation of petroleum hydrocarbons in the contaminated soils: A review of the research. In *Handbook of Biodiesel and Petrodiesel Fuels: Science, Technology, Health, and Environment. Volume 3. Petrodiesel Fuels: Science, Technology, Health, and Environment*, ed. O. Konur. Boca Raton, FL: CRC Press.

Konur, O. 2021ab. Desulfurization of diesel fuels: A review of the research. In *Handbook of Biodiesel and Petrodiesel Fuels: Science, Technology, Health, and Environment. Volume 3. Petrodiesel Fuels: Science, Technology, Health, and Environment*, ed. O. Konur. Boca Raton, FL: CRC Press.

Konur, O. 2021ac. Diesel fuel exhaust emissions: A scientometric review of the research. In *Handbook of Biodiesel and Petrodiesel Fuels: Science, Technology, Health, and Environment. Volume 3. Petrodiesel Fuels: Science, Technology, Health, and Environment*, ed. O. Konur. Boca Raton, FL: CRC Press.

Konur, O. 2021ad. The adverse health and safety impact of diesel fuels: A scientometric review of the research. In *Handbook of Biodiesel and Petrodiesel Fuels: Science, Technology, Health, and Environment. Volume 3. Petrodiesel Fuels: Science, Technology, Health, and Environment*, ed. O. Konur. Boca Raton, FL: CRC Press.

Konur, O. 2021ae. Respiratory illnesses caused by the diesel fuel exhaust emissions: A review of the research. In *Handbook of Biodiesel and Petrodiesel Fuels: Science, Technology, Health, and Environment. Volume 3. Petrodiesel Fuels: Science, Technology, Health, and Environment*, ed. O. Konur. Boca Raton, FL: CRC Press.

Konur, O. 2021af. Cancer caused by the diesel fuel exhaust emissions: A review of the research. In *Handbook of Biodiesel and Petrodiesel Fuels: Science, Technology, Health, and Environment. Volume 3. Petrodiesel Fuels: Science, Technology, Health, and Environment*, ed. O. Konur. Boca Raton, FL: CRC Press.

Konur, O. 2021ag. Cardiovascular and other illnesses caused by the diesel fuel exhaust emissions: A review of the research. In *Handbook of Biodiesel and Petrodiesel Fuels: Science, Technology, Health, and Environment. Volume 3. Petrodiesel Fuels: Science, Technology, Health, and Environment*, ed. O. Konur. Boca Raton, FL: CRC Press.

Lam, M. K., K. T. Tan, K. T. Lee, and A. R. Mohamed. 2009. Malaysian palm oil: Surviving the food versus fuel dispute for a sustainable future. *Renewable and Sustainable Energy Reviews* 13: 1456–1464.

Li, X., H. Y. Hu, K. Gan, and Y. X. Sun. 2010. Effects of different nitrogen and phosphorus concentrations on the growth, nutrient uptake, and lipid accumulation of a freshwater microalga *Scenedesmus* sp. *Bioresource Technology* 101: 5494–5500.

Li, Y. C., Y. F. Chen, and P. Chen, et al. 2011. Characterization of a microalga *Chlorella* sp. well adapted to highly concentrated municipal wastewater for nutrient removal and biodiesel production. *Bioresource Technology* 102: 5138–5144.

Lobell, D. B., M. B. Burke, and C. Tebaldi, et al. 2008. Prioritizing climate change adaptation needs for food security in 2030. *Science* 319: 607–610.

Martinez, M. E., S. Sanchez, J. M. Jimenez, F. El Yousfi, and L. Munoz. 2000. Nitrogen and phosphorus removal from urban wastewater by the microalga *Scenedesmus obliquus*. *Bioresource Technology* 73: 263–272.

Mata, T. M., A. A. Martins, and N. S. Caetano. 2010. Microalgae for biodiesel production and other applications: A review. *Renewable & Sustainable Energy Reviews* 14: 217–232.

McCarthy, J. F. 2010. Processes of inclusion and adverse incorporation: oil palm and agrarian change in Sumatra, Indonesia. *Journal of Peasant Studies* 37: 821–850.

Michalak, A. M., E. J. Anderson, and D. Beletsky, et al. 2013. Record-setting algal bloom in Lake Erie caused by agricultural and meteorological trends consistent with expected future conditions. *Proceedings of the National Academy of Sciences* 110: 6448–6452.

Mohan, D., C. U. Pittman Jr, and P. H. Steele. 2006. Pyrolysis of wood/biomass for bio-oil: A critical review. *Energy & Fuels* 20: 848–889.

Mulbry, W., S. Kondrad, C. Pizarro, and E. Kebede-Westhead. 2008. Treatment of dairy manure effluent using freshwater algae: Algal productivity and recovery of manure nutrients using pilot-scale algal turf scrubbers. *Bioresource Technology* 99: 8137–8142.

Naylor, R. L., A. J. Liska, M. B. Burke, et al. 2007. The ripple effect: Biofuels, food security, and the environment. *Environment: Science and Policy for Sustainable Development* 49: 30–43.

Norsker, N. H., M. J. Barbosa, M. H. Vermue, and R. H. Wijffels. 2011. Microalgal production: A close look at the economics. *Biotechnology Advances* 29: 24–27.

North, D. C. 1991a. *Institutions, Institutional Change and Economic Performance*. Cambridge, Mass.: Cambridge University Press.

North, D.C. 1991b. Institutions. *Journal of Economic Perspectives* 5: 97–112.

O'Neil, J. M., T. W. Davis, M. A. Burford, and C. J. Gobler. 2012. The rise of harmful cyanobacteria blooms: The potential roles of eutrophication and climate change. *Harmful Algae* 14: 313–334.

Obidzinski, K., R. Andriani, H. Komarudin, and A. Andrianto. 2012. Environmental and social impacts of oil palm plantations and their implications for biofuel production in Indonesia. *Ecology and Society* 17: 25.

Paerl, H. W. 1997. Coastal eutrophication and harmful algal blooms: Importance of atmospheric deposition and groundwater as "new" nitrogen and other nutrient sources. *Limnology and Oceanography* 42: 1154–1165.

Park, J., H. F. Jin, B. R. Lim, K. Y. Park, and K. Lee. 2010. Ammonia removal from anaerobic digestion effluent of livestock waste using green alga *Scenedesmus* sp. *Bioresource Technology* 101: 8649–8657.

Park, J. B. K., R. J. Craggs, and A. N. Shilton. 2011. Wastewater treatment high rate algal ponds for biofuel production. *Bioresource Technology* 102: 35–42.

Perron, P. 1989. The great crash, the oil price shock, and the unit root hypothesis. *Econometrica: Journal of the Econometric Society* 57: 1361–1401.

Pittman, J. K., A. P. Dean, and O. Osundeko. 2011. The potential of sustainable algal biofuel production using wastewater resources. *Bioresource Technology* 102: 17–25.

Rawat, I., R. R. Kumar, T. Mutanda, and F. Bux. 2011. Dual role of microalgae: Phycoremediation of domestic wastewater and biomass production for sustainable bio-fuels production. *Applied Energy* 88: 3411–3424.

Rist, L., L. Feintrenie, and P. Levang. 2010. The livelihood impacts of oil palm: Smallholders in Indonesia. *Biodiversity and Conservation* 19: 1009–1024.

Rodolfi, L., G. C. Zittelli, and N. Bassi, et al. 2009. Microalgae for oil: Strain selection, induction of lipid synthesis and outdoor mass cultivation in a low-cost photobioreactor. *Biotechnology and Bioengineering* 102: 100–112.

Rosegrant, M. W. and S. A. Cline. 2003. Global food security: Challenges and policies. *Science* 302: 1917–1919.

Rozendal, R. A., H. V. Hamelers, K. Rabaey, J. Keller, C. J. Buisman. 2008. Towards practical implementation of bioelectrochemical wastewater treatment. *Trends in Biotechnology* 26: 450–459.

Ruiz-Marin, A., L. G. Mendoza-Espinosa, and T. Stephenson. 2010. Growth and nutrient removal in free and immobilized green algae in batch and semi-continuous cultures treating real wastewater. *Bioresource Technology* 101: 58–64.

Schenk, P. M., S. R. Thomas-Hall, and E. Stephens, et al. 2008. Second generation biofuels: High-efficiency microalgae for biodiesel production. *Bioenergy Research* 1: 20–43.

Schmidhuber, J. and F. N. Tubiello. 2007. Global food security under climate change. *Proceedings of the National Academy of Sciences* 104: 19703–19708.

Shi, J., B. Podola, and M. Melkonian. 2007. Removal of nitrogen and phosphorus from wastewater using microalgae immobilized on twin layers: An experimental study. *Journal of Applied Phycology* 19: 417–423.

Speece, R. E. 1983. Anaerobic biotechnology for industrial wastewater treatment. *Environmental Science & Technology* 17: 416A–427A.

Tenenbaum, D. J. 2008. Food vs. fuel: Diversion of crops could cause more hunger. *Environmental Health Perspectives* 116: A254-A2A7.

Wang, L. A., M. Min, and Y. C. Li, et al. 2010a. Cultivation of green algae *Chlorella* sp. in different wastewaters from municipal wastewater treatment plant. *Applied Biochemistry and Biotechnology* 162: 1174–1186.

Wang, L., Y. C. Li, and P. Chen, et al. 2010b. Anaerobic digested dairy manure as a nutrient supplement for cultivation of oil-rich green microalgae *Chlorella* sp. *Bioresource Technology* 101: 2623–2628.

Wilkie, A. C. and W. W. Mulbry. 2002. Recovery of dairy manure nutrients by benthic freshwater algae. *Bioresource Technology* 84: 81–91.

Williams, P. J. L. and L. M. L. Laurens. 2010. Microalgae as biodiesel & biomass feedstocks: Review & analysis of the biochemistry, energetics & economics. *Energy & Environmental Science* 3: 554–590.

Woertz, I., A. Feffer, T. Lundquist, and Y. Nelson. 2009. Algae grown on dairy and municipal wastewater for simultaneous nutrient removal and lipid production for biofuel feedstock. *Journal of Environmental Engineering-ASCE* 135: 1115–1122.

Yang, J., M. Xu, and X. Z. Zhang, et al. 2011. Life-cycle analysis on biodiesel production from microalgae: Water footprint and nutrients balance. *Bioresource Technology* 102: 159–165.

Yun, Y. S., S. B. Lee, J. M. Park, C. I. Lee, and J. W. Yang. 1997. Carbon dioxide fixation by algal cultivation using wastewater nutrients. *Journal of Chemical Technology and Biotechnology* 69: 451–455.

Zhang, Y., M. A. Dube, D. D. McLean, and M. Kates. 2003. Biodiesel production from waste cooking oil. 2. Economic assessment and sensitivity analysis. *Bioresource Technology* 90: 229–240.

Zhang, Q., J. Chang, T. J. Wang, and Y. Xu. 2007. Review of biomass pyrolysis oil properties and upgrading research. *Energy Conversion and Management* 48: 87–92.

Zhou, W. G., Y. C. Li, and M. Min, et al. 2011. Local bioprospecting for high-lipid producing microalgal strains to be grown on concentrated municipal wastewater for biofuel production. *Bioresource Technology* 102: 6909–6919.

Zhu, L. D., Z. M. Wang, and Q. Shu, et al. 2013. Nutrient removal and biodiesel production by integration of freshwater algae cultivation with piggery wastewater treatment. *Water Research* 47: 4294–4302.

36 Algal Lipid Production for Biodiesel Production
A Review of the Research

Ozcan Konur

CONTENTS

36.1 INTRODUCTION

Crude oils have been primary sources of energy and fuels, such as petrodiesel. However, significant public concerns about the sustainability, price fluctuations, and adverse environmental impact of crude oils have emerged since the 1970s (Ahmadun et al., 2009; Atlas, 1981; Babich and Moulijn, 2003; Kilian, 2009; Perron, 1989). Thus, biooils (Bridgwater et al., 1999; Bridgwater and Peacocke, 2000; Czernik and Bridgwater, 2004; Mohan et al., 2006; Zhang et al., 2007) and biooil-based biodiesel fuels (Chisti, 2007; Hill et al., 2006; Hu et al., 2008; Mata et al., 2010; Rodolfi et al., 2009; Schenk et al., 2008) have emerged as alternatives to crude oils and crude oil-based petrodiesel fuels, respectively, in recent decades. Nowadays, although petrodiesel fuels are still used extensively, biodiesel fuels are being used increasingly in the transportation and power sectors (Konur, 2021a–ag). Therefore, there has been great public interest in the development of algal biodiesel fuels as the fourth generation of biodiesel fuels (Chisti, 2007; Hu et al., 2008; Mata et al., 2010; Rodolfi et al., 2009; Schenk et al., 2008). However, it is necessary to reduce the total cost of biodiesel production by reducing the feedstock cost through the improvement of the lipid productivity of algal biomass (Griffiths and Harrison, 2009; Guschina and Harwood, 2006; Hu et al., 2008; Mutanda et al., 2011; Roessler, 1990; Sharma et al., 2012).

Furthermore, for the efficient progression of the research in this field, it is necessary to develop efficient incentive structures for the primary stakeholders and to inform these stakeholders about the research (Konur, 2000, 2002a–c, 2006a–b, 2007a–b; North, 1991a–b).

Although there have been a number of reviews and book chapters in this field (Griffiths and Harrison, 2009; Guschina and Harwood, 2006; Hu et al., 2008; Mutanda et al., 2011; Roessler, 1990; Sharma et al., 2012), there has been no review of the 25-most-cited articles. Thus, this chapter reviews these articles by highlighting the key findings of these most-prolific studies on algal lipid production for algal biodiesel production. Then, it discusses these key findings.

36.2   MATERIALS AND METHODOLOGY

The search for the literature was carried out in the 'Web of Science' (WOS) database in May 2020. It contains the 'Science Citation Index-Expanded' (SCI-E), the 'Social Sciences Citation Index' (SSCI), the 'Book Citation Index-Science' (BCI-S), the 'Conference Proceedings Citation Index-Science' (CPCI-S), the 'Emerging Sources Citation Index' (ESCI), the 'Book Citation Index-Social Sciences and Humanities' (BCI-SSH), the 'Conference Proceedings Citation Index-Social Sciences and Humanities' (CPCI-SSH), and the 'Arts and Humanities Citation Index' (A&HCI).

The keywords for the search of the literature were collated from the screening of abstract pages for the first 500 highly cited papers on algal lipid production. This keyword set is provided in the Appendix.

The 25-most-cited articles are selected for this review and the key findings are presented and discussed briefly.

36.3   RESULTS

36.3.1   Algal Cultivation Engineering in General

There are 12 research papers on the cultivation engineering of algae, mostly microalgae. Irradiance, N deprivation, P deprivation, temperature, mixothropic cultivation, iron repletion, Si deprivation, light–dark cycles, salt repletion, light intensity, glucose repletion, 'fed-batch cultivation', and 'two-phase cultivation' are the key parameters studied in these papers.

Rodolfi et al. (2009) study the effect of N and P deprivation and irradiances on microalgal biomass and lipid productivity in a low-cost photobioreactor in a seminal paper with 1,609 citations. They screened 30 microalgal strains for their biomass productivity and lipid content. They selected four strains with a relatively high lipid content and cultivated them under N deprivation in 0.6 L bubbled tubes. They found that only two of them accumulated lipid under such conditions. They then grew *Nannochloropsis* sp., which attained 60% lipid content after N starvation, in a 20 L flat alveolar panel photobioreactor. They found that fatty acid content increased with high irradiances (up to 32.5% of dry biomass) and following both N and P deprivation (up to about 50%). They then grew this strain outdoors in 110 L green wall panel photobioreactors under nutrient sufficient and deficient conditions. They observed that lipid productivity increased from 117 mg/L/day in nutrient sufficient media (with an average biomass productivity of 0.36 g/L/day and 32% lipid content) to 204 mg/L/day (with an average biomass productivity of 0.30 g/L/day and more than 60% final lipid content) in N-deprived media. In a two-phase cultivation process (a nutrient-sufficient phase to produce the inoculum followed by an N-deprived phase to boost lipid synthesis) they projected the oil production potential as more than 90 kg per hectare per day. They estimated that this microalgal strain has the potential for an annual production of 20 tons of lipid per hectare in the Mediterranean climate and of more than 30 tons of lipid per hectare in sunny tropical areas.

Converti et al. (2009) study the effect of temperature and N concentration on the growth and lipid content of *Nannochloropsis oculata* and *Chlorella vulgaris* for biodiesel production in a paper with 678 citations. They found that the lipid content of microalgae was strongly influenced by the variation of these parameters: an increase in temperature from 20 to 25°C practically doubled the lipid content of *N. oculata* (from 7.90 to 14.92%), while an increase from 25 to 30°C brought about a decrease of the lipid content of *C. vulgaris* from 14.71 to 5.90%. On the other hand, they observed that a 75% decrease of the N concentration in the medium with respect to the optimal values for growth, increased the lipid fractions of *N. oculata* from 7.90 to 15.31% and of *C. vulgaris* from 5.90 to 16.41%.

Li et al. (2008) study the effect of N sources and their concentrations on cell growth and lipid accumulation of *Neochloris oleoabundans* in a paper with 657 citations. They observed that whilst the highest lipid cell content of 0.40 g/g was obtained at the lowest sodium nitrate ($NaNO_3$) concentration (3 mM), a remarkable lipid productivity of 0.133 g L^{-1} day^{-1} was achieved at 5 mM with a lipid cell content of 0.34 g/g and a biomass productivity of 0.40 g L^{-1} day^{-1}. On the other hand, the highest biomass productivity was obtained at 10 mM sodium nitrate, with a biomass concentration of 3.2 g L^{-1} and a biomass productivity of 0.63 g L^{-1} day^{-1}. They observed that cell growth continued after the exhaustion of the external N pool, hypothetically supported by the consumption of intracellular N pools such as chlorophyll molecules.

Liang et al. (2009) study the biomass and lipid productivities of *Chlorella vulgaris* under autotrophic, heterotrophic, and mixotrophic growth conditions in a paper with 530 citations. They observed that whilst autotrophic growth provided higher cellular lipid content (38%), the lipid productivity was much lower compared with those from heterotrophic growth with acetate, glucose, or glycerol. On the other hand, optimal cell growth (2 g L^{-1}) and lipid productivity (54 mg L^{-1} day^{-1}) were attained using glucose at 1% (w/v) whereas higher concentrations were inhibitory. Additionally, the growth of *C. vulgaris* on glycerol had similar dose effects as those from glucose. They conclude that *C. vulgaris* is mixotrophic.

Liu et al. (2008) study the effect of iron on growth and lipid accumulation in *Chlorella vulgaris* in a paper with 467 citations. They found that supplementing the growth media with chelated iron chloride (FeCl$_3$) in the late growth phase increased the final cell density but did not induce lipid accumulation in cells. They collected cells in the late-exponential growth phase collected by centrifugation and reinoculated into new media supplemented with five levels of ferric iron (Fe^{3+}) concentration. They observed that the total lipid content in cultures supplemented with 1.2 × 10^{-5} mol L^{-1} FeCl$_3$ was up to 56.6% biomass by dry weight and was three to sevenfold that in other media supplemented with a lower iron concentration.

Shifrin and Chisholm (1981) study the effects of N and Si deficiencies and cell cycles on lipid accumulation in a survey of 30 algal species in a paper with 405 citations. They observed that during log-phase growth, microalgae contained an average of 17.1% total lipids (percentage of total dry weight), whereas diatoms contained an average of 24.5%. N-deprivation of four to nine days resulted in two to three-fold increases in the lipid content of microalgae, whereas both increases and decreases were noted in diatoms, depending on the species. The greatest lipid content was 72% in *Monallantus salina* which had been deprived of N for nine days. Nitrate replenishment in an N-starved culture of *Oocystis polymorpha* showed that the excess cellular lipids did not rapidly disappear during recovery until cell division occurred. On the other hand, *Cyclotella cryptica* showed an increase in the total cellular lipid fraction from 30 to 42% of dry weight within 6 h of the onset of Si limitation, while the mass of lipid material per cell doubled within 12 h. The total lipid fraction in *O. polymorpha* remained constant over the cell cycle in synchronized cultures regardless of the light regime.

Takagi et al. (2006) study the effect of salt (NaCl) concentration on intracellular accumulation of lipids and triacylglyceride in *Dunaliella* cells in a paper with 359 citations. Although initial NaCl concentration higher than 1.5 M markedly inhibited cell growth, they found that an increase of initial NaCl concentration from 0.5 (equal to sea water) to 1.0 M resulted in a higher intracellular lipid content (67%) in comparison with 60% for the salt concentration of 0.5 M. They then found that the addition of 0.5 or 1.0 M NaCl at mid-log phase or at the end of log phase during cultivation with an initial NaCl concentration of 1.0 M further increased the lipid content (to 70%).

Reitan et al. (1994) study the effect of P limitation on the fatty acid and lipid content of microalgae including diatoms in a paper with 348 citations. They cultured *Phaeodactylum tricornutum*, *Chaetoceros* sp., *Isochrysis galbana*, *Pavlova lutheri*, *Nannochloris atomus*, *Tetraselmis* sp., and *Gymnodinum* sp. at different extents of nutrient-limited growth: 50 and 5% of μ_{max}. They found that the lipid content of the algae was in the range 8.3–29.5% of dry matter and was generally higher in the Prymnesiophyceae than in the Prasinophyceae and the Chlorophyceae. Increasing the extent of P-limitation resulted in increased lipid content in the Bacillariophyceae and Prymnesiophyceae and decreased lipid content in *N. atomus* and *Tetraselmis* sp.

On the other hand, the fatty acid composition of the algae showed taxonomic conformity, especially for the Bacillariophyceae, where the major fatty acids were 14: 0, 16:0, 16:1, and 20:5n-3. These fatty acids were dominant also in the Prymnesiophyceae together with 22:6n-3. An exception was *I. galbana*, in which 18: 1 was the major monounsaturated fatty acid and 20:5n-3 was absent. The fatty acids of *N. atomus* and *Tetraselmis* sp. varied somewhat, but 16:0, 16:1, 18:1, 18:3n-3, and 20:5n-3 were most abundant.

Cheirsilp et al. (2012) study the effect of light intensity, glucose concentration, and 'fed-batch cultivation' on the enhanced growth and lipid production of microalgae under mixotrophic culture conditions in a paper with 325 citations. They screened microalgae in photoautotrophic, heterotrophic, and mixotrophic cultures. They observed that the biomass and lipid production of all tested strains in mixotrophic culture were notably enhanced in comparison with photoautotrophic and heterotrophic cultures. Among the tested strains, *Chlorella* sp. and *Nannochloropsis* sp. were ideal candidates for biodiesel production because of their high lipid production. They then found that although increasing light intensity and initial glucose concentration enhanced the growth of both strains, it reduced their lipid content. They performed a fed-batch cultivation with 'stepwise increasing light intensity' to produce a high amount of biomass with high lipid content. They finally observed that lipid production by this strategy was approximately twice that of conventional batch cultivation. The main fatty acid compositions of the two microalgae were C_{16}-C_{18} (>80%) which were appropriate for biodiesel production.

Ben-Amotz et al. (1985) study the chemical profile of selected species of microalgae with emphasis on lipids in a paper with 307 citations. They found that *Botryococcus braunii* contained the highest lipid content of 45% based on the organic weight, with an increase to 55% under N-deficiency and with no effect of salt (NaCl) stress. *Ankistrodesmus* sp., *Dunaliella* spp., *Isochrysis* sp., *Nannochloris* sp., and *Nitzschia* sp. contained an average of 25% lipids under N-sufficient conditions. N-deficiency resulted in a significant increase in the lipid content in all species but *Dunaliella* spp. There were significant low amounts of acyclic hydrocarbons only in *B. braunii*. The major hydrocarbon fractions in N-deficient *B. braunii*, *Dunaliella salina*, *I.* spp., and *Nannochloris* sp. were cyclic and branched polyunsaturated components. Fatty acid composition was species specific, with changes occurring in the relative amounts of individual acids of cells cultivated under different conditions and growth phases. All species synthesized $C_{14:0}$, $C_{16:0}$, $C_{18:1}$, $C_{18:2}$, and $C_{18:3}$ fatty acids; $C_{16:4}$ in *A.* sp.; $C_{18:4}$ and $C_{22:6}$ in *I.* sp.; $C_{16:2}$, $C_{16:3}$, and $C_{20:5}$ in *Nannochloris* sp.; $C_{16:2}$, $C_{16:3}$, and $C_{20:5}$ in *Nitzschia* sp. N-deficiency and salt stress induced accumulation of $C_{18:1}$ in all treated species and to a lesser extent in *B. braunii*.

Breuer et al. (2012) study the effect of N starvation on the dynamics of 'triacylglycerol' (TAG) accumulation in nine microalgal strains in a paper with 307 citations. They observed that under N-deficient conditions, *Chlorella vulgaris*, *C. zofingiensis*, *Neochloris oleoabundans*, and *Scenedesmus obliquus* accumulated more than 35% of their dry weight as TAGs where palmitic and oleic acid were the major fatty acids produced. The main difference between these strains was the amount of biomass that was produced (a 3.0–7.8-fold increase in dry weight) and the duration for which biomass productivity was retained (two to seven days) after N- depletion. They finally observed that *S. obliquus* and *C. zofingiensis* showed the highest average TAG productivity (322 and 243 mg L^{-1} day^{-1}).

Pal et al. (2011) study the effect of light intensity (170 and 700 µmol photons/ m^2.s), salinity (13, 27, and 40 g/l NaCl), and N availability (0.8 and 1.4 g/l) on lipid production by *Nannochloropsis* sp. in a paper with 279 citations. They observed that on an N-replete medium, increases in light intensity and salinity increased the cellular content of the dry weight (DW) and lipids due to enhanced formation of TAGs. They obtain maximum average productivity of ca. 410 mg TFA/l/d at 700 µmol photons/m^2.s and 40 g/l NaCl within seven days. Under stressful conditions, the content of the major long-chain 'polyunsaturated fatty acids' (LC-PUFA), 'eicosapentaenoic acid' (EPA), was significantly reduced while TAG reached 25% of biomass. In contrast, lower salinity improved major growth parameters, consistent with less variation in EPA contents. Combined higher salinity and light intensity were detrimental to lipid productivity under N-starvation as biomass total fatty acid (TFA) content and lipid productivity amounted for only 33% of DW and ca. 200 mg TFA/l/day, respectively. The highest biomass TFA content (ca. 47% DW) and average lipid productivity of ca. 360 mg TFA/l/day were achieved at 13 g/l NaCl and 700 µmol photons/m^2.s.

36.3.2 ALGAL CULTIVATION ENGINEERING: WASTEWATER AND CARBON DIOXIDE BIOREMEDIATION

36.3.2.1 Algal Cultivation and Wastewater Bioremediation

There are three research papers on the cultivation engineering of microalgae in wastewaters and nutrient removal.

Li et al. (2010a) study the effects of different N and P concentrations on the growth, nutrient uptake, and lipid accumulation of *Scenedesmus* sp. in wastewaters in a paper with 496 citations. They observed that microalgal growth was in accordance with the Monod model, the N- and P-saturated maximum growth rate was 2.21 × 10^6 cells mL^{-1} d^{-1}, and the half-saturation constants of N and P uptake were 12.1 mg L^{-1} and 0.27 mg L^{-1}, respectively. They found that at the N/P ratio of 5:1–12:1, 83–99% N and 99% P could be removed. In conditions of N (2.5 mg L^{-1}) or P (0.1 mg L^{-1}) limitation, microalgae could accumulate lipids to as much as 30 and 53%, respectively, of its algal biomass. However, the lipid productivity/unit volume of culture was not enhanced.

Wang et al. (2010) study 'anaerobic digested dairy manure' as a nutrient supplement for the cultivation of *Chlorella* sp. in a paper with 356 citations. They applied different dilution multiples of 10, 15, 20, and 25 to the digested manure and compared algal growth with regard to growth rate, nutrient removal efficiency, and final algal fatty acid content and composition. They observed slower growth rates with less diluted manure samples with higher turbidities in the initial cultivation days. There was a reverse linear relationship between the average specific growth rate of the first seven days and the initial turbidities. Algae removed ammonia, total N, total P, and 'chemical oxygen demand' (COD) by 100, 75.7–82.5, 62.5–74.7, and 27.4– 38.4%, respectively, in differently diluted dairy manure. COD in digested dairy manure, beside CO_2, was another carbon source for mixotrophic *Chlorella*. Fatty acid profiles derived from triacylglyceride (TAG), phospholipid, and free fatty acids showed that octadecadienoic acid (08:2) and hexadecanoic acid (C16:0) were the two most abundant fatty acids in the algae. The TFA content of the dry weight increased from 9.0 to 13.7% along with the increasing dilution multiples.

Woertz et al. (2009) study the lipid productivity and nutrient removal by microalgae grown during treatment of dairy farm wastewaters and municipal wastewaters supplemented with CO_2 in a paper with 270 citations. Dairy wastewater was treated outdoors in bench-scale batch cultures. They observed that the lipid content of the volatile solids peaked at Day 6, during exponential growth, and declined thereafter. Peak lipid content ranged from 14 to 29%, depending on wastewater concentration. Maximum lipid productivity also peaked at Day 6 of batch growth, with a volumetric productivity of 17 mg/day/L of reactor and an areal productivity of 2.8 g/m²/day, which would be equivalent to 11,000 L/ha/year (1,200 gal/acre/year) if sustained year round. After 12 days, ammonium and orthophosphate removals were 96 and > 99%, respectively. Municipal wastewater was treated in semicontinuous indoor cultures with two to four day 'hydraulic residence times' (HRTs). Maximum lipid productivity for the municipal wastewater was 24 mg/day/L, observed in the three-day HRT cultures. Over 99% removal of ammonium and orthophosphate was achieved.

36.3.2.2 Microalgal Cultivation and Carbon Dioxide Bioremediation

There are six research papers on microalgal lipid production and carbon dioxide bioremediation. CO_2 concentration, growth phases, culture approaches, light intensity, N-deprivation, drying temperature, and harvesting time are the key parameters studied in these papers.

Chiu et al. (2009) study the effects of concentration of CO_2 aeration on the biomass production and lipid accumulation of *Nannochloropsis oculata* in a semicontinuous culture in a paper with 440 citations. They also explored the lipid content of these microalgal cells at different growth phases. They observed that the lipid accumulation from the logarithmic phase to the stationary phase of microalgae was significantly increased from 30.8 to 50.4%. In the microalgal cultures aerated with 2%, 5%, 10%, and 15% CO_2, the maximal biomass and lipid productivity in the semicontinuous system were 0.480 and 0.142 g L^{-1} d^{-1} with 2% CO_2 aeration, respectively. On the other hand, in the microalgae cultured in the semicontinuous system aerated with 15% CO_2, biomass and lipid productivity could reach to 0.372 and 0.084 g L^{-1} d^{-1}, respectively. They next observed that the biomass and lipid productivity of microalgae were 0.497 and 0.151 g L^{-1} d^{-1} in one-day replacements (half the broth was replaced each day), and were 0.296 and 0.121 g L^{-1} d^{-1} in three-day replacements (three-fifths of the broth was replaced every 3 d), respectively. They present the optimized conditions as the growth in the semicontinuous system aerated with 2% CO_2 and operated by one-day replacements.

Yoo et al. (2010) study the selection of microalgae for lipid production under high levels of CO_2 in a paper with 381 citations. They cultivated *Botryococcus braunii*, *Chlorella vulgaris*, and *Scenedesmus* sp. with ambient air containing 10% CO_2 and flue gas. They found that the biomass and lipid productivity for *S.* sp. with 10% CO_2 were 217.50 and 20.65 mg L^{-1} d^{-1} (9% of biomass), while those for *B. braunii* were

26.55 and 5.51 mg L^{-1} d^{-1} (21% of biomass). Furthermore, with flue gas, the lipid productivity for *S*. sp. and *B. braunii* increased 1.9-fold (39.44 mg L^{-1} d^{-1}) and 3.7-fold (20.65 mg L^{-1} d^{-1}), respectively. Oleic acid occupied 55% among the fatty acids in *B. braunii*. They observed that *S.* sp. is appropriate for mitigating CO_2, due to its high biomass productivity and C-fixation ability, whereas *B. braunii* is appropriate for producing biodiesel, due to its high lipid content and oleic acid proportion.

Ho et al. (2012) study the effect of light intensity, N starvation, and CO_2 fixation on the lipid and carbohydrate production of *Scenedesmus obliquus* in a paper with 371 citations. They determined the light intensity that promotes cell growth, carbohydrate/lipid productivity, and CO_2 fixation efficiency. They observed that the highest productivity of biomass lipid and carbohydrate was 840.57 mg L^{-1} d^{-1} and 140.35 mg L^{-1} d^{-1}. The highest lipid and carbohydrate content was 22.4% (five-day N-starvation) and 46.65% (one-day N-starvation), respectively, whereas the optimal CO_2 consumption rate was 1,420.6 mg L^{-1} d^{-1}. They then found that under N starvation, the microalgal lipid was mainly composed of C16/C18 fatty acid (around 90%), which is suitable for biodiesel synthesis. The carbohydrate present in the biomass was mainly glucose, accounting for 77–80% of total carbohydrates.

Tang et al. (2011) study the CO_2 biofixation and fatty acid composition of *Scenedesmus obliquus* and *Chlorella pyrenoidosa* in response to different CO_2 levels in a paper with 370 citations. They cultivated these microalgae with 0.03, 5, 10, 20, 30, and 50% CO_2. They observed that these microalgae could grow at 50% CO_2 (>0.69 g L^{-1}) and grew well (>1.22 g L^{-1}) under CO_2 concentrations ranging from 5 to 20%. However, these microalgae showed best growth potential at 10% CO_2. The maximum biomass concentration and CO_2 biofixation rate were 1.84 g L^{-1} and 0.288 g L^{-1} d^{-1} for *S. obliquus* and 1.55 g L^{-1} and 0.260 g L^{-1} d^{-1} for *C. pyrenoidosa*, respectively. The main fatty acid compositions of these microalgae were fatty acids with C-16-C-18 (>94%) under different CO_2 levels. High CO_2 levels (30–50%) were favorable for the accumulation of total lipids and PUFAs. They propose that these microalgae are appropriate for mitigating CO_2 in the flue gases and biodiesel production.

Widjaja et al. (2009) study the effect of CO_2 concentration, N depletion, and harvesting time as well as the method of extraction on lipid production from *Chlorella vulgaris* in a paper with 318 citations. They observed that the drying temperature during lipid extraction from algal biomass affected not only the lipid composition but also lipid content. Drying at very low temperature under vacuum gave the best result but drying at 60°C still retained the composition of lipid while total lipid content decreased only slightly. Drying at higher temperature decreased the content of triacylglyceride (TAG). As long as enough pulverization was applied to the dried algal sample, ultrasonication made no effect, whether on lipid content or on extraction time. In addition to the increase of total lipid content in microalgal cells as a result of cultivation in N-depletion media, they found that changing from N-replete to N-depleted media gradually changed the lipid composition from free-fatty-acid-rich lipid to lipid mostly containing TAG. Since a higher lipid content was obtained when the growth was very slow due to N starvation, they propose that compromising between lipid content and harvesting time should be taken in order to obtain higher values of both the lipid content and lipid productivity. As the growth was much enhanced by increasing CO_2 concentration, this played an important role in the increase of lipid productivity. They found that from low to moderate CO_2 concentration, the highest lipid productivity could be obtained during N depletion which could surpass productivity during normal nutrition. At high-CO_2 concentration, harvesting at the end of the linear phase during normal nutrition gave the highest lipid productivity. However, they propose that by reducing the incubation time of N depletion, a higher lipid content as well as a higher lipid productivity might still be achieved under this condition.

Francisco et al. (2010) study the CO_2 fixation, lipid production, and biodiesel quality in a paper with 268 citations. They cultivated six strains of algae (three cyanobacteria, two microalgae, and one diatom) photosynthetically in a bubble-column photobioreactor. They observe that *Chlorella vulgaris* was the best strain for use as a feedstock for biodiesel production with a carbon dioxide sequestration rate of 17.8 mg L^{-1} min^{-1}, a biomass productivity of 20.1 mg L^{-1} h^{-1}, a lipid content of 27.0%, and a lipid productivity of 5.3 mg L^{-1} h^{-1}. Qualitative analysis of the fatty acid methyl esters demonstrates the predominance of saturated (43.5%) and monounsaturated (41.9%) fatty acids.

36.3.3 ALGAL LIPID ANALYSIS AND IMAGING

There is only one research paper outlined on lipid analysis and lipid imaging on microalgae. However, the papers containing lipid analysis as well as cultivation engineering and nutrient removal are outlined in the respective sections.

Chen et al. (2009) study the high throughput Nile red fluorescence method for the quantitative measurement of 'neutral lipids' in microalgae in a paper with 434 citations. They introduced the solvent dimethyl sulfoxide (DMSO) to microalgal samples as the stain carrier at an elevated temperature. They then determined the cellular neutral lipids and quantified using a 96-well plate on a fluorescence spectrophotometer with an excitation wavelength of 530 nm and an emission wavelength of 575 nm. They found that an optimized procedure yielded a high correlation coefficient with the lipid standard triolein and repeated measurements of replicates. Application of the improved method to several green algal strains gave very reproducible results with relatively minimal standard errors for repeatability and reproducibility at two concentration levels (2.0 μg/mL and 20 μg/mL), respectively. Moreover, the detection and quantification limits of the improved Nile red staining method were 0.8 μg/mL and 2.0 μg/mL for the neutral lipid standard triolein, respectively. On the other hand, the modified method and a conventional gravimetric determination method provided similar results on replicate samples.

36.3.4 OMICS AND CELL BIOLOGY STUDIES IN ALGAL LIPID PRODUCTION

There are three research papers on the genomics and cell biology of lipid production in microalgae and cyanobacteria. The other parameters related to algal cultivation are N-depletion and salt-repletion.

Siaut et al. (2011) study the lipid (TAG) accumulation in *Chlamydomonas reinhardtii* using various wild-type and starchless strains in a paper with 432 citations.

They found that in response to N deficiency, microalgae produced TAGs enriched in palmitic, oleic, and linoleic acids that accumulated in oil bodies. Oil synthesis was maximal between two and three days following N depletion and reached a plateau around day 5. Furthermore, in the first 48 hours of oil deposition, a similar to 80% reduction in the major plastidial membrane lipids occurred. Upon N resupply, mobilization of TAGs started after starch degradation but was completed within 24 hours. Comparison of the oil content in five common laboratory strains revealed a high variability, from 2 µg TAG per million cells in CC124 to 11 µg in 11-32A. The quantification of TAGs on a cell basis in three mutants affected in starch synthesis (cw15sta1-2, cw15sta6, and cw15sta7-1) showed that blocking starch synthesis did not result in TAG over-accumulation compared to their direct progenitor, the arginine auxotroph strain 330. Moreover, there was no significant correlation between cellular oil and starch levels among the 20 wild-type, mutants, and complemented strains tested. By contrast, cellular oil content increased steeply with salt concentration in the growth medium. At 100 mM NaCl, an oil level similar to N-depletion conditions could be reached in a CC124 strain.

Wang et al. (2009) study the stress induction, purification, and biochemical characterization in wild-type and starchless *Chlamydomonas reinhardtii* in a paper with 373 citations. They noted that when the microalgae were deprived of N after entering a stationary phase in liquid culture, the cells produced abundant cytoplasmic lipid bodies (LBs), as well as abundant starch, via a pathway that accompanies a regulated autophagy program. They observed that after 48 h of N starvation in the presence of acetate, the wild-type LB content increased 15-fold. On the other hand, when starch biosynthesis was blocked in the sta6 mutant, the LB content increased 30-fold, demonstrating that genetic manipulation could enhance LB production. Furthermore, the use of cell-wall-less strains permitted development of a rapid 'popped-cell microscopic assay' to quantify the LB content per cell and permitted gentle cell breakage and LB isolation. They observed that the highly purified LBs contained 90% TAG and 10% free fatty acids (FFA). The fatty acids associated with the TAGs were 50% saturated (C_{16} and C_{18}) fatty acids and 50% unsaturated fatty acids, half of which were in the form of oleic acid ($C_{18:1}$). The FFA were 50% C_{16} and 50% C_{18}. The LB-derived TAG yield from a liter of sta6 cells at 10^7 cells/ml after starvation for 48 h approached 400 mg. The LB fraction also contained low levels of charged glycerolipids, with the same profile as whole-cell charged glycerolipids that presumably formed LB membranes, whereas chloroplast-specific neutral glycerolipids (galactolipids) were absent. Very low levels of protein were also present, but all matrix-assisted laser desorption ionization-identified species were apparent contaminants.

Liu et al. (2011) study fatty acid production in genetically modified cyanobacteria in a paper with 306 citations. They genetically modified cyanobacteria to produce and secrete fatty acids. Starting by introducing an acyl-acyl carrier protein thioesterase gene, they made six successive generations of genetic modifications of *Synechocystis* sp. The fatty acid secretion yield increased to 197 mg/L of culture in one improved strain at a cell density of 1.0 x 10^9 cells/mL by adding codon-optimized thioesterase genes and weakening polar cell wall layers. Although these strains exhibited damaged cell membranes at low cell densities, they grew more rapidly at high cell densities in a late exponential and stationary phase and exhibited less cell damage than cells in wild-type cultures.

36.4 DISCUSSION

The location of these high-impact studies in algal lipid production required a carefully designed keyword set and a collection of databases covering a wide range of academic disciplines. It is apparent that the keyword set for lipid production (keyword set 1 in the Appendix) fairly collects the relevant papers in this area. The keyword set for the algae (keyword set 2 in the Appendix), which is a modified copy of the keyword sets provided in the relevant algal studies (Konur, 2020a–aq), also collects the relevant papers in this area. The final keyword set (keyword set 3 in the Appendix) helps in locating a core set of papers mainly related to lipid production for algal biodiesel production, omitting the papers on lipid production for food, medicine, fisheries, and animal husbandry (Conquer and Holub, 1996; Dawczynski et al., 2007; Patil et al., 2007; Volkman et al., 1989).

Similarly, the papers focusing on algal lipid extraction (Halim et al., 2011, 2012; Lee et al., 2010) and transesterification of algal lipids (Demirbas and Demirbas, 2011; Miao and Wu, 2006; Wahlen et al., 2011) are excluded via the relevant keywords from the set for lipid production (keyword set 1 in the Appendix). More importantly, the papers on the thermochemical conversion of algal biomass to algal biooils through pyrolysis and hydrothermal liquefaction are also excluded in this study (Barreiro et al., 2013; Biller and Ross, 2011; Brown et al., 2010; Miao et al., 2004).

Initially, all the databases contained in the 'Web of Science Core Collection' were searched using these keyword sets. However, it is apparent that the primary database is SCI-E, supported by ESCI and BCI-S.

The core set of papers on algal lipid production were collected using these databases and keyword sets. Hence, the papers presented in this chapter are all relevant to algal lipid production for biodiesel production, rather than for food, biomedicine, fisheries, and animal husbandry. It is notable that the review papers (Griffiths and Harrison, 2009; Guschina and Harwood, 2006; Hu et al., 2008; Mutanda et al., 2011; Roessler, 1990; Sharma et al., 2012) are not presented in this chapter since the focus is on the presentation of the primary findings from experimental and analytical studies. It is notable that these review papers play a critical role for key stakeholders to engage with the key research issues on algal lipid production (North, 1991a–b; Konur, 2000, 2002a–c, 2006a–b, 2007a–b).

It appears that there are five distinct primary research fronts for the research on algal lipid production. These are 'algal cultivation engineering in general', 'algal cultivation and wastewater bioremediation', 'microalgal cultivation and carbon dioxide bioremediation', 'algal lipid analysis and imaging', and 'omics and cell biology studies in algal lipid production' with 12, 3, 6, 1, and 3 papers, respectively. Furthermore, there are sections on lipid analysis in nine papers besides one on lipid imaging. Similarly, there are nine papers which are also related to 'cultivation engineering' besides other respective research fronts.

For the first research front, irradiance, N deprivation, P deprivation, temperature, mixothropic cultivation, iron repletion, Si deprivation, light–dark cycles, salt repletion, light intensity, glucose repletion, 'fed-batch cultivation', and 'two-phase cultivation' are the key parameters studied in these papers.

Nitrogen (N) deficiency is the primary tool to boost lipid content and productivity of microalgae in the cultivation engineering of algae (Rodolfi et al., 2009; Converti et al., 2009; Li et al., 2008; Shifrin and Chisholm, 1981; Ben-Amotz et al., 1985; Breuer et al., 2012; Pal et al., 2011; Li et al., 2010a; Ho et al., 2012; Widjaja et al., 2009; Siaut et al., 2011; Wang et al., 2009). The reported lipid productivity upon N depletion is 0.204 g/l/day for *Nannochloropsis* sp. (Rodolfi et al., 2009), and 0.322 and 0.243 g/l/day for *Scenedesmus obliquus* and *Chlorella zofingiensis*, respectively (Breuer et al., 2012). On the other hand, lipid content following N deprivation is over 60% for *Nannochloropsis* sp. (Rodolfi et al., 2009), 72% for *Monallantus salina* (Shifrin and Chisholm, 1981), and 55% for *Botryococcus braunii* (Ben-Amotz et al., 1985).

The findings of Takagi et al. (2006) on the significant impact of salt repletion for *Dunaliella* producing over 70% lipids are remarkable. The findings of Liu et al. (2008) on the significant impact of iron repletion for *Chlorella vulgaris*, producing up to 56.6% lipid, are similarly remarkable. Furthermore, Pal et al. (2011) obtained a maximum average productivity of 410 mg TFA/l/d obtained at 700 μmol photons/m².s and 40 g/l NaCl within seven days for *Nannochloropsis* sp.

The three studies on wastewater bioremediation confirm the lipid production and nutrient (mainly P and N) removal in wastewaters (Li et al., 2010a; Wang et al., 2010; Woertz et al., 2009). These papers have opened the way for a significant stream of research in this area.

The six studies on algal lipid production and CO_2 bioremediation confirm the positive impact of CO_2 repletion on algal lipid production (Chiu et al., 2009; Francisco et al., 2010; Ho et al., 2012; Tang et al., 2011; Widjaja et al., 2009; Yoo et al., 2010). Similarly, these papers have opened the way for a significant stream of research in this area.

The analysis and imaging of lipids during lipid production have a critical importance as diagnostic and analytical tools (Ben-Amotz et al., 1985; Breuer et al., 2012; Cheirsilp et al., 2012; Chen et al., 2009; Converti et al., 2009; Francisco et al., 2010; Ho et al., 2012; Liu et al., 2008; Reitan et al., 1994; Wang et al., 2009, 2010). The early studies published in the 1980s and 1990s are mostly concerned with the screening and selection of high-lipid algal feedstocks for biodiesel and high-value products for the food and medicinal uses.

Two of the papers on cell biology and omics studies in algal lipid production are concerned with the blocking of a starch biosynthesis, a competitor for the TAG biosynthesis in the *Chlamydomonas reinhardtii* cells (Siaut et al., 2011; Wang et al., 2009). Additionally Liu et al. (2011) enhanced lipid growth in *Synechocystis* sp. through genomic modification. In contrast to Siaut et al. (2011), Wang et al. (2009) reported a significant increase in the lipid body content in *Chlamydomonas reinhardtii* cells after blocking starch biosynthesis.

These studies and other studies on the cell biology and omics technologies in lipid production (Boyle et al., 2012; Kilian et al., 2011; Li et al., 2010b–c; Moellering and

Benning, 2010) provide background studies to understand the biology and omics of microalgal and other algal cells to enhance lipid productivity to make the overall biodiesel production and other related high-value products cost-effective.

These highly cited studies also shape the research papers published in their respective research streams in the 2010s. It is also notable that although there are a number of algae such as microalgae, cyanobacteria, macroalgae, diatoms, dinoflagellates, and coccolithophores (keyword set 2 in the Appendix), these studies mostly use microalgal species for enhancing the lipid content and lipid productivity.

In this context, the prolific microalgal species are *Nannochloropsis* sp. (Rodolfi et al., 2009; Converti et al., 2009; Chiu et al., 2009; Cheirsilp et al., 2012; Pal et al., 2011), *Chlorella* (*C. vulgaris, C. pyrenoidosa, C.* sp., and *C. Zofigensis*) (Converti et al., 2009; Liang et al., 2009; Liu et al., 2008; Yao et al., 2010; Tang et al., 2011; Wang et al., 2010; Cheirsilp et al., 2011; Widjaja et al., 2010; Breuer et al., 2012); *Scenedesmus* (*S.* sp. and *S. obliquus*) (Li et al., 2010a; Yoo et al., 2010; Ho et al., 2012; Tang et al., 2011; Breuer et al., 2012); *Chlamydomonas reinhardtii* (Siaut et al., 2011; Wang et al., 2009); *Neochloris oleoabundans* (Li et al., 2008; Breuer et al., 2012), *Botryococcus braunii* (Yoo et al., 2010; Ben-Amotz et al., 1985); and *Dunaliella* sp. (Takagi et al., 2006; Ben-Amotz et al., 1985).

It is also notable that these papers are published by 12 countries of which the USA, China, Taiwan, and Italy are prolific with seven, three, three, and two papers, respectively.

As around 85% of the total cost of biodiesel production emanates from the feedstock costs (Davis et al., 2011; Norsker et al., 2011; Williams and Laurens, 2010; Zhang et al., 2003), it is helpful to reduce the total cost of biodiesel production by improving lipid productivity and lipid content through engineering the cultivation processes. This would make algal biodiesel fuels more competitive with petrodiesel fuels (Davis et al., 2011; Norsker et al., 2011; Williams and Laurens, 2010; Zhang et al., 2003).

This would also help with the treatment of wastewaters which are used for the cultivation and lipid production of algae (Bigda, 1995; Chen, 2004; Rozendal et al., 2008; Speece, 1983). Another beneficial effect of using wastewaters for algal biomass and lipid accumulation as well as nutrient removal would be the lessening of the volume and harmful effects of algal blooms, which have been a significant ecological disaster in recent years (Anderson et al., 2002; Hallegraeff, 1993; Michalak et al., 2013; O'Neil et al., 2012; Paerl, 1997).

The production of algal biodiesel fuels in competition with petrodiesel fuels would also help in shifting biodiesel production from edible oil-based biodiesel fuels. There has been significant competition between biodiesel production from edible oils and their household consumption, resulting in significant public concerns about food security (Ajanovic, 2011; Godfray et al., 2010; Lam et al., 2009; Lobell et al., 2008; Naylor et al., 2007; Rosegrant and Cline, 2003; Schmidhuber and Tubiello, 2007; Tenenbaum, 2008).

This would also help in dealing with public concerns emanating from the fact that palm plantations expand at the expense of the forests as palm oil plantations have replaced large areas of forests in Southeast Asia and other places (Carlson et al., 2012, 2013; Fitzherbert et al., 2008; Koh and Wilcove, 2008; Koh et al., 2011).

This has resulted in public concerns about the destruction of forests (Fitzherbert et al., 2008; Koh and Wilcove, 2008; Koh et al., 2011), destruction of ecological biodiversity (Fitzherbert et al., 2008; Koh and Wilcove, 2008; Koh et al., 2011), significant increase in CO_2 emissions (Carlson et al., 2012, 2013), and finally, exploitation of local communities (McCarthy, 2010; Obidzinski et al., 2012, Rist et al., 2010).

The basis of these studies is related to the exploration and assessment of structure–processing–property relationships for algal lipid production (Konur and Matthews, 1989; Griffiths and Harrison, 2009; Guschina and Harwood, 2006; Hu et al., 2008; Mutanda et al., 2011; Roessler, 1990; Sharma et al., 2012).

36.5   CONCLUSION

This chapter has presented the key findings of the 25-most-cited article papers in five distinct research streams in this field.

These prolific studies provide valuable evidence on cultivation engineering in general, cultivation engineering in wastewaters and nutrient removal, cultivation engineering and CO_2 bioremediation, algal lipid analysis and imaging, and omics and cell biology studies in lipid production.

All these research streams contribute significantly towards improving lipid content and productivity in algae, mostly high-lipid microalgae. It is apparent that it is highly desirable to enhance both biomass productivity and lipid productivity such as through the use of a two-phase cultivation strategy where first biomass growth is ensured in a nutrient (N, P) sufficient mode and then followed by a nutrient-deficient phase.

It is then apparent that the use of wastewaters and flue gases are of critical importance to reduce production costs. It is also apparent that the studies on lipid analysis and imaging as well as on omics and cell biology complement primary cultivation processes by proving to be valuable tools for understanding and assessing the primary cultivation processes.

The ecological, technoeconomic, and environmental benefits using wastewaters for these purposes have been significant. The use of these wastewaters would make biodiesel production more competitive in relation to petrodiesel fuels by reducing the feedstock costs significantly.

This would also help with the treatment of wastewaters. Another beneficial effect of using wastewaters for algal biomass accumulation and nutrient removal would be the lessening of the volume and harmful effects of algal blooms, which have been a significant ecological disaster in recent years.

The production of algal biodiesel fuels in competition with petrodiesel fuels would also help in shifting biodiesel production from edible oil-based biodiesel fuels relating to public concerns about food security, the destruction of forests for the expansion of oil palm plantations, the destruction of ecological biodiversity, the significant increase in CO_2 emissions, and finally, the exploitation of local communities.

It is recommended that similar studies are carried out for each of the five major research streams. Scientometric studies on each of these streams and for the whole field of algal lipid production would complement these studies, as well as other studies on the scientometrics of algae.

36.A APPENDIX

The keyword set for the search on algal lipid production
Syntax: (I and II) NOT III

36.A.1 LIPID-RELATED KEYWORDS

TI=("bio-oil*" or biocrude* or "bio-crude*" or biooil* or lipid* or glycerolipid* or galactolipid* or "neutral lipid*" or triacylglycerol* or "fatty-acid*" or oil or triglyceride* or triacylglyceride* or lipolysis or "acetyl-CoA" or "malonyl-CoA" or lipogenesis) OR SO=("european journal of lipid*" or "journal of lipid*" or "progress in lipid research" or "biochimica et biophysica acta molecular and cell biology of lipid*" or "chemistry and physics of lipids" or lipids).

36.A.2 ALGAE-RELATED KEYWORDS

TI=(algae or algal or photobioreactor* or alga or "photo-bioreactor*" or "open pond*" or "raceway pond*" or photobioreaction or silicoflagellate* or dictyochale* or dictyocha or "photo-bioreactor*") OR TI=(dinoflagellat* or *coccolith* or dinophy* or alexandrium or emiliania or *gymnodini* or haptophyt* or prymnesi* or zooxanthella* or amphidin* or akashiwo or isochrysis or karenia* or phaeocystis or symbiodin* or chrysophyt* or chrysophyc* or raphidophy* or ochromonas or pfiesteria* or dinocyst* or noctiluca* or aureococcus* or *ceratium or *chattonella or cochlodinium or crypthecodinium or gyrodinium or hematodinium or heterocapsa* or heterosigma or karlodinium or lingulodinium or mallomonas or ostreopsis or oxyrrhis or pleurochrysis or pyrocystis or pyrodinium or scrippsiella or rhodomonas or vaucheria or xanthophyc* or fukuoya or alveolata or mesodinium or dinophysis* or gephyrocapsa or raphidiopsis or synurophycea* or glenodinium or protoperidinium or coolia or ptychodiscus) OR TI=(chlamydomon* or "green alga*" or *chlorella or microalga* or chlorophyt* or chlorophyc* or euglen* or "micro-alga*" or chrysophy* or dunaliella* or haematococcus or nannochloropsis or scenedesmus or cryptophy* or *porphyridium or volvoc* or acetabularia or botryococc* or chlorococc* or phormidium or prototheca or tetraselmis or volvox or prasinophy* or cryptomonad* or desmidia* or eustigmatophy* or selenastr* or streptophy* or trebouxiophy* or ankistrodesmus or aurantiochytr* or chroomonas or coccomyxa or cosmarium or cyanidioschyzon or cyanidium or desmodesmus or galdieria or klebsormid* or micrasterias or micromonas or monoraphid* or nannochloris or neochloris or ostreococcus or pediastrum or platymonas or polytomella or *kirchneriella or pyramimonas or schizochytrium or micractinium or ourococcus or tisochrysis or chlorogonium or volvoca* or odontella or spirogyra or pavlova or cryptomonas) OR TI=(macroalga* or rhodophy* or seaweed* or "red alga*" or "brown alga*" or gracilar* or phaeophy* or *porphyra* or ulva* or caulerp* or corallina* or fucus or gigartina* or laminaria* or saccharina or sargassum or nitell* or characea* or charophyt* or dictyota* or enteromorpha or fucale* or halocynthia* or zygnema* or ascophyllum or bangia* or chondrus or cladophor* or codium or cystoseira* or ecklonia or gelidium or kappaphycus or laurenci* or macrocystis or ectocarp* or ceramiale* or pyropia* or rhodomela* or spirogyra or undaria or "macro-alga*" or "sea-weed*" or bryops* or cryptonemia* or florideophy* or gelidiale* or gelidiella or griffithsia or halimeda* or lessonia* or rhodymeniale* or sargassac* or ulvophyc* or wakame or bangiophy* or "chara vulgaris" or asparagopsis or bifurcaria or bryopsis or ceramium or chaetomorpha or chondracanthus or chondria or cladosiphon or delesseria* or desmarestia* or dictyopteris or durvillaea or "eisenia bicyclis" or eucheuma* or grateloupia or hizikia or hypnea or ishige or lithophyllum or lobophora or lomentaria or monostroma or mougeotia or oedogonium or padina or palmaria* or pelvetia or plocamium or polysiphonia or rhodymenia* or scytosiphon* or solieria* or turbinaria or phyllophora* or charales or streptophyt* or ochrophyt* or halymenia* or bonnemaisonia* or charophyc* or fucacea* or ocrophyt* or furcellaria or pilayella or gracilaria or himanthalia or "alaria esculenta" or nereocystis or osmundea or "sea lettuce" or "irish moss" or derbesia or zonaria or sphaerococcus or colpomenia or cutleria or antithamnion or scinaia or mastocarpus or mazzaella or acanthophora or chordariales or scytosiphonales or ceramiacea* or dasyacea* or pterocladia or characea* or cystophora or lithothamnion or ptilophora or kallymenia or nemaliale* or mesotaenium or symphyocladia or bostrychia or odonthalia or leathesia or lophoclad* or stypopodium or hijiki or kombu or mozuku or delisea) OR TI=(diatoms or bacillarioph* or diatoma* or diatomite* or diatom or thalassio* or *nitzschia or phaeodactylum or chaetoceros* or navicula* or skeletonema or cyclotell* or stephanodisc* or achnanth* or asterionell* or aulacoseira or cocconeis or coscinodisc* or cylindrotheca or cymbella* or didymosphenia or ditylum or eunotia* or fragilaria* or gomphonema* or haslea* or melosira* or rhizosolenia* or stephanodiscus or synedra or coscinodiscus or licmophora* or pseudonitzschia or pleurosigma or achnanthes or frustulia or pinnularia or stephanodiscus or gomphonema) OR TI=(*cyanobact* or *synechoc* or *cylindrospermops* or *microcystis or "blue-green alga*" or *anabaen* or cyanophy* or *nostoc* or *oscillatoria* or spirul* or arthrospira or *lyngbya* or aphanizomenon or planktothrix or prochloro* or trichodesmium or calothrix* or chroococca* or acaryochloris or aphanothece or cyanophora or cyanothece or fischerella or fremyella or gloeobacter or mastigocladus or microcoleus or nodularia or plectonema or scytonem* or tolypothrix or dolichospermum or geitlerinema or prochlorothrix or prochloron or westiellopsis or stigonematale* or glaucophyt* or glaucocystis or symploca or hapalosiphon or "moorea producens") OR SO=("algal research*" or "european journal of phycology" or "harmful algae" or "journal of applied phycology" or "journal of phycology" or phycologia or "british phycological journal" or "diatom research" or "phycological research" or algae or "cryptogamie algologie" or fottea* or alga* or microalga* or cyanobacter* or macroalga* or diatoms or dinoflagellate* or coccolithophor* or phytoplankton or photobiroreactor* or phyco* or seaweed* or lipid*). It is a modified copy of the keyword list contained in Konur (2020a–q).

36.A.3 EXCLUDING KEYWORD SET

TI=(photosyst* or extraction or symbiogenesis or "phosporus lipid*" or "membrane-lipid*" or chilling or bleaching or supercapacitor* or crassostr* or zooplank* or alkenone* or cold or garlic or biomarker* or daphnia or peroxidation or alkane* or "essential oil*" or transester* or *deoxygenation or "fish oil" or liquefaction or pyrolysis or upgrading or hydrothermal or pecten* or diet* or "oil shale*" or hydro-cracking or pyrolytic or desatur* or materials or thermochemical or sediment* or membrane* or kerogen* or gasoline or bloom* or hens or antibacteria* or *toxic* or coral* or rotif* or protein* or hydrotreatment or "vegetable oil*" or extracted or "olive-oil*" or hydroprocess* or food* or arachidonic or eicosapentaenoic or doco-sahexaenoic or pufa* or aquaculture or "lipid recovery" or health* or "omega-3" or "polyunsaturated fatty acid*" or "unsaturated fatty acid*" or pufa* or "oil well*" or "palm oil*" or conversion or "oil sand" or "oil and gas" or "crop oil*" or "cooking oil*") or WC=(geo* or ocean* or food* or fish* or nutr* or ecol* or pharm* or "chemistry med*" or toxic* or limnol* or veter* or med*).

ACKNOWLEDGMENTS

The contribution of the highly cited researchers in this field is greatly acknowledged.

REFERENCES

Ahmadun, F. R., A. Pendashteh, and L. C. Abdullah, et al. 2009. Review of technologies for oil and gas produced water treatment. *Journal of Hazardous Materials* 170:530–551.

Ajanovic, A. 2011. Biofuels versus food production: Does biofuels production increase food prices? *Energy* 36: 2070–2076.

Anderson, D. M. P. M. Glibert, and J. M. Burkholder. 2002. Harmful algal blooms and eutro-phication: Nutrient sources, composition, and consequences. *Estuaries* 25: 704–726.

Atlas, R. M. 1981. Microbial degradation of petroleum hydrocarbons: An environmental per-spective. *Microbiological Reviews* 45:180–209.

Babich, I. V. and J. A. Moulijn. 2003. Science and technology of novel processes for deep desulfurization of oil refinery streams: A review. *Fuel* 82:607–631.

Barreiro, DL, W. Prins, F. Ronsse, and W. Brilman. 2013. Hydrothermal liquefaction (HTL) of microalgae for biofuel production: State of the art review and future prospects. *Biomass & Bioenergy* 53: 113–127.

Ben-Amotz, A., T. G. Tornabene, W. H. Thomas. 1985. Chemical profile of selected species of microalgae with emphasis on lipids. *Journal of Phycology* 21: 72–81.

Bigda, R. J. 1995. Consider Fenton's chemistry for wastewater treatment. *Chemical Engineering Progress* 91: 62–66.

Biller, P. and A. B. Ross. 2011. Potential yields and properties of oil from the hydrothermal liquefaction of microalgae with different biochemical content. *Bioresource Technology* 102: 215–225.

Boyle, N. R., M. D. Page, and B. S. Liu, et al. 2012. Three acyltransferases and nitrogen-responsive regulator are implicated in nitrogen starvation-induced triacylglycerol accu-mulation in *Chlamydomonas. Journal of Biological Chemistry* 287: 15811–15825.

Breuer, G., P. P. Lamers, D. E. Martens, R. B. Draaisma, and R. H. Wijffels. 2012. The impact of nitrogen starvation on the dynamics of triacylglycerol accumulation in nine microal-gae strains. *Bioresource Technology* 124: 217–226.

Bridgwater, A. V., D. Meier, and D. Radlein. 1999. An overview of fast pyrolysis of biomass. *Organic Geochemistry* 30: 1479–1493.

Bridgwater, A. V. and G. V. C. Peacocke. 2000. Fast pyrolysis processes for biomass. *Renewable & Sustainable Energy Reviews* 4: 1–73.

Brown, T. M., P. G. Duan, and P. E. Savage. 2010. Hydrothermal liquefaction and gasification of *Nannochloropsis* sp. *Energy & Fuels* 24: 639–646.

Carlson, K. M., L. M. Curran, and G. P. Asner, et al. 2013. Carbon emissions from forest conversion by Kalimantan oil palm plantations. *Nature Climate Change* 3: 283–287.

Carlson, K. M., L. M. Curran, and D. Ratnasari, et al. 2012. Committed carbon emissions, deforestation, and community land conversion from oil palm plantation expansion in West Kalimantan, Indonesia. *Proceedings of the National Academy of Sciences of the United States of America* 109: 7559–7564.

Cheirsilp, B. and S. Torpee. 2012. Enhanced growth and lipid production of microalgae under mixotrophic culture condition: Effect of light intensity, glucose concentration and fed-batch cultivation. *Bioresource Technology* 110:510–516.

Chen, G. 2004. Electrochemical technologies in wastewater treatment. *Separation and Purification Technology* 38(1), 11–41.

Chen, W., C. W. Zhang, L. R. Song, M. Sommerfeld, and Q. Hu. 2009. A high throughput Nile red method for quantitative measurement of neutral lipids in microalgae. *Journal of Microbiological Methods* 77: 41–47.

Chisti, Y. 2007. Biodiesel from microalgae. *Biotechnology Advances* 25: 294–306.

Chiu, S. Y., C. Y. Kao, and M. T. Tsai, et al. 2009. Lipid accumulation and CO_2 utilization of *Nannochloropsis oculata* in response to CO_2 aeration. *Bioresource Technology* 100: 833–838.

Conquer, J. A. and B. J. Holub. 1996. Supplementation with an algae source of docosahexaenoic acid increases (3-3) fatty acid status and alters selected risk factors for heart disease in vegetarian subjects. *Journal of Nutrition* 126: 3032–3039.

Converti, A., A. A. Casazza, E. Y. Ortiz, P. Perego, and M. del Borghi. 2009. Effect of temperature and nitrogen concentration on the growth and lipid content of *Nannochloropsis oculata* and *Chlorella vulgaris* for biodiesel production. *Chemical Engineering and Processing-Process Intensification* 48:1146–1151.

Czernik, S. and A. V. Bridgwater. 2004. Overview of applications of biomass fast pyrolysis oil. *Energy & Fuels* 18: 590–598.

Davis, R., A. Aden, and P. T. Pienkos. 2011. Techno-economic analysis of autotrophic microalgae for fuel production. *Applied Energy* 88:3524–3531.

Dawczynski, C., R. Schubert, and G. Jahreis. 2007. Amino acids, fatty acids, and dietary fibre in edible seaweed products. *Food Chemistry* 103: 891–899.

Demirbas, A. and M. F. Demirbas. 2011. Importance of algae oil as a source of biodiesel. *Energy Conversion and Management* 52:163–170.

Fitzherbert, E. B., M. J. Struebig, and A. Morel, et al. 2008. How will oil palm expansion affect biodiversity? *Trends n Ecology & Evolution* 23: 538–545.

Francisco, E. C., D. B. Neves, E. Jacob-Lopes, and T. T. Franco. 2010. Microalgae as feedstock for biodiesel production: Carbon dioxide sequestration, lipid production and biofuel quality. *Journal of Chemical Technology and Biotechnology* 85: 395–403.

Godfray, H. C. J., J. R. Beddington, and I. R. Crute, et al., 2010. Food security: The challenge of feeding 9 billion people. *Science* 327: 812–818.

Griffiths, M. J. and S. T. L. Harrison. 2009. Lipid productivity as a key characteristic for choosing algal species for biodiesel production. *Journal of Applied Phycology* 21: 493–507.

Guschina, I. A. and J. L. Harwood. 2006. Lipids and lipid metabolism in eukaryotic algae. *Progress in Lipid Research* 45:160–186.

Halim, R., M. K. Danquah, and P. A. Webley. 2012. Extraction of oil from microalgae for biodiesel production: A review. *Biotechnology Advances* 30: 709–732.

Halim, R., B. Gladman, M. K. Danquah, and P. A. Webley. 2011. Oil extraction from microalgae for biodiesel production. *Bioresource Technology* 102:178–185.

Hallegraeff, G. M. 1993. A review of harmful algal blooms and their apparent global increase. *Phycologia* 32: 79–99.

Hill, J., E. Nelson, D. Tilman, S. Polasky, and D. Tiffany. 2006. Environmental, economic, and energetic costs and benefits of biodiesel and ethanol biofuels. *Proceedings of the National Academy of Sciences of the United States of America* 103: 11206–11210.

Ho, S. H., C. Y. Chen, and J. S. Chang. 2012. Effect of light intensity and nitrogen starvation on CO_2 fixation and lipid/carbohydrate production of an indigenous microalga *Scenedesmus obliquus* CNW-N. *Bioresource Technology* 113: 244–252.

Hu, Q., M. Sommerfeld, and E. Jarvis, et al. 2008. Microalgal triacylglycerols as feedstocks for biofuel production: Perspectives and advances. *Plant Journal* 54: 621–639.

Kilian, L. 2009. Not all oil price shocks are alike: Disentangling demand and supply shocks in the crude oil market. *American Economic Review* 99: 1053–1069.

Kilian, O., C. S. E. Benemann, K. K. Niyogi, and B. Vick. 2011. High-efficiency homologous recombination in the oil-producing alga *Nannochloropsis* sp. *Proceedings of the National Academy of Sciences of the United States of America* 108:21265–21269.

Koh, L. P., J. Miettinen, S. C. Liew, and J. Ghazoul. 2011. Remotely sensed evidence of tropical peatland conversion to oil palm. *Proceedings of the National Academy of Sciences of the United States of America* 108: 5127–5132.

Koh, L. P. and D. S. Wilcove. 2008. Is oil palm agriculture really destroying tropical biodiversity? *Conservation Letters* 1: 60–64.

Konur, O. 2000. Creating enforceable civil rights for disabled students in higher education: An institutional theory perspective. *Disability & Society* 15:1041–1063.

Konur, O. 2002a. Access to Nursing Education by disabled students: Rights and duties of nursing programs. *Nurse Education Today* 22: 364–374.

Konur, O. 2002b. Assessment of disabled students in higher education: Current public policy issues. *Assessment and Evaluation in Higher Education* 27: 131–152.

Konur, O. 2002c. Access to employment by disabled people in the UK: Is the Disability Discrimination Act working? *International Journal of Discrimination and the Law* 5: 247–279.

Konur, O. 2006a. Participation of children with dyslexia in compulsory education: Current public policy issues. *Dyslexia* 12: 51–67.

Konur, O. 2006b. Teaching disabled students in Higher Education. *Teaching in Higher Education* 11: 351–363.

Konur, O. 2007a. A judicial outcome analysis of the Disability Discrimination Act: A windfall for the employers? *Disability & Society* 22:187–204.

Konur, O. 2007b. Computer-assisted teaching and assessment of disabled students in higher education: The interface between academic standards and disability rights. *Journal of Computer Assisted Learning* 23: 207–219.

Konur, O., 2020a. The scientometric analysis of the research on the algal science, technology, and medicine. In *Handbook of Algal Science, Technology and Medicine*, ed. O. Konur, 3–18. London: Academic Press.

Konur, O., 2020b. 100 citation classics in the algal science, technology, and medicine: a scientometric analysis. In *Handbook of Algal Science, Technology and Medicine*, ed. O. Konur, 19–38. London: Academic Press.

Konur, O., 2020c. The scientometric analysis of the research on the algal structures. In In *Handbook of Algal Science, Technology and Medicine*, ed. O. Konur, 41–60. London: Academic Press.

Konur, O., 2020d. The scientometric analysis of the research on the algal genomics. In *Handbook of Algal Science, Technology and Medicine*, ed. O. Konur, 105–125. London: Academic Press.

Konur, O., 2020e. The scientometric analysis of the research on the algal photosystems and phtosythesis. In *Handbook of Algal Science, Technology and Medicine*, ed. O. Konur, 195–215. London: Academic Press.

Konur, O., 2020f. The pioneering research on the cyanobacterial photosystems and photosynthesis. In *Handbook of Algal Science, Technology and Medicine*, ed. O. Konur, 231–343. London: Academic Press.

Konur, O., 2020g. The scientometric analysis of the research on the algal ecology. In *Handbook of Algal Science, Technology and Medicine*, ed. O. Konur, 257–276. London: Academic Press.

Konur, O., 2020h. The scientometric analysis of the research on the algal bioenergy and biofuels. In *Handbook of Algal Science, Technology and Medicine*, ed. O. Konur, 319–341. London: Academic Press.

Konur, O., 2020i. The scientometric analysis of the research on the bioethanol production from green macroalgae. In *Handbook of Algal Science, Technology and Medicine*, ed. O. Konur, 385–401. London: Academic Press.

Konur, O., 2020j. The scientometric analysis of the research on the algal biomedicine. In *Handbook of Algal Science, Technology and Medicine*, ed. O. Konur, 405–427. London: Academic Press.

Konur, O., 2020k. The pioneering research on the wound care by alginates. In *Handbook of Algal Science, Technology and Medicine*, ed. O. Konur, 467–481. London: Academic Press.

Konur, O., 2020l. The scientometric analysis of the research on the algal foods. In *Handbook of Algal Science, Technology and Medicine*, ed. O. Konur, 485–506. London: Academic Press.

Konur, O., 2020m. The scientometric analysis of the research on the algal toxicology. In *Handbook of Algal Science, Technology and Medicine*, ed. O. Konur, 521–541. London: Academic Press.

Konur, O., 2020n. The scientometric analysis of the research on the algal bioremediation. In *Handbook of Algal Science, Technology and Medicine*, ed. O. Konur, 607–627. London: Academic Press.

Konur, O., 2020o. Algal drugs: The state of the research. In *Encyclopedia of Marine Biotechnology*. Vol. 1, ed. S.K. Kim. Oxford: Wiley-Blackwell.

Konur, O., 2020p. Algal genomics. In *Encyclopedia of Marine Biotechnology*, Vol. 3, ed. S.K. Kim. Oxford: Wiley-Blackwell.

Konur, O., ed., 2020q. *Handbook of Algal Science, Technology and Medicine*. London: Academic Press.

Konur, O., ed. 2021a. *Handbook of Biodiesel and Petrodiesel Fuels: Science, Technology, Health, and Environment*. Boca Raton, FL: CRC Press.

Konur, O., ed. 2021b. *Handbook of Biodiesel and Petrodiesel Fuels: Science, Technology, Health, and Environment. Volume 1. Biodiesel Fuels: Science, Technology, Health, and Environment*. Boca Raton, FL: CRC Press.

Konur, O., ed. 2021c. *Handbook of Biodiesel and Petrodiesel Fuels: Science, Technology, Health, and Environment. Volume 2. Biodiesel Fuels based on the Edible and Nonedible Feedstocks, Wastes, and Algae: Science, Technology, Health, and Environment*. Boca Raton, FL: CRC Press.

Konur, O., ed. 2021d. *Handbook of Biodiesel and Petrodiesel Fuels: Science, Technology, Health, and Environment. Volume 3. Petrodiesel Fuels: Science, Technology, Health, and Environment*. Boca Raton, FL: CRC Press.

Konur, O. 2021e. Biodiesel and petrodiesel fuels: Science, technology, health, and environment. In *Handbook of Biodiesel and Petrodiesel Fuels: Science, Technology, Health,* and *Environment. Volume 1. Biodiesel Fuels: Science, Technology, Health, and Environment*, ed. O. Konur. Boca Raton, FL: CRC Press.

Konur, O. 2021f. Biodiesel and petrodiesel fuels: A scientometric review of the research. In *Handbook of Biodiesel and Petrodiesel Fuels: Science, Technology, Health, and Environment. Volume 1. Biodiesel Fuels: Science, Technology, Health, and Environment*, ed. O. Konur. Boca Raton, FL: CRC Press.

Konur, O. 2021g. Biodiesel and petrodiesel fuels: A review of the research. In *Handbook of Biodiesel and Petrodiesel Fuels: Science, Technology, Health, and Environment. Volume 1. Biodiesel Fuels: Science, Technology, Health, and Environment*, ed. O. Konur. Boca Raton, FL: CRC Press.

Konur, O. 2021h Nanotechnology applications in the diesel fuels and the related research fields: A review of the research. In *Handbook of Biodiesel and Petrodiesel Fuels: Science, Technology, Health, and Environment. Volume 1. Biodiesel Fuels: Science, Technology, Health, and Environment*, ed. O. Konur. Boca Raton, FL: CRC Press.

Konur, O. 2021i. Biooils: A scientometric review of the research. In *Handbook of Biodiesel and Petrodiesel Fuels: Science, Technology, Health, and Environment. Volume 1. Biodiesel Fuels: Science, Technology, Health, and Environment*, ed. O. Konur. Boca Raton, FL: CRC Press.

Konur, O. 2021j. Characterization and properties of biooils: A review of the research. In *Handbook of Biodiesel and Petrodiesel Fuels: Science, Technology, Health, and Environment. Volume 1. Biodiesel Fuels: Science, Technology, Health, and Environment*, ed. O. Konur. Boca Raton, FL: CRC Press.

Konur, O. 2021k. Biomass pyrolysis and pyrolysis oils: A review of the research. In *Handbook of Biodiesel and Petrodiesel Fuels: Science, Technology, Health, and Environment. Volume 1. Biodiesel Fuels: Science, Technology, Health, and Environment*, ed. O. Konur. Boca Raton, FL: CRC Press.

Konur, O. 2021l. Biodiesel fuels: A scientometric review of the research. In *Handbook of Biodiesel and Petrodiesel Fuels: Science, Technology, Health, and Environment. Volume 1. Biodiesel Fuels: Science, Technology, Health, and Environment*, ed. O. Konur. Boca Raton, FL: CRC Press.

Konur, O. 2021m. Glycerol: A scientometric review of the research. In *Handbook of Biodiesel and Petrodiesel Fuels: Science, Technology, Health, and Environment. Volume 1. Biodiesel Fuels: Science, Technology, Health, and Environment*, ed. O. Konur. Boca Raton, FL: CRC Press.

Konur, O. 2021n. Propanediol production from glycerol: A review of the research. In *Handbook of Biodiesel and Petrodiesel Fuels: Science, Technology, Health, and Environment. Volume 1. Biodiesel Fuels: Science, Technology, Health, and Environment*, ed. O. Konur. Boca Raton, FL: CRC Press.

Konur, O. 2021o. Edible oil-based biodiesel fuels: A scientometric review of the research. In *Handbook of Biodiesel and Petrodiesel Fuels: Science, Technology, Health, and Environment. Volume 2. Biodiesel Fuels based on the Edible and Nonedible Feedstocks, Wastes, and Algae: Science, Technology, Health, and Environment*, ed. O. Konur. Boca Raton, FL: CRC Press.

Konur, O. 2021p. Palm oil-based biodiesel fuels: A review of the research. In *Handbook of Biodiesel and Petrodiesel Fuels: Science, Technology, Health, and Environment. Volume 2. Biodiesel Fuels based on the Edible and Nonedible Feedstocks, Wastes, and Algae*, ed. O. Konur. Boca Raton, FL: CRC Press.

Konur, O. 2021q. Rapeseed oil-based biodiesel fuels: A review of the research. In *Handbook of Biodiesel and Petrodiesel Fuels: Science, Technology, Health, and Environment. Volume 2. Biodiesel Fuels based on the Edible and Nonedible Feedstocks, Wastes, and Algae*, ed. O. Konur. Boca Raton, FL: CRC Press.

Konur, O. 2021r. Nonedible oil-based biodiesel fuels: A scientometric review of the research. In *Handbook of Biodiesel and Petrodiesel Fuels: Science, Technology, Health, and*

Environment. Volume 2. Biodiesel Fuels based on the Edible and Nonedible Feedstocks, Wastes, and Algae: Science, Technology, Health, and Environment, ed. O. Konur. Boca Raton, FL: CRC Press.

Konur, O. 2021s. Waste oil-based biodiesel fuels: A scientometric review of the research. In *Handbook of Biodiesel and Petrodiesel Fuels: Science, Technology, Health, and Environment. Volume 2. Biodiesel Fuels based on the Edible and Nonedible Feedstocks, Wastes, and Algae: Science, Technology, Health, and Environment*, ed. O. Konur. Boca Raton, FL: CRC Press.

Konur, O. 2021t. Algal biodiesel fuels: A scientometric review of the research. In *Handbook of Biodiesel and Petrodiesel Fuels: Science, Technology, Health, and Environment. Volume 2. Biodiesel Fuels based on the Edible and Nonedible Feedstocks, Wastes, and Algae: Science, Technology, Health, and Environment*, ed. O. Konur. Boca Raton, FL: CRC Press.

Konur, O. 2021u. Algal biomass production for biodiesel production: A review of the research. In *Handbook of Biodiesel and Petrodiesel Fuels: Science, Technology, Health, and Environment. Volume 2. Biodiesel Fuels based on the Edible and Nonedible Feedstocks, Wastes, and Algae*, Ed. O. Konur. Boca Raton, FL: CRC Press.

Konur, O. 2021v. Algal biomass production in wastewaters for biodiesel production: A review of the research. In *Handbook of Biodiesel and Petrodiesel Fuels: Science, Technology, Health, and Environment. Volume 2. Biodiesel Fuels based on the Edible and Nonedible Feedstocks, Wastes, and Algae*, ed. O. Konur. Boca Raton, FL: CRC Press.

Konur, O. 2021x. Algal lipid production for biodiesel production: A review of the research. In *Handbook of Biodiesel and Petrodiesel Fuels: Science, Technology, Health, and Environment. Volume 2. Biodiesel Fuels based on the Edible and Nonedible Feedstocks, Wastes, and Algae*, ed. O. Konur. Boca Raton, FL: CRC Press.

Konur, O. 2021y. Crude oils: A scientometric review of the research. In *Handbook of Biodiesel and Petrodiesel Fuels: Science, Technology, Health, and Environment. Volume 3. Petrodiesel Fuels: Science, Technology, Health, and Environment*, ed. O. Konur. Boca Raton, FL: CRC Press.

Konur, O. 2021z. Petrodiesel fuels: A scientometric review of the research. In *Handbook of Biodiesel and Petrodiesel Fuels: Science, Technology, Health, and Environment. Volume 3. Petrodiesel Fuels: Science, Technology, Health, and Environment*, ed. O. Konur. Boca Raton, FL: CRC Press.

Konur, O. 2021aa. Bioremediation of petroleum hydrocarbons in the contaminated soils: A review of the research. In *Handbook of Biodiesel and Petrodiesel Fuels: Science, Technology, Health, and Environment. Volume 3. Petrodiesel Fuels: Science, Technology, Health, and Environment*, ed. O. Konur. Boca Raton, FL: CRC Press.

Konur, O. 2021ab. Desulfurization of diesel fuels: A review of the research. In *Handbook of Biodiesel and Petrodiesel Fuels: Science, Technology, Health, and Environment. Volume 3. Petrodiesel Fuels: Science, Technology, Health, and Environment*, ed. O. Konur. Boca Raton, FL: CRC Press.

Konur, O. 2021ac. Diesel fuel exhaust emissions: A scientometric review of the research. In *Handbook of Biodiesel and Petrodiesel Fuels: Science, Technology, Health, and Environment. Volume 3. Petrodiesel Fuels: Science, Technology, Health, and Environment*, ed. O. Konur. Boca Raton, FL: CRC Press.

Konur, O. 2021ad. The adverse health and safety impact of diesel fuels: A scientometric review of the research. In *Handbook of Biodiesel and Petrodiesel Fuels: Science, Technology, Health, and Environment. Volume 3. Petrodiesel Fuels: Science, Technology, Health, and Environment*, ed. O. Konur. Boca Raton, FL: CRC Press.

Konur, O. 2021ae. Respiratory illnesses caused by the diesel fuel exhaust emissions: A review of the research. In *Handbook of Biodiesel and Petrodiesel Fuels: Science, Technology, Health, and Environment. Volume 3. Petrodiesel Fuels: Science, Technology, Health, and Environment*, ed. O. Konur. Boca Raton, FL: CRC Press.

Konur, O. 2021af. Cancer caused by the diesel fuel exhaust emissions: A review of the research. In *Handbook of Biodiesel and Petrodiesel Fuels: Science, Technology, Health, and Environment. Volume 3. Petrodiesel Fuels: Science, Technology, Health, and Environment*, ed. O. Konur. Boca Raton, FL: CRC Press.

Konur, O. 2021ag. Cardiovascular and other illnesses caused by the diesel fuel exhaust emissions: A review of the research. In *Handbook of Biodiesel and Petrodiesel Fuels: Science, Technology, Health, and Environment. Volume 3. Petrodiesel Fuels: Science, Technology, Health, and Environment*, ed. O. Konur. Boca Raton, FL: CRC Press.

Konur, O. and F. L. Matthews. 1989. Effect of the properties of the constituents on the fatigue performance of composites: A review. *Composites* 20: 317–328.

Lam, M. K., K. T. Tan, K. T. Lee, and A. R. Mohamed. 2009. Malaysian palm oil: Surviving the food versus fuel dispute for a sustainable future. *Renewable and Sustainable Energy Reviews* 13: 1456–1464.

Lee, J. Y., C. Yoo, S. Y. Jun, C. Y. Ahn, and H. M. Oh. 2010. Comparison of several methods for effective lipid extraction from microalgae. *Bioresource Technology* 101: S75–S77.

Li, X., H. Y. Hu, K. Gan, and Y. X. Sun. 2010a. Effects of different nitrogen and phosphorus concentrations on the growth, nutrient uptake, and lipid accumulation of a freshwater microalga *Scenedesmus* sp. *Bioresource Technology* 101: 5494–5500.

Li, Y. Q., M. Horsman, B. Wang, N. Wu, and C. Q. Lan. 2008. Effects of nitrogen sources on cell growth and lipid accumulation of green alga *Neochloris oleoabundans*. *Applied Microbiology and Biotechnology* 81: 629–636.

Li, Y. T., D. X. Han, and G. R. Hu, et al. 2010b. *Chlamydomonas* starchless mutant defective in ADP-glucose pyrophosphorylase hyper-accumulates triacylglycerol. *Metabolic Engineering* 12: 387–391.

Li, Y. T., D. X. Han, G. R. Hu, M. Sommerfeld, and Q. A. Hu. 2010c. Inhibition of starch synthesis results in overproduction of lipids in *Chlamydomonas reinhardtii*. *Biotechnology and Bioengineering* 107: 258–268.

Liang, Y. N., N. Sarkany, and Y. Cui. 2009. Biomass and lipid productivities of *Chlorella vulgaris* under autotrophic, heterotrophic and mixotrophic growth conditions. *Biotechnology Letters* 31: 1043–1049.

Liu, X. Y., J. Sheng, and R. Curtiss. 2011. Fatty acid production in genetically modified cyanobacteria. *Proceedings of the National Academy of Sciences of the United States of America* 108: 6899–6904.

Liu, Z. Y., G. C. Wang, and B. C. Zhou. 2008. Effect of iron on growth and lipid accumulation in *Chlorella vulgaris*. *Bioresource Technology* 99: 4717–4722.

Lobell, D. B., M. B. Burke, and C. Tebaldi, et al. 2008. Prioritizing climate change adaptation needs for food security in 2030. *Science* 319: 607–610.

Mata, T. M., A. A. Martins, and N. S. Caetano. 2010. Microalgae for biodiesel production and other applications: A review. *Renewable & Sustainable Energy Reviews* 14: 217–232.

McCarthy, J. F. 2010. Processes of inclusion and adverse incorporation: Oil palm and agrarian change in Sumatra, Indonesia. *Journal of Peasant Studies* 37: 821–850.

Miao, X. L. and Q. Y. Wu 2006. Biodiesel production from heterotrophic microalgal oil. *Bioresource Technology* 97: 841–846.

Miao, X. L., Q. Y. Wu, and C. Y. Yang. 2004. Fast pyrolysis of microalgae to produce renewable fuels. *Journal of Analytical and Applied Pyrolysis* 71: 855–863.

Michalak, A. M., E. J. Anderson, and D. Beletsky, et al. 2013. Record-setting algal bloom in Lake Erie caused by agricultural and meteorological trends consistent with expected future conditions. *Proceedings of the National Academy of Sciences* 110: 6448–6452.

Moellering, E. R. and C. Benning. 2010. RNA interference silencing of a major lipid droplet protein affects lipid droplet size in *Chlamydomonas reinhardtii*. *Eukaryotic Cell* 9: 97–106.

Mohan, D., C. U. Pittman Jr, and P. H. Steele. 2006. Pyrolysis of wood/biomass for bio-oil: A critical review. *Energy & Fuels* 20: 848–889.

Mutanda, T., D. Ramesh, and S. Karthikeyan, et al. 2011. Bioprospecting for hyper-lipid pro-
    ducing microalgal strains for sustainable biofuel production. *Bioresource Technology*
    102: 57–70.
Naylor, R. L., A. J. Liska, M. B. Burke, et al. 2007. The ripple effect: Biofuels, food security,
    and the environment. *Environment: Science and Policy for Sustainable Development* 49:
    30–43.
Norsker, N. H., M. J. Barbosa, M. H. Vermue, and R. H. Wijffels. 2011. Microalgal produc-
    tion: A close look at the economics. *Biotechnology Advances* 29: 24–27.
North, D. C. 1991a. *Institutions, Institutional Change and Economic Performance*. Cambridge,
    MA: Cambridge University Press.
North, D.C. 1991b. Institutions. *Journal of Economic Perspectives* 5: 97–112.
Obidzinski, K., R. Andriani, H. Komarudin, and A. Andrianto. 2012. Environmental and social
    impacts of oil palm plantations and their implications for biofuel production in Indonesia.
    Ecology and Society 17: 25.
O'Neil, J. M., T. W. Davis, M. A. Burford, and C. J. Gobler. 2012. The rise of harmful cyano-
    bacteria blooms: The potential roles of eutrophication and climate change. *Harmful
    Algae* 14: 313–334.
Paerl, H. W. 1997. Coastal eutrophication and harmful algal blooms: Importance of atmo-
    spheric deposition and groundwater as "new" nitrogen and other nutrient sources.
    Limnology and Oceanography 42:1154–1165.
Pal, D., I. Khozin-Goldberg, Z. Cohen, and S. Boussiba. 2011. The effect of light, salinity, and
    nitrogen availability on lipid production by *Nannochloropsis* sp. *Applied Microbiology
    and Biotechnology* 90: 1429–1441.
Patil, V., T. Kallqvist, E. Olsen, G. Vogt, and H. R. Gislerod. 2007. Fatty acid composition of
    12 microalgae for possible use in aquaculture feed. *Aquaculture International* 15:1–9.
Perron, P. 1989. The great crash, the oil price shock, and the unit root hypothesis. *Econometrica:
    Journal of the Econometric Society* 57: 1361–1401.
Reitan, K. I., J. R. Rainuzzo, and Y. Olsen. 1994. Effect of nutrient limitation on fatty-acid and
    lipid-content of marine microalgae. *Journal of Phycology* 30: 972–979.
Rist, L., L. Feintrenie, and P. Levang. 2010. The livelihood impacts of oil palm: Smallholders
    in Indonesia. *Biodiversity and Conservation* 19: 1009–1024.
Rodolfi, L., G. C. Zittelli, and N. Bassi, et al. 2009. Microalgae for oil: Strain selection, induc-
    tion of lipid synthesis and outdoor mass cultivation in a low-cost photobioreactor.
    Biotechnology and Bioengineering 102:100–112.
Roessler, P. G. 1990. Environmental control of glycerolipid metabolism in microalgae -
    Commercial implications and future research directions. *Journal of Phycology* 26:
    393–399.
Rosegrant, M. W. and S. A. Cline. 2003. Global food security: Challenges and policies. *Science*
    302:1917–1919.
Rozendal, R. A., H. V. Hamelers, K. Rabaey, J. Keller, C. J. Buisman. 2008. Towards practical
    implementation of bioelectrochemical wastewater treatment. *Trends in Biotechnology*
    26: 450–459.
Schenk, P. M., S. R. Thomas-Hall, and E. Stephens, et al. 2008. Second generation biofuels:
    High-efficiency microalgae for biodiesel production. *Bioenergy Research* 1: 20–43.
Schmidhuber, J. and F. N. Tubiello. 2007. Global food security under climate change.
    Proceedings of the National Academy of Sciences 104: 19703–19708.
Sharma, K. K., H. Schuhmann, and P. M. Schenk. 2012. High lipid induction in microalgae for
    biodiesel production. *Energies* 5:1532–1553.
Shifrin, N. S. and S. W. Chisholm. 1981. Phytoplankton lipids - Interspecific differences and
    effects of nitrate, silicate and light-dark cycles. *Journal of Phycology* 17: 374–384.
Siaut, M., S. Cuine, and C. Cagnon, et al. 2011. Oil accumulation in the model green alga
    Chlamydomonas reinhardtii: Characterization, variability between common laboratory
    strains and relationship with starch reserves. *BMC Biotechnology* 11: 7.

Speece, R. E. 1983. Anaerobic biotechnology for industrial wastewater treatment. *Environmental Science & Technology* 17: 416A–427A.

Takagi, M., Karseno, and T. Yoshida. 2006. Effect of salt concentration on intracellular accumulation of lipids and triacylglyceride in marine microalgae *Dunaliella* cells. *Journal of Bioscience and Bioengineering* 101: 223–226.

Tang, D. H., W. Han, P.L. Li, X. L. Miao, and J. J. Zhong. 2011. CO_2 biofixation and fatty acid composition of *Scenedesmus obliquus* and *Chlorella pyrenoidosa* in response to different CO_2 levels. *Bioresource Technology* 102: 3071–3076.

Tenenbaum, D. J. 2008. Food vs. fuel: Diversion of crops could cause more hunger. *Environmental Health Perspectives* 116: A254-A2A7.

Volkman, J. K., S. W. Jeffrey, P. D. Nichols, G. I. Rogers, and C. D. Garland. 1989. Fatty-acid and lipid-composition of 10 species of microalgae used in mariculture. *Journal of Experimental Marine Biology and Ecology* 128 219–240.

Wahlen, B. D., R. M. Willis, and L. C. Seefeldt. 2011. Biodiesel production by simultaneous extraction and conversion of total lipids from microalgae, cyanobacteria, and wild mixed-cultures. *Bioresource Technology* 102: 2724–2730.

Wang, L., Y. C. Li, and P. Chen, et al. 2010. Anaerobic digested dairy manure as a nutrient supplement for cultivation of oil-rich green microalgae *Chlorella* sp. *Bioresource Technology* 101: 2623–2628.

Wang, Z. T., N. Ullrich, S. Joo, S. Waffenschmidt, and U. Goodenough. 2009. Algal lipid bodies: Stress induction, purification, and biochemical characterization in wild-type and starchless *Chlamydomonas reinhardtii*. *Eukaryotic Cell* 8:1856–1868.

Widjaja, A., C. C. Chien, and Y. H. Ju. 2009. Study of increasing lipid production from fresh water microalgae *Chlorella vulgaris*. *Journal of the Taiwan Institute of Chemical Engineers* 40: 13–20.

Williams, P. J. L. and L. M. L. Laurens. 2010. Microalgae as biodiesel & biomass feedstocks: Review & analysis of the biochemistry, energetics & economics. *Energy & Environmental Science* 3: 554–590.

Woertz, I., A. Feffer, T. Lundquist, and Y. Nelson. 2009. Algae grown on dairy and municipal wastewater for simultaneous nutrient removal and lipid production for biofuel feedstock. *Journal of Environmental Engineering-ASCE* 135:1115–1122.

Yoo, C., S. Y. Jun, J. Y. Lee, C. Y. Ahn, and H. M. Oh. 2010. Selection of microalgae for lipid production under high levels carbon dioxide. *Bioresource Technology* 101: S71–S74.

Zhang, Q., J. Chang, T. J. Wang, and Y. Xu. 2007. Review of biomass pyrolysis oil properties and upgrading research. *Energy Conversion and Management* 48: 87–92.

Zhang, Y., M. A. Dube, D. D. McLean, and M. Kates. 2003. Biodiesel production from waste cooking oil. 2. Economic assessment and sensitivity analysis. *Bioresource Technology* 90: 229–240.

37 Biooil Production from Microalgae
Current Status and Future Perspectives

Shankha Koley

Sashi Sonkar

Nirupama Mallick

CONTENTS

37.1   BIOCRUDE/BIOOIL

Biocrude (biooil), a clean and renewable energy source, is derived from biomass and can be proposed as a probable alternative feedstock for the production of diesel and gasoline (Long et al., 2016). HTL or pyrolysis is needed to convert the biomass slurry into biooil. Pyrolysis (Hsu, 2012; Kamarudin et al., 2013) and hydrothermal processes (Selvaratnam et al., 2015; Yin et al., 2010) have the capacity to produce biooil from wood waste,

TABLE 37.1
Biocrude (Biooil) Yield from Various Feedstocks using Different Solvents

Feedstock	Solvent Used	Biocrude/Biooil Yield (wt%)
Microalgae	Dichloromethane	58
Sewage	Ethanol	55
Cattle manure	Dichloromethane	49
Rice husk	Tetrahydrofuran	40
Beech wood	Acetone	34
Coconut shell	Acetone	34
Acacia mangium wood	Acetone	32
Corn stalk	Acetone	28
Tea waste	Acetone	23
Oil palm	Diethyl ester	19

Source: Baloch et al. (2018).

agricultural straw, algae, animal and vegetable oils, animal waste, and municipal sludge. Besides the use of biomass, microalgae can also be used as an alternative feedstock for the production of biooil (Yang, 2014). The biooil yield from various biological feedstocks is shown in Table 37.1, which depicts a higher potential for microalgae.

37.2 MICROALGAE: VIABLE FEEDSTOCK FOR PRODUCTION OF BIOCRUDE (BIOOIL)

Microalgae are microscopic organisms, which are found both in aquatic and terrestrial ecosystems. They characterize a diversity of species, living in a wide array of environmental conditions. The three foremost components for microalgal growth include water, sunlight, and carbon dioxide. The doubling time of microalgae is much less during the exponential phase. Microalgal oil yield was projected to be in the range of 58,700–136,900 L ha^{-1} year^{-1}, subject to the variation of the oil content (Chisti, 2007). Microalgae have been cultivated for utilization in various products, which includes chemicals, oils, proteins, and for the production of bioethanol and biomethane (Coustets et al., 2013; Klok et al., 2014; Santos et al., 2014). High oil-accumulating microalgal species can also work as a viable alternative to vegetable oils used for biodiesel production (Girard et al., 2014). Various known microalgal species like *Chlorella protothecoids*, *C. zofingiensis*, *Scenedesmus obliquus*, and *Schizochytrium limacinum* are used as viable oil-producing sources as they have the potential to accumulate oil at the rate of 50% of their dry body weight under specific growth conditions (Johnson and Wen, 2010; Mata et al., 2010). Microalgae have a major advantage over agricultural crops due to their faster growth rate and non-requirement of arable land for cultivation. Additionally, carbon sequestration and clean burning are interesting aspects of microalgae utilization for biofuel production (Rawat et al., 2013). In the current scenario, microalgae are known for being a superior feedstock for biocrude production, for their high rate of growth, and for lipid accumulation capability under favorable culture conditions (Tredici, 2010). Despite many advantages of microalgae, there are certain major bottlenecks in the process of biooil production, such as the low-cost of cultivation of microalgae for biomass production and harvesting of the microscopic organisms, which are discussed below.

37.2.1 LOW-COST BIOMASS GENERATION

A major limitation of microalgal cultivation under large-scale practices is the availability of low-cost nutrient sources. Analytical grade inorganic chemicals are mainly used as nutrients for the preparation of the culture medium of microalgae (Bold and Wynne, 1978), but the cost goes high enough with the price of these inorganic salts. Therefore, an alternative is required to overcome this cultivation cost. A short review of various types of alternative media for microalgae cultivation, developed in the last few years, is presented in Table 37.2.

Agricultural fertilizers can be a viable alternative to analytical grade inorganic salts. They are cheap and have previously been used in microalgal cultures, demonstrating their nutritional efficiency in microalgal growth with respect to the F/2 medium (Valenzuela-Espinoza et al., 1999). Various alternative culture media like wastewater, manures, and agricultural wastes were used to find a suitable microalgal growth medium; but everything has its own constraints. Wastewater is rich in nitrate and phosphate concentrations, making it useful for microalgal growth (Kong et al., 2010), but the availability of wastewater and a proper concentration of nutrients are the major problems in its use. Cow dung manure was used in the past to serve as an algal growth medium and was found efficient with respect to an F/2 medium due to its nutritional efficiency (Machuca and Ramirez, 1984).

Reports suggested that swine manure medium for algal growth is in need of upgrading to make the process more cost effective (Hu et al., 2012). Previous reports suggested that a $1.68–3.47 \text{ g L}^{-1}$ algal biomass yield can be obtained by addition of swine manure (Deng et al., 2018). Productivity of $2.5–25.0 \text{ gm}^{-2}\text{d}^{-1}$ with the use of diluted dairy manure effluents, at a varying total N loading rate of $0.3–2.5 \text{ gm}^{-2}\text{d}^{-1}$, was reported by Mulbry et al. (2008) and Johnson and Wen (2010). An agricultural fertilizer such as nitrogen, phosphorus, and potassium (NPK) (10 26 26) was used by various researchers as an algal growth medium and found to provide comparable and efficient results with respect to an inorganic salt medium (Valenzuea-Espinoza et al., 1999). Agricultural fertilizers with respect to various other culture media are much more easily available at lower cost. Thus, the use of agricultural fertilizers like urea, potash, and NPK can be effective for the large-scale production of microalgal biomass.

An extensive study was carried out by Koley et al. (2020) at the Indian Institute of Technology Kharagpur, West Bengal, India, to formulate an alternative culture medium by using locally available agricultural fertilizers for cost reduction. The chlorophycean microalga, *Scenedesmus obliquus* (Trup.) Kutz. (SAG 276-3a), was grown in various agricultural fertilizer media and the growth pattern was compared with an N11 medium taken as the control. The best growth was found to be by the use of urea as an N source, namely urea 1x medium. The results showed a maximum yield of

TABLE 37.2
Reports on Various Alternative Media Developed for Microalgae Cultivation

Species	Medium	Outcome	Reference
Tetraselmis suecica	Agricultural fertilizer Igromurtonik (Murphy Ltd., England)	Maximum yield (33×10^5 cells cm^{-3}) was almost as high as in control F-medium (34×10^5 cells cm^{-3}) (Guillard and Ryther, 1962).	Corsini and Karydis (1990)
Chondracanthus squarrulosus	Ammonium nitrate, ammonium sulphate, and urea (fertilizer grade)	Better growth in fertilizer medium (4% per day) than control seawater medium (3% per day).	Pacheco-Ruiz et al. (2004)
Nannochloropsis sp.	Combination of agricultural fertilizers such as ammonium sulfate, urea, calcium superphosphate, micronutrients, and vitamin solutions.	The average cell density of 69×10^6 cells mL^{-1} was obtained in fertilizer medium with respect to 25×10^6 cells mL^{-1} in F/2 medium (Guillard and Ryther, 1962).	El Nabris (2012)
Chlorella vulgaris	Organic fertilizer medium (details not available regarding composition)	*C. vulgaris* biomass yield of 0.55 g L^{-1} and lipid content of 18% (dcw) was obtained in organic fertilizer medium with respect to 0.37 g L^{-1} with inorganic nutrient medium (Bolds Basal Medium) (Bischoff and Bold, 1963)	Lam and Lee (2012)
Nannochloropsis gaditana	Fertilizer-based medium with varying nitrogen (2–16 M) and phosphorous concentrations of (0.11–1.9 mM)	Maximum biomass productivity of 0.51 g L^{-1} day^{-1} was obtained at an N concentration of 11.3 mM, a phosphorus concentration of 0.16 mM, 50% higher yield with respect to nitrogen concentration of 2.0 M, and phosphorous 0.11 mM	Camacho-Rodríguez et al. (2013)
Nannochloropsis sp.	NPK-10:26:26	Biomass yield of 0.76 g L^{-1} in NPK medium with a comparison to 0.65 g L^{-1} in standard ASW medium (Darley and Volcani, 1969)	Banerjee et al. (2016)
Ankistrodesmus gracili	NPK-20:5:20	Increase in biomass productivity by ~7% in NPK medium with a comparison to commercial CHU medium (Chu, 1943)	Sipauba-Tavares et al. (2017)
Desmodesmus subspicatus	NPK solution	Average biomass productivity of 28 g m^{-2} d^{-1} in NPK solution compared to 17 g m^{-2} day^{-1} in wastewater	De Souza Schneider et al. (2018)
Chlorella vulgaris	Organic liquid fertilizer (Syaron Co., Korea)	The specific growth rate in BG-11 (Stanier et al., 1971) and fertilizer medium were 0.143 and 0.165 d^{-1}, respectively	Dang and Lee (2018)
Scenedesmus accuminatus	Agricultural fertilizer-based medium (UZn)	Maximum biomass productivity of 12.4 g m^{-2} d^{-1} and lipid productivity of 1.3 g m^{-2} d^{-1}	Koley et al. (2019)

2.21 in terms of guaranteed minimum yield (OD) at 540 nm in urea 1x and 2.01 in an N 11 medium, for an incubation period of 21 days. The lipid yield was also found to be comparable to an N 11 medium.

In an effort to further optimize the concentrations of the major nutrients of the urea 1x medium, a 'central composite rotary design' (CCRD) coupled with a response surface methodology was performed, which predicted a yield of 3.01 in terms of OD by decreasing the contents of the major nutrient in the medium, i.e. potash, magnesium sulfate, and superphosphate along with an increase in urea concentration (Koley et al., 2020). The validation of the predicted model was confirmed by experimentation and a maximum OD of 2.95 was recorded in comparison to 2.21 OD in urea 1x medium. The optimized medium was thus named 'OUR medium' (optimized urea medium) and was also found to be cheaper by 50 times in comparison to an N 11 medium.

The final OUR medium was further evaluated for its suitability to other microalgal species, i.e. *Scenedesmus accuminatus*, *S. armatus*, *Chlorella vulgaris*, *C. sorokiniana*, *C. minutissima*, *Chlamydomonas* sp., *Selenastrum* sp., and *Haematococcus pluvialis* (Koley et al., 2020). Among these microalgae, *S. armatus*, *C. vulgaris*, *C. minuttisima*, *Chlamydomonas* sp., *Selenastrum* sp., and *H. pluvialis* showed comparable growth performance in the OUR medium with respect to an N 11 medium. For *S. accuminatus* and *C. sorokiniana*, the maximum OD values were, however, significantly higher for the optimized OUR medium than an N 11 medium.

A pilot-scale study was also conducted for two years with *S. obliquus* in open raceway ponds, varying the cultivation parameters, i.e. culture depth, impeller speed, and seasons (Figure 37.1). The culture depth was found to be suitable at 30 cm with an impeller speed of 60 rpm. Among all seasons, winter was found to be more

FIGURE 37.1 Pictorial view of *S. obliquus* cultivation in open raceway ponds under ambient conditions.

productive, with a maximum yield of 1.23 gL^{-1} on the 30th day of incubation. The lipid yield was found to be 125.3 mgL^{-1} with a lipid content of 10.4% of the dry cell weight (dcw). The areal biomass productivity was recorded to be 12.61 gm^{-2}day^{-1} with a lipid productivity of 1.2 g m^{-2}day^{-1} for the winter season. An approximately 8–14% reduction in biomass and lipid productivity was recorded during the summer season as compared to the winter season.

37.2.2 MICROALGAL HARVESTING

Recovery of the biomass from the dilute algal suspension is another challenging task for the scientist working in the algal biotechnology field. The negative zeta potential of microalgal cells helps them to maintain a stable suspension in the medium they grow (Becker, 1994). For harvesting, four techniques are mainly practiced world-wide, i.e. filtration, centrifugation, flocculation, and gravity settling. In addition to the 'flocculation efficiency' (FE) and economy of the method used, reusability of the culture medium is another important aspect to be considered for partially reducing the cost of the culture medium and lowering the pressure on water resources. Flocculation can be considered as a suitable method for large volumes of cultures, which can be achieved by various flocculants such as chitosan and the use of various inorganic salts (Morales et al., 1985; Papazi et al., 2010). Gravity settling or sedimentation is also another technique to handle large volumes of cultures, but it is a slow process and usually recommended for biomass with low commercial values (Munoz and Guieysse, 2006).

Harvesting experiments were carried out by Koley et al. (2017) at the Indian Institute of Technology Kharagpur, West Bengal, India, both in the laboratory as well as for large-scale setups. Various flocculating methods, i.e. flocculation with inorganic chemicals, cationic polymers, pH-induced flocculation, 'dissolved air flotation' (DAF), and electroflotation, were compared to find out the most pertinent one to be used for field-scale setups. The highest FE was achieved in electroflotation (99%), followed by alum-induced flocculation (93%), chitosan-induced flocculation (92%), DAF (90%), pH-induced flocculation (83%), and ferric chloride-induced flocculation (80%). Pictorial views of DAF and electroflotation are presented in Figures 37.2 and 37.3.

Although, alum- and chitosan-induced flocculation displayed the highest efficiency, the cost was found to be inhibitory for application in large-scale setups. On the other hand, the pH-induced flocculation was found to be the most efficient with a cost of only US$0.02 per kg dry wt. biomass. Considering the reusability of the supernatant and the cost incurred, pH-induced flocculation by NaOH, DAF, and electroflotation seemed to be the best for large-scale microalgal harvesting. pH-induced flocculation experiments with NaOH displayed an FE value of 75% after a time duration of 1 h, which increased up to 90% at the end of 12 h, and resembled the laboratory scale study in terms of efficiency (Figure 37.4). In electroflotation, a 24 V supply was found to be inadequate under the field-scale studies and an FE of only 50% was observed after a duration of 24 h. FE was found to increase with an increase in voltage supply and reached 90% after a time duration of 24 h by the application of 60 V. In DAF, it was observed that a 90% FE was achieved after a

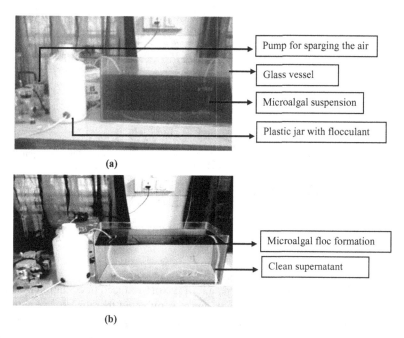

FIGURE 37.2 Pictorial view of the flocculation experiments conducted by DAF (a) before and (b) with algal floc formation at the surface.

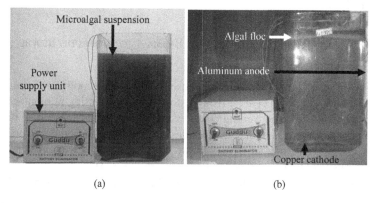

FIGURE 37.3 Electroflotation setup in laboratory showing flocculation of *S. obliquus* (a) at the initiation of the experiment and (b) with algal floc formation at the surface after 1 h. (*Source:* (Koley et al. (2017)).

retention time of 7 h with 10 mgL^{-1} of alum followed by an 85% FE with an alum concentration of 8 mgL^{-1}. Thus, the requirement of alum was significantly high as compared to laboratory scale studies, where a 99% FE was recorded with only a 1 mgL^{-1} alum concentration. The cost effectiveness of pH-induced flocculation by NaOH was found to be the most pertinent, with an estimated value of US$0.03/kg of biomass, followed by US$0.79/kg of biomass in DAF and US$5.61/kg of biomass in

0.25 hp motor for stirring the algal culture

RPM regulator for controlling the rpm of stirrer

FIGURE 37.4 pH-induced flocculation at pH 12 set up in a 1,000 L tank showing initiation of experiment. (*Source:* Koley et al. (2017)).

electroflotation. Above all, the supernatant of the pH-induced flocculation by NaOH was reused and the time-course growth analysis reflected that the growth pattern in both the freshly prepared N 11 medium, OUR medium, and the adjusted medium (supernatant obtained from the pH-induced flocculation experiments, pH normalized and nutrient resupplemented) were comparable. Thus, the pH-induced flocculation by NaOH was recommended for the harvesting of microalgal biomass for large-scale setups (Koley et al., 2017).

37.2.3 HTL: EFFICIENTLY CONVERTING BIOMASS TO BIOCRUDE/BIOOIL

37.2.3.1 HTL

Various thermochemical procedures are used to produce biooil or biocrude from microalgal biomass (Silva et al., 2016). Hydrothermal processing of biomass among others is a technique for converting complex organic materials to biocrude or biooil. The biomass is composed of macromolecules, which under hydrothermal processing are broken down into smaller molecules by hydrolysis or degradation. The oxygen content is removed in the form of H_2O by dehydration. In HTL, various components of the biomass, i.e. protein, carbohydrate, and lipids, are processed together to form biocrude (Toor et al., 2014). Biomass is treated at high temperature and pressure in the presence of water with or without the assistance of a catalyst in a hydrothermal reactor (Figure 37.5). This can be elaborated as physical and chemical conversions occurring at high temperature (200–500°C) and high pressure (5–40 MPa). Microalgae are fed as slurry to the HTL reactor with a biomass concentration varying between 5 and 20% out of the total reactor feed, while the other is just a reaction medium, i.e. water. The generated biooil or biocrude is of a liquid blend due to the water medium used, and is more deoxygenated and viscous than biooil from pyrolysis (Mathimani et al., 2019). The biooil could be comparable with crude oil, which after refining could be utilized in various applications.

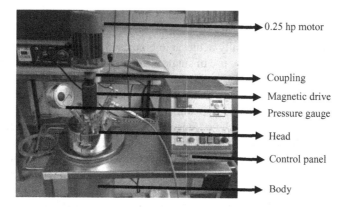

FIGURE 37.5 Pictorial view of 10 L capacity continuous stirred tank reactor at Agro-Environment Laboratory (Make: Amar Equipment Pvt. Ltd., Mumbai, India).

37.2.3.2 The Need for HTL

Microalgal-based biodiesel production involves five major steps: cultivation of resilient strains, followed by dewatering or harvesting of the biomass, drying, lipid extraction, and biodiesel production through transesterification (Han et al., 2015; Mathimani et al., 2017). Among these processes, the drying of biomass, oil extraction, and transesterification are the major limiting steps to marketing the biodiesel with energy balance and positive economy (Figure 37.6). Extraction of complete lipid from microalgal cells through an appropriate cell-rupturing technique is dominant since it governs the lipid content of the strain. The most commonly used disruption methods are homogenization, oil expelling, sonication, soxhlet extraction, and supercritical fluid extraction (Martinez-Guerra et al., 2014; Shwetharani and Balakrishna, 2016; Wang et al., 2017). However, the common issue for all these

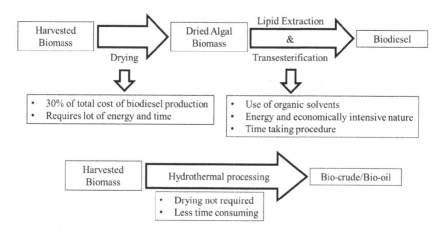

FIGURE 37.6 Schematic representation of the need for HTL for biocrude production from microalgal biomass.

methods is the requirement for dried biomass for oil extraction; microalgal drying is estimated to cost ~30% of the overall biofuel production cost (Becker, 1994). Thus, commercially viable, technically achievable biodiesel production is a nightmare at the present time. Moreover, the use of single or binary solvent systems are hazardous for the environment and are also highly expensive (Mubarak et al., 2015). Due to the issues mentioned above, though a handful of techniques are being used, lipid extraction is certainly a key economic impediment towards lucrative fuel production. In a line of technicity barriers, catalytic transesterification (with acid or alkali) of lipids into biodiesel with alcohol or hydrogenation of lipids should be performed after the extraction of lipids (Guo et al., 2015; Mandal et al., 2013). In this perspective, processing of wet algal slurry directly for production of biooil or biocrude, avoiding the major energy and cost-intensive processes, seems to be a lucrative option.

Currently, various thermochemical processes such as gasification, liquefaction, and pyrolysis are in use for the processing of microalgal biomass (Demirbas, 2009; Wang et al., 2017). Among the thermochemical methods, liquefaction is carried out at low temperature (250–350°C) and high pressure (5–40 MPa), whereas pyrolysis is operated at higher temperatures (400–600°C) and atmospheric pressures (Jena and Das, 2011; Kumar et al., 2018), and therefore liquefaction is more advantageous than pyrolysis in terms of energy efficiency, due to the requirement of a lower temperature. As reported by Peterson et al. (2008), HTL is an energy efficient thermochemical method compared to pyrolysis, as HTL is carried out at a subcritical water temperature compared to pyrolysis (supercritical temperature). In addition, pyrolysis can only be conducted with less than 45% moisture content, thus it requires drying, whereas liquefaction can be done at a higher moisture content. In gasification, the operating temperature was above 700°C, which is highly cost intensive. Moreover, the high moisture content of microalgae makes it unsuitable for the gasification process (Barreiro et al., 2013). Among these methods, HTL is a promising technique by which the wet algal biomass with ~90% moisture content can be converted to biooil/biocrude at a high temperature and pressure.

37.2.3.3 Biooil (Biocrude)

Biooil or biocrude is defined as in an organic liquid phase, comprising broken simpler units from macromolecules such as carbohydrate, protein, and lipid; it has been foreseen as a plausible substitute for petrocrude (Saber et al., 2016). The process flowchart is as depicted in Figure 37.7.

Reviewing the available literature has shown that the development of HTL is at its infancy and that very few reports are available on the microalgal biocrude production process. Table 37.3 presents the methodologies adapted for biocrude preparation with the yield from various microalgal species. A critical analysis of the results shows that a temperature range of 200–350°C and a pressure of ~10 MPa were mostly used by various researchers working with different microalgal species. A maximum yield of 55% of dry weight was recorded for *Nannochloropsis* sp. at 260°C and 20 MPa with a reaction time of 60 min in a 100 ml reactor without the addition of any catalyst (Li et al., 2014), followed by 50% of dry weight for *Auxenochlorella protothecoides* at 350°C and 25 MPa with a 15 min holding time in a 10 ml micro-autoclave reactor without any catalyst (Guo et al. 2019). Similarly, a 49% yield was

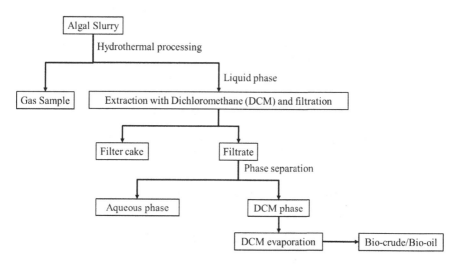

FIGURE 37.7 Flowchart of biocrude (biooil) preparation from algal slurry through HTL.

reported for *Desmodesmus* sp. at 375°C and 10 MPa in a 45 ml batch reactor without any catalyst (Alba et al., 2012). However, all these studies were conducted in a very small volumed reactor, thus its validity for large scale systems needs to be explored. Moreover, most of the studies did not report on the proximate analysis, thus corelating with the biocrude yield remains uncertain.

37.2.4 FACTORS AFFECTING HTL

The products following HTL of microalgal biomass, e.g. biooil yield, composition, and biooil characteristics, are majorly influenced by the operational parameters of HTL, such as temperature, retention time, reactor pressure, and the presence of a catalyst (Koley et al., 2018; Mathimani and Mallick, 2019).

37.2.4.1 Temperature

The reaction temperature is one of the prime factors which controls the production of biooil along with other end products in HTL (Reddy et al., 2016; Ruiz et al., 2013). Temperature plays the part of providing the heat required for the fragmentation of the components present in the biomass (Ji et al., 2017). To be more specific, within the temperature range of 0 to 100°C, macromolecules present in the biomass hydrolyse and amino acids are produced from protein; lipid hydrolyses to form glycerol and fatty acids, whereas carbohydrate hydrolyses into sugars (Gai et al., 2015). Increasing the temperature above 100°C up to 200°C, decarboxylation, deamination, and degradation processes takes place in amino acids, fatty acids, and sugars (Barreiro et al., 2013). Fatty acids convert themselves to esters and amide products, above the temperature of 200°C. Above 250°C products seems to be hydrophilic and thus retard the generation of biooil (Alba et al., 2012). However, the maximum temperature of 300°C is administered, keeping in view biooil generation with optimized energy consumption (Ji et al., 2017).

TABLE 37.3

Reports on the Hydrothermal Processing of Microalgal Biomass for Production of Biocrude (Biooil)

Species	Method	Outcome	Reference
Spirulina sp.	Processing at 300°C at 10–12 MPa for 30 min without catalyst	Biocrude yield was 33% of dry weight	Vardon et al. (2011)
Nannochloropsis sp.	Processing at 350°C at 10 MPa in 31 ml reactor for 60 min without catalyst	Biocrude yield was 39% of dry weight	Valdez et al. (2011)
Scenedesmus sp.	Processing at 300°C and 10–12 MPa in 500 ml reactor for 30 min without catalyst	Biocrude yield of 45% of dry wt.	Vardon et al. (2012)
Desmodesmus sp.	Processing over a wide range of temperature (175–450°C) and reaction time of 60 min at 10 MPa in a 45 ml batch reactor system, without catalyst	Biocrude yield of 49% of the dry weight was obtained at 375°C	Alba et al. (2012)
Nannochloropsis sp.	Processing at 260°C, 60 min at 20 MPa in 100 ml reactor, without catalyst	Maximum yield of 55% of dry weight	Li et al. (2014)
Chlorella sp.	Liquefaction at 350°C, 10 MPa pressure and at varying flow rate of 10 ml/min and 40 ml/min, in a 98 ml reactor without catalyst	Maximum yield of 40% of dry weight at 1.4 min residence time	Biller et al. (2015)
Nannochloropsis sp.	Liquefaction at 200–250°C, with three different types of catalysts i.e. Nano-Ni/SiO$_2$, zeolite, Na$_2$CO$_3$	30% biooil yield was obtained at 250°C with Nano Ni/SiO2.	Saber et al. (2016)
Chlorella vulgaris	Liquefaction at a temperature range of 220–340°C and 5 MPa pressure with NaOH as a catalyst	Maximum biooil yield of 25% was obtained at 300°C for 60 min residence time	Arun et al. (2017)
Scenedesmus obliquus	Processing at 200–300°C, 30–60 min, 100–300 bar pressure, and catalysts in a 2 L reactor	Maximum biooil yield of 45.1% at 300°C, 60 min residence time, 200 bar pressure with acetic acid as catalyst	Koley et al. (2018)
Chlorella pyrenoidosa	Processing at 250–400°C, 10–50 min and 30 KPa pressure, following two temperature steps	Maximum yield of 32% at 350°C and 30 min residence time	He et al. (2018)
Auxenochlorella protothecoides	Processing at 10 ml micro-autoclave reactor at 250 and 350°C, 25 MPa, and 15 min holding time	Maximum yield of 50% at 350°C	Guo et al. (2019)
Nannochloropsis sp.	Processing at 50 mL stainless steel autoclave batch reactor at 200–300°C and 30 min time	Biocrude yield maximum of 30.2% at 250°C	Tang et al. (2020)

37.2.4.2 Reaction Time

Reaction time inside the chamber, among others, is an important factor determining the efficiency of the process (Barreiro et al., 2013). The biooil and biochar yield by cracking; the repolymerization of algal biomass is controlled by the residence time in the chamber (Ji et al., 2017). Biooil generation is favored by a short holding time, as longer duration increases the rate of secondary reactions, which in turn leads to the production of gases, char, and aqueous products (Guo et al., 2015). The composition of biooil is not affected by the reaction time (Cheng et al., 2017). Research is still at pace to find out an optimum reaction time for HTL, as it is interdependent on other reaction parameters like biomass type, reaction temperature, and catalyst (Dimitriadis and Bezergianni, 2017).

37.2.4.3 Pressure

The reaction pressure helps to maintain the water balance in the liquid phase (Brown et al., 2010). It also maintains the single phase of the reaction which accelerates the degradation of biomass under super/sub-critical temperatures (Akhtar and Amin, 2011). The biomass slurry transportation inside the hydrothermal reactor is facilitated under higher temperature with higher pressure (Elliott et al., 2015). The reaction pressure signifies high water density, which controls the release of excess H^+ in compressed water, and enhances acid-catalyzed reactions in biomass liquefaction (Akiya and Savage, 2001).

37.2.4.4 Catalysts

The catalysts used can be broadly distinguished into homogeneous and heterogeneous classes, and their usage enhances the quality of microalgal biooil (Xiu and Shahbazi, 2012). Catalysts help in removing N, oxygen, and sulfur from biocrude. They also upgrade the quality of biooil by uplifting the H/C ratio and decreasing the viscosity (Guo et al., 2015). The char formation in HTL is decreased and the biooil yield is increased by the help of catalysts (Dimitriadis and Bezergianni, 2017). The catalysts also have the prime function of improving the ester bonds of lipids, the peptide bonds of proteins, and the glycosidic ether linkages of carbohydrates in hydrothermal water (Yeh et al., 2013).

37.3 CONCLUSION

Various studies have been undertaken to realize microalgal biofuel as an alternative source for petrofuels. The economics of microalgal biofuel production on the other hand has been a major constraint to this aim. To overcome this bottleneck, direct conversion of biomass to biocrude could be a viable solution. It enables biooil produced to meet the characteristics of petrocrude, and can be refined in established petroleum refineries. To further enhance the economics of biofuel production, the by-products of direct conversion, i.e. the aqueous phase and biochar, are rich in nutrients and can be used as fertilizers. Biooil production can also be helpful if simultaneous production of biodiesel and biocrude are done from the same biomass. The defatted biomass after the extraction of lipids can be again subjected to HTL for the production of biocrude (Koley et al., 2020). Nevertheless, there is enough space to fill the gaps and improve the economics of biooil production from microalgal biomass for obtaining a sustainable substitute for fossil fuels.

ACKNOWLEDGMENTS

The authors thank NASF, New Delhi, India for financial aid, and the Indian Institute of Technology Kharagpur, India for providing research facilities. The authors also wish to acknowledge the contribution of Miss Saumita Chakravarty, Ph.D. Scholar, Indian Institute of Technology Kharagpur, for preparation of the manuscript.

REFERENCES

Akhtar, J. and N. A. S. Amin. 2011. A review on process conditions for optimum bio-oil yield in hydrothermal liquefaction of biomass. *Renewasble and Sustainable Energy Reviews* 15: 1615–1624.

Akiya, N. and P. E. Savage. 2001. Kinetics and mechanism of cyclohexanol dehydration in high-temperature water. *Industrial & Engineering Chemistry Research* 40: 1822–1831.

Alba, L. G., C. Torri, and C. Samori, et al. 2012. Hydrothermal treatment (HTT) of microalgae: Evaluation of the process as conversion method in an algae biorefinery concept. *Energy & Fuels* 26: 642–657.

Arun, J., S. J. Shreekanth, and R. Sahana, et al. 2017. Studies on influence of process parameters on hydrothermal catalytic liquefaction of microalgae (Chlorella vulgaris) biomass grown in wastewater. *Bioresource Technology* 244: 963–968.

Baloch, H. A., S. Nizamuddin, and M. T. H. Siddiqui, et al. 2018. Recent advances in production and upgrading of bio-oil from biomass: A critical overview. *Journal of Environmental Chemical Engineering* 6: 5101–5118.

Banerjee, C., K. K. Dubey, and P. Shukla. 2016. Metabolic engineering of microalgal based biofuel production: Prospects and challenges. *Frontiers in Microbiology* 7: 432.

Barreiro, D. L., W. Prins, F. Ronsse, and W. Brilman. 2013. Hydrothermal liquefaction (HTL) of microalgae for biofuel production: State of the art review and future prospects. *Biomass and Bioenergy* 53: 113–127.

Becker, E. W. 1994. *Microalgae: Biotechnology and Microbiology*. Cambridge: Cambridge University Press.

Biller, P., B. K. Sharma, B. Kunwar, and A. B. Ross. 2015. Hydroprocessing of bio-crude from continuous hydrothermal liquefaction of microalgae. *Fuel* 159: 197–205.

Bischoff, H. W. and H. C. Bold. 1963. *Phycological Studies IV. Some Soil Algae from Enchanted Rock and Related Algal Species*. Austin, TX: University of Texas.

Bold, H. C. and M. J. Wynne. 1978. *Introduction to Algae: Structure and Reproduction*. Upper Saddle River, NJ: Prentice-Hall.

Brown, T. M., P. Duan, and P. E. Savage. 2010. Hydrothermal liquefaction and gasification of *Nannochloropsis* sp. *Energy & Fuels* 24: 3639–3646.

Camacho-Rodriguez, J., M. C. Ceron-Garcia, and C. V. Gonzalez-Lopez, et al. 2013. A low-cost culture medium for the production of *Nannochloropsis gaditana* biomass optimized for aquaculture. *Bioresource Technology* 144: 57–66.

Cheng, F., Z. Cui, and L. Chen, et al. 2017. Hydrothermal liquefaction of high- and low-lipid algae: Biocrude oil chemistry. *Applied Energy* 206: 278–292.

Chisti, Y. 2007. Biodiesel from microalgae. *Biotechnology Advances* 25: 294–306.

Chu, S. P. 1943. The influence of the mineral composition of the medium on the growth of planktonic algae: part II. *The influence of the concentration of inorganic nitrogen and phosphate phosphorus. Journal of Ecology* 31: 109–148.

Corsini, M. and M. Karydis. 1990. An algal medium based on fertilizers and its evaluation in mariculture. *Journal of Applied Phycology* 2: 333–339.

Coustets, M., N. Al-Karablieh, C. Thomsen, and J. Teissie. 2013. Flow process for electroextraction of total proteins from microalgae. *Journal of Membrane Biology* 246: 751–760.

Dang, N. M. and K. Lee. 2018. Utilization of organic liquid fertilizer in microalgae cultivation for biodiesel production. *Biotechnology and Bioprocess Engineering* 23: 405–414.

Darley, W. M. and B. E. Volcani. 1969. Role of silicon in diatom metabolism: a silicon requirement for deoxyribonucleic acid synthesis in the diatom *Cylindrotheca fusiformis* Reimann and Lewin. *Experimental Cell Research* 58: 334–342.

Demirbas, A. 2009. Progress and recent trends in biodiesel fuels. *Energy Conversion and Management* 50: 14–34.

de Souza Schnedier, R. D. C., M. de Moura Lima, and M. Hoeltz, et al. 2018. Life cycle assessment of microalgae production in a raceway pond with alternative culture media. *Algal Research* 32: 280–292.

Deng, X. Y., K. Gao, and M. Addy, et al. 2018. Cultivation of *Chlorella vulgaris* on anaerobically digested swine manure with daily recycling of the post-harvest culture broth. *Bioresource Technology* 247: 716–723.

Dimitriadis, A. and S. Bezergianni. 2017. Hydrothermal liquefaction of various biomass and waste feedstocks for biocrude production: A state of the art review. *Renewable and Sustainable Energy Reviews* 68: 113–125.

El Nabris, K. J.-A. 2012. Development of cheap and simple culture medium for the microalgae *Nannochloropsis* sp. based on agricultural grade fertilizers available in the local market of Gaza strip (Palestine). *Journal of Al Azhar University-Gaza (Natural Sciences)* 14: 61–76.

Elliott, D. C., P. Biller, A. B. Ross, A. J. Schmidt, and S. B. Jones. 2015. Hydrothermal liquefaction of biomass: Developments from batch to continuous process. *Bioresource Technology* 178: 147–156.

Gai, C., Y. Zhang, W.-T. Chen, P. Zhang, and Y. Dong. 2015. An investigation of reaction pathways of hydrothermal liquefaction using *Chlorella pyrenoidosa* and *Spirulina platensis*. *Energy Conversion and Management* 96: 330–339.

Girard, J. M., M. L. Roy, and M. ben Hafsa, et al. 2014. Mixotrophic cultivation of green microalgae *Scenedesmus obliquus* on cheese whey permeate for biodiesel production. *Algal Research* 5: 241–248.

Guillard, R. R. and J. H. Ryther. 1962. Studies of marine planktonic diatoms: I. Cyclotella nana Hustedt, and Detonula confervacea (Cleve) Gran. *Canadian Journal of microbiology* 8: 229–239.

Guo, B., B. Yang, and A. Silve, et al. 2019. Hydrothermal liquefaction of residual microalgae biomass after pulsed electric field-assisted valuables extraction. *Algal Research* 43: 101650.

Guo, M., W. Song, and J. Buhain. 2015. Bioenergy and biofuels: History, status, and perspective. *Renewable and Sustainable Energy Reviews* 42: 712–725.

Han, S. F., W. B. Jin, R. J. Tu, and W. M. Wu. 2015. Biofuel production from microalgae as feedstock: Current status and potential. *Critical Reviews in Biotechnology* 35: 255–268.

He, Z., D. Xu, and L. Liu, et al. 2018. Product characterization of multi-temperature steps of hydrothermal liquefaction of *Chlorella* microalgae. *Algal Research* 33: 8–15.

Hsu, D. D. 2012. Life cycle assessment of gasoline and diesel produced via fast pyrolysis and hydroprocessing. *Biomass and Bioenergy* 45: 41–47.

Hu, B., M. Min, and W. Zhou, et al., 2012. Enhanced mixotrophic growth of microalga *Chlorella* sp. on pretreated swine manure for simultaneous biofuel feedstock production and nutrient removal. *Bioresource Technology* 126: 71–79.

Jena, U. and K. C. Das. 2011. Comparative evaluation of thermochemical liquefaction and pyrolysis for bio-oil production from microalgae. *Energy & Fuels* 25: 5472–5482.

Ji, C., Z. He, and Q. Wang, et al. 2017. Effect of operating conditions on direct liquefaction of low-lipid microalgae in ethanol-water co-solvent for bio-oil production. *Energy Conversion and Management* 141: 155–162.

Johnson, M. B. and Z. Wen. 2010. Development of an attached microalgal growth system for biofuel production. *Applied Microbiology and Biotechnology* 85: 525–534.

Kamarudin, S. K., N. S. Shamsul, and J. A. Ghani, et al. 2013. Production of methanol from biomass waste via pyrolysis. *Bioresource Technology* 129: 463–468.

Klok, A. J., P. P. Lamers, D. E. Martens, R. B. Draaisma, and R. H. Wijffels. 2014. Edible oils from microalgae: Insights in TAG accumulation. *Trends in Biotechnology* 32: 521–528.

Koley, S., M. S. Khadase, T. Mathimani, H. Raheman, and N. Mallick. 2018. Catalytic and non-catalytic hydrothermal processing of *Scenedesmus obliquus* biomass for bio-crude production: A sustainable energy perspective. *Energy Conversion and Management* 163: 111–121.

Koley, S., T. Mathimani, S. K. Bagchi, S. Sonkar, and N. Mallick. 2019. Microalgal biodiesel production at outdoor open and polyhouse raceway pond cultivations: A case study with *Scenedesmus accuminatus* using low-cost farm fertilizer medium. *Biomass and Bioenergy* 120: 156–165.

Koley, S., S. Prasad, S. K. Bagchi, and N. Mallick. 2017. Development of a harvesting technique for large-scale microalgal harvesting for biodiesel. *RSC Advances* 7: 7227–7237.

Koley, S., S. Sonkar, S. K. Bagchi, and N. Mallick. 2020. Simultaneous production of bio-diesel and bio-crude from the chlorophycean microalga *Tetradesmus obliquus*: A renewable energy prospective. *Journal of Cleaner Production*.

Kong, Q. X., L. Li, B. Martinez, P. Chen, and R. Ruan. 2010. Culture of microalgae *Chlamydomonas reinhardtii* in wastewater for biomass feedstock production. *Applied Biochemistry and Biotechnology* 160: 9.

Kumar, M., A. O. Oyedun, and A. Kumar. 2018. A review on the current status of various hydrothermal technologies on biomass feedstock. *Renewable and Sustainable Energy Reviews* 81: 1742–1770.

Lam, M. K. and K. T. Lee. 2012. Potential of using organic fertilizer to cultivate *Chlorella vulgaris* for biodiesel production. *Applied Energy* 94: 303–308.

Li, H., Z. Liu, and Y. Zhang, et al. 2014. Conversion efficiency and oil quality of low-lipid high-protein and high-lipid low-protein microalgae via hydrothermal liquefaction. *Bioresource Technology* 154: 322–329.

Long, J., Y. Li, and X. Zhang, et al. 2016. Comparative investigation on hydrothermal and alkali catalytic liquefaction of bagasse: Process efficiency and product properties. *Fuel* 186: 685–693.

Machuca, C. G. and L. F. B. Ramirez. 1984. Cultivo de las microalgas *Monochrysis lutheri* y *Skeletonema costatum* con nutrientes producidos por estiercoles digeridos. [Cultivation of *Monochrysis lutheri* and *Skeletonema costatum* microalgae with nutrients produced by digested estiercoles]. *Anales del Instituto de Ciencias del mar y Limnologia* 11: 241–256.

Mandal, S., R. Patnaik, A. K. Singh, and N. Mallick. 2013. Comparative assessment of various lipid extraction protocols and optimization of transesterification process for microalgal biodiesel production. *Environmental Technology* 34: 2009–2018.

Martinez-Guerra, E., V. G. Gude, A. Mondala, W. Holmes, and R. Hernandez. 2014. Microwave and ultrasound enhanced extractive-transesterification of algal lipids. *Applied Energy* 129: 354–363.

Mata, T. M., A. A. Martins, and N. S. Caetano. 2010. Microalgae for biodiesel production and other applications: A review. *Renewable and Sustainable Energy Reviews* 14: 217–232.

Mathimani, T., A. Baldinelli, and K. Rajendran, et al. 2019. Review on cultivation and thermo-chemical conversion of microalgae to fuels and chemicals: Process evaluation and knowledge gaps. *Journal of Cleaner Production* 208: 1053–1064.

Mathimani, T., T. S. Kumar, M. Chandrasekar, L. Uma, and D. Prabaharan. 2017. Assessment of fuel properties, engine performance and emission characteristics of outdoor grown marine *Chlorella vulgaris* BDUG 91771 biodiesel. *Renewable Energy* 105: 637–646.

Mathimani, T. and N. Mallick. 2019. A review on the hydrothermal processing of microalgal biomass to bio-oil: Knowledge gaps and recent advances. *Journal of Cleaner Production* 217: 69–84.

Morales, J., J. de la Noue, and G. Picard. 1985. Harvesting marine microalgae species by chitosan flocculation. *Aquacultural Engineering* 4: 257–270.

Mubarak, M., A. Shaija, and T. V. Suchithra. 2015. A review on the extraction of lipid from microalgae for biodiesel production. *Algal Research* 7: 117–123.

Mulbry, W., S. Kondrad, C. Pizarro, and E. Kebede-Westhead. 2008. Treatment of dairy manure effluent using freshwater algae: Algal productivity and recovery of manure nutrients using pilot-scale algal turf scrubbers. *Bioresource Technology* 99: 8137–8142.

Munoz, R. and B. Guieysse. 2006. Algal-bacterial processes for the treatment of hazardous contaminants: A review. *Water Research* 40: 2799–2815.

Pacheco-Ruiz, I., J. A. Zertuche-Gonzalez, E. Arroyo-Ortega, and E. Valenzuela-Espinoza. 2004. Agricultural fertilizers as alternative culture media for biomass production of *Chondracanthus squarrulosus* (Rhodophyta, Gigartinales) under semi-controlled conditions. *Aquaculture* 240201–9.

Papazi, A., P. Makridis, and P. Divanach. 2010. Harvesting *Chlorella minutissima* using cell coagulants. *Journal of Applied Phycology* 22: 349–355.

Peterson, A. A., F. Vogel, and R. P. Lachance, et al. 2008. Thermochemical biofuel production in hydrothermal media: A review of sub-and supercritical water technologies. *Energy & Environmental Science* 1: 32–65.

Rawat, I., R. R. Kumar, T. T. Mutanda, and F. Bux. 2013. Biodiesel from microalgae: A critical evaluation from laboratory to large scale production. *Applied Energy* 103: 444–467.

Reddy, H. K., T. Muppaneni, and S. Ponnusamy, et al. 2016. Temperature effect on hydrothermal liquefaction of *Nannochloropsis gaditana* and *Chlorella* sp. *Applied Energy* 165: 943–951.

Ruiz, H. A., R. M. Rodriguez-Jasso, B. D. Fernandes, A. A. Vicente, and J. A. Teixeira. 2013. Hydrothermal processing, as an alternative for upgrading agriculture residues and marine biomass according to the biorefinery concept: A review. *Renewable and Sustainable Energy Reviews* 21: 35–51.

Saber, M., B. Nakhshiniev, and K. Yoshikawa. 2016. A review of production and upgrading of algal bio-oil. *Renewable and Sustainable Energy Reviews* 58: 918–930.

Santos, C. A. and A. Reis. 2014. Microalgal symbiosis in biotechnology. *Applied Microbiology and Biotechnology* 98: 5839–5846.

Selvaratnam, T., A. K. Pegallapati, and H. Reddy, et al. 2015. Algal biofuels from urban wastewaters: Maximizing biomass yield using nutrients recycled from hydrothermal processing of biomass. *Bioresource Technology* 182: 232–238.

Shwetharani, R. and R. G. Balakrishna. 2016. Efficient algal lipid extraction via photocatalysis and its conversion to biofuel. *Applied Energy* 168: 364–374.

Silva, C. M., A. F. Ferreira, A. P. Dias, and M. Costa. 2016. A comparison between microalgae virtual biorefinery arrangements for bio-oil production based on lab-scale results. *Journal of Cleaner Production* 130: 58–67.

Sipauba-Tavares, L. H., A. M. D. L. Segali, F. A. Berchielli-Morais, and B. Scardoeli-Truzzi. 2017. Development of low-cost culture media for *Ankistrodesmus gracilis* based on inorganic fertilizer and macrophyte. *Acta Limnologica Brasiliensia* 29: 5.

Stanier, R. Y., R. Kunisawa, M. Mandel, and G. Cohen-Bazire. 1971. Purification and properties of unicellular blue-green algae (order Chroococcales). *Bacteriological Reviews* 35: 171–205.

Tang, X., C. Zhang, and X. Yang. 2020. Optimizing process of hydrothermal liquefaction of microalgae via flash heating and isolating aqueous extract from bio-crude. *Journal of Cleaner Production* 258: 120660.

Toor, S. S., L. A. Rosendahl, and J. Hoffmann, et al. 2014. Hydrothermal liquefaction of biomass. In: *Application of Hydrothermal Reactions to Biomass Conversion*, ed. F. Jin, 189–217. Berlin: Springer.

Tredici, M. R. 2010. Photobiology of microalgae mass cultures: Understanding the tools for the next green revolution. *Biofuels* 1: 143–162.

Valdez, P. J., J. G. Dickinson, and P. E. Savage. 2011. Characterization of product fractions from hydrothermal liquefaction of *Nannochloropsis* sp. and the influence of solvents. *Energy & Fuels* 25: 3235–3243.

Valenzuela-Espinoza, E., R. Millan-Nunez, and F. Nunez-Cebrero. 1999. Biomass production and nutrient uptake by *Isochrysis* aff. *galbana* (Clone T-ISO) cultured with a low cost alternative to the f/2 medium. *Aquacultural Engineering* 20: 135–147.

Vardon, D. R., B. K. Sharma, G. V. Blazina, K. Rajagopalan, and T. J. Strathmann. 2012. Thermochemical conversion of raw and defatted algal biomass via hydrothermal liquefaction and slow pyrolysis. *Bioresource Technology* 109: 178–187.

Vardon, D. R., B. K. Sharma, and J. Scott, et al. 2011. Chemical properties of biocrude oil from the hydrothermal liquefaction of *Spirulina* algae, swine manure, and digested anaerobic sludge. *Bioresource Technology* 102: 8295–8303.

Wang, X., F. Guo, Y. Li, and X. Yang. 2017. Effect of pretreatment on microalgae pyrolysis: Kinetics, biocrude yield and quality, and life cycle assessment. *Energy Conversion and Management* 132: 161–171.

Xiu, S. and A. Shahbazi. 2012. Bio-oil production and upgrading research: A review. *Renewable and Sustainable Energy Reviews* 16: 4406–4414.

Yang, W., X. Li, S. Liu, and L. Feng. 2014. Direct hydrothermal liquefaction of undried macroalgae *Enteromorpha prolifera* using acid catalysts. *Energy Conversion and Management* 87: 938–945.

Yeh, T. M., J. G. Dickinson, and A. Franck, et al. 2013. Hydrothermal catalytic production of fuels and chemicals from aquatic biomass. *Journal of Chemical Technology and Biotechnology* 88: 13–24.

Yin, S., R. Dolan, M. Harris, and Z. Tan. 2010. Subcritical hydrothermal liquefaction of cattle manure to bio-oil: Effects of conversion parameters on bio-oil yield and characterization of bio-oil. *Bioresource Technology* 101 3657–3664.

38 Extraction of Algal Neutral Lipids for Biofuel Production

Saravanan Krishnan

Jaison Jeevanandam

Caleb Acquah

Michael K. Danquah

CONTENTS

38.1   INTRODUCTION

Fuel is generally considered as a chemical compound with the ability to oxidize under rapid combustion to release exothermal energy. The heat generated from the combustion of fuels is used as energy to run machines and vehicles, power plants, and produce electricity (Ritchie and Roser, 2017). Fossil fuels with hydrocarbons are the main source of fuel to generate energy for significant applications (Gerasimchuk et al., 2019). Solid fuels such as wood, coal, liquid fuels, including crude petrol, and natural gas as gaseous fuel are distinct types of fossil fuels (Theodosiou et al., 2007). Although these fuels generate potentially higher heat energy they contain N, free carbon, and sulfur as impurities, which affects their calorific value and causes hazardous effects to living organisms and adds to the greenhouse effect (Nyashina et al., 2018). This has led to the refinement of fossil fuels to produce synthetic fuels which are more efficient than the former due to the elimination of impurities and which are beneficial to meeting the growing demand for fuel (Maass and Mockel, 2020). Nevertheless, synthetic fuels also have a high carbon footprint and are detrimental to the environment due to their greenhouse effect (Artan et al., 2015; Hooftman et al., 2016). The emergence of alternative sources of energy such as solar, wind, and hydropower, which are renewable, has revealed green ways of meeting the ever-increasing energy demand; however, they are still not efficient compared to non-renewable fuels, resulting in the need for increased efficiency (Pirjola et al., 2019).

Biofuels are produced from biological organisms such as bacteria, algae, fungi, and plants by extracting biomolecules that can undergo rapid oxidation and combustion to produce energy (Kumar et al., 2018; Srivastava et al., 2018; Voloshin et al., 2016). Among these organisms, algae are widely employed for the production of biofuel due to their enhanced capability to scale up production, generate essential commercial biomolecules, and the ability to grow with minimum resources (Halim et al., 2011, 2012; Mu et al., 2017). Biodiesel is a form of biofuel with similar properties to conventional diesel; it is extensively produced from algal extracts (Nam et al., 2017). It has been reported by the Center for Sustainable Systems, University of Michigan, that 5,020 gallons of biodiesel per acre were produced in the United States by using algae as the feedstock in the year 2018 (Joo and Kumar, 2019). Biodiesels produced from algae are similar to conventional diesel in their efficiency and are currently being considered as a mixture with commercial diesel to meet energy demands (Edeseyi et al., 2015; Singh and Jaiswal, 2019).

Lipids are the most important algal biomolecule to be used as a fuel. The quantity, as well as the quality, of extracted lipids from algae determines the efficiency of biodiesel (Karatay et al., 2019). Further, neutral lipids are extensively reported to be beneficial for the combustion, oxidation, and calorific value of algae-based biodiesel (You et al., 2020). Hence, this chapter provides an overview of the diverse conservative and progressive extraction methods used as efficient approaches to neutral lipid extraction from algae. The impact of nanotechnology for enhanced lipid extraction to improve biodiesel production is also discussed.

38.2 OVERVIEW OF MICROALGAL LIPIDS IN BIOFUEL PRODUCTION

Microalgae are oleaginous microorganisms that grow in aquatic habitats like marine water or freshwater ecosystems. Microalgae are a diverse group of living species and are either prokaryotic or eukaryotic. They can utilize visible light and carbon (autotrophic) from the sun and carbon dioxide, respectively, to produce biomass, which entails polysaccharides, proteins, lipids, and hydrocarbons. Of note, some microalgae are heterotrophic and mixotrophic. Prokaryotic microalgae lack a nucleus, and they include cyanobacteria (blue-green algae). On the other hand, eukaryotic microalgae, such as green algae, red algae, and diatoms, have a true nucleus within the cell (Brennan and Owende, 2010).

Microalgal lipids are the preferred feedstock source for producing biofuel. Relative to vegetative sources such as corn or oil palms, microalgae act as better feedstocks for biofuel production (Takisawa et al., 2014) in many aspects, such as enhanced efficiency of photosynthesis, rapid biomass production due to elevated growth rate (Chisti, 2013), enhanced oil content with 20–50% dry weight of TAG (Hu et al., 2008), short cultivation time, improved solar energy to biomass conversion (10–50 times), less land and low water consumption for cultivation, and requiring only marginal lands (Georgianna and Mayfield, 2012). They also utilize essential nutrients for growth from wastewater sources (Mohan et al. 2015) and carbon dioxide from the atmosphere, while generating by-products which are of high value (Brennan and Owende, 2010; Pal et al., 2019). Algae-based biofuels can be readily utilized in modern engines without any special requirements or modifications and can be directly mixed with petroleum diesel in any proportion.

38.2.1 MICROALGAL LIPIDS AND THEIR PHYSIOLOGICAL ROLES

Lipids are essential macromolecules in living systems, having the ability to carry out significant functions at the cellular level. Microalgal lipids are divided into two types based on the structure, namely neutral (non-polar) and polar lipids, as outlined in Figure 38.1. Accumulation of polar and non-polar lipids varies with the growth phase of microalgae. For instance, in the case of *Chlorella ellipsoidea*, polar lipids accumulate in large proportion during the early growth phase, whereas non-polar lipids start accumulating before cell division, thus acting as an energy reservoir (Otsuka and Morimura, 1966). The biological membrane of algae is composed of structural components with polar lipids such as phosphoglycerides, glycosylglycerides, and sterols that can act as a passageway or permeable membrane for the mobility of various molecules, ions, and solutes required for metabolic processes within the cell. Some of the polar lipids, namely inositol, sphingolipids, and oxidative polyunsaturated fatty acid products, act as significant intermediates in cellular signaling pathways.

Mono-, di-, or TAGs, 'free fatty acids' (FFAs), sterols, steryl esters, and wax are some examples of neutral (non-polar) lipids. TAG is the main storage lipid which is catabolized to provide the energy required for cellular metabolism (Gurr et al., 2002).

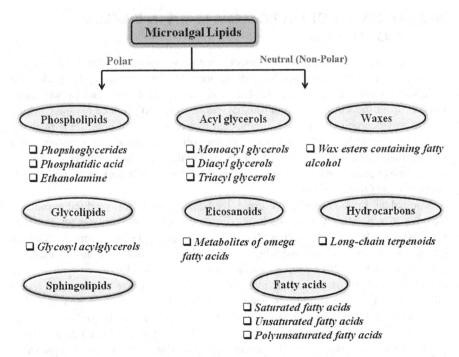

FIGURE 38.1   Classification of microalgal lipids with examples. (*Source:* Adapted from Kumar et al. (2015)).

TAG functions to assist microalgae to adapt in various environmental conditions. Under complex (stress) conditions, TAG is deposited inside the microalgae in cytoplasmic oil forms (Solovchenko, 2012). Mono- or diacylglycerols are the fatty acid mono- or di-esters of glycerol, which act as intermediates in metabolism. Additionally, wax acts as an energy store to protect the organism from extremely low temperatures by forming separate layers on the extracellular surface of different parts of higher plants (Guschina and Harwood, 2007).

38.2.2   MICROALGAL NEUTRAL LIPIDS FOR BIOFUEL PRODUCTION

Microalgal oil is a rich source of neutral lipids and considered a viable alternative to other crop oils that are used for biofuel production. Neutral lipids are storage lipids whose fatty acids are amphiphilic in nature. Triglycerides and hydrocarbons are the core neutral lipid components. Triglycerides are the source of the parent oil in microalgae that is utilized as a significant feedstock for biodiesel production. Further, TAGs derived from microalgae have 16–18 carbons esterified to glycerol moiety, which is regarded as chemically equivalent to diesel (10–15 carbons per molecule) (Liu et al., 2013). Moreover, biofuel obtained from microalgae is renewable and ecofriendly, compared to fossil-derived fuels. The most preferred algal species for renewable biodiesel production is *Chlorella*, which is a green alga that belongs to the family Chlorophyta.

TABLE 38.1
Significant Oil-Producing Microalgae and their Neutral Lipid Content

Micro Algae	Family	Neutral Lipid (% Dry Weight)	Reference
Nitzschia laevis	Bacillariaceae	79.2	Chen et al. (2007)
Pavlova lutheri	Pavlovaceae	56.5	Meireles et al. (2003)
Chlorella sorokiniana	Chlorellaceae	78.9	Zheng et al. (2013)
Dunaliella viridis	Dunaliellaceae	0.5–21.5	Gordillo et al. (1998)
Scenedesmus sp.	Scenedesmaceae	81.3–82.3	Yang et al. (2014)
Gymnodinium sp.	Gymnodiniaceae	7.5–28.8	Mansour et al. (2003)
Schizochytrium Limacinum	Thraustochytriaceae	69.0	Wang and Wang (2012)
Nannochloropsis sp.	Eustigmataceae	41.4	Wang and Wang (2012)

The content of oil in microalgae varies, depending on the species type, the growth conditions, and the microenvironment. Table 38.1 shows that microalgae such as *Botryococcus braunii*, *Schizochytrium* sp., and *Nanochloropsis* sp. are rich in oil content (% dry weight of biomass) (Dong et al., 2016). Ideally, microalgae with high oil productivity, especially a large proportion of TAGs, are the preferred choice for biodiesel production.

Microalgal lipid accumulation can be enhanced by: (i) culturing algae under nutrient deprived conditions, e.g. N starvation; (ii) altering culture parameters (growth phase); (iii) changing physicochemical parameters (temperature, light intensity, salinity); (iv) biomass concentration; and (v) harvesting method (Guschina and Harwood, 2013; Zhu et al., 2017). By engineering the nutritional or cultivation conditions it is possible to regulate the accumulation of microalgal lipids. During N deprived circumstances, the rate of lipid accumulation in microalgae increases and the excess carbon assimilation is channeled to form triacylglycerides, which can be used as a precursor for producing biofuel (Solovchenko, 2012). The concept of metabolic engineering to elevate neutral lipid accumulation in microalgae for biofuel production has also been well-recognized in recent times (Tan and Lee, 2016).

38.3   EXTRACTION METHODS OF ALGAL NEUTRAL LIPIDS

Total lipid, mechanical, and solvent-free extraction are the three main approaches that are widely used for the extraction of algal neutral lipids. Each of these methods is further subclassified into several approaches, as shown in Figure 38.2, based on the equipment type, solvent, or the process utilized for the extraction process.

38.3.1   TOTAL LIPID EXTRACTION METHODS

This method is beneficial for retrieving total algal lipid content; the neutral lipids are later separated via other approaches such as chromatography. Folch, as well as Bligh and Dyer, are superior solvent extraction approaches, whereas *in situ* lipid hydrolysis and supercritical transesterification are subclasses of total lipid extraction approaches applicable to algae.

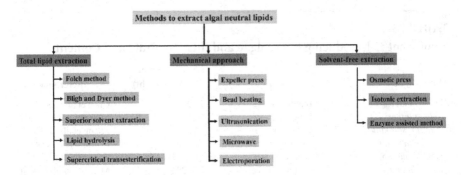

FIGURE 38.2 Classification of various methods for neutral lipid extraction from algal biomass.

38.3.1.1 Folch Method

In this method, a 1: 2 volume ratio of methanol-chloroform is utilized to retrieve lipids from endogenous cells. The mixture of alcohol and chloroform will lead to two phases during extraction; the lipids will be present in the upper phase (Folch et al., 1957). Jones et al. (2012) evaluated the efficiency of the Folch method in extracting total lipids from salt water *Chlorella* KAS603 algae. Two hundred mg of a concentrated pellet of algae was subjected to disruption in a methanol-chloroform solution at an optimized volume via a Dounce homogenizer. The study emphasized that the method is highly beneficial in extracting various significant lipids from the algae compared to other methods. However, 'high-performance liquid chromatography' (HPLC) analysis showed that the Folch method yielded a lower concentration of TAGs and hydrocarbon which are required for biofuel production. Further, the study also recommended that 2-ethoxyethanol would be beneficial as an alcohol solvent for lipid extraction via the Folch approach, instead of methanol (Jones et al., 2012).

Likewise, the Folch approach was used to extract the lipids from microalgal species such as *Schizochytrium limacinum* and *Nannochloropsis* and compared with aqueous and isopropanol homogenization. The study showed that the Folch method is suitable in retrieving significant lipids with a low concentration of TAG and hydrocarbons. The study also stated that extraction efficiency depends on the solvents, algae type (lipid content), and the tedious lipid quantification method (Wang and Wang, 2012). Recently, Donot et al. (2016) utilized the Folch method for neutral lipid extraction and the extraction of FFAs from algal species such as *Botryococcus braunii*, *Spirulina plantensis*, *Chlorella vulgaris*, and *Porphyridium cruentum*. HPLC-Evaporative light scattering detection analysis reported the existence of neutral lipid classes and fatty acids along with high hydrocarbon percentages that are useful for enhanced biofuel production (Donot et al., 2016). The crucial benefit of this approach is its capacity to support the rapid processing of several samples with ease in a short time (Kumar et al., 2015). However, this method is useful for extracting lipids from animal cells, though they require modification to use them for algal lipid extractions, which increases the cost involved and is a time-consuming process (Iverson et al., 2001).

38.3.1.2 Bligh and Dyer Method

This method is comparable to the Folch approach, whereby a chloroform, methanol, and water combination is utilized for algal lipid extraction in two phases by suspending homogenized algal cells in the solvent mixture (Bligh and Dyer, 1959). The solvent mixture extraction is followed by evaporation of the extract at 30°C in a water bath, to eliminate solvent, with subsequent column chromatography to separate neutral lipids from the crude extract (Alonzo and Mayzaud, 1999; Chen et al., 2009). Chen et al. (2011) extracted lipids of a neutral nature from 13 algal species, including *Chlorella zofingiensis*, *Pseudochlorococcum*, and *Scenedesmus dimorphus*, using the Bligh and Dyer approach. The study introduced a dual step of staining and a microwave-mediated method for neutral lipid quantification present in green algae, which showed that this approach is highly useful in lipid extraction from oleaginous algal species (Chen et al., 2011).

Recently, Terme et al. (2017) used the Bligh and Dyer method for specific neutral lipid extraction from *Solieria chordalis* and *Sargassum muticum*. The study revealed that this approach helps in 100% lipid extraction from *S. chordalis* in which 37% are neutral lipids and 38% are glycolipids, compared to a supercritical carbon dioxide method that yields 60% of only glycolipids (Terme et al., 2017). Further, this extraction approach is improved by doping acids, such as phosphoric acid, hydrochloric acid, or acetic acid, with sodium chloride salts at an optimized molarity, instead of water, to extract acidic phospholipids (Jensen, 2008; Weerheim et al., 2002). This helps to prevent the interaction of acidic lipids with denatured lipids. Recently, Saroya et al. (2018) extracted lipids from *Chlorella* species using Bligh and Dyer as well as the wet lipid extraction approach. The modified Bligh and Dyer method was described as yielding 25% of lipids from the algae, which is less than the lipids (20%) extracted from the wet approach (Saroya et al., 2018).

38.3.1.3 Superior Solvent Extraction Method

This method utilizes less-toxic solvents, such as butanol, isopropanol, hexane, and ethanol, as an alternative to toxic chloroform and methanol that are used in the methods mentioned in the previous sections for the extraction of lipids from algae without causing any adverse effect on the environment (Kumar et al., 2017). However, the less-toxic solvents are less effective for lipid extraction and thus a combination of these solvents is used, depending on the type of lipids to be extracted (Sheng et al., 2011). Recently, the accelerated extraction of a solvent has been utilized to retrieve neutral lipids from algal species such as *C. vulgaris*, *C. zofingiensis*, *C. sorokiniana*, and *Nannochloropsis gaditana* via pressure or heat. The study stated that this extraction method yielded 1.3–2.7 more dry biomass content of lipids and total FAME, compared to conventional chloroform-methanol based solvent extraction methods (Tang et al., 2016).

Further, Yang et al. (2017) utilized an amphiphilic amine solvent as a replacement for alcohol for neutral lipid retrieval from the slurry of microalgae. There, N-methylcyclohexylamine (MCHA) was utilized as a switchable solvent that is easy to retrieve significant compounds from the algae while $[C_4\text{-mim}]$ $[PF_6]$

(1-butyl-3-methylimidazolium hexafluorophosphate) was utilized as a recovering ionic liquid for lipid separation from the extracted compounds. The study stated that 77% of algal lipids were recovered via these novel green solvents, which are non-toxic to the environment and are beneficial for biofuel production (Yang et al., 2017). However, this approach is not recommended for large-scale extraction of algal lipids for producing biofuel, as organic solvents may reduce the biofuel efficiency (Kumar et al., 2015).

38.3.1.4 *In situ* Lipid Hydrolysis and Supercritical *in situ* Transesterification

Levine et al. (2010) introduced novel *in situ* hydrolysis and transesterification at supercritical conditions to retrieve lipids from the wet biomass of algae, which was converted into solids via a concurrent dehydration and drying process for the accurate loading of solids. Later, the solids were cooled to separate the aqueous phase via specific Whatman 934-AH filter paper under vacuum conditions (Levine et al., 2010). Recently, the two *in situ* methods were separately utilized for the extraction of algal neutral lipids. Patil et al. (2017) disclosed that rapid hydrolysis of lipids using supercritical methyl acetate technology was useful for the total lipid conversion of *Nannochloropsis salina* into triacetin and FAMEs, which can be a readily useable biofuel (Patil et al., 2017). Further, Shirazi et al. (2017) stated that biodiesel can be formed from the feedstock of *Spirulina* by extracting neutral lipids, hydrocarbons, and fatty acids via direct *in situ* transesterification using methanol at the near supercritical condition. The study showed that the biodiesel possessed significant efficiency of 0.44–99.32% under different near supercritical conditions (Shirazi et al., 2017). However, this method requires further stringent feasibility tests to render it commercially applicable and for large-scale biofuel production via lipid extraction.

38.3.2 MECHANICAL APPROACH

Expeller press, bead beating, ultrasonication, microwave-assisted extraction, and electroporation are the types of mechanical approaches used for algal neutral lipid extraction; they are classified based on the type of equipment used for the disruption of algal cells.

38.3.2.1 Expeller Press

This mechanical method is the oldest approach for the extraction of oils from seeds and was later utilized for lipid extraction from algae biomass (Demirbas, 2008). In this method, dried algal biomass with lipid content is pressed to break the algal cells, using an expeller with high mechanical pressure to retrieve lipids present in the cells. Optimization of pressure is crucial in this method as inappropriate pressure may result in the disruption of the molecular integrity of extracts due to choking and generation of heat (Ramesh, 2013). It was reported by Topare et al. (2011) that 75% of oil, including lipids, can be retrieved from algae with filamentous structures via a screw expeller press (Topare et al., 2011). Moreover, Li et al. (2019) demonstrated that pretreatment of *Nannochloropsis oceanica* biomass with a twin-screw extruder was useful in increasing the extraction of neutral lipids, exclusively for biofuel production (Li et al., 2019). The study revealed that the temperature, water content of biomass, and screw rotation speed were significant parameters that controlled the extraction of lipids via this screw-based pressing method and elevated the presence of polyunsaturated fatty acid in the crude extract to about 30% (Li et al., 2019). However, the existence of a firm algal cell wall and pigments along with lipids in the crude extract are the major limitation of this method (Johnson and Wen, 2009). Thus, solvent extraction must be performed in addition to the expeller method to extract significant neutral lipids from algae for biofuel applications which will also lead to less efficiency, high cost, and skilled labor requirement (Kumar et al., 2015).

38.3.2.2 Bead Beating

In this approach, the algal cells are disordered by spinning them at high speed as a slurry of biomass with fine beads to extract the lipids from them (Geciova et al., 2002). The combination of collision, grinding, and agitation in the bead beating process is responsible for the effective disruption of algal cells (Lee et al., 2012). Generally, beads with a 0.5 mm diameter that are made up of zirconium oxide, zirconia-silica, and titanium carbide with optimized density and hardness are utilized for neutral lipid extraction from algae (Hopkins, 1991). Sydney et al. (2018) recently compared the efficiency of bead beating and high-intensity ultraviolet (UV) light for neutral lipid extraction from *Chlamydomonas reinhardtii*, *Micractinium inermum*, and *Dunaliella salina*. The results revealed that the bead beating approach yielded 45.3 mg/L of neutral lipid, including FAMEs, whereas UV light exposure-mediated disruption of algal cells yielded 79.9 mg/L of neutral lipids in lab-scale tests (Sydney et al., 2018). The core advantage of this approach was the non-requirement of the algal slurry dewatering process relative to the expeller press. This was due to the heat created during the spinning process being enough to dewater the slurry, thus drastically reducing the operating cost. Since heat was expelled during the spinning process, mechanical cooling or a cooling agent was required to avoid damage of the extracted liquid in large scale production (Kumar et al., 2015). Thus, this method is highly recommended only for lab-based small-scale algal neutral lipid extraction.

38.3.2.3 Ultrasonication

Ultrasound was also employed for the disruption of algal cells to extract lipids as an alternative to the conventional mechanical approach. Acoustic streaming and cavitation are the two significant mechanisms that lead to the breakage of algal cells in the liquid culture to ease the retrieving process of lipids (Khanal et al., 2007; Suslick and Flannigan, 2008). Microbubbles are produced to create pressure in the cavitation procedure for algal cell disruption (Adam et al., 2012) and the acoustic streaming is utilized to blend the culture of algae for proper disruption and extrusion of lipids. Moreover, the efficiency of the lipid yield extraction process relies on the heightened mass transfer and microstreaming processes of cavitation and the bursting of microbubbles (Kurokawa et al., 2016). Recently, Patel et al. (2019) introduced a rapid and novel treatment via microwave and ultrasonication for total lipid extraction from the biomass of oleaginous yeasts to produce sustainable biodiesel. The approach generated 70.86% w/w of total lipid content which was higher compared to the ultrasonication method.

Further, the extracted lipids contained 76.5% of monounsaturated, 10.3% of saturated, and 11.5% of polyunsaturated fatty acids which are essential for quality biodiesel production (Patel et al., 2019). Also, Garoma and Janda (2016) recently investigated the effect of pretreatment approaches such as electroporation, microwave, and ultrasonication to disrupt *C. vulgaris* algal cells and extract neutral lipids. The study showed that the yield of lipids increased to 26.4% while using ultrasonication, which was less than microwave (28.9%) and higher than the electroporation method (5.3%) (Garoma and Janda, 2016). The advantages of this method were high performance with simple working conditions, yielded lipids with high purity, reduced waste generation, eco-friendliness, low cost, and less energy input with enhanced reproducibility (Chemat and Khan, 2011). However, prolonged exposure of algae to ultrasound may lead to the formation of free radicals which may affect the extracted neutral lipid quality (Mason et al., 1994).

38.3.2.4 Microwave-Assisted Extraction

Since the 1980s, microwaves have been employed for effective algal cell disruption to retrieve lipids (Ganzler et al., 1986). This method is also simple, rapid, economic, and safe in retrieving lipids from algal cells without the need for the dewatering process as heat generated from microwaves evaporates residual water in the slurry (Pare et al., 1997). Pan et al. (2016) stated that microwaves can be utilized to heat the algal biomass to disrupt them and increase the quantity of lipids to be extracted. The study revealed that the microwave approach with ionic '1-butyl-3-methylimidazolium' hydrogen sulfate in liquid form enhanced the lipid extraction rate by over 370% from algal species such as *Chlorella sorokiniana*, *Galdieria sulphuraria*, and *Nannochloropsis salina*, compared to the solvent extraction method (Pan et al., 2016). Similarly, microwave-assisted extraction was used to retrieve lipids from *Nannochloropsis* algal species using a brine solution by Zghaib et al. (2019). The study emphasized that the addition of 10% w/v of brine will lead to a 16.1% higher lipid yield at 100°C and 30 min of microwave irradiation.

Furthermore, the extracted lipid contained 43% of omega-3 fatty acids and 44.5% of fatty acids with polyunsaturated compounds which are useful for food and biofuel applications (Zghaibi et al., 2019). Likewise, microwaves have been used to retrieve lipids from the wet microalgal paste. A study by de Moura et al. (2018) showed that a higher lipid yield of 33.6% was achieved via 1 min of microwave irradiation time and was relatively more efficient than an ultrasonication approach that required 20 min of reaction time (de Moura et al., 2018). However, this method is not recommended for commercial purposes due to the requirement of high maintenance costs (Kumar et al., 2015).

38.3.2.5 Electroporation

This membrane phenomenon-based method utilizes cytoplasm permeability and electrical conductivity via an external electric field to damage algal cells for the efficient retrieval of lipids. It was reported by Sommerfeld et al. (2008) that 92% of total lipids, including neutral lipids, can be extracted from algae via electroporation methods, whereas extraction via the conventional method yields about 62% of total lipids (Sommerfeld et al., 2008). Recently, an irreversible electroporation approach was utilized for retrieving lipids from *Lipomyces starkeyi* yeast. The result showed that 63 mg/g (31.8%) of lipids were extracted from the yeast, after electroporation for about 10 min, which was higher than ultrasound (11.8%) and Fenton's reagent-mediated (16.8%) lipid extraction (Karim et al., 2018). Elersek et al. (2016) demonstrated the non-polar compounds from *C. vulgaris* can be extracted using the electroporation approach. However, this approach is not widely employed for commercial biofuel production as a high electric field may disrupt the molecular integrity of lipids and other essential compounds, thereby reducing the quality of the extracted lipids.

38.3.3 Solvent-Free Extraction Method

Neutral lipids from algae can also be extracted without the help of solvents such as osmotic pressure and the isotonic and enzyme-assisted extraction approach, based on the type of equipment or compounds used in the extraction process.

38.3.3.1 Osmotic Pressure Method

This method disrupts the algal cells by altering the concentration of salt in aqueous media to interrupt the osmotic pressure equilibrium via hyper- and hypo-osmosis in both the internal and peripheral portion of the algal cells (Lee et al., 2010). Recently, Adetya and Hadiyanto (2018) reported that the osmotic shock method – coupled with ultrasound – is beneficial in lipid extraction from *Spirulina platensis* microalgae. The study revealed that the presence of 16.9% of sodium chloride concentration as an osmotic agent yielded 6.39% of lipids after 36 min and 10 seconds of ultrasound exposure of the algal cells (Adetya and Hadiyanto, 2018). Further, Vo et al. (2017) revealed that osmotic stress leads to more lipid accumulation which can be extracted via osmotic pressure from *Dunaliella salina* A9 algal species (Vo et al., 2017). Besides, Wang et al. (2017) demonstrated that hyper-osmotic pressure will lead to an enhanced yield of lipids from *Chlorella protothecoides* algae (Wang et al., 2017). Despite these advantages, this method is not feasible for pilot and large-scale extraction of lipids and is not recommended for commercial applications (Kumar et al., 2015).

38.3.3.2 Isotonic Extraction Method

This method utilizes ionic liquids at certain specific conditions as a replacement of toxic solvents for the disruption of algal cells and the extraction of lipids. Ionic liquids are made up of non-aqueous salt solutions with the synthetic flexibility of anion and cation combination to exhibit enhanced conductivity, polarity, solubility, and hydrophobicity, based on the algal biomass from which the lipids should be extracted (Cooney et al., 2009). Recently, Zhang et al. (2018) reported that 'imidazolium 1-ethyl-3-methylimidazolium ethylsulfate' and phosphonium (tetrabutylphosphonium propanoate) were used as ionic liquids for docosahexaenoic acid (DHA)-rich lipid extraction and FAME from *Thraustochytrium* species (Zhang et al., 2018). More so, Orr et al. (2016) showed that '1-ethyl-3-methylimidazolium ethylsulfate' was useful for retrieving around 82% w/w of lipids from dewatered *Chlorella vulgaris* along with methanol at room temperature after 75 min of reaction time. Furthermore, $[TMAm][SO_4]$ has been reported to be beneficial as an ionic liquid for lipid extraction, especially eicosapentaenoic acid and docosahexaenoic acid, from microalgal biomass for biomedical and biofuel applications (Motlagh et al., 2019). However, the technical and economic viability of this method is still under investigation regarding whether it would be useful to produce large-scale biofuels.

38.3.3.3 Enzyme-Assisted Extraction Method

This method utilizes enzymes instead of solvents or ionic liquids for algal cell disruption to extract lipids. Cellulase and trypsin are the most common enzymes that are used in this approach to degrade cell surface polymers of algae to easily retrieve intracellular lipids (Taher et al., 2014). Recently, Wu et al. (2017) demonstrated that the combination of aqueous enzymes such as lysozyme, cellulase, protease, and pectinase is beneficial in the disruption of *Nannochloropsis* algal species cell walls at pH 4 and 50°C of temperature for 30 min to yield 90% of lipids, including neutral lipids. Likewise, Guo et al. (2017) showed that the lignocellulolytic enzymes from eight biomass-degrading bacterial strains containing laccase, xylanase, carboxymethyl cellulase, and filter paper activity enzymes can be utilized for algal cell wall disruption. This enzymatic approach has led to the extraction of 10.4–43.9% of neutral lipids from algae based on the utilization of enzyme type and bacterial strain. Although this method is rapid, specific, and utilizes less energy compared to mechanical methods, the quality of extracted lipids is dependent on the algal class and is prohibitively costly for large-scale processes in order to purify the extracted lipids (Liang et al., 2012).

38.3.4 NANOFILTRATION METHOD FOR FRACTIONATION OF ALGAL LIPIDS

Solvent resistant nanofiltration (SRNF) is a membrane separation approach with molecular cut-off (MWCO) in the range 200–1000 Da (Sereewatthanawut et al., 2018). Here, the solvent molecules pass through the membrane, whereas the oil molecules get rejected. Most of the natural product purification and fractionation involves a solvent extraction step. Hence, it is important to employ a solvent stable membrane in the extraction process. Polymers of elevated molecular weight such as polyimide and poly (amideimide) are used to fabricate a solvent resistant nanofiltration membrane that is suitable for non-aqueous systems. The polymer material used to make SRNF should have excellent chemical or thermal stability and compaction resistance. For example, STARMEM™ SRNF membrane (W.R. Grace & Co., USA) is a polyimide polymer based on SRNF with a molecular weight of 200–400 Da which is compatible with a wide range of organic solvents. More significantly, PI-crosslinked membranes are stable in common organic solvents such as dimethylformamide (DMF), dichloromethane (DCM), tetrahydrafuran (THF), and n-methyl-2-pyrolidone (NMP).

Some of the recent examples include DuraMem and PuraMem, polyimide-based membranes suitable for organic solvent nanofiltration for extensive practical and industrial applications (Gurr et al., 2002). These membranes have an MWCO in the 150–900 Da range, tolerate aggressive organic solvents, a maximum pressure up to 60 bar, and a maximum temperature of 50°C. In a separate study, an organic solvent nanofiltration technique to extract the microalgal oil, employing two commercial membranes (hydrophobic and amphiphilic) in the presence of different organic solvents, has been reported (Lopresto et al., 2017). In this study, membrane performance for oil rejection using organic solvents like n-Hexane, isopropanol, ethanol, methanol, acetone, binary hexane/isopropanol (3: 2 v/v) mixtures, and hexane/ethanol (1,2.5 v/v) were examined. Thus, it is noteworthy that the nanoparticles serve as a superior catalyst for retrieving algal lipids in the future for extensive biodiesel production.

38.4 CONCLUSION

Biodiesel generated from algae has been proven to be a better substitute in the search for sustainable and effective sources of fuel to meet the current demand. This chapter has discussed several methods for neutral lipid extraction from algal cells and their capability to be effectively used as biodiesel. Additionally, the emergence of nano-technology-based flocculants and membranes for enhancing neutral lipid extraction and increase their energy production efficiency by acting as a biodiesel were also discussed. It is noteworthy that the emergence of nanotechnology and its positive incorporation in an algal lipid extraction process and biodiesel production shows promise to replace conventional lipid extraction approaches in the future. Further, it is possible to mitigate the maximum carbon release via an algal neutral lipid-based biodiesel with the help of nanosized flocculants and membranes to improve their quality and efficiency.

REFERENCES

Adam, F., M. Abert-Vian, G. Peltier, and F. Chemat. 2012. Solvent-free ultrasound-assisted extraction of lipids from fresh microalgae cells: A green, clean and scalable process. *Bioresource technology* 114: 457–465.

Adetya, N. P., and H. Hadiyanto. 2018. Improvement of lipid yield from microalgae *Spirulina platensis* using ultrasound assisted osmotic shock extraction method. *IOP Conference Series: Earth and Environmental Science* 102: 012012.

Alonzo, F. and P. Mayzaud. 1999. Spectrofluorometric quantification of neutral and polar lipids in zooplankton using Nile red. *Marine chemistry* 67: 289–301.

Artan, A., C. Acquah, M. K. Danquah, and C. M. Ongkudon. 2015. Process analysis of microalgae biomass thermal disruption for biofuel production. In *Advances in Bioprocess Technology*, ed. P. Ravindra, 113–131. Cham: Springer International Publishing.

Bligh, E. G., and W. J. Dyer. 1959. A rapid method of total lipid extraction and purification. *Canadian Journal of Biochemistry and Physiology* 37: 911–917.

Brennan, L., and P. Owende. 2010. Biofuels from microalgae: A review of technologies for production, processing, and extractions of biofuels and co-products. *Renewable and Sustainable Energy Reviews* 14: 557–577.

Chemat, F. and M. K. Khan. 2011. Applications of ultrasound in food technology: Processing, preservation and extraction. *Ultrasonics Sonochemistry* 18: 813–835.

Chen, G.-Q., Y. Jiang, and F. Chen. 2007. Fatty acid and lipid class composition of the eicosa-pentaenoic acid-producing microalga, *Nitzschia laevis*. *Food Chemistry* 104: 1580–1585.

Chen, W., C. Zhang, L. Song, M. Sommerfeld, and Q. Hu. 2009. A high throughput Nile red method for quantitative measurement of neutral lipids in microalgae. *Journal of Microbiological Methods* 77: 41–47.

Chen, W., M. Sommerfeld, and Q. Hu. 2011. Microwave-assisted Nile red method for *in vivo* quantification of neutral lipids in microalgae. *Bioresource Technology* 102: 135–141.

Chisti, Y. 2013. Constraints to commercialization of algal fuels. *Journal of biotechnology* 167: 201–204.

Cooney, M., G. Young, and N. Nagle. 2009. Extraction of bio-oils from microalgae. *Separation & Purification Reviews* 38: 291–325.

de Moura, R. R., B. J. Etges, and E. O. dos Santos, et al. 2018. Microwave-assisted extraction of lipids from wet microalgae paste: A quick and efficient method. *European Journal of Lipid Science and Technology* 120: 1700419.

Demirbas, A. 2008. Production of biodiesel from algae oils. *Energy Sources, Part A: Recovery, Utilization, and Environmental Effects* 31: 163–168.

Dong, T., E. P. Knoshaug, P. T. Pienkos, and L. M. L. Laurens. 2016. Lipid recovery from wet oleaginous microbial biomass for biofuel production: A critical review. *Applied Energy* 177: 879–895.

Donot, F., C. Strub, and A. Fontana, et al. 2016. Rapid analysis and quantification of major neutral lipid species and free fatty acids by HPLC-ELSD from microalgae. *European Journal of Lipid Science and Technology* 118: 1550–1556.

Edeseyi, M. E., A. Y. Kaita, and R. Harun, et al. 2015. Rethinking sustainable biofuel marketing to titivate commercial interests. *Renewable and Sustainable Energy Reviews* 52: 781–792.

Elersek, T., A. Kapun, J. Golob, K. Flisar, and D. Miklavcic. 2016. Extraction of non-polar molecules from green alga *Chlorella vulgaris* by electroporation. *IFMBE Proceedings* 53: 379–383.

Folch, J., M. Lees, and G. H. S. Stanley. 1957. A simple method for the isolation and purification of total lipides from animal tissues. *Journal of Biological Chemistry* 226: 497–509.

Ganzler, K., A. Salgo, and K. Valko. 1986. Microwave extraction: A novel sample preparation method for chromatography. *Journal of Chromatography A* 371:299–306.

Garoma, T., and D. Janda. 2016. Investigation of the effects of microalgal cell concentration and electroporation, microwave and ultrasonication on lipid extraction efficiency. *Renewable Energy* 86: 117–123.

Geciova, J., D. Bury, and P. Jelen. 2002. Methods for disruption of microbial cells for potential use in the dairy industry: A review. *International Dairy Journal* 12: 541–553.

Georgianna, D. R. and S. P. Mayfield. 2012. Exploiting diversity and synthetic biology for the production of algal biofuels. *Nature* 488: 329–335.

Gerasimchuk, I., K. Kuhne, and J. Roth, et al. 2019. *Beyond Fossil Fuels: Fiscal Transition in BRICS*. Winnipeg, Manitoba: International Institute for Sustainable Development.

Gordillo, F. J. L., M. Goutx, F. L. Figueroa, and F. X. Niell. 1998. Effects of light intensity, CO_2 and nitrogen supply on lipid class composition of *Dunaliella viridis*. *Journal of Applied Phycology* 10: 135–144.

Guo, H., H. Chen, and L. Fan, et al. 2017. Enzymes produced by biomass-degrading bacteria can efficiently hydrolyze algal cell walls and facilitate lipid extraction. *Renewable Energy* 109: 195–201.

Gurr, M. I., J. L. Harwood, and K. N. Frayn. 2002. *Lipid Biochemistry*. 5th edition. Oxford: Blackwell Science.

Guschina, I. A., and J. L. Harwood. 2007. Complex lipid biosynthesis and its manipulation in plants. In *Improvement of Crop Plants for Industrial End Uses*, ed. P. Ranalli, 253–279. Dordrecht: Springer.

Guschina, I. A., and J. L. Harwood. 2013. Algal lipids and their metabolism. In *Algal Lipids and Their Metabolism*, ed. M. Borowitzka and N. Moheimani, 17–36. Dordrecht: Springer.

Halim, R., M. K. Danquah, and P. A. Webley. 2012. Extraction of oil from microalgae for biodiesel production: A review. *Biotechnology Advances* 30: 709–732.

Halim, R., B. Gladman, M. K. Danquah, and P. A. Webley. 2011. Oil extraction from microalgae for biodiesel production. *Bioresource Technology* 102: 178–185.

Hooftman, N., L. Oliveira, M. Messagie, T. Coosemans, and J. van Mierlo. 2016. Environmental analysis of petrol, diesel and electric passenger cars in a Belgian urban setting. *Energies* 9: 84.

Hopkins, T. R. 1991. Physical and chemical cell disruption for the recovery of intracellular proteins. *Bioprocess Technology* 12: 57–83.

Hu, Q., M. Sommerfeld, and E. Jarvis, et al. 2008. Microalgal triacylglycerols as feedstocks for biofuel production: Perspectives and advances. *Plant journal* 54: 621–639.

Iverson, S. J., S. L. C. Lang, and M. H. Cooper. 2001. Comparison of the Bligh and Dyer and Folch methods for total lipid determination in a broad range of marine tissue. *Lipids* 36: 1283–1287.

Jensen, S. K. 2008. Improved Bligh and Dyer extraction procedure. *Lipid Technology* 20: 280–281.

Johnson, M. B., and Z. Wen. 2009. Production of biodiesel fuel from the microalga *Schizochytrium limacinum* by direct transesterification of algal biomass. *Energy & Fuels* 23: 5179–5183.

Jones, J., S. Manning, M. Montoya, K. Keller, and M. Poenie. 2012. Extraction of algal lipids and their analysis by HPLC and mass spectrometry. *Journal of the American Oil Chemists' Society* 89: 1371–1381.

Joo, H., and A. Kumar. 2019. *World Biodiesel Policies and Production*. Boca Raton, FL: CRC Press.

Karatay, S. E., E. Demiray, and D. Donmez. 2019. Efficient approaches to convert *Coniochaeta hoffmannii* lipids into biodiesel by in-situ transesterification. *Bioresource technology* 285: 121321.

Karim, A., A. Yousuf, M. A. Islam, and Y. H. Naif, et al. 2018. Microbial lipid extraction from Lipomyces starkeyi using irreversible electroporation. *Biotechnology Progress* 34: 838–845.

Khanal, S. K., D. Grewell, S. Sung, and J. van Leeuwen. 2007. Ultrasound applications in wastewater sludge pretreatment: A review. *Critical Reviews in Environmental Science and Technology* 37: 277–313.

Kumar, M., S. Sundaram, E. Gnansounou, C. Larroche, and I. S. Thakur. 2018. Carbon dioxide capture, storage and production of biofuel and biomaterials by bacteria: A review. *Bioresource Technology* 247: 1059–1068.

Kumar, R. R., P. H. Rao, and M. Arumugam. 2015. Lipid extraction methods from microalgae: A comprehensive review. *Frontiers in Energy Research* 2: 61.

Kumar, S. P. Jeevan, G. V. Kumar, et al. 2017. Sustainable green solvents and techniques for lipid extraction from microalgae: A review. *Algal Research* 21: 138–147.

Kurokawa, M., P. M. King, and X. Wu, et al. 2016. Effect of sonication frequency on the disruption of algae. *Ultrasonics Sonochemistry* 31: 157–162.

Lee, A. K., D. M. Lewis, and P. J. Ashman. 2012. Disruption of microalgal cells for the extraction of lipids for biofuels: Processes and specific energy requirements. *Biomass and Bioenergy* 46: 89–101.

Lee, J. Y., C. Yoo, S. Y. Jun, C. Y. Ahn, and H. M. Oh. 2010. Comparison of several methods for effective lipid extraction from microalgae. *Bioresource Technology* 101: S75–SS7.

Levine, R. B., T. Pinnarat, and P. E. Savage. 2010. Biodiesel production from wet algal biomass through *in situ* lipid hydrolysis and supercritical transesterification. *Energy & Fuels* 24: 5235–5243.

Li, Q., Z. Zhou, D. Zhang, Z. Wang, and W. Cong. 2019. Lipid extraction from *Nannochloropsis oceanica* biomass after extrusion pretreatment with twin-screw extruder: Optimization of processing parameters and comparison of lipid quality. *Bioprocess and Biosystems Engineering* 2019: 02263.

Liang, K., Q. Zhang, and W. Cong. 2012. Enzyme-assisted aqueous extraction of lipid from microalgae. *Journal of Agricultural and Food Chemistry* 60: 11771–11776.

Liu, B., A. Vieler, C. Li, A. D. Jones, and C. Benning. 2013. Triacylglycerol profiling of micro-algae *Chlamydomonas reinhardtii* and *Nannochloropsis oceanica*. *Bioresource Technology* 146: 310–316.

Lopresto, C. G., S. Darvishmanesh, and A. Ehsanzadeh, et al. 2017. Application of organic solvent nanofiltration for microalgae extract concentration. *Biofuels, Bioproducts and Biorefining* 11: 307–324.

Maass, H. J. and H. O. Mockel. 2020. Combined decarbonization of electrical energy genera-tion and production of synthetic fuels by renewable energies and fossil fuels. *Chemical Engineering & Technology* 43: 111–118.

Mansour, M. P., J. K. Volkman, and S. I. Blackburn. 2003. The effect of growth phase on the lipid class, fatty acid and sterol composition in the marine dinoflagellate, *Gymnodinium* sp. in batch culture. *Phytochemistry* 63: 145–153.

Mason, T. J., J. P. Lorimer, D. M. Bates, and Y. Zhao. 1994. Dosimetry in sonochemistry: The use of aqueous terephthalate ion as a fluorescence monitor. *Ultrasonics Sonochemistry* 1: S91-SS5.

Meireles, L. A., A. C. Guedes, and F. X. Malcata. 2003. Lipid class composition of the micro-alga *Pavlova lutheri*: Eicosapentaenoic and docosahexaenoic acids. *Journal of Agricultural and Food Chemistry* 51:2237–2241.

Mohan, S. V., M. V. Rohit, P. Chiranjeevi, R. Chandra, and B. Navaneeth. 2015. Heterotrophic microalgae cultivation to synergize biodiesel production with waste remediation: Progress and perspectives. *Bioresource Technology* 184: 169–178.

Motlagh, S. R., R. Harun, and R. D. A. Biak, et al. 2019. Screening of suitable ionic liquids as green solvents for extraction of eicosapentaenoic acid (EPA) from microalgae biomass using COSMO-RS Model. *Molecules* 24: 713.

Mu, D., R. Ruan, and M. Addy, et al. 2017. Life cycle assessment and nutrient analysis of vari-ous processing pathways in algal biofuel production. *Bioresource Technology* 230: 33–42.

Nam, K., H. Lee, S.-W. Heo, Y. K. Chang, and J.-I. Han. 2017. Cultivation of *Chlorella vul-garis* with swine wastewater and potential for algal biodiesel production. *Journal of Applied Phycology* 29: 1171–1178.

Nyashina, G. S., K. Yu Vershinina, M. A. Dmitrienko, and P. A. Strizhak. 2018. Environmental benefits and drawbacks of composite fuels based on industrial wastes and different ranks of coal. *Journal of Hazardous Materials* 347: 359–370.

Orr, V. C. A., N. V. Plechkova, K. R. Seddon, and L. Rehmann. 2016. Disruption and wet extraction of the microalgae *Chlorella vulgaris* using room-temperature ionic liquids. *ACS Sustainable Chemistry & Engineering* 4: 591–600.

Otsuka, H. and Y. Morimura. 1966. Change of fatty acid composition of *Chlorella ellipsoidea* during its cell cycle. *Plant and Cell Physiology* 7: 663–670.

Pal, P., K. W. Chew, and H. W. Yen, et al. 2019. Cultivation of oily microalgae for the produc-tion of third-generation biofuels. *Sustainability* 11: 5424.

Pan, J., T. Muppaneni, and Y. Sun, et al. 2016. Microwave-assisted extraction of lipids from microalgae using an ionic liquid solvent [BMIM][HSO$_4$]. *Fuel* 178: 49–55.

Pare, J. R., G. Matni, and J. M. Belanger, et al. 1997. Use of the microwave-assisted process in extraction of fat from meat, dairy, and egg products under atmospheric pressure con-ditions. *Journal of AOAC International* 80: 928–933.

Patel, A., N. Arora, V. Pruthi, and P. A. Pruthi. 2019. A novel rapid ultrasonication-microwave treatment for total lipid extraction from wet oleaginous yeast biomass for sustainable biodiesel production. *Ultrasonics Sonochemistry* 51: 504–516.

Patil, P. D., H. Reddy, T. Muppaneni, and S. Deng. 2017. Biodiesel fuel production from algal lipids using supercritical methyl acetate (glycerin-free) technology. *Fuel* 195: 201–207.

Pirjola, L., H. Kuuluvainen, and H. Timonen, et al. 2019. Potential of renewable fuel to reduce diesel exhaust particle emissions. *Applied Energy* 254: 113636.

Ramesh, D. 2013. Lipid identification and extraction techniques. In: *Biotechnological Applications of Microalgae: Biodiesel and Value-Added Products*, ed. F. Bux, 89–97. Boca Raton, FL: CRC Press.

Ritchie, H. and M. Roser. 2017. *Fossil fuels: Our World in Data.* Oxford: Oxford University.

Saroya, S., V. Bansal, R. Gupta, A. S. Mathur, and P. Mehta. 2018. *Comparison of lipid extraction from algae (Chlorella species) using wet lipid extraction procedure and Bligh and dry method.* In *2018 IEEE International Students' Conference on Electrical, Electronics and Computer Science (SCEECS),* 24–25 Feb. 2018, Bhopal, India.

Sereewatthanawut, I., F. C. Ferreira, and J. Hirunlabh. 2018. Advances in green engineering for natural products processing: Nanoseparation membrane technology. *Journal of Engineering Science & Technology Review* 11: 196–214.

Sheng, J., R. Vannela, and B. E. Rittmann. 2011. Evaluation of methods to extract and quantify lipids from *Synechocystis* PCC 6803. *Bioresource Technology* 102: 1697–1703.

Shirazi, H. M., J. Karimi-Sabet, and C. Ghotbi. 2017. Biodiesel production from *Spirulina* microalgae feedstock using direct transesterification near supercritical methanol condition. *Bioresource Technology* 239: 378–386.

Singh, P. and P. Jaiswal. 2019. A review of production, properties and advantages of biodiesel. *International Journal of Innovative Research in Technology* 6: 40–45.

Solovchenko, A. E. 2012. Physiological role of neutral lipid accumulation in eukaryotic microalgae under stresses. *Russian Journal of Plant Physiology* 59: 167–176.

Sommerfeld, M., W. Chen, and Q. Hu, et al. 2008. Application of electroporation for lipid extraction from microalgae. In *8th International Conference on Algal Biomass, Biofuels and Bioproducts,* June 11–12, Seattle, WA.

Srivastava, N., M. Srivastava, and P. K. Mishra, et al. 2018. Applications of fungal cellulases in biofuel production: Advances and limitations. *Renewable and Sustainable Energy Reviews* 82: 2379–2386.

Suslick, K. S. and D. J. Flannigan. 2008. Inside a collapsing bubble: sonoluminescence and the conditions during cavitation. *Annual Review of Physical Chemistry* 59: 659–683.

Sydney, T., J.-A. Marshall-Thompson, V. R. Kapoore, et al. 2018. The effect of high-intensity ultraviolet light to elicit microalgal cell lysis and enhance lipid extraction. *Metabolites* 8: 65.

Taher, H., S. Al-Zuhair, A. H. Al-Marzouqi, Y. Haik, and M. Farid. 2014. Effective extraction of microalgae lipids from wet biomass for biodiesel production. *Biomass and Bioenergy* 66: 159–167.

Takisawa, K., K. Kanemoto, M. Kartikawati, and Y. Kitamura. 2014. Overview of biodiesel production from microalgae. *Journal of Developments in Sustainable Agriculture* 9: 120–128.

Tan, K. W. M. and Y. K. Lee. 2016. The dilemma for lipid productivity in green microalgae: importance of substrate provision in improving oil yield without sacrificing growth. *Biotechnology for Biofuels* 9: 255.

Tang, Y., Y. Zhang, and J. N. Rosenberg, et al. 2016. Efficient lipid extraction and quantification of fatty acids from algal biomass using accelerated solvent extraction (ASE). *RSC Advances* 6: 29127–29134.

Terme, N., R. Boulho, and M. Kendel, et al. 2017. Selective extraction of lipid classes from *Solieria chordalis* and *Sargassum muticum* using supercritical carbon dioxide and conventional solid-liquid methods. *Journal of Applied Phycology* 29: 2513–2519.

Theodosiou, G., C. Koroneos, and N. Moussiopoulos. 2007. Alternative scenarios analysis concerning different types of fuels used for the coverage of the energy requirements of a typical apartment building in Thessaloniki, Greece. Part I: Fuel consumption and emissions. *Building and Environment* 42: 1522–1530.

Topare, N. S., S. J. Raut, and V. C. Renge, et al. 2011. Extraction of oil from algae by solvent extraction and oil expeller method. *International Journal of Chemical Sciences* 9: 1746–1750.

Vo, T., T. Mai, and H. Vu, et al. 2017. Effect of osmotic stress and nutrient starvation on the growth, carotenoid and lipid accumulation in *Dunaliella salina* A9. *Research in Plant sciences* 5: 1–8.

Voloshin, R. A., M. V. Rodionova, S. K. Zharmukhamedov, T. N. Veziroglu, and S. I. Allakhverdiev. 2016. Biofuel production from plant and algal biomass. *International Journal of Hydrogen Energy* 41: 17257–17273.

Wang, G. and T. Wang. 2012. Characterization of lipid components in two microalgae for biofuel application. *Journal of the American Oil Chemists' Society* 89: 135–143.

Wang, T., X. Tian, and T. Liu, et al. 2017. A two-stage fed-batch heterotrophic culture of *Chlorella protothecoides* that combined nitrogen depletion with hyperosmotic stress strategy enhanced lipid yield and productivity. *Process Biochemistry* 60: 74–83.

Weerheim, A. M., A. M. Kolb, A. Sturk, and R. Nieuwland. 2002. Phospholipid composition of cell-derived microparticles determined by one-dimensional high-performance thin-layer chromatography. *Analytical Biochemistry* 302: 191–198.

Wu, C., Y. Xiao, and W. Lin, et al. 2017. Aqueous enzymatic process for cell wall degradation and lipid extraction from *Nannochloropsis* sp. *Bioresource Technology* 223: 312–316.

Yang, F., L. Long, and X. Sun, et al. 2014. Optimization of medium using response surface methodology for lipid production by *Scenedesmus* sp. *Marine Drugs* 12: 1245–1257.

Yang, H. Y., W. J. Lu, and Y. C. Chen, et al. 2017. New algal lipid extraction procedure using an amphiphilic amine solvent and ionic liquid. *Biomass and Bioenergy* 100: 108–115.

You, J., K. Mallery, and D. Mashek, et al. 2020. Microalgal swimming signatures and neutral lipids production across growth phases. *Biotechnology and Bioengineering* 2020: 27271.

Zghaibi, N., R. Omar, and M. S. M. Kamal, et al. 2019. Microwave-assisted brine extraction for enhancement of the quantity and quality of lipid production from microalgae *Nannochloropsis* sp. *Molecules* 24: 3581.

Zhang, Y., V. Ward, and D. Dennis, et al. 2018. Efficient extraction of a docosahexaenoic acid (DHA)-rich lipid fraction from *Thraustochytrium* sp. using ionic liquids. *Materials* 11: 1986.

Zheng, Y., T. Li, X. Yu, P. D. Bates, T. Dong, and S. Chen. 2013. High-density fed-batch culture of a thermotolerant microalga *Chlorella sorokiniana* for biofuel production. *Applied Energy* 108: 281–287.

Zhu, L., N. Gao, and R. G. Cong. 2017. Application of biotechnology for the production of biomass-based fuels. *Bio Med Research International* 2017: 3896505.

39 Microalgal Biodiesel Production

Ramachandran Sivaramakrishnan

Aran Incharoensakdi

CONTENTS

39.1   INTRODUCTION

Fossil fuels and coal are non-renewable sources, which serve conventional energy requirements. Maximum oil production will be reached by 2022 after which oil resources will be depleted (Bankovic-Ilic et al., 2012). Nuclear, coal, wind, solar energy, etc. serve as energy resources. However, around 86% of primary energy is produced from fossil fuels (Correa and Arbilla, 2008). Due to the depletion of fuel reserves and environmental issues, researchers have directed the focus towards renewable resources or alternative fuels.

Biodiesel is primarily considered as the alternative fuel which has the potential to replace fossil fuels. Biodiesel is nothing but methyl esters and it can be produced from edible, nonedible, and waste oils (Singh et al., 2019). It can be used directly or as a blend with petrodiesel (Hu et al., 2008; Qi et al., 2009). The first and second-generation fuels cannot be commercialized due to food vs. fuel conflicts and sustainability issues. However, microalgae are considered as efficient feedstocks for biodiesel production due to their rapid growth rate and high lipid content. Moreover, the use of microalgae does not compete with the food supply. The main advantages of biodiesel are that it is sulfur free, renewable, eco-friendly, and biodegradable (Shi et al., 2013). Engine performance can be improved by biodiesel and the modified engine system can reduce engine emissions by recirculating them. Biodiesel almost shows similar properties to diesel fuel (Shameer et al., 2017).

However, some drawbacks need to be addressed, such as high viscosity, and high cloud and pour point (Demirbas and Demirbas, 2011). Various sources are available for biodiesel production. Biodiesel can be produced from pyrolysis, cracking, supercritical methods, and transesterification. However, the important route of biodiesel production is through transesterification which converts triglycerides into methyl esters with the help of alcohol and a catalyst, whereas glycerol is obtained as a by-product (Bet-Moushoul et al., 2016). This chapter describes the downstream processing, including extraction methods, biodiesel production, and fuel quality, to understand and explore microalgal biodiesel production in detail.

39.2   HISTORY OF BIODIESEL PRODUCTION

In general, biodiesel is produced from the lipid which is obtained by the photosynthesis process in plants. According to the ASTM, biodiesel is the mono alkyl ester which is produced by the transesterification of waste oils, and nonedible and edible oils with the help of a catalyst and methanol (Singh and Singh, 2010). Various research improvements were undertaken for biodiesel production, such as development of new feedstocks, novel biodiesel production methods, quality of fuel properties, and carbon-neutral challenges (Kumar et al., 2010). Hence, depending on the types of feedstocks, biodiesel production has been classified into first, second, and third generation biodiesel. However, man-made biological tools are considered as the fourth generation of biodiesel which is still at a premature stage (Singh et al., 2019).

39.2.1 First-Generation Biodiesel

First generation biodiesel is derived from edible oils such as palm oil, soybean oil, rapeseed oil, sunflower oil, olive oil, rice bran oil, coconut oil, corn oil, and mustard oil (Mahdavi et al., 2015). At the beginning of biodiesel research these edible oils were commonly considered for biodiesel production. The readily available edible oils and simple production methods make first-generation biodiesel production very convenient. However, this affects the food supply and increases oil costs and regular food products costs. These plant-based oils have a limitation regarding cultivation area. The high costs are also the major drawback of using edible oils for biodiesel production. Hence, researchers moved to other oil sources for biodiesel production which do not affect the food supply (Tariq et al., 2012).

39.2.2 Second-Generation Biodiesel

Nonedible oil resources were considered for second-generation biodiesel. The prominent nonedible oils used were Jatropha oil, karanja oil, rubber seed oil, neem oil, *Mahua indica* oil, and nagchampa oil (Shameer et al., 2017; Shameer and Ramesh, 2017). The main drawback of first-generation biodiesel was food competition. The second-generation nonedible feedstock does not compete with food and is eco-friendly in nature. In addition, the production cost is lower when compared to first-generation biodiesel. The major advantage of second-generation biodiesel is that it does not require arable lands and neither does it rely on food-based plants. However, nonarable lands showed very low yields of plants with Jatropha, karanja, and jojoba oil. Hence, it is necessary to cultivate plants in arable land which directly affects food production and social economic issues (Tariq et al., 2012).

39.2.3 Third-Generation Biodiesel

Microalgal feedstock is mainly considered for third-generation biodiesel production. The important advantage of microalgae-based biodiesel is the high growth rate of the microalgae, its high lipid productivity, the reduction of greenhouse gas effects, and competition with food productivity. However, the major drawback of using microalgal biodiesel is the requirement of a huge investment, the requirement of continuous sunlight, as well as the harvesting and oil extraction procedures, which are difficult. Other important sources like waste cooking oil, animal fats, and fish oils are also considered for third-generation biodiesel (Verma et al., 2016). However, excluding microalgae, other sources also faced the same issues as second-generation biodiesel. On the other hand, waste oils or other cooking oils have the advantage of recycling the used oils for biodiesel purposes, which indirectly contributes to the diminishing of water pollution (Bhatti et al., 2008). However, microalgae are considered as promising feedstock for third-generation biodiesel due to their high growth rate and potential to survive in various environmental conditions (Mata et al., 2010). Nowadays, electro-fuels and photobiological solar fuels can produce biodiesel by converting solar energy and raw materials, and this is considered as fourth-generation biodiesel (Cameron et al., 2014).

39.3   FEEDSTOCKS FOR BIODIESEL PRODUCTION

Various feedstocks are available for biodiesel production, for example, plant or vegetable oils, microbial oil, and animal fats. Different oils have different lipid compositions and purity and exhibit different fuel properties (Mahdavi et al., 2015). The selection of feedstock is very important for biodiesel production because it may affect the production cost, biodiesel yield, and fuel properties. Mostly, the current feedstocks for biodiesel production are classified according to their nature, such as edible, nonedible, and waste oil (Demirbas and Demirbas, 2011). Feedstock selection also depends on region and environmental conditions, as well as on availability and a country's economic aspects. For instance, Canada used canola oil as biodiesel feedstock, whereas the USA and Brazil prefer soybean oil for the biodiesel feedstock. European countries like the UK, Finland, Italy, and Germany used rapeseed oil for biodiesel production. Jatropha and karanja oils are preferably used as biodiesel feedstock in India, whereas Malaysia and Indonesia prefer palm and coconut oils.

However, serious issues were raised due to the utilization of food plant oils and edible oils for biodiesel production which may affect the food chain (Kumar et al., 2010). Hence, it is recommended to consider nonedible oils for biodiesel feedstock, which also has the advantages of low sulfur emission, is biodegradable, and does not disrupt the food chain. Biodiesel feedstocks are classified according to the biodiesel generations. Coconut oil, palm oil, soybean oil, and sunflower oil are considered first-generation feedstocks. In the second generation, cotton seed oil, Jatropha oil, jojoba oil, karanja oil, linseed oil, mahua oil, neem oil, rubber seed oil, and tobacco oil were considered for biodiesel feedstocks. The third-generation biodiesel feedstocks are animal fat, waste-based oil or waste cooking oil, and microalgae (Demirbas and Demirbas, 2011).

39.4   POTENTIAL OF MICROALGAL BIODIESEL PRODUCTION

In the current age, microalgae receive much attention for biodiesel production due to their rapid growth rate and high lipid content. Microalgae also produce various valuable chemicals or products other than lipids, such as enzymes, proteins, vitamins, antioxidants, pigments, and omega fatty acids, whereas spent biomass can also be used for animal feed, biogas production, or ethanol production (Coustets et al., 2015; Klok et al., 2014; Santos et al., 2014; Solana et al., 2014; Sivaramakrishnan and Incharoensakdi, 2020). Microalgae have the capability of producing a higher yield of oil than that obtained by plants and which is suitable for biodiesel production (Girard et al., 2014; Nan et al., 2015; Tan et al., 2016). *Chlorella* sp. and *Schizochytrium limacium* sp. are known for their high oil production, and such microalgae contain more than 50% dry cell weight (DCW) lipids (Gao et al., 2010; Johnson and Wen, 2009; Mata et al., 2010; Sharma and Singh, 2017).

The noteworthy advantage of microalgae over plants is that they have a rapid growth rate and their cultivation requires no arable land (Nan et al., 2015; Singh et al., 2014). Besides, microalgae also sequestrate carbon dioxide from the environment, which is the beneficial aspect of microalgal biodiesel production. In nature, microalgae produce a certain amount of lipids. Moreover, when they are subjected to environmental stress they can produce more lipids, which is an advantage (Sivaramakrishnan and Incharoensakdi, 2017b). Microalgae have a high photosynthetic efficiency which results in more carbon dioxide fixation (Kiran and Thanasekaran, 2011). In addition, upon cultivation of microalgae in nutrient rich wastewater, the heavy metals present in the wastewater can be removed (Kiran and Thanasekaran, 2011). Microalgae have the ability to produce 19,000 to 57,000 l oil/acre/year which is far better than other oil sources (Demirbas and Demirbas, 2011).

39.5   LIPID EXTRACTION FROM MICROALGAE

39.5.1   ORGANIC SOLVENT EXTRACTION

After the harvesting of microalgae, organic solvents are used for the extraction of lipids. Unlike other terrestrial plants, mechanical pressing is not suitable for microalgal extraction due to the size of the microalgae and the toughness of the cell wall. Hence, organic solvents are preferred for oil extraction from microalgae (Cooney et al., 2009). Lipid extraction depends on the type of solvent used, for instance neutral lipids are extracted by non-polar solvents, whereas cell membrane associated polar lipids are extracted using polar solvents (Martinez-Guerra et al., 2014). In addition, a suitable physical or mechanical cell wall disruption method increases the efficiency of lipid extraction. Hence, it is necessary to choose a suitable solvent and cell disruption method for low cost microalgal oil extraction.

39.5.2   ULTRASOUND-MEDIATED OIL EXTRACTION

Ultrasound mediated oil extraction is a mechanical method in which high energy acoustic waves are used. The ultrasonic waves not only break the strong microalgal cell walls, but also generate an attractive force between solvent and lipid molecules through cavitation effects. During ultrasonic bubble collapse, very high pressures and temperatures are generated which cause dispersive effects on cell walls and bring the intracellular lipids to the solvent medium (Araujo et al., 2013). Hence, the overall extraction process becomes easy. However, the energy requirement for the ultrasound is high which affects the overall processing cost.

39.5.3   MICROWAVE-MEDIATED OIL EXTRACTION

Microwave mediated oil extraction is brought about by the conduction of dipole moments between lipid molecules and ions. During treatment, the electric and magnetic field of microwaves change their direction rapidly, whereas the polar molecules adapt to this change – the so-called dipole moment. The response to this effect causes friction and localized heating. Polar molecules like solvents or water easily get heated when exposed to electromagnetic waves. The high temperature generated at molecule level due to the dipole moment increases the dissolution efficiency, resulting in high extraction efficiency (Martinez-Guerra et al., 2014; Mubarak et al., 2015). Although microwave mediated microalgal lipid extraction has high efficiency, the energy consumption of the method is very high and affects the overall energy consumption of the process (Nogueira et al., 2018).

39.5.4 SUPERCRITICAL FLUID (SF) EXTRACTION

Due to the short reaction time and absence of a catalyst, SF is used for biodiesel production (Saka and Kusdiana, 2001). However, due to the benefits of SF, it is also used in microalgal lipid extraction. The common SFs used for lipid extraction are CO_2, ethanol, and methanol. In addition, the polarity of the solvent can be reduced during SF treatment, which is also suitable for the neutral lipids (Patil et al., 2012). The major advantage of SF is that the separation of the solvent after extraction is easy and that pure lipids are obtained. Among the various SFs, CO_2 is widely considered for microalgal lipid extraction due to its high extraction efficiency (Cheng et al., 2011; Halim et al., 2011). Among the various extraction methods, SF extraction shows high efficiency for oil extraction. The initial set-up cost is high for SF mediated extraction. However, it is more suitable for triglyceride extraction as well as for large-scale operations (Cheng et al., 2011; Lorenzen et al., 2017).

39.5.5 HYDROTHERMAL LIQUEFACTION

Microalgal 'hydrothermal liquefaction' (HTL) is the process of converting raw biomass into biooil or valuable components at subcritical conditions using high water activity. In this method, wet microalgae can be used for the process and there is no need for drying of the biomass. In addition, apart from the lipid extraction, the HTL process also converts proteins and carbohydrates into biooil which in turn reduces the overall production cost (Biller and Ross, 2011; Brown et al., 2010; Sawayama et al., 1999). A previous study reported that the HTL of *Chlorella vulgaris* and *Spirulina* produce satisfactory biooil, which is higher than the microalgal lipid extraction yield (Biller and Ross, 2011; Yang et al., 2004). Yu et al. (2011) slightly modified the HTL method and converted microalgae into biocrude oil. The results showed that the increasing HTL temperature increased the biooil yield, carbon, and nitrogen fraction. Despite the high energy consumption by HTL, the process can also produce multiple valuable products which can compensate for the energy investment.

39.6 MICROALGAL BIODIESEL PRODUCTION METHODS

39.6.1 BASE-CATALYZED TRANSESTERIFICATION

Transesterification is the reaction where triglycerides react with alcohol to produce methyl esters and glycerol with the help of catalysts. The catalysts used in transesterification are base, acid, or enzymes. However, the prominent catalysts for biodiesel production are base catalysts such as KOH or NaOH. Base catalyzed transesterification can be achieved under lower pressure and temperatures with 98% biodiesel conversion (Taher et al., 2011). However, when the free fatty acid content of oil exceeds 2–5% saponification will occur, which badly affects the biodiesel yield (Banani et al., 2015). When comparing base catalyzed two-step transesterification and one-step transesterification, the results showed that both transesterifications achieved a satisfactory yield of biodiesel with acceptable fuel properties (Sivaramakrishnan and Incharoensakdi, 2017d, 2019a).

39.6.2 Acid-catalyzed Transesterification

Comparing to the base catalysts, acid catalysts show lower efficiency in transesterification. However, acid catalysts are suitable for oil containing high free fatty acid. Acid catalyzed transesterification has poor performance on biodiesel yield. However, acid catalyst biodiesel yield can be increased by applying high temperature and pressure, though the process is not feasible for large scale biodiesel production (Sivaramakrishnan and Incharoensakdi, 2018b). Biodiesel from acid catalyzed transesterification also causes engine corrosion. The biodiesel yield of 85% could be obtained after the acid catalyzed transesterification of non-edible *Spirulina platensis* oil under optimum conditions (El-Shimi et al., 2013). There are some limitations using chemical (base or acid) catalysts due to the drawbacks in downstream processing, such as catalyst separation, soap formation during biodiesel washing, difficulty in glycerol separation, engine corrosion issues, and high energy input. Addressing these limitations may enhance the possibility of microalgal biodiesel production on a large scale.

39.6.3 Enzyme-catalyzed Transesterification

Due to the limitations in chemical catalyst-based transesterification, enzyme catalysts are considered as promising. The advantages are that free fatty acid or water has no effect on the reaction, the reaction can be performed at lower temperatures, the enzyme catalysts can be easily recovered and reused, there is a low alcohol ratio, and less power consumption (Sivaramakrishnan and Muthukumar, 2012a; Sivaramakrishnan and Incharoensakdi, 2016). However, the main disadvantages are due to the enzyme cost and slow reaction rate. Both intracellular and extracellular enzymes are considered for biodiesel production, mostly in the form of immobilized lipase (Rawat et al., 2011). However, the enzyme activity can be seriously inhibited by excess alcohol and the glycerol by-product. Hence, prior purification is required before the reuse of enzyme which affects the overall production cost. Many studies successfully utilized enzyme catalysts for microalgal biodiesel production and produced satisfactory biodiesel yields (Li et al., 2007; Taher et al., 2011; Xiong et al., 2008).

39.6.4 *In situ* Transesterification

Conventional microalgal transesterification consumes high energy due to the dewatering of microalgae, drying, and lipid extraction. Hence, *in situ* transesterification was developed in order to achieve biodiesel production in one step (Shirazi et al., 2017). In *in situ* transesterification, microalgal biomass reacts with alcohol and an extraction solvent or reacts with a single solvent which has both extraction and transesterification capacity to produce biodiesel (Zhu et al., 2017). *In situ* transesterification eliminates lipid loss and converts maximum lipids into biodiesel. In addition, valuable by-products such as ethyl formate, diethyl ether, ethyl levulinate, glycerol carbonate, and glycerol can be produced during *in situ* transesterification (Chisti, 2007). Previous studies reported algal *in situ* transesterification by using dimethyl

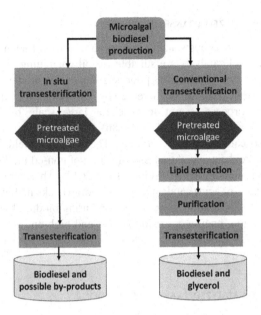

FIGURE 39.1   Comparison of *in situ* and conventional transesterification.

carbonate (DMC) as a solvent (Sivaramakrishnan and Muthukumar, 2014; Sivaramakrishnan and Incharoensakdi, 2017a). DMC acts as both extraction and transesterification reagent and efficiently produces biodiesel with a higher yield than that by two-step transesterification with satisfactory fuel properties. Besides, DMC mediated transesterification also produces valuable glycerol carbonate as a by-product. The comparison of conventional and *in situ* transesterification is represented in Figure 39.1.

39.6.5   Ultrasound-assisted Transesterification

The transesterification process can be enhanced by exposing the transesterification reaction mixture to ultrasonic waves. A previous study reported that the intensification of ultrasound on enzyme catalyzed algal oil transesterification enhances biodiesel production and reduces production time considerably (Sivaramakrishnan and Muthukumar, 2012b). On the other hand, ultrasound assisted *in situ* transesterification of *Botryococcus* sp. using an enzyme catalyst showed high biodiesel yield with a shorter reaction time and efficient by-product synthesis (Sivaramakrishnan and Incharoensakdi, 2017a). Martinez-Guerra et al. (2014) reported that the ultrasound assisted transesterification of *Spirulina* showed a higher conversion rate than the regular conventional method. The application of ultrasound to the transesterification reaction mixture decreases the immiscibility of catalysts and reactants and increases the mass transfer properties, resulting in a higher reaction rate and short reaction duration. Reducing the reaction time by ultrasound reduces the overall production cost of microalgal biodiesel production (Sivaramakrishnan and Incharoensakdi, 2018b).

39.6.6 SUPERCRITICAL TRANSESTERIFICATION

Transesterification can also be performed under supercritical conditions without any catalyst (Jazzar et al., 2015). Supercritical transesterification of various sources have already been reported, such as rapeseed oil (Lim et al., 2010), palm oil (Song et al., 2008), Jatropha oil (Lim et al., 2010), and waste vegetable oil (Ghoreishi and Moein, 2013). In the case of microalgae, the rigid microalgal cell gets damaged during supercritical methanol reaction and allows solvent diffusion into the cells. The microalgae then act as a reactant and produce methyl ester by utilizing microalgal lipids. A study reported that the replacement of methanol with ethanol under supercritical conditions produces a similar ethyl ester yield (Lemoes et al., 2016). The high temperature during supercritical methanol reaction can be reduced by the addition of carbon dioxide (Saka and Isayama, 2009). Due to the reason that methanol is toxic, other longer chain alcohols such as isopropanol, butanol, and ethanol were considered as potential replacements for the supercritical transesterification (Huang et al., 2015). Although the supercritical transesterification shows a high biodiesel yield, the operating method consumes more energy which lowers the possibility of commercialization.

39.7 FUEL PROPERTIES OF MICROALGAL BIODIESEL

It is necessary to select the appropriate microalgae with suitable fatty acid compositions, ensuring acceptable fuel properties, as mentioned in international standards such as EN14214 or ASTM D675. It is important to achieve an economical and environmentally friendly biodiesel as an alternative for diesel fuel. The important biodiesel properties required to ensure the quality of biodiesel are the saponification value, iodine number, cetane number, cold filter plugging point, cloud point, degree of unsaturation, oxidative stability, kinematic viscosity, density, and higher heating value (Sivaramakrishnan and Incharoensakdi, 2017c). Most of the microalgal oils have a polyunsaturated fatty acid content higher than the standard which is not suitable for biodiesel quality. To ensure a high quality of biodiesel, the microalgae should be rich in saturated and monounsaturated fatty acid content (Ramirez-Verduzco et al., 2012).

39.8 ECONOMIC FEASIBILITY OF MICROALGAL BIODIESEL PRODUCTION

It was believed that commercial biodiesel production would be very costly. Several factors are involved when embarking on the building of a biodiesel plant, such as feedstock selection, location, design plan, capacity, and type of equipment (Khanna, 2015). However, microalgae have several attractive advantages like rapid growth, no arable land requirement, no competition with food, utilizing wastewater for

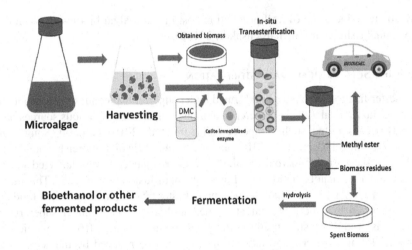

FIGURE 39.2 Biorefinery approach using microalgae feedstock for multiple products.

culturing, carbon sequestration, additional valuable compounds, and biorefinery possibilities (Sivaramakrishnan and Incharoensakdi, 2018c; Sivaramakrishnan et al., 2019) (Figure 39.2). In order to achieve high lipid content microalgae, strain development, genetic engineering strategies, physical or chemical stress, and heterotrophic cultivation can be used to cut the overall production cost (Sivaramakrishnan and Incharoensakdi 2018a, 2019b). To achieve commercial biodiesel production, the dry biomass should be produced at a cost below USD300/ton (Schenk et al., 2008). A study reported that the cost of biomass from open ponds, flat panel photobioreactors, and horizontal tubular photobioreactors are 4.95, 5.96, and 4.15 €/kg respectively (Norsker et al., 2011). The biomass production cost could be reduced by 0.68 €/kg if proper optimized conditions were developed, and this would allow for a reasonable price for biodiesel commercialization considerations (Norsker et al., 2011).

39.9   CONCLUSIONS AND PERSPECTIVES

Microalgal biodiesel has many advantages compared to petrodiesel fuels. Nevertheless, some technological bottlenecks still prevail with respect to the commercialization of microalgal biodiesel production. The primary concern is the cost requirement on initial set-up and continuous energy requirement for the microalgal biodiesel production steps. Hence it is necessary to further develop the existing methods to reduce the production cost. For instance, lipid extraction from wet biomass or direct transesterification eliminates the drying process and reduces production cost. To achieve cost-effective microalgal biodiesel production it is necessary to develop direct transesterification methods or transesterification of biomass with supercritical fluids. Utilization of nutrient-rich wastewater is another important strategy to cut the cost. However, scaling up is feasible when other valuable products can be simultaneously obtained from microalgae in addition to lipids. At present, the existing current technologies for biomass cultivation and biodiesel conversion at large scale are still at the premature stage. To achieve a greater breakthrough in microalgal biodiesel production, it is necessary to integrate more research with sufficiently wide collaboration, including government and private sectors.

ACKNOWLEDGMENTS

R. Sivaramakrishnan is grateful to the Graduate School and Faculty of Science, Chulalongkorn University (CU), for a senior post-doctoral fellowship from the Ratchadaphiseksomphot Endowment Fund. A.I. acknowledges the research grant from CU on the Frontier Research Energy Cluster (CU-59-048-EN) and from the Thailand Research Fund (IRG 5780008).

REFERENCES

Araujo, G. S., L. J. B. L. Matos, and J. O. Fernandes, et al. 2013. Extraction of lipids from microalgae by ultrasound application: Prospection of the optimal extraction method. *Ultrasonics Sonochemistry* 20:95–98.

Banani, R., S. Youssef, M. Bezzarga, and M. Abderrabba. 2015. Waste frying oil with high levels of free fatty acids as one of the prominent sources of biodiesel production. *Journal of Materials and Environmental Science* 6:1178–1185.

Bankovic-Ilic, I. B., O. S. Stamenkovic, and V. B. Veljkovic. 2012. Biodiesel production from non-edible plant oils. *Renewable and Sustainable Energy Reviews* 16:3621–3647.

Bet-Moushoul, E., K. Farhadi, and Y. Mansourpanah, et al. 2016. Application of CaO-based/ Au nanoparticles as heterogeneous nanocatalysts in biodiesel production. *Fuel* 164:119–127.

Bhatti, H. N., M. A. Hanif, M. Qasim, and A.-U. Rehman. 2008. Biodiesel production from waste tallow. *Fuel* 87:2961–2966.

Biller, P. and A. B. Ross. 2011. Potential yields and properties of oil from the hydrothermal liquefaction of microalgae with different biochemical content. *Bioresource Technology* 102:215–225.

Brown, T. M., P. Duan, and P. E. Savage. 2010. Hydrothermal liquefaction and gasification of *Nannochloropsis* sp. *Energy & Fuels* 24:3639–3646.

Cameron, D. E., C. J. Bashor, and J. J. Collins. 2014. A brief history of synthetic biology. *Nature Reviews Microbiology* 12:381–390.

Cheng, C. H., T. B. Du, and H. C. Pi, et al. 2011. Comparative study of lipid extraction from microalgae by organic solvent and supercritical CO_2. *Bioresource Technology* 102:10151–10153.

Chisti, Y. 2007. Biodiesel from microalgae. *Biotechnology Advances* 25:294–306.

Cooney, M., G. Young, and N. Nagle. 2009. Extraction of bio-oils from microalgae. *Separation & Purification Reviews* 38:291–325.

Correa, S. M. and G. Arbilla. 2008. Carbonyl emissions in diesel and biodiesel exhaust. *Atmospheric Environment* 42:769–775.

Coustets, M., V. Joubert-Durigneux, and J. Herault, et al. 2015. Optimization of protein electroextraction from microalgae by a flow process. *Bioelectrochemistry* 103:74–81.

Demirbas, A., and M. F. Demirbas. 2011. Importance of algae oil as a source of biodiesel. *Energy Conversion and Management* 52:163–170.

El-Shimi, H. I., N. K. Attia, S. T. El-Sheltawy and G. I. El-Diwani. 2013. Biodiesel production from *Spirulina platensis* microalgae by *in-situ* transesterification process. *Journal of Sustainable Bioenergy Systems.* 3:224–233.

Gao, C., Y. Zhai, Y. Ding, and Q. Wu. 2010. Application of sweet sorghum for biodiesel production by heterotrophic microalga *Chlorella protothecoides*. *Applied Energy* 87:756–761.

Ghoreishi, S. M. and P. Moein. 2013. Biodiesel synthesis from waste vegetable oil via transesterification reaction in supercritical methanol. *Journal of Supercritical Fluids* 76:24–31.

Girard, J. M., M. L. Roy, and M. ben Hafsa, et al. 2014. Mixotrophic cultivation of green microalgae *Scenedesmus obliquus* on cheese whey permeate for biodiesel production. *Algal Research* 5:241–248.

Halim, R., B. Gladman, M. K. Danquah, and P. A. Webley. 2011. Oil extraction from microalgae for biodiesel production. *Bioresource Technology* 102:178–185.

Hu, Z., P. Tan, X. Yan, and D. Lou. 2008. Life cycle energy, environment and economic assessment of soybean-based biodiesel as an alternative automotive fuel in China. *Energy* 33:1654–1658.

Huang, R., J. Cheng, and Y. Qiu, et al. 2015. Using renewable ethanol and isopropanol for lipid transesterification in wet microalgae cells to produce biodiesel with low crystallization temperature. *Energy Conversion and Management* 105:791–797.

Jazzar, S., P. Olivares-Carrillo, A. P. de los Rios, et al. 2015. Direct supercritical methanolysis of wet and dry unwashed marine microalgae (*Nannochloropsis gaditana*) to biodiesel. *Applied Energy* 148:210–219.

Johnson, M. B. and Z. Wen. 2009. Production of biodiesel fuel from the microalga *Schizochytrium limacinum* by direct transesterification of algal biomass. *Energy & Fuels* 23:5179–5183.

Khanna, N. 2015. Perspectives on algal engineering for enhanced biofuel production. In *Algal Biorefinery: An Integrated Approach*, ed. D. Das, 73–101. Cham: Springer International Publishing.

Kiran, B. and K. Thanasekaran. 2011. Metal tolerance of an indigenous cyanobacterial strain, *Lyngbya putealis*. *International Biodeterioration & Biodegradation* 65:1128–1132.

Klok, A. J., P. P. Lamers, D. E. Martens, R. B. Draaisma, and R. H. Wijffels. 2014. Edible oils from microalgae: Insights in TAG accumulation. *Trends in biotechnology* 32:521–528.

Kumar, A., K. Kumar, N. Kaushik, S. Sharma, and S. Mishra. 2010. Renewable energy in India: Current status and future potentials. *Renewable and Sustainable Energy Reviews* 14:2434–2442.

Lemoes, J. S., R. C. M. A. Sobrinho, and S. P. Farias, et al. 2016. Sustainable production of biodiesel from microalgae by direct transesterification. *Sustainable Chemistry and Pharmacy* 3:33–38.

Li, X., H. Xu, and Q. Wu. 2007. Large-scale biodiesel production from microalga *Chlorella protothecoides* through heterotrophic cultivation in bioreactors. *Biotechnology and Bioengineering* 98:764–771.

Lim, S., S. S. Hoong, L. K. Teong, and S. Bhatia. 2010. Supercritical fluid reactive extraction of *Jatropha curcas* L. seeds with methanol: A novel biodiesel production method. *Bioresource Technology* 101:7169–7172.

Lorenzen, J., N. Igl, and M. Tippelt, et al. 2017. Extraction of microalgae derived lipids with supercritical carbon dioxide in an industrial relevant pilot plant. *Bioprocess and Biosystems Engineering* 40:911–918.

Mahdavi, M., E. Abedini, and A. H. Darabi. 2015. Biodiesel synthesis from oleic acid by nanocatalyst (ZrO_2/Al_2O_3) under high voltage conditions. *RSC Advances* 5:55027–55032.

Martinez-Guerra, E., V. G. Gude, A. Mondala, W. Holmes, and R. Hernandez. 2014. Microwave and ultrasound enhanced extractive-transesterification of algal lipids. *Applied Energy* 129:354–363.

Mata, T. M., A. A. Martins, and N. S. Caetano. 2010. Microalgae for biodiesel production and other applications: A review. *Renewable and Sustainable Energy Reviews* 14:217–232.

Mubarak, M., A. Shaija, and T. V. Suchithra. 2015. A review on the extraction of lipid from microalgae for biodiesel production. *Algal Research* 7:117–123.

Nan, Y., J. Liu, R. Lin, and L. L. Tavlarides. 2015. Production of biodiesel from microalgae oil (*Chlorella protothecoides*) by non-catalytic transesterification in supercritical methanol and ethanol: process optimization. *Journal of Supercritical Fluids* 97:174–182.

Nogueira, D. A., J. M. da Silveira, E. M. Vidal, N. T. Ribeiro, and C. A. V. Burkert. 2018. Cell disruption of *Chaetoceros calcitrans* by microwave and ultrasound in lipid extraction. *International Journal of Chemical Engineering* 2018:9508723.

Norsker, N. H., M. J. Barbosa, M. H. Vermue, and R. H. Wijffels. 2011. Microalgal production: A close look at the economics. *Biotechnology Advances* 29:24–27.

Patil, P. D., V. G. Gude, and A. Mannarswamy, et al. 2012. Comparison of direct transesterification of algal biomass under supercritical methanol and microwave irradiation conditions. *Fuel* 97:822–831.

Qi, D. H., L. M. Geng, and H. Chen, et al. 2009. Combustion and performance evaluation of a diesel engine fueled with biodiesel produced from soybean crude oil. *Renewable Energy* 34:2706–2713.

Ramirez-Verduzco, L. P., J. E. Rodriguez-Rodriguez, and A. del R. Jaramillo-Jacob. 2012. Predicting cetane number, kinematic viscosity, density and higher heating value of biodiesel from its fatty acid methyl ester composition. *Fuel* 91:102–111.

Rawat, I., R. R. Kumar, T. Mutanda, and F. Bux. 2011. Dual role of microalgae: Phycoremediation of domestic wastewater and biomass production for sustainable biofuels production. *Applied Energy* 88:3411–3424.

Saka, S. and Y. Isayama. 2009. A new process for catalyst-free production of biodiesel using supercritical methyl acetate. *Fuel* 88:1307–1313.

Saka, S. and D. Kusdiana. 2001. Biodiesel fuel from rapeseed oil as prepared in supercritical methanol. *Fuel* 80:225–231.

Santos, N. O., S. M. Oliveira, L. C. Alves, and M. C. Cammarota. 2014. Methane production from marine microalgae *Isochrysis galbana*. *Bioresource Technology* 157:60–67.

Sawayama, S., T. Minowa, and S. Y. Yokoyama. 1999. Possibility of renewable energy production and CO_2 mitigation by thermochemical liquefaction of microalgae. *Biomass and Bioenergy* 17:33–39.

Schenk, P. M., S. R. Thomas-Hall, and E. Stephens, et al. 2008. Second generation biofuels: High-efficiency microalgae for biodiesel production. *BioEnergy Research* 1:20–43.

Shameer, P. M. and K. Ramesh. 2017. Green technology and performance consequences of an eco-friendly substance on a 4-stroke diesel engine at standard injection timing and compression ratio. *Journal of Mechanical Science and Technology* 31:1497–1507.

Shameer, P. M., K. Ramesh, R. Sakthivel, and R. Purnachandran. 2017. Effects of fuel injection parameters on emission characteristics of diesel engines operating on various biodiesel: A review. *Renewable and Sustainable Energy Reviews* 67:1267–1281.

Sharma, Y. C. and V. Singh. 2017. Microalgal biodiesel: A possible solution for India's energy security. *Renewable and Sustainable Energy Reviews* 67:72–88.

Shi, W., J. Li, and B. He, et al. 2013. Biodiesel production from waste chicken fat with low free fatty acids by an integrated catalytic process of composite membrane and sodium methoxide. *Bioresource Technology* 139:316–322.

Shirazi, H. M., J. Karimi-Sabet, and C. Ghotbi. 2017. Biodiesel production from *Spirulina* microalgae feedstock using direct transesterification near supercritical methanol condition. *Bioresource Technology* 239:378–386.

Singh, B., A. Guldhe, I. Rawat, and F. Bux. 2014. Towards a sustainable approach for development of biodiesel from plant and microalgae. *Renewable and Sustainable Energy Reviews* 29:216–245.

Singh, D., D. Sharma, S. L. Soni, S. Sharma, and D. Kumari. 2019. Chemical compositions, properties, and standards for different generation biodiesels: A review. *Fuel* 253:60–71.

Singh, S. P. and D. Singh. 2010. Biodiesel production through the use of different sources and characterization of oils and their esters as the substitute of diesel: A review. *Renewable and Sustainable Energy Reviews* 14:200–216.

Sivaramakrishnan, R. and A. Incharoensakdi. 2016. Purification and characterization of solvent tolerant lipase from *Bacillus* sp. for methyl ester production from algal oil. *Journal of Bioscience and Bioengineering* 121:517–522.

Sivaramakrishnan, R. and A. Incharoensakdi. 2017a. Direct transesterification of *Botryococcus* sp. catalysed by immobilized lipase: Ultrasound treatment can reduce reaction time with high yield of methyl ester. *Fuel* 191:363–370.

Sivaramakrishnan, R. and A. Incharoensakdi. 2017b. Enhancement of lipid production in *Scenedesmus* sp. by UV mutagenesis and hydrogen peroxide treatment. *Bioresource Technology* 235:366–370.

Sivaramakrishnan, R. and A. Incharoensakdi. 2017c. Enhancement of total lipid yield by nitrogen, carbon, and iron supplementation in isolated microalgae. *Journal of Phycology* 53:855–868.

Sivaramakrishnan, R. and A. Incharoensakdi. 2017d. Production of methyl ester from two microalgae by two-step transesterification and direct transesterification. *Environmental Science and Pollution Research* 24:4950–4963.

Sivaramakrishnan, R. and A. Incharoensakdi. 2018a. Enhancement of lipid production in *Synechocystis* sp. PCC 6803 overexpressing glycerol kinase under oxidative stress with glycerol supplementation. *Bioresource Technology* 267:532–540.

Sivaramakrishnan, R. and A. Incharoensakdi. 2018b. Microalgae as feedstock for biodiesel production under ultrasound treatment: A review. *Bioresource Technology* 250:877–887.

Sivaramakrishnan, R. and A. Incharoensakdi. 2018c. Utilization of microalgae feedstock for concomitant production of bioethanol and biodiesel. *Fuel* 217:458–466.

Sivaramakrishnan, R. and A. Incharoensakdi. 2019a. Enhancement of lipid extraction for efficient methyl ester production from *Chlamydomonas* sp. *Journal of Applied Phycology* 31:2365–2377.

Sivaramakrishnan, R. and A. Incharoensakdi. 2019b. Low power ultrasound treatment for the enhanced production of microalgae biomass and lipid content. *Biocatalysis and Agricultural Biotechnology* 20:101230.

Sivaramakrishnan, R. and A. Incharoensakdi. 2020. Plant hormone induced enrichment of *Chlorella* sp. omega-3 fatty acids. *Biotechnology for Biofuels* 13:7.

Sivaramakrishnan, R. and K. Muthukumar. 2012a. Isolation of thermo-stable and solvent-tolerant *Bacillus* sp. lipase for the production of biodiesel. *Applied Biochemistry and Biotechnology* 166:1095–1111.

Sivaramakrishnan, R. and K. Muthukumar. 2012b. Production of methyl ester from *Oedogonium* sp. oil using immobilized isolated novel *Bacillus* sp. lipase. *Energy & Fuels* 26:6387–6392.

Sivaramakrishnan, R. and K. Muthukumar. 2014. Direct transesterification of *Oedogonium* sp. oil be using immobilized isolated novel *Bacillus* sp. lipase. *Journal of Bioscience and Bioengineering* 117:86–91.

Sivaramakrishnan, R., S. Suresh, and A. Incharoensakdi. 2019. *Chlamydomonas* sp. as dynamic biorefinery feedstock for the production of methyl ester and ε-polylysine. *Bioresource Technology* 272:281–287.

Solana, M., C. S. Rizza, and A. Bertucco. 2014. Exploiting microalgae as a source of essential fatty acids by supercritical fluid extraction of lipids: Comparison between *Scenedesmus obliquus*, *Chlorella prototothecoides* and *Nannochloropsis salina*. *Journal of Supercritical Fluids* 92:311–318.

Song, E. S., J. Won Lim, H. S. Lee, and Y. W. Lee. 2008. Transesterification of RBD palm oil using supercritical methanol. *Journal of Supercritical Fluids* 44:356–363.

Taher, H., S. Al-Zuhair, A. H. Al-Marzouqi, Y. Haik, and M. M. Farid. 2011. A review of enzymatic transesterification of microalgal oil-based biodiesel using supercritical technology. *Enzyme Research* 2011:468292.

Tan, C. H., C.-Y. Chen, and P. L. Show, et al. 2016. Strategies for enhancing lipid production from indigenous microalgae isolates. *Journal of the Taiwan Institute of Chemical Engineers* 63:189–194.

Tariq, M., S. Ali, and N. Khalid. 2012. Activity of homogeneous and heterogeneous catalysts, spectroscopic and chromatographic characterization of biodiesel: A review. *Renewable and Sustainable Energy Reviews* 16:6303–6316.

Verma, P., M. P. Sharma, and G. Dwivedi. 2016. Impact of alcohol on biodiesel production and properties. *Renewable and Sustainable Energy Reviews* 56:319–333.

Xiong, W., X. Li, J. Xiang, and Q. Wu. 2008. High-density fermentation of microalga *Chlorella prototothecoides* in bioreactor for microbio-diesel production. *Applied Microbiology and Biotechnology* 78:29–36.

Yang, Y. F., C. P. Feng, Y. Inamori, and T. Maekawa. 2004. Analysis of energy conversion characteristics in liquefaction of algae. *Resources, Conservation and Recycling* 43:21–33.

Yu, G., Y. Zhang, L. Schideman, T. L. Funk, and Z. Wang. 2011. Hydrothermal liquefaction of low lipid content microalgae into bio-crude oil. *Transactions of the ASABE* 54:239–246.

Zhu, L., Y. K. Nugroho, and S. R. Shakeel, et al. 2017. Using microalgae to produce liquid transportation biodiesel: What is next? *Renewable and Sustainable Energy Reviews* 78:391–400.

Index

Page numbers in *Italics* refers figures; **bold** refers table

Printed in the United States
by Baker & Taylor Publisher Services